Algorithms and Computation in Mathematics · Volume 21

Editors

Arjeh M. Cohen   Henri Cohen
David Eisenbud Michael F. Singer Bernd Sturmfels

Dmitry Kozlov

Combinatorial Algebraic Topology

With 115 Figures and 1 Table

Springer

Author

Dmitry Kozlov

Fachbereich 3 - Mathematik
University of Bremen
28334 Bremen
Germany
E-mail: dfk@math.uni-bremen.de

Library of Congress Control Number: 2007933072

Mathematics Subject Classification (2000): 55U10, 06A07, 05C15.

ISSN 1431-1550

ISBN   978-3-540-71961-8 Springer Berlin Heidelberg New York

Springer is a part of Springer Science+Business Media
springer.com
© Springer-Verlag Berlin Heidelberg 2008

Typesetting by the author and SPi using a Springer LATEX macro package
Cover design: *WMXDesign,* Heidelberg

Printed on acid-free paper   SPIN: 11936756   46/SPi   5 4 3 2 1 0

Dedicated to my family

Preface

The intent of this book is to introduce the reader to the beautiful world of Combinatorial Algebraic Topology. While the main purpose is to describe the modern research tools and latest applications of this field, an attempt has been made to keep the presentation as self-contained as possible.

A book to teach from

The text is divided into three major parts, which provide several options for adoption for course purposes, depending on the time available to the instructor.

The first part furnishes a brisk walk through some of the basic concepts of algebraic topology. While it is in no way meant to replace a standard course in that field, it could prove helpful at the beginning of the lectures, in case the audience does not have much prior knowledge of algebraic topology or would like to focus on refreshing those notions that will be needed in subsequent chapters. The first part can be read by itself, or used as a blueprint with a standard textbook in algebraic topology such as [Mun84] or [Hat02] as additional reading. Alternatively, it could also be used for an independent course or for a student seminar.

If the audience is sufficiently familiar with algebraic topology, then one could start directly with the second part. This is suitable for a graduate or advanced undergraduate course whose purpose would be to learn contemporary tools of Combinatorial Algebraic Topology and to see them in use on some examples. At the end of the course, a successful student should be able to conduct independent research on this topic.

The third and last part of the book is a foray into one specific realm of a present-day application: the topology of complexes of graph homomorphisms. It fits well at the end of the envisioned graduate course, and is meant as a source of illustrations of various techniques developed in the second part. Another possibility would be to use it as material for a reading seminar.

What is different in our presentation

In the second part we lay the foundations of Combinatorial Algebraic Topology. In particular, we survey many of the tools that have been used in research in topological combinatorics over the last 20 years. However, our approach is at times quite different from the one prevailing in some of the literature.

Perhaps the major novelty is the general shift of focus from the category of posets to the category of acyclic categories. Correspondingly, the entire Chapter 10 is devoted to the development of the fundamental theory of acyclic categories and of the topology of their nerves, which in turn are no longer abstract simplicial complexes, but rather regular triangulated spaces.

Also, Chapter 11 is designed to give quite a different take on discrete Morse theory. The theory is broken into three major branches: combinatorial, topological, and algebraic; each one with its own specifics. A very new feature here is the recasting of discrete Morse theory for posets in terms of *poset maps with small fibers*. This, together with the existence of a universal object associated to every acyclic matching and the Patchwork Theorem allows for a structural understanding of the techniques that have been used until now.

There are further novelties scattered in the remaining four chapters of the second part. In Chapter 13 we connect the notion of evasiveness with monotone poset maps, and introduce the notion of NE-reduction. After that, the importance of colimits in Combinatorial Algebraic Topology is emphasized. We look at regular colimits and their relation with group actions in Chapter 14, and at homotopy colimits in Chapter 15. We provide complete proofs for all the statements in Chapter 15, based on the previous groundwork pertaining to cofibrations in Chapter 7. Finally, in Chapter 16, we take a daring step of counting the machinery of spectral sequences to the core methods of Combinatorial Algebraic Topology.

Let us also comment briefly on our citation policy. As far as possible we have tried to avoid citations directly in the text, choosing to present material in the way that appeared to us to be most coherent from the contemporary point of view. Instead, each chapter in the second and third parts ends with a detailed bibliographic account of the contents of that chapter. Since the mathematics of the first part is much more classical, we skip bibliographic information there almost completely, giving only general references to the existing textbooks. An exception is provided by Chapter 8, where the material is slightly less standard, thus justifying making some reading suggestions.

Acknowledgments

Many organizations as well as individuals have made it possible for this book project to be completed. To start with, it certainly would not have materialized without the generous financial support of the Swiss National Science Foundation and of the Institute of Theoretical Computer Science at the Swiss Federal Institute of Technology in Zürich. Furthermore, a major part of this work has been done while the author was in residence as a research professor at the Mathematical Science Research Institute in Berkeley, whose hospitality, as well as the collegiality of the organizers of the special program during the fall term 2006, is warmly appreciated. The last academic institution that the author would like to thank is the University of Bremen, which has generously granted him a research leave, so that in particular this project could be completed.

The staff at Springer has been most encouraging and helpful indeed. Many thanks go to Martin Peters and Ruth Allewelt, who have managed to circumvent all the clever excuses that I kept fabricating for not being able to finish the writing.

This text has grown from a one-year graduate course that I have given at ETH Zürich to an enthusiastic group of students. Their comments have been most welcome and have led to substantial improvements. Special thanks go to Peter Csorba, whose additional careful proofreading of the text has revealed many inconsistencies both notational and mathematical.

The head of the Institute of Theoretical Computer Science at ETH Zürich, Emo Welzl, has been a congenial host and a thoughtful mentor during my sojourn as an assistant professor here. For this he has my deepest gratitude.

Much of the material in this book is based on joint research with my long-time collaborator Eric Babson, from UC Davis. Without him there would be no book. He is the spiritual coauthor and I thank him for this.

Writing a text of this length can be a daunting task, and it is invaluable when someone's support is guaranteed come rain or come shine. During this work, I was in a singularly fortunate situation of having my mathematical collaborator and my wife, Eva-Maria Feichtner, by my side, to help me to persevere when it seemed all but futile. There is no way I can thank her enough for all the advice, comfort, and reassurance that she lent me.

Finally, I would like to thank my daughter, Esther Yael Feichtner, who was born in the middle of this project and immediately introduced an element of randomness into the timetable. The future looks bright for her, as the opportunities for this welcome sabotage abound.

ETH Zürich, Switzerland   *Dmitry N. Kozlov*
                                                              March 2007

Contents

1

Overture

The subject of *Combinatorial Algebraic Topology* is in a certain sense a classical one, since modern algebraic topology derives its roots from dealing with various combinatorially defined complexes and with combinatorial operations on them. Yet the aspects of the theory that we consider here and that we distinguish under the title of this book are far from classical and have been brought to the attention of the general mathematical public fairly recently.

If one asks oneself the question

What does Combinatorial Algebraic Topology do?

then the answer will be the same of for regular algebraic topology: one computes various algebraic invariants of topological spaces, for example homology groups, or special cohomology elements such as characteristic classes; at times, one is even able to determine the homotopy type. The discriminating feature is provided not by *what* one is computing, but by *how* and for *which classes* of topological spaces it is done.

More precisely, in this book the focus will be on the algebraic topology of cellular complexes, which are combinatorial both *locally*, meaning that the cell attachments are simple, and *globally*. Being combinatorial locally usually means that we have simplicial complexes, though more and more, further classes of complexes, such as cubical and prodsimplicial ones, find their application in Combinatorial Algebraic Topology. The word "globally" here refers to the fact that the cells themselves are combinatorially enumerated. Of course, the meaning of being combinatorially enumerated is open to interpretation, and probably cannot formally be pinned down without the loss of the desired flexibility. Typically this alludes to the fact that one has a bijection between cells and some objects that are universally perceived as combinatorial, for example graphs, partitions, permutations, and various combinations and enrichments (e.g., by labelings) of these.

Additionally, though the cell attachment maps are easy, the cell inclusions themselves indicate some combinatorial relationship between the objects that are indexing the cells in question. Normally, to obtain the combinatorial objects that are indexing the cells on the boundary of a given cell σ, one would need to perform some combinatorial operation on the object that is indexing σ itself.

Such complexes arise in all sorts of contexts. Sometimes the complexes are simply given directly, though more often they are induced implicitly. For example, frequently one happens to consider a topological space that allows additional structure, such as some kind of stratification. The combinatorial data that can be extracted from such a stratification is the partially ordered set of strata. This is of course a serious trivialization of the space, since only the bare incidence structure is left. There are then standard ways, such as taking the nerve, to associate a simplicial complex to this poset, with the idea that some of the algebro-topological invariants of this complex will reflect something about the initial stratification.

This is an example of a procedure that constitutes the first of perhaps the three major venues of Combinatorial Algebraic Topology: being able to derive new interesting combinatorial objects by building suitable models for topological questions. A classical example of this is the so-called Goresky–MacPherson formula, which we mention in Subsection 9.1.2. In short, given a collection of linear subspaces, this formula provides a way to calculate the cohomology groups of the complement of the union of these subspaces, in terms of a certain "combinatorial model," namely homology groups of the so-called order complex of a combinatorial object associated to this family of subspaces, the *intersection lattice*, see Subsection 9.1.2 and (9.1) for precise details.

The second major venue is that the methods of computation that are established as standard in algebraic topology lead to the unearthing of new discrete structures. For example, spectral sequences are such a tool, and once the filtration on the studied complex is chosen, the calculation, though possibly technically challenging, is nonetheless uniquely determined. The subsequent steps in the computation will unveil new combinatorial objects on a constant basis. As an example, we refer to the lengthy computation performed in Chapter 20 of this book. The primary goal there is to calculate the homology groups of certain standard prodsimplicial complexes associated to cycle graphs. However, in performing the actual calculation along the lines prescribed by the spectral sequence, one uncovers the important Hom$_+$ *construction* and witnesses the appearance of other classical instances of combinatorial complexes. This can trace its genesis to the original work of Eric Babson and the author on the resolution of the Lovász Conjecture.

Finally, the third major venue is that the combinatorial properties of the indexing objects from discrete mathematics get distinguished by the topology, providing a deeper insight both into the structure theory of these objects and into which part of it is relevant for topology. For example, there are many operations on graphs. However, it is specifically the operation of *fold* that has been singled out in the study of the Hom *complexes*, based solely on the fact that it is extremely well behaved from the topological point of view.

As another example, we refer to a combinatorial computation of some concrete Stiefel-Whitney characteristic classes in Theorem 19.13 in Subsection 19.2.2. There, once the combinatorial description of the characteristic classes has been found, the entire calculation hinges on one combinatorial lemma, namely Lemma 19.14.

In addition to the beautiful fundamental theory, Combinatorial Algebraic Topology has numerous applications. Classically, these lie in discrete mathematics, as well as in theoretical computer science. As we shall see in Chapter 9, there are many constructions that take some combinatorial data as input and produce some cell complex as output. This time, the idea is that the algebro-topological invariants of this complex should have a bearing on the combinatorial properties of the initial data.

One famous application is the so-called Evasiveness Conjecture; see Chapter 13. To get a rough idea, the reader should consider the set of all graphs on n vertices, where n is fixed, satisfying a certain graph property that is preserved by deletion of edges (e.g., planarity). The Evasiveness Conjecture then says that if one is trying to determine whether a given graph has this property or not, by using simple edge oracle, i.e., by asking whether a certain edge is in the graph or not, then one may have to ask all $\binom{n}{2}$ questions. Curiously, this conjecture is still open, although it has been settled using topological methods for some special cases, including when n is a prime power. Chapter 13 provides more insight into this and related problems.

Let us consider a further example here at some length. Assume that we have a graph G and we are asking ourselves whether it is possible to break the vertices of G into 3 disjoint groups such that no two vertices in the same group are connected by an edge. Even though this question sounds very elementary, in practice it is extremely difficult to answer. In fact, using the language of computational complexity theory, this problem is NP-complete. In concrete terms, this implies that any multiplicative increase of the computer speed by a constant factor will yield only an additive increase in the size of the instances (here meaning the number of vertices of the graph) for which the problem can be solved in the lifetime of the universe.

Of course, sometimes one can say right away that such a partition is impossible. This happens, for example, if G has four vertices all of which are connected to each other. However, this is a very restricted set of configurations, since this argument would be local in nature, we would find a small subgraph that is already not possible to partition. Most interesting cases have some sort of global obstructions, where one has to pay attention to the whole graph before concluding whether such a partition is possible.

There is an indirect way to try to answer this question, though. It is as follows. Assume that we have constructed in some clever way a cellular complex $X(G)$, and that our construction is functorial, meaning that graph homomorphisms induce continuous maps. The partition of the set of vertices of G into three groups as desired could be phrased by saying that we are taking a graph homomorphism from G to the complete graph on three vertices, called

K_3. Therefore, if such a partition were possible, we would have a map from $X(G)$ to $X(K_3)$.

Of course, by itself, this conclusion does not contradict anything yet. In fact, for any two nonempty topological spaces there is a continuous map from one to the other: simply map the entire first space to some chosen point in the second one. Assume, however, that there is an additional structure on the space $X(G)$, namely a free involution, and assume further that the induced map from $X(G)$ to $X(K_3)$ must commute with the respective involutions. This is already more restrictive, since for two given spaces which free involutions, it may happen that no such map exists.

For example, by the Borsuk-Ulam theorem it is impossible to map a higher dimensional sphere to a lower dimensional one in a way that commutes with the antipodal maps. Or here, if $X(G)$ is connected and $X(K_3)$ consists of several connected components and no single component is mapped to itself by the involution, then again no such involution-preserving map is possible. Note that to use this statement for different graphs, we would need each time to worry only about the connectivity of $X(G)$, since the complex $X(K_3)$ could be analyzed and understood once and for all.

We will expand on this line of argument in the last part of the book, where we will actually look at applications of somewhat more sophisticated invariants, which are routinely used in obstruction theory. Namely, we explore the use of Stiefel-Whitney classes associated to involutions (or to free $\mathbb{Z}$-actions, or to line bundles) for this sort of question. However, before an application of such complexity can be properly put in context, it is important to develop the toolbox of Combinatorial Algebraic Topology.

We finish this introductory chapter by stressing that the field of Combinatorial Algebraic Topology is both large and fast-growing. New connections to other areas are found continuously, reaching currently even to such distant subjects as computer vision and statistical analysis. We hope that this text will serve the function of helping to fill the void between the standard course in topology and this forefront of research.

Concepts of Algebraic Topology

2

Cell Complexes

In this chapter we consider all sorts of cell complexes that make an appearance in the combinatorial context. Following the tradition, we begin in Section 2.1 with the abstract simplicial complexes, which have long been the main workhorse applications to discrete mathematics. After dealing with them, we proceed in Section 2.2 to look at polyhedral complexes, including generalized simplicial complexes, cubical complexes, and, more generally, prodsimplicial complexes, which have all proved important in Combinatorial Algebraic Topology.

Section 2.3 is entirely devoted to triangulated spaces. These will be of crucial importance when we study nerves of acyclic categories in Chapter 10, and will also appear in various combinatorial quotient constructions in Chapter 14. Finally, the last section of this chapter considers the general CW complexes.

2.1 Abstract Simplicial Complexes

2.1.1 Definition of Abstract Simplicial Complexes and Maps Between Them

We start by recalling a basic and simple yet versatile way to describe topological spaces by means of purely combinatorial data.

Definition 2.1. *A finite* **abstract**[1] **simplicial complex** *is a finite set A together with a collection Δ of subsets of A such that if $X \in \Delta$ and $Y \subseteq X$, then $Y \in \Delta$.*

We denote the abstract simplicial complex described in Definition 2.1 simply by Δ. The element $v \in A$ such that $\{v\} \in \Delta$ is called a *vertex* of Δ. We denote the set of all vertices of Δ by $V(\Delta)$. When Δ consists of *all* subsets of A, it is called a *simplex*, and is denoted by Δ^A. In the same spirit, the sets

---

[1] Sometimes for brevity the word *abstract* is omitted.

$\delta \in \Delta$ are called *simplices of Δ*. Those simplices $\delta \in \Delta$ that are contained in no other simplex of Δ are called *maximal*.

Example 2.2. The collection of sets $\{\emptyset, \{1\}, \{2\}, \{3\}, \{1,2\}, \{1,3\}, \{2,3\}\}$ is an abstract simplicial complex. To obtain the simplex $\Delta^{\{1,2,3\}}$ one would need to add to this collection the set $\{1,2,3\}$.

Given two finite abstract simplicial complexes Δ_1 and Δ_2 such that $\sigma \in \Delta_1$ implies $\sigma \in \Delta_2$, we say that Δ_1 is an *abstract simplicial subcomplex* of Δ_2, and write $\Delta_1 \subseteq \Delta_2$. If in addition there exists $\sigma \in \Delta_2$ such that $\sigma \notin \Delta_1$, we say that Δ_1 is a *proper* subcomplex of Δ_2.

One also talks about the *dimension* of each simplex, which is 1 less than its cardinality as a set. When a simplex has dimension d, one says *d-simplex*. Vertices have dimension 0. The dimension is defined for finite abstract simplicial complexes as well: it is equal to the maximum of the dimensions of its simplices. The dimension is denoted by dim. If Δ_1 is an abstract simplicial subcomplex of Δ_2, then $\dim \Delta_1 \leq \dim \Delta_2$.

For any finite abstract simplicial complex Δ, the collection of all simplices of Δ up to dimension d is called the *d-skeleton* of Δ, and is denoted by $\Delta^{(d)}$ or $\mathrm{Sk}_d(\Delta)$.

Remark 2.3. The void and the empty.
Some words about degeneracies occurring in this context are in order. In Definition 2.1 we have allowed the empty collection of sets. The corresponding abstract simplicial complex is called the *void* abstract simplicial complex, and is denoted by $\emptyset$, or by $\{\}$. The void abstract simplicial complex is the only one that does not contain the empty set as one of its simplices.

Alternatively, we may also have the collection consisting of an empty set. The corresponding abstract simplicial complex is called the *empty* abstract simplicial complex, and is denoted by $\{\emptyset\}$. Notice that the sets of vertices of the void and of the empty complexes are empty sets.

Remark 2.4. The dimensions of the empty simplex and of the void complex.
When determining the dimension, one should keep in mind the degenerate case of the empty simplex $\emptyset \in \Delta$. By definition, since the cardinality of the empty set is 0, the dimension of this simplex is equal to -1. Correspondingly, the dimension of the empty simplicial complex is equal to -1. We set the dimension of the void simplicial complex to be equal to $-\infty$.

Remark 2.5. The term simplex.
The meticulous reader may have noticed that the term *simplex* was used by us in two ways, first, to denote any set that is in the collection of sets defining an abstract simplicial complex, and second to denote the abstract simplicial complex whose collection consists of all sets. The distinction is of course purely formal, and the reason one usually uses the same term for these two notions is that one can associate to each $\delta \in \Delta$ the simplex Δ^δ, which in turn will be a subcomplex of Δ.

Definition 2.6. *Let Δ_1 and Δ_2 be two finite abstract simplicial complexes. A* **simplicial map** *from Δ_1 to Δ_2 is a set map $f : V(\Delta_1) \to V(\Delta_2)$ such that if σ is a simplex of Δ_1, then $f(\sigma)$ is a simplex of Δ_2. In such a situation, we shall simply write $f : \Delta_1 \to \Delta_2$.*

In this definition we have used the notation $f(S) := \{f(s) \mid s \in S\}$ for any set S and any function f on S.

Example 2.7. Let $\Delta_1 = \Delta_2 = \{\emptyset, \{1\}, \{2\}, \{3\}, \{1,2\}, \{1,3\}, \{2,3\}\}$. Then any $f : [3] \to [3]$ is a simplicial map, whereas the same is not true for $\Delta_1 = \Delta_2 = \{\emptyset, \{1\}, \{2\}, \{3\}, \{1,2\}, \{1,3\}\}$.

Let us make some observations on the properties of simplicial maps:

- The identity map on the set of vertices is always a simplicial map from the abstract simplicial complex onto itself.
- The composition of two simplicial maps is again a simplicial map, since a simplex maps to a simplex, which again maps to a simplex.
- Even if the function f is bijective and simplicial, its inverse does not have to be simplicial.
- For any abstract simplicial complex Δ and for any finite set A, an arbitrary set map $f : V(\Delta) \to A$ induces a simplicial map $f : \Delta \to \Delta^A$.
- Whenever Δ_1 and Δ_2 are abstract simplicial complexes such that Δ_1 is a subcomplex of Δ_2, we have a natural simplicial inclusion map $\Delta_1 \hookrightarrow \Delta_2$.
- For $\Delta = \Delta^{[1]}$ and any abstract simplicial complex Λ, there exists a unique simplicial map from Λ to Δ; this map takes all vertices of Λ to the vertex of Δ.
- For any abstract simplicial complex Δ, there exists a unique simplicial map from the void abstract simplicial complex to Δ. Additionally, if Δ is not the void abstract simplicial complex, then there exists a unique simplicial map from the empty abstract simplicial complex to Δ.

Definition 2.8. *Let Δ_1 and Δ_2 be two abstract simplicial complexes, and let $f : \Delta_1 \to \Delta_2$ be a simplicial map between them. Then f is called an* **isomorphism of abstract simplicial complexes** *if the induced map $f : V(\Delta_1) \to V(\Delta_2)$ is a bijection and its inverse induces a simplicial map as well. If such an isomorphism exists, then Δ_1 and Δ_2 are said to be* **isomorphic as abstract simplicial complexes.**

Clearly, the isomorphism is an equivalence relation. It is the "equality" relation for the abstract simplicial complexes.

An important special class of simplicial maps are isomorphisms $f : \Delta \to \Delta$, where Δ is an abstract simplicial complex. These maps are called *automorphisms* of Δ. The composition of two automorphisms is again an automorphism, and an inverse of each automorphism is also an automorphism; hence the set of automorphisms forms a group by composition. This is the group of "symmetries" of the abstract simplicial complex Δ, and we denote it by

Aut (Δ). For example, the group of automorphisms of the abstract simplicial complex in Example 2.2 is the full symmetric group $\mathcal{S}_3$.

Until now, we have restricted ourselves to considering the finite abstract simplicial complexes. The natural question is, what happens if we drop the condition that the ground set should be finite. We invite the reader to check that all the definitions make sense, and that all the statements hold just as they do for the finite ones. For reasons that will become clear once we look at the notion of geometric realization, we keep the condition that the simplices have finite cardinality. We restate Definition 2.1 for future reference.

Definition 2.9. *An* **abstract simplicial complex** *is a set A together with a collection Δ of finite subsets of A such that if $X \in \Delta$ and $Y \subseteq X$, then $Y \in \Delta$.*

Example 2.10. Let A be the set of natural numbers. Then we obtain an abstract simplicial complex by taking all finite subsets $\sigma \in A$ such that for any two elements from σ, one of them must divide the other one.

Remark 2.11. In Chapter 4 we shall see that the finite abstract simplicial complexes together with simplicial maps actually form a category, and that the same is true if one takes all abstract simplicial complexes.

2.1.2 Deletion, Link, Star, and Wedge

Since the abstract simplicial complexes are some of the main characters of Combinatorial Algebraic Topology, there is a large variety of concepts and constructions pertaining to them. We shall now describe some of these.

Definition 2.12. *Let Δ be an abstract simplicial complex, and let τ be a simplex of Δ. The* **deletion** *of τ is the abstract simplicial subcomplex of Δ, denoted by $\mathrm{dl}_\Delta(\tau)$, defined by*

$$\mathrm{dl}_\Delta(\tau) := \{\sigma \in \Delta \,|\, \sigma \not\supseteq \tau\}.$$

In the degenerate cases, the deletion of the vertex v from the abstract simplicial complex $\{\emptyset, \{v\}\}$ will give the empty simplex, whereas the deletion of the empty set from any abstract simplicial complex will give the void simplex. Furthermore, if S is a set of simplices of Δ, then we define the deletion of S by setting

$$\mathrm{dl}_\Delta(S) := \{\sigma \in \Delta \,|\, \sigma \not\supseteq \tau, \text{ for all } \tau \in S\} = \bigcap_{\tau \in S} \mathrm{dl}_\Delta(\tau). \tag{2.1}$$

Another important concept in the context of abstract simplicial complexes is that of a link of a simplex.

Definition 2.13. *Let Δ be an abstract simplicial complex, and let τ be a simplex of Δ. The **link** of τ is the abstract simplicial subcomplex of Δ, denoted by $\mathrm{lk}_\Delta(\tau)$, defined by*

$$\mathrm{lk}_\Delta(\tau) := \{\sigma \in \Delta \mid \sigma \cap \tau = \emptyset, \text{ and } \sigma \cup \tau \in \Delta\}.$$

For example, in a *hollow tetrahedron*, the abstract simplicial complex Δ consisting of all subsets of $\{1, 2, 3, 4\}$ except for the set $\{1, 2, 3, 4\}$ itself, a link of an edge consists of the two vertices that do not belong to that edge. Note also that the void simplex can never be the link of anything, since any link contains an empty set.

In analogy to the definition above, if S is a set of simplices of Δ, then we define the link of S by setting

$$\mathrm{lk}_\Delta(S) := \{\sigma \in \Delta \mid \sigma \cap \tau = \emptyset, \text{ and } \sigma \cup \tau \in \Delta, \text{ for all } \tau \in S\} \qquad (2.2)$$
$$= \bigcap_{\tau \in S} \mathrm{lk}_\Delta(\tau).$$

Closely related to the notion of link is the notion of star.

Definition 2.14. *Let Δ be an abstract simplicial complex, and let τ be a simplex of Δ.*

*(1) The **closed star** of τ is the abstract simplicial subcomplex of Δ, denoted by $\mathrm{star}_\Delta(\tau)$, defined by*

$$\mathrm{star}_\Delta(\tau) := \{\sigma \in \Delta \mid \sigma \cup \tau \in \Delta\}.$$

*(2) The **open star** of τ is the set of simplices of Δ, denoted by $\mathrm{ostar}_\Delta(\tau)$, defined by*

$$\mathrm{ostar}_\Delta(\tau) := \{\sigma \in \Delta \mid \sigma \supseteq \tau\}.$$

For any simplex $\tau \in \Delta$, we have

$$\mathrm{lk}_\Delta(\tau) = \mathrm{star}_\Delta(\tau) \cap \mathrm{dl}_\Delta(V(\tau))$$

and

$$\Delta = \mathrm{ostar}_\Delta(\tau) \cup \mathrm{dl}_\Delta(\tau),$$

where the latter union is actually disjoint. Furthermore, for a vertex $v \in \Delta$, we have a simple but important for subsequent chapters decomposition

$$\Delta = \mathrm{star}_\Delta(v) \cup \mathrm{dl}_\Delta(v). \qquad (2.3)$$

Definition 2.15. *Given two abstract simplicial complexes Δ_1 and Δ_2, with vertices $v_1 \in V(\Delta_1)$ and $v_2 \in V(\Delta_2)$, the **wedge** of Δ_1 and Δ_2, with respect to the vertices v_1 and v_2, is the abstract simplicial complex $\Delta_1 \vee \Delta_2$ defined by*

- $V(\Delta_1 \vee \Delta_2) = (V(\Delta_1) \setminus \{v_1\}) \cup (V(\Delta_2) \setminus \{v_2\}) \cup \{v\}$;
- $\sigma \subseteq V(\Delta_1 \vee \Delta_2)$ *is a simplex of* $\Delta_1 \vee \Delta_2$ *if and only if either* $\sigma \subseteq V(\Delta_1) \cup \{v\}$ *and* σ *is a simplex of* Δ_1 *once* v *is replaced with* v_2, *or* $\sigma \subseteq V(\Delta_2) \cup \{v\}$ *and* σ *is a simplex of* Δ_2 *once* v *is replaced with* v_2.

Even though the abstract simplicial complex $\Delta_1 \vee \Delta_2$ depends on the choice of vertices v_1 and v_2, in practice these are usually suppressed from notations.

2.1.3 Simplicial Join

The following is one of the most fundamental constructions that allow one to produce new abstract simplicial complexes from old ones.

Definition 2.16. *Let* Δ_1 *and* Δ_2 *be two abstract simplicial complexes whose vertices are indexed by disjoint sets. The* **join** *of* Δ_1 *and* Δ_2 *is the abstract simplicial complex* $\Delta_1 * \Delta_2$ *defined as follows: the set of vertices of* $\Delta_1 * \Delta_2$ *is* $V(\Delta_1) \cup V(\Delta_2)$, *and the set of simplices is given by*

$$\Delta_1 * \Delta_2 = \{\sigma \subseteq V(\Delta_1) \cup V(\Delta_2) \,|\, \sigma \cap V(\Delta_1) \in \Delta_1 \text{ and } \sigma \cap V(\Delta_2) \in \Delta_2\}.$$

Clearly, we have commutativity: for arbitrary abstract simplicial complexes Δ_1 and Δ_2, the joins $\Delta_1 * \Delta_2$ and $\Delta_2 * \Delta_1$ are isomorphic. The join is also associative; namely, for arbitrary abstract simplicial complexes Δ_1, Δ_2, and Δ_3, the joins $(\Delta_1 * \Delta_2) * \Delta_3$ and $\Delta_1 * (\Delta_2 * \Delta_3)$ are isomorphic.

Another important property of the join is that for any abstract simplicial complex Δ and any simplex $\tau \in \Delta$, the abstract simplicial complexes $\mathrm{lk}_\Delta(\tau) * \tau$ and $\mathrm{star}(\tau)$ are isomorphic.

Joining with the abstract simplicial complex consisting of a single vertex is also called *coning*. One can also take a join with the abstract simplicial complex with n vertices and no simplices of dimension 1 and higher. This is the *n-coning*, giving the same result as a sequence of n single conings.

Example 2.17.

(1) We have $\Delta^A * \Delta^B = \Delta^{A \cup B}$.
(2) The join of an arbitrary abstract simplicial complex Δ with the empty simplex is equal to Δ.
(3) The join of an arbitrary abstract simplicial complex with the void simplex is equal to the void simplex.

2.1.4 Face Posets

A standard combinatorial gadget that one associates to an abstract simplicial complex is that of a face poset. To start with, we have the following definition.

Definition 2.18. *A* **partially ordered set**, *or simply* **poset**, P *is a set together with a relation* $\geq$ *that satisfies the following three axioms:*

(1) idempotency: for any $x \in P$, we have $x \geq x$;
(2) antisymmetry: for any $x, y \in P$, if $x \geq y$ and $y \geq x$, then $x = y$;
(3) transitivity: for any $x, y, z \in P$, if $x \geq y$ and $y \geq z$, then $x \geq z$.

Given a poset P, we let $\succ$ denote the *covering relation* in P, i.e., for $x, y \in P$, we write $x \succ y$ if $x > y$ and there is no $z \in P$ such that $x > z > y$. The following poset is of a particular importance.

Definition 2.19. *Let Δ be an arbitrary abstract simplicial complex. A **face poset** of Δ is the poset $\mathcal{F}(\Delta)$ whose set of elements consists of all nonempty simplices of Δ and whose partial order relation is the inclusion relation on the set of simplices.*

The following is a classical notion in order theory.

Definition 2.20. *Let $(P, >)$ be a poset. A total order $>_L$ on the set of elements of P is called a **linear extension** L of P if for any two elements x, y of P we have $x >_L y$ whenever $x > y$.*

For example, for an arbitrary abstract simplicial complex Δ, a standard linear extension of the face poset $\mathcal{F}(\Delta)$ is obtained by setting $\sigma >_L \tau$ whenever $\dim \sigma > \dim \tau$, and choosing an arbitrary order within each set of simplices of the same dimension.

2.1.5 Barycentric and Stellar Subdivisions

There are two standard ways to subdivide abstract simplicial complexes.

Definition 2.21. *Let Δ be an abstract simplicial complex. The **barycentric subdivision** of Δ is also an abstract simplicial complex, which is denoted by $\mathrm{Bd}\,\Delta$ and defined by*

$$\mathrm{Bd}\,\Delta = \{\{\sigma_1, \ldots, \sigma_t\} \mid \sigma_1 \supset \sigma_2 \supset \cdots \supset \sigma_t, \sigma_i \in \Delta, t \geq 1\} \cup \{\emptyset\}.$$

In particular, the set of vertices of $\mathrm{Bd}\,\Delta$ is indexed by the nonempty simplices of Δ.

While taking barycentric subdivision is useful in many situations, sometimes it just produces too many simplices. The next definition provides a more economic, local construction.

Definition 2.22. *Let Δ be an abstract simplicial complex, and let σ be a simplex of Δ. The **stellar subdivision**[2] of Δ at σ is the abstract simplicial complex $\mathrm{sd}_\Delta(\sigma)$ defined by the following:*

---

[2] The stellar subdivision is a special case of *combinatorial blowups*; see [FK04] for the definition and [CD06, Del06, FK05, FM05, FS05, FY04, Fei05, Fei06] for further applications of the latter concept.

- *For the set of vertices we have $V(\mathrm{sd}_\Delta(\sigma)) = V(\Delta) \cup \{\hat\sigma\}$, where $\hat\sigma$ simply denotes the new vertex "indexed by σ," and in case σ itself is a vertex, we have $\hat\sigma = \sigma$, and no new vertex is introduced.*
- *The simplex $\tau \in \Delta$ is a simplex of $\mathrm{sd}_\Delta(\sigma)$ if and only if τ does not contain σ as a subset. Additionally, the abstract simplicial complex $\mathrm{sd}_\Delta(\sigma)$ has simplices of the form $\tau \cup \{\hat\sigma\}$, where $\tau \in \Delta$ such that $\tau \cup \sigma \in \Delta$ and τ does not contain σ as a subset.*

For example, if σ is a vertex itself, then $\mathrm{sd}_\Delta(\sigma)$ is isomorphic to Δ. Next we show that taking the barycentric subdivision can be accomplished by a sequence of stellar subdivisions.

Proposition 2.23. *Let Δ be an arbitrary finite abstract simplicial complex, and let L be an arbitrary linear extension of the face poset $\mathcal{F}(\Delta)$. Then, the barycentric subdivision $\mathrm{Bd}\,\Delta$ is isomorphic to the abstract simplicial complex obtained from Δ by a sequence of stellar subdivisions, consisting of one stellar subdivision for every nonempty simplex of Δ, taking the simplices in decreasing order with respect to the given linear extension.*

Proof. Let $\{\sigma_1, \ldots, \sigma_t\}$ be the set of all nonempty simplices of Δ, indexed along the given linear extension, i.e., $\sigma_i >_L \sigma_{i-1}$, for every $i = 2, \ldots, t$. Furthermore, we set $\sigma_{t+1} := \emptyset$. For every $i = 1, \ldots, t$ we let Δ_i denote the abstract simplicial complex obtained as a result of the sequence of stellar subdivisions of $\sigma_1, \sigma_2, \ldots, \sigma_i$, and we use the convention that $\Delta_0 = \Delta$. We need to show that the complex Δ_t is isomorphic to $\mathrm{Bd}\,\Delta$. This follows from the following more general claim.

Claim. *For every $k = 0, \ldots, t$, the simplices of the abstract simplicial complex Δ_k are indexed by the l-tuples $(\hat\sigma_{i_1}, \ldots, \hat\sigma_{i_{l-1}}, \sigma_{i_l})$, satisfying conditions*

- $l \geq 1$;
- $\sigma_{i_1} \supset \cdots \supset \sigma_{i_l}$, where all the set inclusions are strict, which implies $i_1 < \cdots < i_l$;
- $i_{l-1} \leq k < i_l$, in particular $i_l = t + 1$ is allowed, whereas we always have $i_{l-1} \leq t$.

The simplex indexed by $(\hat\sigma_{i_1}, \ldots, \hat\sigma_{i_{l-1}}, \sigma_{i_l})$ has dimension $l + |\sigma_{i_l}| - 2$, and its boundary simplices are obtained by either deleting any of the elements in the tuple except for the last one, or replacing the last element by its proper subset.

Proof of the claim. We use induction on k. For $k = 0$ we have only the simplices with $l = 1$, and the conditions of the claim describe precisely the abstract simplicial complex Δ. Consider now what happens when we pass from the complex Δ_{k-1} to the complex Δ_k, for $k \geq 1$.

First, the simplices of Δ_{k-1} that contain the simplex indexed by (σ_k) get removed. Since we follow a linear extension, the simplex σ_k is maximal in $\mathrm{dl}_\Delta(\{\sigma_1, \ldots, \sigma_{k-1}\})$. Hence, according to the induction assumption, the removed tuples are indexed by $(\hat{\sigma}_{i_1}, \ldots, \hat{\sigma}_{i_{l-1}}, \sigma_k)$, satisfying the condition $\sigma_{i_1} \supset \cdots \supset \sigma_{i_{l-1}} \supset \sigma_k$.

After that, the new simplices are added. By the definition of the stellar subdivision together with our induction assumption, these are indexed by the pairs $((\hat{\sigma}_{i_1}, \ldots, \hat{\sigma}_{i_{l-2}}, \sigma_{i_l}), \hat{\sigma}_k)$ such that the abstract simplicial complex Δ_{k-1} has a simplex that contains both $(\hat{\sigma}_{i_1}, \ldots, \hat{\sigma}_{i_{l-2}}, \sigma_{i_l})$ and σ_k, but the simplex $(\hat{\sigma}_{i_1}, \ldots, \hat{\sigma}_{i_{l-2}}, \sigma_{i_l})$ itself does not contain σ_k. These conditions are equivalent to $\sigma_{i_{l-2}} \supset \sigma_k \supset \sigma_{i_l}$, and therefore, one can think of the added simplices as being indexed with the tuples $(\hat{\sigma}_{i_1}, \ldots, \hat{\sigma}_{i_{l-2}}, \hat{\sigma}_k, \sigma_{i_l})$ such that $\sigma_{i_1} \supset \cdots \supset \sigma_{i_{l-2}} \supset \sigma_k \supset \sigma_{i_l}$.

All in all, we see that this description of the simplices that are added or removed when taking the stellar subdivision of (σ_k) in Δ_{k-1}, together with the induction assumption that the conditions of our claim describe the abstract simplicial complex Δ_{k-1}, implies that the conditions of our claim describe the abstract simplicial complex Δ_k as well, providing the induction step. This finishes the proof of our claim. $\square$

In particular, for $k = t$ we have only the simplices $(\hat{\sigma}_{i_1}, \ldots, \hat{\sigma}_{i_{l-1}}, \emptyset)$, i.e., $i_l = t + 1$. The rule for the simplex inclusions given in the claim is the same as that for Bd Δ, and hence the abstract simplicial complex Δ_t is isomorphic to the abstract simplicial complex Bd Δ. $\square$

In particular, Proposition 2.23 suggests one standard way to view the barycentric subdivision as a sequence of stellar ones: simply start by taking the stellar subdivisions of the simplices of top dimension, then take the stellar subdivisions of the simplices of dimension one less, and so on, until reaching the vertices.

For dealing with subdivisions in later chapters, it is practical to introduce the following shorthand notation: when a simplicial complex Δ_2 subdivides another simplicial complex Δ_1, we shall write $\Delta_2 \rightsquigarrow \Delta_1$. We shall also use the same notation for more general classes of complexes.

2.1.6 Pulling and Pushing Simplicial Structures

Let A and B be two sets, and let $f : A \to B$ be a set map.

Definition 2.24. *Assume that Δ is an abstract simplicial complex on A. We define the* **pushforward** *abstract simplicial complex $f(\Delta)$ by setting*

$$f(\Delta) := \{f(\sigma) \mid \sigma \in \Delta\}.$$

The pushforward abstract simplicial complex satisfies the following universal property: whenever Ω is an abstract simplicial complex on B such that f is a simplicial map, the complex $f(\Delta)$ must be an abstract simplicial subcomplex of Ω. Phrased colloquially, $f(\Delta)$ is obtained by pushing the simplicial structure Δ forward, and it is the minimal one that makes the set map f simplicial.

Definition 2.25. *Assume that Ω is an abstract simplicial complex on B. We define the **pullback** abstract simplicial complex $f^{-1}(\Omega)$ by setting*

$$f^{-1}(\Omega) := \{\sigma \subseteq A \,|\, f(\sigma) \in \Omega, |\sigma| < \infty\}.$$

The pullback abstract simplicial complex satisfies the following universal property: any abstract simplicial complex on A such that f is a simplicial map must be an abstract simplicial subcomplex of $f^{-1}(\Omega)$. Again, one could say that $f^{-1}(\Omega)$ is obtained by pulling the simplicial structure Ω back, and it is the maximal one that makes the set map f simplicial.

Note that $f^{-1}(\Omega)$ does not denote two different things, since if the set map f is bijective, then $f(\Delta)$ is isomorphic (as an abstract simplicial complex) to Δ, and $f^{-1}(\Omega)$ is isomorphic to Ω.

The reader is invited to check that the following properties of pullbacks and pushforwards hold for any abstract simplicial complex Δ:

$$\mathrm{id}(\Delta) = \Delta, \quad f(g(\Delta)) = (f \circ g)(\Delta), \quad f^{-1}(g^{-1}(\Delta)) = (g \circ f)^{-1}(\Delta).$$

2.2 Polyhedral Complexes

2.2.1 Geometry of Abstract Simplicial Complexes

It is now time to describe the geometric picture that is encoded by the combinatorial data of an abstract simplicial complex.

Various definitions of the geometric realization

Definition 2.26.

(1) *A geometric n-**simplex** σ is the convex hull of the set A of $n + 1$ affine independent points in $\mathbb{R}^N$, for some $N \geq n$. The convex hulls of the subsets of A are called **subsimplices** of σ.*

(2) *The **standard** n-simplex is the convex hull of the set of the endpoints of the standard unit basis $(1, 0, \ldots, 0), (0, 1, 0, \ldots, 0), \ldots, (0, \ldots, 0, 1)$ in $\mathbb{R}^{n+1}$.*

More generally, given any finite set A, we have the vector space $\mathbb{R}^A$, whose coordinates are indexed by the elements of A; and correspondingly, for any subset $B \subseteq A$, we can define a standard B-simplex in $\mathbb{R}^A$ as the one that is spanned by the endpoints of the part of the standard unit basis indexed by elements in B in that vector space. In this language, the simplex described in Definition 2.26(2) would be called the standard $[n + 1]$-simplex in $\mathbb{R}^{[n+1]}$.

Definition 2.27. *Given a finite abstract simplicial complex Δ, we define its **standard geometric realization** to be the topological space obtained by taking the union of standard σ-simplices in $\mathbb{R}^{V(\Delta)}$, for all $\sigma \in \Delta$.*

*Any topological space that is homeomorphic to the standard geometric realization of Δ is called the **geometric realization** of Δ, and is denoted by $|\Delta|$.*

Very often, for the sake of brevity, in case no confusion arises, we shall talk about the topological properties of the abstract simplicial complex Δ, always having in mind the properties of the topological space $|\Delta|$. We note that the geometric realizations of the void and of the empty complexes are both empty sets.

It is possible to give a similar definition of the geometric realization for the infinite case. However, one would need a careful treatment of the resulting infinite-dimensional vector space and the topology involved. Instead, we use this as an opportunity to introduce a *gluing process*, which will be used to construct various classes of cell complexes.

Given a nonempty abstract simplicial complex Δ, the constructive definition of the geometric realization of Δ goes as follows:

- Start with an arbitrary vertex of Δ, and then add new simplices one by one, in any order, with the only condition being that all the proper subsimplices of the simplex that is being added have already been glued on at this point.
- Assume that we are at the situation in which we would like to glue the simplex $\sigma \in \Delta$ onto the part of the realization X that we have so far. The new space is $X \cup_{\partial \Delta^\sigma} \Delta^\sigma$, which is obtained by identifying the boundary of Δ^σ with the subspace of X, which is the result of gluing the simplices corresponding to the proper subsimplices of σ. In shortly, the simplex Δ^σ is glued onto X along its boundary in the natural way.

One way to think of this gluing process is the following. We have a collection of simplices $\{\Delta^\sigma\}_{\sigma \in \Delta}$, together with inclusion maps $i_{\sigma,\tau} : \Delta^\sigma \hookrightarrow \Delta^\tau$, whenever σ is a proper subset of τ. The space $|\Delta|$ is obtained from the disjoint union of the simplices by one extra condition: we would like to identify two points whenever one of them maps to the other one by one of these inclusion maps.

Of course, one easy way to satisfy such a condition is simply to identify all points, and to obtain just a point as the resulting quotient space. The additional requirement for the gluing process is that this identification should be in some sense "minimal"; in other words, no identification is done unless it is a consequence of the prescribed identifications. We shall see how similar universality conditions appear in further definitions of the cell complexes. Furthermore, in Chapter 4 it will be demonstrated how all of these are just special instances of the general colimit construction.

Yet another alternative to define the geometric realization of an abstract simplicial complex would be to give a direct description of the set of points together with topology. This is what we do next.

Definition 2.28. *Given a set S, a* **convex combination** *of the elements in S is a function $f : S \to \mathbb{R}$ such that*

- $f(x) \geq 0$, *for all $x \in S$;*
- $f(x) \neq 0$, *for only finitely many $x \in S$;*

- $\sum_{x \in S} f(x) = 1$.

The finite set $\{x \in S \mid f(x) \neq 0\}$ is called the support of the convex combination and is denoted by supp.

The convex combination is usually written as an algebraic expression $\sum_{x \in S} f(x)x$. These expressions can be added and multiplied by real numbers in the usual way. In this notation, two algebraic expressions differing only in terms with zero coefficients will be identified.

The crucial observation is that the points of the geometric realization of an abstract simplicial complex Δ are in 1-to-1 correspondence with the set of all convex combinations whose support is a simplex of Δ. In fact, the support of each convex combination tells us precisely to which simplex it belongs.

Furthermore, we can define a distance function on the set of all convex combinations by setting

$$d\left(\sum_{x \in S} f(x)x, \sum_{x \in S} g(x)x\right) := \sum_{x \in S} |f(x) - g(x)|. \tag{2.4}$$

It is easy to check that (2.4) indeed defines a metric on our space; hence one can take the topology induced by this metric. This is exactly the topology of the geometric realization.

Intuitively, one can say that the points that are near to a point in the geometric realization can be obtained by a small deformation of the coefficients of the corresponding convex combination. If the deformation is sufficiently small, then the nonzero coefficients will stay positive. However, even under a very small deformation it may happen that the zero coefficients become nonzero. This is allowed as long as the support set remains a simplex of the initial abstract simplicial complex. Geometrically, this corresponds to entering the interior of an adjacent higher-dimensional cell. An illustration is provided in Figure 2.1.

Geometry of simplicial maps

Proposition 2.29. Let Δ_1 and Δ_2 be arbitrary abstract simplicial complexes. A simplicial map $f : \Delta_1 \to \Delta_2$ induces a continuous map $|f| : |\Delta_1| \to |\Delta_2|$.

Proof. We define $|f|$ by setting

$$|f| : \sum_{v_i \in V(\Delta_1)} t_i v_i \mapsto \sum_{v_i \in V(\Delta_1)} t_i f(v_i). \tag{2.5}$$

Since simplices are required to map to simplices, this map is well-defined. We leave the proof of the continuity as an exercise. $\square$

We note that when both Δ_1 and Δ_2 are finite, the map $|f|$ is actually a restriction of a linear map $|f| : \mathbb{R}^{V(\Delta_1)} \to \mathbb{R}^{V(\Delta_2)}$, whose matrix is the

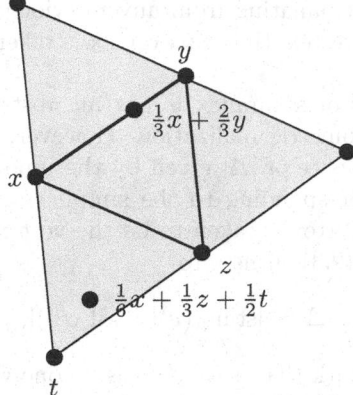

Fig. 2.1. Coordinate description of points of a geometric realization of an abstract simplicial complex.

standard 0-1 matrix associated to the set map $f : V(\Delta_1) \to V(\Delta_2)$: it simply maps a basis vector e_i to the basis vector $e_{f(i)}$.

In general, it is clear from (2.5) that

$$|f \circ g| = |f| \circ |g| \qquad (2.6)$$

whenever f and g are simplicial maps, and that

$$|f^{-1}| = |f|^{-1} \qquad (2.7)$$

whenever f has an inverse, that is also a simplicial map.

It follows that an automorphism of an abstract simplicial complex induces a continuous automorphism of its geometric realization, and that the geometric realizations of two isomorphic abstract simplicial complexes are homeomorphic, with homeomorphisms induced by the isomorphism maps. On the other hand, the existence of a homeomorphism between $|\Delta_1|$ and $|\Delta_2|$ does not imply the isomorphism of Δ_1 and Δ_2. If a topological space allows a simplicial structure, then it allows infinitely many nonisomorphic simplicial structures.

2.2.2 Geometric Meaning of the Combinatorial Constructions

Geometry of deletion, link, star, and wedge

Geometrically, the deletion operation does precisely what it is supposed to do: it deletes from Δ all the simplices that contain the simplex τ.

For a vertex v, the link is modeling the simplicial complex that one obtains if one cuts the given simplicial complex by a sphere of small radius with the center in v. For a general simplex τ, one thinks of the link of τ intuitively as the *space of directions* emanating from any interior point of τ, transversally to τ itself. If τ is just a vertex, then this vertex is taken in place of the interior point.

Since the open star of a simplex is not an abstract simplicial complex, one cannot take its geometric realization. However, sometimes one instead considers the open subspace of $|\Delta|$ given by the union of the interiors of the geometric simplices corresponding to the simplices in the open star, where again the interior of a vertex is taken to be the vertex itself.

The decomposition (2.3) turns into

$$|\Delta| = |\text{star}_\Delta(v)| \cup |\text{dl}_\Delta(v)|,$$

where we explicitly remark that $|\text{star}_\Delta(v)|$ is a cone with apex v.

It is easy to see that Definition 2.15 carries over to the case of topological spaces with a selected point (the so-called *based spaces*) essentially without change.

Definition 2.30. *Given topological spaces X_1 and X_2, with points $x_1 \in X_1$ and $x_2 \in X_2$, we define the* **wedge** *of X_1 and X_2 to be the quotient space $X_1 \vee X_2 = (X_1 \cup X_2)/ \sim$, with the equivalence relation given by $x_1 \sim x_2$.*

Clearly, for two abstract simplicial complexes Δ_1 and Δ_2, we have

$$|\Delta_1 \vee \Delta_2| \cong |\Delta_1| \vee |\Delta_2|, \tag{2.8}$$

where the wedges are taken with respect to the same pair of points.

Geometry of the simplicial join

One can define the join of arbitrary topological spaces. Let I denote the closed unit interval.

Definition 2.31. *Let X and Y be two topological spaces. The* **join** *of X and Y is the topological space $X * Y$ defined as follows:*

$$X * Y = I \times X \times Y/ \sim,$$

where the equivalence relation $\sim$ is given by

- $(0, x, y) \sim (0, x, \tilde{y})$, *for all* $y, \tilde{y} \in Y$;
- $(1, x, y) \sim (1, \tilde{x}, y)$, *for all* $x, \tilde{x} \in X$.

We note that for two abstract simplicial complexes Δ_1 and Δ_2 one has

$$|\Delta_1| * |\Delta_2| \cong |\Delta_1 * \Delta_2|, \tag{2.9}$$

where on the righthand side we take the simplicial join.

Given geometric realizations of Δ_1 in $\mathbb{R}^m$ and Δ_2 in $\mathbb{R}^n$, a geometric realization of $\Delta_1 * \Delta_2$ in $\mathbb{R}^{m+n+1}$ can be obtained as follows. Identify $\mathbb{R}^m$

with the coordinate subspace of $\mathbb{R}^{m+n+1}$ given by $(x_1, \ldots, x_m, 0, \ldots, 0)$, and identify $\mathbb{R}^n$ with the translated coordinate subspace of $\mathbb{R}^{m+n+1}$ given by $(0, \ldots, 0, y_1, \ldots, y_n, 1)$. Take the induced embeddings of $|\Delta_1|$ and $|\Delta_2|$ into $\mathbb{R}^{m+n+1}$ and let $|\Delta_1 * \Delta_2|$ be the union of convex hulls of pairs of simplices: one from Δ_1 and one from Δ_2.

First, if σ is a simplex in $|\Delta_1|$ and τ is a simplex in $|\Delta_2|$, then the union of the sets of vertices of σ and τ is a linearly independent set. Second, these newly spanned simplices will not overlap (other than along their boundaries), since the convex combination of $x \in |\Delta_1|$ and $y \in |\Delta_2|$, $x = (x_1, \ldots, x_m, 0, \ldots, 0)$, $y = (0, \ldots, 0, y_1, \ldots, y_n, 1)$, has the coordinates $((1-t)x_1, \ldots, (1-t)x_m, ty_1, \ldots, ty_n, t)$, defining x, y, and t uniquely.

Example 2.32.

(1) For an arbitrary topological space X, the space $X * point$ is called **a cone over** X, and the space $X * \mathbb{S}^0$ is called a **suspension of** X; the latter terminology comes from the fact that one can view the space X as being suspended on threads between the two opposite apexes inside the space $\operatorname{susp} X$.

(2) Let m and n be arbitrary nonnegative integers. Then we have

$$\mathbb{S}^m * \mathbb{S}^n = \mathbb{S}^{m+n+1}. \tag{2.10}$$

To see (2.10), note that in general, $(X * Y) * Z = X * (Y * Z)$. Therefore

$$\mathbb{S}^m * \mathbb{S}^n = \underbrace{\mathbb{S}^0 * \cdots * \mathbb{S}^0}_{m+1} * \underbrace{\mathbb{S}^0 * \cdots * \mathbb{S}^0}_{n+1} = \underbrace{\mathbb{S}^0 * \cdots * \mathbb{S}^0}_{m+n+2} = \mathbb{S}^{m+n+1}.$$

In parallel with the explicit embedding described above, the points of a join of two spaces can be described as follows. The equivalence class of the point $(t, x, y) \in I \times X \times Y$ is denoted by $((1-t)x, ty)$. This notation helps encode the fact that when $t = 0$, then y does not matter, and when $t = 1$, then x does not matter. More generally, a point on the join of finitely many topological spaces $X_1 * \cdots * X_k$ can be described as $(t_1 x_1, t_2 x_2, \ldots, t_k x_k)$, such that $t_1 + t_2 + \cdots + t_k = 1$, and $t_i \geq 0$ for all $i = 1, \ldots, k$.

We also remark that instead of taking the successive joins, one can think of $X_1 * \cdots * X_k$ as the quotient space

$$X_1 * \cdots * X_k = \Delta^{[k]} \times X_1 \times \cdots \times X_k / \sim,$$

where the equivalence relation $\sim$ is given by $(\alpha, x_1, \ldots, x_k) \sim (\alpha, x_1', \ldots, x_k')$ if tuples $(x_1, \ldots, x_k)$ and $(x_1', \ldots, x_k')$ coincide on the support simplex of α (where the support simplex of α is the minimal subsimplex of $\Delta^{[k]}$ containing α).

Geometry of barycentric subdivision

The geometric realizations of the abstract simplicial complexes $\mathrm{Bd}\,\Delta$ and Δ are related in a fundamental way.

Proposition 2.33. *For any abstract simplicial complex Δ, the topological spaces $|\Delta|$ and $|\mathrm{Bd}\,\Delta|$ are homeomorphic.*

Proof. The explicit point description of the geometric realization of an abstract simplicial complex tells us that the points of $|\Delta|$ are indexed by convex combinations $a_1 v_1 + \cdots + a_s v_s$ such that $\{v_1, v_2, \ldots, v_s\} \in \Delta$, whereas the points of $|\mathrm{Bd}\,\Delta|$ are indexed by convex combinations $b_1 \sigma_1 + \cdots + b_t \sigma_t$ such that $\sigma_1 \supset \sigma_2 \supset \cdots \supset \sigma_t$. Let us now give the procedure for translating between the two point descriptions.

(1) To define a continuous function $f : |\Delta| \to |\mathrm{Bd}\,\Delta|$, take a point $x = a_1 v_1 + \cdots + a_s v_s$ in $|\Delta|$. Choose a permutation $(i_1, i_2, \ldots, i_s)$ that orders the coefficients $0 \leq a_{i_1} \leq a_{i_2} \leq \cdots \leq a_{i_s} \leq 1$. For $k = 1, \ldots, s$ we set

$$b_k := (s - k + 1)(a_{i_k} - a_{i_{k-1}}) \text{ and } \sigma_k := \{v_{i_k}, v_{i_{k+1}}, \ldots, v_{i_s}\}, \qquad (2.11)$$

where the convention $a_{i_0} = 0$ is used. The point $f(x)$ is encoded by the convex combination $b_1 \sigma_1 + \cdots + b_s \sigma_s$.

While the formulae (2.11) give a succinct algebraic definition of the function f, it is also possible to see an intuitive algorithmic picture. To start with, by the properties of our point encoding, we can assume that $a_i \neq 0$, for all $i \in [s]$. Let $a := \min(a_1, \ldots, a_s)$, and set $b_1 := sa$ and $\sigma := \{v_1, \ldots, v_s\}$. After that, we proceed recursively with the linear (though no longer convex) combination $(a_1 - a)v_1 + \cdots + (a_s - a)v_s$ as follows.

- Delete all the zero terms, say k terms remain.
- Set b_2 to be equal to ka', where a' is the minimum of the remaining coefficients, and set σ_2 to be the set of the remaining vertices.
- Subtract a' from the remaining coefficients and repeat the whole procedure to find the next pair b_j and σ_j.

Note, that in the end we shall get the same convex combination as using (2.11), just with the zero terms already removed. This algorithm is illustrated by the pictogram-like Figure 2.2.

(2) To define a continuous function $g : |\mathrm{Bd}\,\Delta| \to |\Delta|$, take a point $y = b_1 \sigma_1 + \cdots + b_t \sigma_t$ in $|\mathrm{Bd}\,\Delta|$. The point $g(y)$ can be encoded by the convex combination $a_1 v_1 + \cdots + a_s v_s$, where $\{v_1, \ldots, v_s\} = \sigma_1$. To determine the coefficient of v_i, let k be the maximal index such that $v_i \in \sigma_k$. Then we set $a_i := b_1/|\sigma_1| + b_2/|\sigma_2| + \cdots + b_k/|\sigma_k|$. In words, one could say that each simplex σ_i distributes its coefficient in a fair way to its vertices.

It follows immediately from our description of topology on geometric realizations of abstract simplicial complexes that the maps f and g are continuous. We leave it as an exercise for the reader to verify that these two maps are actually inverses of each other. $\square$

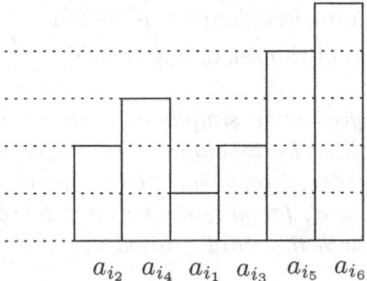

$$a_{i_2} \quad a_{i_4} \quad a_{i_1} \quad a_{i_3} \quad a_{i_5} \quad a_{i_6}$$

Fig. 2.2. The horizontal levels correspond to the successive steps of the algorithm computing the value of the function f.

One way to think about the barycentric subdivision, which can come in handy in certain situations, is the following. First, we define the barycentric subdivision in the standard n-simplex. By definition, it is the simplicial complex obtained by stratifying the standard n-simplex with the intersections with the hyperplanes $x_i = x_j$, for $1 \le i < j \le n+1$. The new vertices will be in barycenters (also called the *centers of gravity*) of the subsimplices of the standard n-simplex, hence the name *barycentric*.

When the abstract simplicial complex Δ is finite, then we can take its standard geometric realization and take the barycentric subdivisions of the individual n-simplices as just described. This gives the geometric realization of the barycentric subdivision of Δ. When, on the other hand, the abstract simplicial complex Δ is infinite, we can take the barycentric subdivisions of its simplices before the gluing and then observe that the gluing process is compatible with the new cell structure; hence we will obtain the geometric realization of the barycentric subdivision of Δ as well.

2.2.3 Geometric Simplicial Complexes

Sometimes the embedding of the geometric realization in the ambient space is prescribed from the beginning and is of importance. For this reason, many texts in algebraic topology introduce the following concept.

Definition 2.34. *A **geometric simplicial complex** K in $\mathbb{R}^N$ is a collection of simplices in $\mathbb{R}^N$ such that every subsimplex of a simplex of K is a simplex of K and the intersection of any two simplices of K is a subsimplex of each of them.*

As for abstract simplicial complexes, the collection of all simplices of K of dimension less than or equal to d is called the d-skeleton of K, and is denoted by $K^{(d)}$ or $\mathrm{Sk}_d(K)$. For example, K_0 is the set of vertices, K_1 is the edge graph, etc.

Example 2.35. Define a simplicial complex K as follows: the set of vertices is $K_0 = \{0\} \cup \left\{ \frac{1}{n} \mid n \in \mathbb{N} \right\}$, and the set of edges is $K_1 = \left\{ \left[\frac{1}{n+1}, \frac{1}{n} \right] \mid n \in \mathbb{N} \right\}$.

Definition 2.36. *For a geometric simplicial complex K, let $|K|$ denote the union of all simplices of K. The topology on $|K|$ is defined as follows: every simplex σ of K has the induced topology, and in general $A \subseteq |K|$ is open if and only if $A \cap \sigma$ is open in σ, for all simplices $\sigma \in K$ (equivalently, the word "open" could be replaced with the word "closed").*

We note here that if K has finitely many simplices, then the intrinsic topology of $|K|$ is the same as the topology induced from the encompassing space $\mathbb{R}^n$. Otherwise, the two topologies may differ. To illustrate that effect, consider the simplicial complex in Example 2.35. The induced topology on the union of simplices is that of a unit interval; in particular, the set $\{0\}$ is not open. On the other hand,

$$\{0\} \cap \sigma = \begin{cases} \{0\}, & \text{if } \sigma = \{0\}; \\ \emptyset, & \text{otherwise.} \end{cases}$$

In particular, $\{0\}$ is open in $|K|$. The topological space $|K|$ is not connected, whereas it would be connected if we simply took the induced topology.

In general, a geometric realization of an abstract simplicial complex is a geometric simplicial complex, while the combinatorial incidence structure of the geometric simplicial complex will give an abstract one.

The notions of a subcomplex and simplicial maps are defined for the geometric simplicial complexes in full analogy with the abstract simplicial complexes. These notions then coincide under the described correspondence between the two families of simplicial complexes. Note that when two complexes K and L have dimension less than 2, i.e., can be viewed as graphs, the graph homomorphisms (see Definition 9.20) between K and L are simplicial maps, but not vice versa.

To illustrate the use of the topology described in Definition 2.36, we prove the following result.

Proposition 2.37. *Let L be a subcomplex of a geometric simplicial complex K; then $|L|$ is a closed subspace of $|K|$.*

Proof. For an arbitrary simplex $\sigma \in K$, the intersection $\sigma \cap |L|$ is a union of those subsimplices of σ that belong to L; in particular, it is closed. $\square$

By our discussion above, Proposition 2.37 implies that whenever Δ_1 and Δ_2 are abstract simplicial complexes and Δ_1 is a subcomplex of Δ_2, we know that $|\Delta_1|$ is a closed subspace of $|\Delta_2|$. In fact, if Δ_1 is also finite, we see that $|\Delta_1|$ is a compact subspace of $|\Delta_2|$.

2.2.4 Complexes Whose Cells Belong to a Specified Set of Polyhedra

In this subsection we shall describe a general procedure that allows one to consider families of complexes whose cells are sampled from some specified set of polyhedra. In this way we will construct both familiar families as well as complexes that only recently have proved to be of importance for combinatorial computations.

Polyhedral complexes

Recall that a *convex polytope* P is a bounded subset of $\mathbb{R}^d$ that is the solution of a finite number of linear inequalities and equalities. Recall that $F \subseteq P$ is called a *face* of P if there exists a linear functional f on $\mathbb{R}^d$ such that $f(s) = 0$, for all $s \in S$, and $f(p) \geq 0$, for all $p \in P$.

Definition 2.38. *A geometric polyhedral complex Γ in $\mathbb{R}^N$ is a collection of convex polytopes in $\mathbb{R}^N$ such that*

(i) every face of a polytope in Γ is itself a polytope in Γ;
(ii) the intersection of any two polytopes in Γ is a face of each of them.

Most of the terminology, such as *skeleton, subcomplex, join,* carries over from the simplicial situation. One new property worth observing is that a direct product of two geometric polyhedral complexes is again a geometric polyhedral complex, whereas the same is not true for the geometric simplicial complexes.

Let C be a geometric polyhedral complex. For every face F of C we insert a new vertex b_F in the barycenter of F, i.e., $b_F = \frac{1}{k} \sum_{i=1}^{k} v_i$, where $v_1, \ldots, v_k$ are the vertices of F. We can now subdivide each face of C into simplices in the inductive manner: the edges are simply divided into two, and at each step, the face F is subdivided by spanning a cone from b_F to the subdivided boundary of F. Clearly, this will give a homeomorphic geometric simplicial complex.

Let us now adopt the gluing process from Subsection 2.2.1 to define a more general family of complexes.

- We start with a discrete set of points. This is our 0-skeleton, and we proceed by induction on the dimension of the attached faces.
- At step d we attach the d-dimensional faces, all at once. Each face is represented by some convex polytope P in the sample space $\mathbb{R}^d$. To attach it we need a continuous map $f : \partial P \to X$, where X denotes the part of the complex created in the first $d - 1$ steps. The attaching map must satisfy an additional condition: we request that it should be a homeomorphism between ∂P and $f(\partial P)$, and that this homeomorphism should preserve the cell structures, where the cell structure on ∂P is simply the given polytopal structure, and the cell structure on $f(\partial P)$ is induced from the previous gluing process.

When a cell of the complex was obtained by gluing on the polyhedron P; we shall sometimes simply say that this cell *is* the polyhedron P.

Definition 2.39. *A topological space X is called a* **polyhedral complex** *if it can be obtained by the above gluing procedure.*

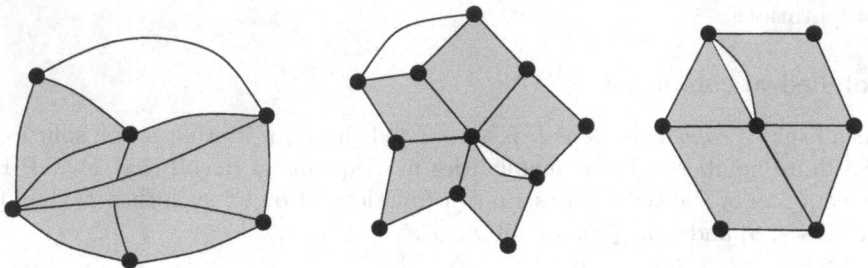

Fig. 2.3. Examples of polyhedral, cubical, and prodsimplicial complexes.

An example of a polyhedral complex is shown on the left of Figure 2.3. Again, all the basic operations and terminology of the simplicial complex extend to the polyhedral ones. This includes the barycentric subdivision, since it can be done on the polyhedra before the gluing, and then one can observe that the obtained complexes will glue to a simplicial complex in a compatible way.

For future reference we introduce the following notion.

Definition 2.40. *Let X be a polyhedral complex, and let S be the set of some of its vertices. We let $X[S]$ denote the polyhedral complex that consists of all cells whose set of vertices is a subset of S. This complex is called the* **induced subcomplex**.

The notion of the link does not generalize in a straightforward fashion. We shall not need it in this generality, so rather than indulging in technical details, we provide some intuition behind the various constructions that exist in the literature. The notion of a link in triangulated spaces, which we will actually need, will be defined rigorously in the next section.

To start with, notice that every polyhedral complex is embeddable into $\mathbb{R}^N$ for a sufficiently large number N. Indeed, all we need to verify is that we can attach each cell to an embeddable complex X so that it stays embeddable, possibly increasing the dimension. Assume that $X \subseteq \mathbb{R}^N$ and add one more dimension. Place a vertex on the new coordinate axis and span a cone over the subspace of X along which the new cell is to be glued. Since the attaching map of the new cell is a homeomorphism, we see that adding this cone is the same up to homeomorphism as actually attaching the new cell to X. Hence the resulting complex is embeddable as well.

Once the complex is embedded into $\mathbb{R}^N$, a usual way to define the link of a vertex v is to place a sphere of sufficiently small radius with its center in v. The intersection of this sphere with the complex is the link. To see that this is actually a polyhedral complex, it is enough to notice that up to face-preserving homeomorphism, the intersection of a small sphere with the polytope can be replaced by the intersection with a hyperplane. This hyperplane can be found as follows: take the hyperplane H whose intersection with our polytope is equal to the considered vertex, and consider the parallel translation of this hyperplane by a sufficiently small number in the direction of the polytope.

Finally, one can also define the link of an arbitrary face σ of the polyhedral complex. To do that, take any point x in the interior of σ. It has a small closed neighborhood that can be represented as a direct product $B^d \times Q$, where $d = \dim \sigma$, B^d is a closed ball of dimension d, and Q is some polyhedral complex, which can be thought of as the transversal complex of σ. The face σ is replaced by the vertex x in Q, and we can take the link of x in Q. This is the link of σ in our polyhedral complex.

Complexes glued from simplices

It is now possible to define whole classes of complexes by specifying the set of allowed polyhedra. The first and most natural set of polyhedra that one could choose is the set of all simplices.

Definition 2.41. *A polyhedral complex whose cells are simplices is called a* **generalized simplicial complex**.[3]

A generalized simplicial complex may be not representable as a geometric realization of an abstract simplicial complexes. Perhaps the simplest example is the complex with two vertices and two edges connecting these vertices. However, all constructions involving abstract simplicial complexes, including simplicial maps, can also be done with the generalized simplicial complexes in the sense of Definition 2.41. The generalizations are straightforward and left to the reader.

Cubical complexes

In many contexts, such as geometric group theory, another family of complexes is of fundamental importance.

Definition 2.42. *A polyhedral complex whose cells are cubes of various dimensions is called a* **cubical complex**.

The join operation does not work well for the cubical complexes. Instead, we see that a direct product of two cubical complexes is again a cubical complex. An example of a cubical complex is shown in the middle of Figure 2.3.

---

[3] We choose not to call it a *simplicial complex* to avoid confusion with the terminology in other literature.

Prodsimplicial complexes

The next class of complexes provides a hybrid of the cubical and simplicial complexes.[4]

Definition 2.43. *A polyhedral complex whose cells are direct products of simplices is called a* **prodsimplicial complex**.

This is a class of polyhedral complexes that contains the generalized simplicial complexes and is closed under direct products. An example of a prodsimplicial complex is shown on the right of Figure 2.3.

2.3 Trisps

Given an abstract simplicial complex Δ with the set of vertices $[n]$, we see that the natural order on the set $[n]$ induces an order on the set of vertices of each simplex. More generally, an abstract simplicial complex equipped with a partial order on the set of vertices such that it induces total orders on the sets of vertices of every simplex is called an *ordered abstract simplicial complex*.

In this section we define a class of spaces, the so-called *triangulated spaces*, which in this book will simply be called *trisps*,[5] whose open cells are simplices with coherently ordered vertex sets, but which are more general than ordered abstract simplicial complexes.

2.3.1 Construction Using the Gluing Data

A trisp is described in a purely combinatorial way by its *gluing data*.

Definition 2.44. *The* **gluing data** *for a trisp Δ comprise the following parts:*

- *a sequence of sets $S_0(\Delta), S_1(\Delta), \ldots$, where the set $S_i(\Delta)$ indexes the i-simplices of Δ, for all $i \geq 0$, and the sets $S_i(\Delta)$ are not required to be finite;*
- *for each $m \leq n$, and for each order-preserving injection $f : [m+1] \hookrightarrow [n+1]$, we are given a set map $B_f : S_n(\Delta) \to S_m(\Delta)$ satisfying two additional properties:*

  (1) for any pair of composable order-preserving injections $[k+1] \xhookrightarrow{g} [m+1] \xhookrightarrow{f} [n+1]$, we have $B_{f \circ g} = B_g \circ B_f$;

  (2) for any identity map $\mathrm{id}_n : [n+1] \hookrightarrow [n+1]$, we have $B_{\mathrm{id}_n} = \mathrm{id}_{S_n(\Delta)}$.

---

[4] This class of complexes has recently gained in importance in Combinatorial Algebraic Topology; see [Ko05a].

[5] Our terminology follows Gelfand & Manin [GeM96], though the abbreviation *trisp* appears to be used here for the first time. The same objects have appeared in the literature under other names, most notably as *semisimplicial sets* in Eilenberg & Zilber [EZ50], and, more recently, as *Δ-complexes* in Hatcher [Hat02].

Let us say a few words to illuminate the second point in Definition 2.44. When thinking through it, the reader should identify for her- or himself the order-preserving injections into the set $[n + 1]$ with the subsets of $[n + 1]$. Therefore, each such order-preserving injection $f : [m + 1] \hookrightarrow [n + 1]$ should be thought of as indexing certain m-dimensional subsimplices of the n-dimensional simplices. For example, when $m = 1$ and $n = 2$, there are precisely three order-preserving injections $[2] \hookrightarrow [3]$, corresponding to the fact that a triangle has three edges. For each such map f we have a set map $B_f : S_n(\Delta) \to S_m(\Delta)$, and for each $\sigma \in S_n(\Delta)$, the value $B_f(\sigma)$ simply tells us onto which m-simplex we should glue the corresponding m-simplex from the boundary of σ.

The fact that the gluing data are discrete and are given by set maps only is very important. The precise gluing procedure is as follows. Note that an order-preserving injection $f : [m + 1] \hookrightarrow [n + 1]$ induces a linear inclusion map $M_f : \mathbb{R}^{m+1} \hookrightarrow \mathbb{R}^{n+1}$, which takes the ith vector of the normal orthogonal basis in $\mathbb{R}^{m+1}$ to the $f(i)$th vector of the normal orthogonal basis in $\mathbb{R}^{n+1}$. In particular, it can be restricted to a homeomorphism from the standard m-simplex to a certain m-subsimplex of the n-standard simplex. This is the map that glues this subsimplex of a simplex $\sigma \in S_n$ to the m-simplex $B_f(\sigma) \in S_m(\Delta)$. The condition $B_{f \circ g} = B_g \circ B_f$ ensures that the gluing procedure is consistent.

Definition 2.45. *A **trisp** Δ is the complex that is obtained from the gluing data from Definition 2.44 by the gluing procedure above.*

It may be informative to observe that in order to specify a trisp, we just need to list these order-preserving injections for the codimension-1 pairs, i.e., when $m = n - 1$. We also see that when an n-simplex is attached, each of its proper open subsimplices is attached by a homeomorphism to a simplex that is already in the complex.

In the structure of trisps, the "vertices" of each simplex have a prescribed order. In particular, all edges in Δ are directed. When v and w are vertices of Δ, and $e = (v, w)$ is an edge directed from v to w, we shall often write $(v \to w)$ instead of e.

Remark 2.46. Notice that we have not excluded the possibility that all of the simplex sets $S_i(\Delta)$ are empty. The trisp for which this happens is called the *void trisp*.

Recall furthermore that for the nonvoid abstract simplicial complexes we have also had an empty simplex. It is practical to have it for the trisps as well. Therefore we adopt the convention that we also have the set $S_{-1}(\Delta)$. If this set is empty, then also all other sets $S_i(\Delta)$ are required to be empty; otherwise, we require that $|S_{-1}(\Delta)| = 1$. Since by convention $[0] = \emptyset$, there is a unique order-preserving injection $f : [0] \hookrightarrow [n]$ for each n, and the map $B_f : S_n(\Delta) \to S_{-1}(\Delta)$ is the unique map that takes everything to one element.

In a trisp there could be several simplices with the same set of vertices, and furthermore, the boundary of every simplex may have self-identifications. In the combinatorial applications, we shall usually have spaces without such self-identifications. Therefore, it is useful to distinguish this special case by a separate definition.

Definition 2.47. *Let Δ be a trisp with the gluing data $(\{S_i(\Delta)\}_i, \{B_f\}_f)$. Then Δ is called a **regular trisp** if the vertices of every $\sigma \in S_k(\Delta)$ are distinct, i.e., the values $B_{f_i}(\sigma)$, for $i = 1, \ldots, k+1$, are all distinct; here each $f_i : [1] \hookrightarrow [k+1]$ is the injection defined by $f_i(1) := i$.*

Clearly, if the vertices of $\sigma \in S_k(\Delta)$ are distinct, then in fact, all the subsimplices of the attached simplex are glued over distinct subsimplices. In particular, the simplex is glued over a homeomorphic copy of its boundary. In this book, regular trisps will appear as nerves of acyclic categories.

Given a generalized simplicial complex, see Definition 2.41, choose an arbitrary order on its set of vertices. This will induce an order on vertices of every simplex, and one can see that the gluing maps will be order-preserving. Hence we will get a regular trisp. Conversely, once the orderings of vertices in every simplex are forgotten in a regular trisp, we get a generalized simplicial complex.

However, it is of course not true that an abstract simplicial complex with already chosen orders of vertices in simplices can be realized as a trisp. The simplest example is provided by the hollow triangle in which the directions on edges are chosen so that these go in a circle.

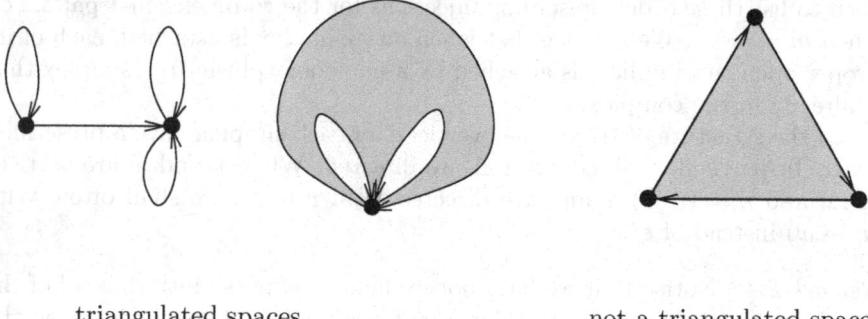

triangulated spaces   not a triangulated space

Fig. 2.4. Examples.

2.3.2 Constructions Involving Trisps

Essentially all constructions on the abstract simplicial complexes generalize to the trisps. It might be instructive to follow some of this in further detail.

To start with, the *deletion* of a vertex, or, more generally of a higher-dimensional simplex, is straightforward. When Δ is a trisp and $\sigma \in S_m(\Delta)$ is an m-simplex of Δ, the gluing data of the trisp $\mathrm{dl}_\Delta(\sigma)$ are obtained from the gluing data for Δ by deleting the element σ from $S_m(\Delta)$ and deleting all elements of $S_n(\Delta)$, for $n > m$ that map to σ under some map B_f. In other words, we delete all simplices that contain σ as a part of their boundary.

Defining a *link* is only slightly more difficult. Let again Δ be a trisp, given by its gluing data as specified by Definition 2.44, and let $\sigma \in S_m(\Delta)$ be an m-simplex of Δ. We shall now describe the gluing data of a trisp $\mathrm{lk}_\Delta(\sigma)$. To start with, the sets indexing simplices are given by

$$S_{n-m-1}(\mathrm{lk}_\Delta(\sigma)) := \{(f, \tau) \mid \tau \in S_n(\Delta), \, f : [m+1] \hookrightarrow [n+1], \, \sigma = B_f(\tau)\},$$

where the injection f is assumed to be order-preserving. Let us define the gluing maps. Assume that $n > k > m$, and we have an order-preserving injection $\varphi : [k - m] \hookrightarrow [n - m]$. Take $\tau \in S_n(\Delta)$ such that $B_f(\tau) = \sigma$, where f is an order-preserving injection $f : [m+1] \hookrightarrow [n+1]$. The maps φ and f give rise to order-preserving injections $g : [m+1] \hookrightarrow [k+1]$ and $h : [k+1] \hookrightarrow [n+1]$ as follows. First of all, identify the target set of φ with the complement of the image of f. Now, to find h simply take the union of images of f and of φ. We also find the map g by taking the image of f inside that union. We can now define

$$B_\varphi(\mathrm{lk}_\Delta(\sigma))(f, \tau) = (g, B_h(\Delta)(\tau)).$$

We leave the verification of the compatibility condition to the reader. The geometric meaning of the link is the same as for generalized simplicial and polyhedral complexes.

For a trisp Δ and a simplex σ, the *star* of σ, denoted by $\mathrm{star}_\Delta(\sigma)$, is a cone with apex, which we denote by σ as well, over the link of σ in Δ. In particular, it is always contractible.

It is also possible to define *joins of trisps*. Assume that the triangulated spaces Δ_1 and Δ_2 are given by their gluing data and let us describe the gluing data for the trisp $\Delta_1 * \Delta_2$. The sets indexing simplices of $\Delta_1 * \Delta_2$ are given by

$$S_n(\Delta_1 * \Delta_2) := \bigcup_{i+j+1=n} \{(\sigma_1, \sigma_2) \mid \sigma_1 \in S_i(\Delta_1), \sigma_2 \in S_j(\Delta_2)\}.$$

Let us describe the gluing maps. Assume that we have an order-preserving injection $f : [m+1] \hookrightarrow [n+1]$, and let $(\sigma_1, \sigma_2) \in S_n(\Delta_1 * \Delta_2)$, say $\sigma_1 \in S_i(\Delta_1)$ and $\sigma_2 \in S_j(\Delta_2)$. Note that $n + 1 = (i + 1) + (j + 1)$, and set

$$\alpha := |\mathrm{Im} f \cap [i + 1]| - 1 \text{ and } \beta := |\mathrm{Im} f \cap \{i + 2, \ldots, n + 1\}| - 1.$$

We have $\alpha + \beta + 1 = m$. The map f can then be represented by two order-preserving injections $f_1 : [\alpha + 1] \hookrightarrow [i + 1]$ and $f_2 : [\beta + 1] \hookrightarrow [j + 1]$. Correspondingly, we have two gluing maps $B_{f_1}(\Delta_1) : S_i[\Delta_1] \hookrightarrow S_\alpha[\Delta_1]$ and $B_{f_2}(\Delta_2) : S_j[\Delta_2] \hookrightarrow S_\beta[\Delta_2]$. Thus we can define

$$B_f(\Delta_1 * \Delta_2) : (\sigma_1, \sigma_2) \mapsto (B_{f_1}(\sigma_1), B_{f_2}(\sigma_2)).$$

We leave it to the reader to verify the gluing compatibility condition. Note that as a topological space, $\Delta_1 * \Delta_2$ is homeomorphic to the topological join of Δ_1 and Δ_2.

The maps between trisps are defined in the natural way as maps between their sets of simplices that commute with the gluing maps.

Definition 2.48. *Let Δ_1 and Δ_2 be two trisps. A **trisp map** F between Δ_1 and Δ_2 is the following collection of data: for every simplex $\sigma \in S_n(\Delta_1)$ we have a triple $(n(\sigma), p(\sigma), F(\sigma))$, where*

- *$n(\sigma) \leq n$ is a nonnegative integer,*
- *$p(\sigma) : [n+1] \longrightarrow [n(\sigma)+1]$ is an order-preserving surjection,*
- *$F(\sigma) \in S_{n(\sigma)}(\Delta_2)$; this is the simplex in Δ_2 of dimension $n(\sigma)$ to which σ is mapped by F following the map $p(\sigma)$.*

These data are required to satisfy the following axiom: for any order-preserving injection $f : [m+1] \hookrightarrow [n+1]$, there exists an order-preserving injection $g : [n(B_f(\sigma))+1] \hookrightarrow [n(\sigma)+1]$ such that

$$g \circ p(B_f(\sigma)) = p(\sigma) \circ f \tag{2.12}$$

and

$$B_g(F(\sigma)) = F(B_f(\sigma)), \tag{2.13}$$

where the gluing maps are taken in the corresponding trisps.

The composition of trisp maps is defined by composing the structure data in the natural way. One can see that the composition is well-defined by concatenating the corresponding commutative diagrams. It is easy to see that the identity map is a trisp map, and that the composition of trisp maps is associative.

Assume that Δ is a trisp and that F is an automorphism of Δ, i.e., the trisp map F takes Δ to itself, and it is a trisp isomorphism. Then Definition 2.48 specifies the following:

- for every simplex $\sigma \in S_n(\Delta)$, we have $n(\sigma) = n$, $p(\sigma) = \mathrm{id}_{[n+1]}$, and hence $F(\sigma) \in S_n(\Delta)$;
- for any order-preserving injection $f : [m+1] \hookrightarrow [n+1]$, the map g whose existence is stipulated by Definition 2.48 is equal to f;
- condition (2.13) translates into the commutation relation

$$B_f(F(\sigma)) = F(B_f(\sigma)). \tag{2.14}$$

Given a trisp Δ, we can define its *barycentric subdivision* $\mathrm{Bd}\,\Delta$, which is actually a regular trisp. The set of vertices of $\mathrm{Bd}\,\Delta$ is equal to the union $S_0(\Delta) \cup S_1(\Delta) \cup \cdots$. The d-simplices of $\mathrm{Bd}\,\Delta$ are indexed by $(d+1)$-tuples $(\sigma, f_1, \ldots, f_d)$, where $\sigma \in S_{n_d}$, and the f_i's are order-preserving injections

$f_i : [n_{i-1}] \hookrightarrow [n_i]$, for some $n_0 < n_1 < \cdots < n_d$. The boundary simplices of such a d-simplex are obtained by either replacing two injections f_i and f_{i+1} with their composition or by deleting the map f_d and at the same time replacing the simplex σ with $B_{f_d}(\sigma)$.

Since the barycentric subdivision of the generalized simplicial complex is a geometric realization of an abstract simplicial complex, we can be sure that after taking the barycentric subdivision twice, the trisp will turn into the geometric realization of an abstract simplicial complex. On the other hand, with many trisps, taking the barycentric subdivision once would not suffice for that purpose.

2.4 CW Complexes

In this section we define the most general class of cell complexes that we use, the *CW complexes*. Roughly speaking, a CW complex is made up of balls of different dimensions, glued to each other, possibly in a fairly complicated manner.

2.4.1 Gluing Along a Map

First we need some terminology. Let an *m-cell* be a topological space homeomorphic to an m-dimensional closed unit ball $B^m = \{v \in \mathbb{R}^m \,|\, \|v\| \leq 1\}$. An *open m-cell* is a topological space homeomorphic to the interior of a ball $\text{Int}\, B^m$. The CW complexes are constructed from such cells by means of a general gluing procedure. We have already seen several instances of that. Since the gluings we need now are more general, we would like to be more pedantic about precise definitions. To start with, the actual gluing step is what is called "attaching by a continuous map." More precisely, we have the following definition.

Definition 2.49. *Let X and Y be topological spaces, let $A \subseteq X$ be a closed subspace, and let $f : A \to Y$ be a continuous map. $Y \cup_f X$ denotes the quotient space $X \coprod Y / \sim$, where the equivalence relation $\sim$ is generated by $a \sim f(a)$, for all $a \in A$. We say that the space $Y \cup_f X$ is obtained from Y by attaching X along f.*

Note that the mapping cone and the mapping cylinder, which will be defined in Section 6.3, are examples of the space attachment constructions. Attaching a cell along its boundary is another such example, in this case $X = B^m$, $A = \partial B^m$, and the attachment map is an arbitrary continuous map $f : \partial B^m \to Y$.

Observe that the space $Y \cup_f X$ is equipped with the quotient topology, which means that $S \subseteq Y \cup_f X$ is open if and only if $q^{-1}(S)$ is open in $X \coprod Y$, where $q : X \coprod Y \to Y \cup_f X$ is the quotient map.

2.4.2 Constructive and Intrinsic Definitions

We shall now give the first definition of a CW complex; this one is along the same "gluing" lines that we have followed several times already.

Definition 2.50. (Constructive).
A **CW complex** X *is obtained by the following inductive construction of the skeletons:*

(1) The 0-skeleton $X^{(0)}$ is a discrete set.
(2) Construct the n-skeleton $X^{(n)}$ by the simultaneous attachment of the n-cells to $X^{(n-1)}$ along their boundaries. In particular, $X^{(n)}$ gets the quotient topology as described above.
(3) Equip the space $X = \bigcup_{n=0}^{\infty} X^{(n)}$ with the weak topology: $A \subseteq X$ is open if and only if $A \cap X^{(n)}$ is open for any n.

Unwinding definitions we get that $A \subseteq X$ is open if and only if $f_\alpha^{-1}(A)$ is open for any cell α, where $f_\alpha : B_\alpha^n \to X$ is the attachment map (also called the *characteristic map*).

Note that for an arbitrary CW complex X, the space $X^{(n)}/X^{(n-1)}$ is homeomorphic to a wedge of n-dimensional spheres, one for each n-cell of X.

Definition 2.51. *Let X be a CW complex; $A \subseteq X$ is called a* **subcomplex** *if A is a union of open cells such that if $e \subseteq A$, then $\bar{e} \subseteq A$.*

An important property of CW complexes is expressed by the following proposition.

Proposition 2.52. *A compact subspace of a CW complex is contained in a finite subcomplex.*

We leave the verification of this proposition to the reader.

Corollary 2.53. *For each open cell e in a CW complex X, its closure $\bar{e}$ is contained in finitely many open cells.*

This proves a part of the equivalence of the two definitions of CW complex. Historically, the CW structure was defined intrinsically, as follows.

Definition 2.54. (J.H.C. Whitehead).
Let X be a Hausdorff topological space, and assume that it is represented as a disjoint union of open cells e_α. Then, the pair (X, the collection of the cells $\{e_\alpha\}_\alpha$) is called a **CW complex** *if the following two conditions are satisfied:*

(1) For any α, there exists a continuous map $f_\alpha : B^m \to X$ (m is the dimension of e_α) such that
  • the restriction of f_α to Int B^m is a homeomorphism onto e_α;
  • $f_\alpha(\partial B^m)$ is a subset of a union of finitely many cells of dimension less than m.
(2) The subset $A \subseteq X$ is closed in X if and only if $A \cap \bar{e}_\alpha$ is closed for any α.

2.4.3 Properties and Examples

CW complexes enjoy several properties:

- CW complexes are normal (meaning that disjoint closed subspaces can be encapsulated in disjoint open subspaces);
- a CW complex is connected if and only if it is path connected;
- all CW complexes are locally contractible.

A nicer class of complexes is obtained by imposing an extra condition.

Definition 2.55. *A CW complex X is called* **regular** *if for each cell α, the restriction of the characteristic map $f_\alpha : \partial B_\alpha \to f_\alpha(\partial B_\alpha)$ is a homeomorphism.*

For example, all the cell spaces, except for the trisps that we have defined up to now are regular CW complexes. Regular trisps are also regular CW complexes.

Example 2.56.

(1) There are many representations of the sphere $\mathbb{S}^n$ as a CW complex. Two popular ones are as follows:
   (a) There is one 0-cell and one n-cell. There is only one choice of the attachment map, and the obtained complex is not regular.
   (b) There are 2 cells in each dimension. The n-cells are attached to $\mathbb{S}^{n-1}$ by the identity maps along their boundaries. This is a regular CW complex, and the cell structure is invariant under the antipodal map.
(2) A torus can be economically represented by one 0-cell, one 2-cell, and two 1-cells.
(3) The real projective space $\mathbb{RP}^n$ can be represented as a CW complex with one cell in each dimension from 0 to n. This cell structure is the $\mathbb{Z}_2$-quotient, with respect to the antipodal map, of the cell structure on the sphere $\mathbb{S}^n$, which we described in (1)(b) above.
(4) The complex projective space $\mathbb{CP}^n$ can be represented as a CW complex with one cell in every even dimension from 0 to $2n$. The interior of the $2n$-cell corresponds to all the points $(1 : z_1 : \cdots : z_n)$ identifying it with $\mathbb{C}^n$. The characteristic map identifies the ray generated by the vector v (the boundary of $\mathbb{C}^n$ can be thought of as the set consisting of all the rays emanating from the origin) with the point $(0 : v)$.

3

Homology Groups

The general idea of homology theory is to have computable algebraic invariants of topological spaces. One basic set of invariants is given by the Betti numbers. These can be further enriched: for example, one can consider homology groups or cohomology rings. In the opposite direction, one can also simplify, and consider such invariants as the Euler characteristic, which could be much easier to compute.

As usual, these invariants are used to distinguish between topological spaces. In fact, it will turn out that most of the invariants that we will define are preserved by a coarser equivalence relation than homeomorphism, called homotopy equivalence. Thus the invariants may also help in distinguishing the homotopy types of spaces.

In the subsequent sections we shall define homology in stages, proceeding from special to increasingly general settings. For example, the boundary map will be introduced four times in various contexts: as a set map (Betti numbers over $\mathbb{Z}_2$ of abstract simplicial complexes), as a linear transformation (Betti numbers of trisps), as an abelian group homomorphism (homology with integer coefficients), and finally using the winding numbers (cellular homology).

3.1 Betti Numbers of Finite Abstract Simplicial Complexes

In the first section of this chapter we would like to introduce the reader to the basic invariants as quickly as possible. We therefore restrict our attention to finite abstract simplicial complexes for now.

Let Δ be a finite abstract simplicial complex, and let $S_n(\Delta)$ be the set of n-simplices of Δ, for all $n \geq 0$. Let $C_n(\Delta)$ be the set of all subsets of $S_n(\Delta)$, in particular $|C_n(\Delta)| = 2^{|S_n(\Delta)|}$. We define a function, called the *boundary operator* $\partial_n : S_n(\Delta) \to C_{n-1}(\Delta)$, as follows:

$$\{v_0, \ldots, v_n\} \mapsto \bigcup_{i=0}^{n} \{\{v_0, \ldots, \hat{v}_i, \ldots, v_n\}\},$$

where $\{v_0, \ldots, \hat{v}_i, \ldots, v_n\}$ means *the subset of* $\{v_0, \ldots, v_n\}$ *obtained by removing the element* v_i, and the double bracket $\{\}$ comes from the fact that we are dealing with a set of sets. In words, the set $\{v_0, , \ldots, v_n\}$ is mapped to the set of all the subsets that can be obtained from it by deleting one of its elements.

The definition of the boundary operator can be extended to all subsets of $S_n(\Delta)$, which we still denote by $\partial_n : C_n(\Delta) \to C_{n-1}(\Delta)$, by setting

$$\partial_n(\Sigma) := \bigoplus_{S \in \Sigma} \partial_n(S),$$

for all $\Sigma \in C_n(\Delta)$, where $\oplus$ denotes the exclusive or. Spelled out in words: the set Σ of n-simplices is mapped to the set of all $(n-1)$-simplices that can be obtained from an odd number of simplices in Σ by deleting one of their vertices.

Let $\mathrm{Im}(\partial_n)$ denote the image of the boundary map $\partial_n : C_n(\Delta) \to C_{n-1}(\Delta)$. Furthermore, let us call the subset of $C_n(\Delta)$ consisting of the families of subsets of $S_n(\Delta)$ that are mapped by ∂_n to the empty set the *kernel* of ∂_n, and let us denote it by $\mathrm{Ker}(\partial_n)$. It is convenient to adopt the convention that $C_{-1}(\Delta)$ consists of the empty set alone, and that $\mathrm{Ker}(\partial_0) = C_0(\Delta)$.

Before we proceed, we need the following combinatorial lemma.

Lemma 3.1. *Let S be a finite set, and let Σ be a family of subsets of S that contains the empty set and that is closed under the operation $\oplus$. Then we have $|\Sigma| = 2^d$, for some $0 \leq d \leq |S|$.*

Remark 3.2. Lemma 3.1 is an easy consequence of linear algebra over the finite field $\mathbb{Z}_2$. Indeed, the elements of the set S can be identified with the coordinates of $\mathbb{Z}_2^{|S|}$, which is an $|S|$-dimensional vector field over $\mathbb{Z}_2$. Under this identification, the subsets of S are vectors in that vector space, and the condition that the family Σ is closed under the operation $\oplus$ translates to the fact that Σ is a linear subspace of $\mathbb{Z}_2^{|S|}$. Say this subspace has dimension d. Then by linear algebra over $\mathbb{Z}_2$, we see that $|\Sigma| = 2^d$.

Alternatively, one can also give a combinatorial proof that does not use linear algebra.

Proof of Lemma 3.1. If Σ contains no other sets except for the empty one, the statement trivially holds with $d = 0$. Let us proceed by induction on $|\Sigma|$. We can assume without loss of generality that $S = [n]$, and that there exists $A \in \Sigma$ such that $n \in A$. Let Σ_1 be set of the sets in Σ that do not contain n, and let Σ_2 be set of the sets in Σ that do.

The operation of taking the symmetric difference with A gives a bijection between Σ_1 and Σ_2. On the other hand, Σ is a disjoint union of Σ_1 and Σ_2, hence $|\Sigma| = |\Sigma_1| + |\Sigma_2| = 2|\Sigma_1|$. By the induction assumption, since Σ_1 is a family of the subsets of $[n-1]$ that is closed under $\oplus$, we have $|\Sigma_1| = 2^d$, for some d. We conclude that $|\Sigma| = 2^{d+1}$. $\square$

Examples of families of sets that satisfy the conditions of Lemma 3.1 are the kernel and the image of every boundary map ∂_n, since $\partial_n(\Sigma_1 \oplus \Sigma_2) = \partial_n(\Sigma_1) \oplus \partial_n(\Sigma_2)$ for arbitrary $\Sigma_1, \Sigma_2 \in C_n(\Delta)$. We are now ready to state the main definition of this section.

Definition 3.3. *Let Δ be a finite abstract simplicial complex. Let $n \geq 0$, and assume that $|\mathrm{Im}(\partial_{n+1})| = 2^i$, $|\mathrm{Ker}(\partial_n)| = 2^k$, for some i and k. We define the nth **Betti number** of Δ with coefficients in $\mathbb{Z}_2$ to be $\beta_n(\Delta) := k - i$.*

An important property of this boundary operator is that when applied twice it gets reduced to the trivial map.

Lemma 3.4. *Let Δ be a finite abstract simplicial complex. Then for all $n \geq 0$, the composition map $\partial_n \circ \partial_{n+1}$ maps every set to the empty one.*

Proof. For any $\Sigma \in C_{n+1}(\Delta)$, the simplices in $\partial_n(\partial_{n+1}(\Sigma))$ are obtained by taking some simplex in Σ and deleting from it two different vertices. This can be done in two different ways, depending on the order in which the two vertices are removed. In the total symmetric difference these two simplices give the combined contribution $\emptyset$, and hence the total value of $\partial_n(\partial_{n+1}(\Sigma))$ is equal to $\emptyset$ as well. $\square$

One way to rephrase Lemma 3.4 is to say that $\mathrm{Ker}(\partial_n)$ contains $\mathrm{Im}(\partial_{n+1})$ for all $n \geq 0$. In particular, we see that the Betti numbers are always nonnegative.

3.2 Simplicial Homology Groups

3.2.1 Homology Groups of Trisps with Coefficients in $\mathbb{Z}_2$

We shall now upgrade our discussion from Section 3.1 in two ways: first we replace abstract simplicial complexes with arbitrary trisps; second we now phrase our invariants algebraically as groups.

Let Δ be a trisp. For $n \geq 0$, we let $C_n(\Delta)$ be vector spaces over $\mathbb{Z}_2$ with basis indexed by the elements of $S_n(\Delta)$. The *boundary operator* $\partial_n : C_n(\Delta) \to C_{n-1}(\Delta)$ is a linear transformation that can be defined as follows. For $n \geq 1$ and an n-simplex $\sigma \in S_n(\Delta)$ we set

$$\partial_n(\sigma) := \sum_{i=1}^{n+1} B_{f_i}(\sigma), \tag{3.1}$$

where for i ranging from 1 to $n+1$, $f_i : [n] \hookrightarrow [n+1]$ is the order-preserving injection whose image does not contain the element i. The definition is then extended linearly over $\mathbb{Z}_2$ to the whole vector space $C_n(\Delta)$.

We would like to point out that in the special case that Δ is actually an abstract simplicial complex, formula (3.1) reads as

$$\partial_n(\{v_0,\ldots,v_n\}) := \sum_{i=0}^{n} \{v_0,\ldots,\hat{v}_i,\ldots,v_n\}, \qquad (3.2)$$

where $\{v_0,\ldots,v_n\}$ is an n-simplex of Δ.

An element of $C_n(\Delta)$ is called an n-chain of Δ. We use the convention that $C_{-1}(\Delta) = \{0\}$, and hence $\partial_0 : C_0(\Delta) \to C_{-1}(\Delta)$ is a 0-map.

Lemma 3.5. *Let Δ be a trisp, and let the boundary operator ∂_n be as defined by (3.1). Then we have*

$$\partial_{n-1} \circ \partial_n = 0, \qquad (3.3)$$

for all $n \geq 0$.

We remark that frequently the identity (3.3) is simply stated as $\partial^2 = 0$.

Proof of Lemma 3.5. For $n = 0$ the statement is obvious. Let $n \geq 1$, and take $\sigma \in S_n(\Delta)$. Since $B_g \circ B_f = B_{f \circ g}$, we see that to each simplex in the sum $\partial_{n-1}(\partial_n \sigma)$ we can associate an order-preserving injection $[n-2] \hookrightarrow [n]$. On the other hand, each such order-preserving injection can be represented as a composition of order-preserving injections $[n-2] \hookrightarrow [n-1] \hookrightarrow [n]$ in exactly two ways. This means that there will be precisely two identical simplices associated to each order-preserving injection $[n-2] \hookrightarrow [n]$. This implies that the total sum $\partial_{n-1}(\partial_n \sigma)$ is 0. $\square$

Let $Z_n(\Delta; \mathbb{Z}_2)$ denote the kernel of ∂_n, and let $B_n(\Delta; \mathbb{Z}_2)$ denote the image of ∂_{n+1}. The elements of the vector space $Z_n(\Delta; \mathbb{Z}_2)$ are called n-cycles with $\mathbb{Z}_2$-coefficients, and the elements of the vector space $B_n(\Delta; \mathbb{Z}_2)$ are called n-boundaries with $\mathbb{Z}_2$-coefficients. Again, reformulating Lemma 3.5 we obtain the following corollary.

Corollary 3.6. *For an arbitrary trisp Δ and any integer $n \geq 0$, we have $B_n(\Delta; \mathbb{Z}_2) \subseteq Z_n(\Delta; \mathbb{Z}_2)$.*

Now we have all the notions in hand to give an algebraic analogue of Definition 3.3.

Definition 3.7. *For $n \geq 0$, the nth **simplicial homology group with $\mathbb{Z}_2$-coefficients** of Δ is defined by*

$$H_n(\Delta; \mathbb{Z}_2) := Z_n(\Delta; \mathbb{Z}_2) / B_n(\Delta; \mathbb{Z}_2).$$

We shall skip the coefficients when these are clear.

Remark 3.8. Clearly, $H_n(\Delta; \mathbb{Z}_2)$ actually has a structure of a $\mathbb{Z}_2$-vector space, since it is a quotient of a vector space by a vector subspace. It is, however, customary to call it *homology group*.

3.2.2 Orientations

To move beyond the case of $\mathbb{Z}_2$-coefficients, we need to introduce an additional structure: the choice of orientations of all the simplices.

Assume that Δ is an abstract simplicial complex, and let $\sigma = (v_0, \ldots, v_n)$ be an n-dimensional simplex of Δ. We have a permutation action of the symmetric group $\mathcal{S}_{n+1}$ on the set of the vertices of σ. Clearly, this action is transitive, i.e., it has only one orbit. When restricted to the action of the subgroup $\mathcal{A}_{n+1}$ of even permutations, we shall have two orbits, since $\mathcal{A}_{n+1}$ is a subgroup of index 2.

Definition 3.9. *The above-defined orbits of $\mathcal{A}_{n+1}$-action are called* **orientations** *of σ.*

We have a total of two possible orientations, which geometrically correspond to choosing an orientation on the linear subspace spanned by the simplex. We shall use the term *oriented simplex* to denote a simplex with a chosen orientation, and we shall denote it by $[v_0, \ldots, v_n]$.

Example 3.10. We have $[v_0, v_1, v_2] = [v_1, v_2, v_0] \neq [v_0, v_2, v_1]$.

Note that an ordering on the set of vertices $V(A)$ gives an order within each specific simplex of Δ as well. In particular, this implies that such an ordering gives an induced orientation on all the simplices in our complex.

When Δ is an arbitrary trisp, the simplices in the gluing data come with a standard orientation. Sometimes, however, it can be advantageous to consider some nonstandard orientations. It is therefore useful to have the following notion.

Definition 3.11. *Let Δ be an arbitrary trisp. An* **orientation** *of Δ is a collection of functions $\{\epsilon_n\}_{n=0}^{\infty}$, where $\epsilon_n : S_n(\Delta) \to \mathbb{Z}_2$, for all $n \geq 0$.*

Intuitively we think that $\epsilon_n(\sigma) = 0$ corresponds to the orientation induced by the order of the simplices, whereas $\epsilon_n(\sigma) = 1$ corresponds to changing that orientation to the opposite one.

3.2.3 Homology Groups of Trisps with Integer Coefficients

We shall now combine the orientations of Subsection 3.2.2 with the algebraic procedure of Subsection 3.2.1.

Let Δ be an arbitrary trisp, with the standard orientation on its simplices. For $n \geq 0$, let $C_n(\Delta)$ denote the free abelian group generated by the oriented simplices of Δ of dimension n, that is,

$$C_n(\Delta) := \left\{ \sum_{\sigma \in S_n(\Delta)} c_\sigma \sigma \,\middle|\, c_\sigma \in \mathbb{Z} \right\}.$$

The algebraic convention that we use is that whenever the simplex changes its orientation, the corresponding generator changes its sign. Put formally, we generate the free abelian group by all orientations of simplices, and then take the quotient by the relation stating that the sum of the opposite orientations of the same underlying simplex is equal to 0.

In line with the previous terminology, a linear combination $\sum_{\sigma \in S_n(\Delta)} c_\sigma \sigma$ is called an *n-chain with integer coefficients*, and the group $C_n(\Delta)$ is called an *n-chain group with integer coefficients*.

Just as in previous cases, one can define the *boundary operator*, which this time is an abelian group homomorphism

$$\partial_n : C_n(\Delta) \to C_{n-1}(\Delta),$$

as follows: on oriented simplices, which are the group generators here, it is defined by

$$\partial_n(\sigma) := \sum_{i=1}^{n+1} (-1)^{i+1} B_{f_i}(\sigma), \tag{3.4}$$

and in the case of the abstract simplicial complex, by

$$[v_0, \dots, v_n] \mapsto \sum_{i=0}^{n} (-1)^i [v_0, \dots, \hat{v}_i, \dots, v_n], \tag{3.5}$$

which are then extended linearly to the whole abelian group $C_n(\Delta)$. We see that equations (3.4) and (3.5) differ from equations (3.1) and (3.2) only by sign.

More generally, when some other orientation ϵ on Δ is chosen, equation (3.4) changes to

$$\partial_n(\sigma) := \sum_{i=1}^{n+1} (-1)^{i+1} \epsilon_n(\sigma) \epsilon_{n-1}(B_{f_i}(\sigma)) B_{f_i}(\sigma). \tag{3.6}$$

Example 3.12. We have $\partial_2([v_0, v_1, v_2]) = [v_1, v_2] - [v_0, v_2] + [v_0, v_1]$.

Lemma 3.13. *Let Δ be an arbitrary trisp, and let Δ be the boundary operator defined by (3.4). Then for $n \geq 0$ we have*

$$\partial_{n-1} \circ \partial_n = 0. \tag{3.7}$$

Proof. In the case that Δ is an abstract simplicial complex, we see that the oriented simplex $[v_0, \dots, \hat{v}_i, \dots, \hat{v}_j, \dots, v_n]$ appears in $\partial_{n-1}(\partial_n[v_0, \dots, v_n])$ two times. This is the same as in our previous proofs. The additional ingredient here is "chasing signs": the two copies come once with the coefficient $i + (j-1)$, when v_i is deleted first, and once with the coefficient $j + i$, when v_i is deleted second. The total contribution is therefore 0, and hence we can conclude that $\partial_{n-1}(\partial_n[v_0, \dots, v_n]) = 0$.

The case that Δ is a trisp is completely analogous. $\square$

As before, $Z_n(\Delta; \mathbb{Z}) := \mathrm{Ker}(\partial_n)$ are the *n-cycles with integer coefficients*, and $B_n(\Delta; \mathbb{Z}) := \mathrm{Im}(\partial_{n+1})$ are the *n-boundaries with integer coefficients*. For all $n \geq 0$, we have $B_n(\Delta; \mathbb{Z}) \subseteq Z_n(\Delta; \mathbb{Z})$.

Definition 3.14. *Let Δ be an arbitrary trisp, and let $n \geq 0$. The nth* **simplicial homology group with integer coefficients** *of Δ is defined by*

$$H_n(\Delta; \mathbb{Z}) := Z_n(\Delta; \mathbb{Z}) / B_n(\Delta; \mathbb{Z}). \tag{3.8}$$

In particular, $H_n(\Delta; \mathbb{Z})$ is an abelian group, and $H_i(\Delta; \mathbb{Z}) = 0$, whenever $n > \dim \Delta$. To denote the totality of the homology groups one uses the notation $H_*(\Delta; \mathbb{Z})$, where $*$ stands for a variable to be assigned. We shall also omit the coefficients when it does not lead to ambiguity.

Example 3.15.
(1) Let us calculate the homology groups of a hollow triangle, i.e., let Δ be the abstract simplicial complex whose set of vertices is $\{v_0, v_1, v_2\}$ and whose set of edges is $\{\{v_0, v_1\}, \{v_1, v_2\}, \{v_0, v_2\}\}$. Its chain groups are

- $C_0(\Delta) = \langle [v_0], [v_1], [v_2] \rangle = \mathbb{Z}^3$,
- $C_1(\Delta) = \langle [v_0, v_1], [v_1, v_2], [v_0, v_2] \rangle = \mathbb{Z}^3$.

The boundary operator is $\partial_1([v_i, v_j]) = -[v_i] + [v_j]$, for all $i, j \in \{0, 1, 2\}$, $i \neq j$. Therefore, for the homology groups we have

$$H_0(\Delta) = Z_0(\Delta)/B_0(\Delta) = C_0(\Delta)/\langle [v_0] - [v_1], [v_0] - [v_2] \rangle = \langle v_0 + B_0(\Delta) \rangle = \mathbb{Z}$$

and

$$H_1(\Delta) = Z_1(\Delta)/B_1(\Delta) = Z_1(\Delta) = \langle [v_0, v_1] + [v_1, v_2] - [v_0, v_2] \rangle = \mathbb{Z}.$$

Clearly, $H_i(\Delta) = 0$ for $i \notin \{0, 1\}$.

In general, we see by a similar calculation that when Δ is the boundary of an *n*-simplex, i.e., $V(\Delta) = \{0, 1, \ldots, n\}$, and all subsets of $\{0, 1, \ldots, n\}$ except for $\{0, 1, \ldots, n\}$ itself are simplices of Δ, we have $H_i(\Delta) = 0$ for $i \neq n - 1$, and $H_{n-1}(\Delta) = \mathbb{Z}$.

(2) For any trisp that is homeomorphic to a 2-dimensional torus T, we have $H_0(T) = \mathbb{Z}$, $H_1(T) = \mathbb{Z}^2$, $H_2(T) = \mathbb{Z}$, and $H_i(T) = 0$ for $i \neq \{0, 1, 2\}$. We shall not do a complete computation here, since it is tedious and will be done more succinctly later in the book.

In general it is easy to see by a computation similar to the above that $H_0(\Delta) = \mathbb{Z}^c$, where c is the number of connected components of Δ.

Remark 3.16. Let us point out two problems with this definition:

(1) It is unclear why the homology groups are independent of the particular triangulation. This can be resolved, for example, by passing to the *singular homology*, which is defined later on in this chapter. We shall simply assume the fact, and refer the interested reader to any of the standard textbooks in algebraic topology.
(2) Simplicial computations, even for such small spaces as the torus, can be rather tedious. A quicker way to compute is provided by the *cellular homology*, which we introduce in Section 3.6.

3.3 Invariants Connected to Homology Groups

Classically, one studied invariants of topological spaces that were actually numbers, not groups. In this section we shall extract these numbers from $H_*(\Delta)$.

3.3.1 Betti Numbers and Torsion Coefficients

To start with, we need a classical result from group theory.

Theorem 3.17. (Basis theorem for abelian groups)
Every finitely generated abelian group can be written as a direct product of cyclic groups of prime-power order with a finite number of infinite cyclic groups. In this presentation, the summands are uniquely determined up to isomorphism and order.

Let Δ be a trisp such that $H_n(\Delta; \mathbb{Z})$ is finitely generated. This holds, for example, if Δ has only finitely many simplices in each dimension. By Theorem 3.17, the group $H_n(\Delta; \mathbb{Z})$ can be represented as a direct sum $F \oplus T$, where F is a free group and T is a finite group.

The rank of F (i.e., the number of generators of F) is called the *nth Betti number of* Δ and is denoted by $\beta_n(\Delta)$. The elements of T are called the *torsion elements* of $H_n(\Delta; \mathbb{Z})$. When T is trivial, one says that the nth homology is *torsion-free*, or simply that *there is no torsion*.

The torsion part can be represented as a direct product of cyclic groups of prime power order. The exponents of this groups are called the *torsion coefficients*. The nth Betti number and the torsion coefficients are uniquely determined by the group $H_n(\Delta; \mathbb{Z})$.

Finally, let us remark that the Betti numbers can be defined also when the groups $H_n(\Delta; \mathbb{Z})$ are not finitely generated. One simply takes the rank of this group as the definition. In this generality it is possible that a Betti number is equal to infinity.

3.3.2 Euler Characteristic and the Euler–Poincaré Formula

The following is perhaps the most basic numerical invariant of a topological space considered in algebraic topology.

Definition 3.18. *Let Δ be a trisp with finitely many simplices. The **Euler characteristic** of Δ is defined by*

$$\chi(\Delta) := \sum_{\sigma \in \Delta} (-1)^{\dim \sigma} = \sum_{n \geq 0} (-1)^n |S_n(\Delta)|. \tag{3.9}$$

The crucial property of the Euler characteristic is that passing to the Betti numbers simplifies the righthand side of (3.9) tremendously, due to major cancellations that take place in the process.

Theorem 3.19. (Euler–Poincaré formula).
Let Δ be a trisp with finitely many simplices. Then we have

$$\chi(\Delta) = \sum_{n \geq 0} (-1)^n \beta_n(\Delta). \tag{3.10}$$

Before we proceed with the proof, let us recall that a sequence

$$\cdots \to A_{n+1} \to A_n \to A_{n-1} \to \cdots$$

of abelian groups is called *exact* if for every n, the kernel of the map from A_n to A_{n-1} is equal to the image of the map from A_{n+1} to A_n.

Proof of Theorem 3.19. We have exact sequences

$$0 \longrightarrow Z_n(\Delta) \hookrightarrow C_n(\Delta) \longrightarrow B_{n-1}(\Delta) \longrightarrow 0$$

and

$$0 \longrightarrow B_n(\Delta) \hookrightarrow Z_n(\Delta) \longrightarrow H_n(\Delta) \longrightarrow 0.$$

By general abstract algebra we have $\operatorname{rk} C_n(\Delta) = \operatorname{rk} B_{n-1}(\Delta) + \operatorname{rk} Z_n(\Delta)$, and $\operatorname{rk} Z_n(\Delta) = \operatorname{rk} B_n(\Delta) + \operatorname{rk} H_n(\Delta)$. Combining these, we get

$$\operatorname{rk} C_n(\Delta) = \operatorname{rk} B_{n-1}(\Delta) + \operatorname{rk} H_n(\Delta) + \operatorname{rk} B_n(\Delta).$$

Summing over all n yields

$$\chi(\Delta) = \sum_{n \geq 0} (-1)^n \operatorname{rk} C_n(\Delta) = \sum_{n \geq 0} (-1)^n \operatorname{rk} H_i(\Delta) = \sum_{n \geq 0} (-1)^n \beta_n(\Delta),$$

which finishes the proof. $\square$

We remark that formula (3.10) allows us to extend the definition of the Euler characteristic to any trisp whose homology groups vanish above a certain dimension: simply take (3.10) as the definition of $\chi(\Delta)$.

3.4 Variations

3.4.1 Augmentation and Reduced Homology Groups

A useful variation of the definitions above is that of *reduced homology groups*. To define these, let us take another look at dimension -1. Until now, we have always set $C_{-1}(\Delta) := 0$. However, as the reader will recall from our discussion in Chapter 2, we do have an extra simplex in dimension -1, unless of course our complex is void.

Let us define the *reduced chain groups* $\{\widetilde{C}_n(\Delta)\}_{n=-1}^{\infty}$ by setting $\widetilde{C}_n(\Delta) := C_n(\Delta)$, for $n \geq 0$, and

$$\widetilde{C}_{-1}(\Delta) := \begin{cases} \mathbb{Z}, & \text{if } \Delta \text{ is not void;} \\ 0, & \text{otherwise.} \end{cases}$$

Assume that Δ is not void, and let ε denote a chosen generator of $\widetilde{C}_{-1}(\Delta)$. For $n \geq 1$, set $\widetilde{\partial}_n := \partial_n$. Set $\widetilde{\partial}_0(v) := \varepsilon$ for all vertices v, and extend $\widetilde{\partial}_0$ linearly to the whole group $C_0(\Delta)$. The process of adding the group $\widetilde{C}_{-1}(\Delta)$ together with the boundary operator $\widetilde{\partial}_0$ is called *augmentation*.

Definition 3.20. *Let Δ be an arbitrary trisp. Set $\widetilde{Z}_n(\Delta) := \mathrm{Ker}(\widetilde{\partial}_n)$ and $\widetilde{B}_n(\Delta) := \mathrm{Im}(\widetilde{\partial}_{n+1})$. The **reduced homology groups** of Δ are defined by $\widetilde{H}_*(\Delta) := \widetilde{Z}_n(\Delta)/\widetilde{B}_n(\Delta)$.*

The difference between the reduced homology groups and the nonreduced ones is not large. In fact,

$$\widetilde{H}_i(\Delta) = H_i(\Delta), \text{ for } i \geq 1,$$

$$H_0(\Delta) = \begin{cases} \widetilde{H}_0(\Delta) \times \mathbb{Z}, & \text{if } \Delta \text{ is not void;} \\ \widetilde{H}_0(\Delta) = 0, & \text{otherwise,} \end{cases}$$

$$\widetilde{H}_{-1}(\Delta) = \begin{cases} 0, & \text{if } \Delta \text{ is not the empty complex;} \\ \mathbb{Z}, & \text{otherwise.} \end{cases}$$

In the same vein, we get *reduced Betti numbers* $\widetilde{\beta}_i(\Delta)$. If the complex is not empty and not void, then its reduced Betti numbers are equal to the nonreduced ones in all dimensions except for 0, where we have $\beta_0(\Delta) = \widetilde{\beta}_0(\Delta) + 1$. For the empty complex we see that $\widetilde{\beta}_{-1} = 1$, and all other reduced Betti numbers are equal to 0, whereas for the void complex all of its reduced Betti numbers are equal to 0.

Finally, the *reduced Euler characteristic* is defined by setting

$$\widetilde{\chi}(\Delta) := \begin{cases} \chi(\Delta) - 1, & \text{if } \Delta \text{ is not void,} \\ 0, & \text{otherwise.} \end{cases}$$

3.4.2 Homology Groups with Other Coefficients

The complete algebraic apparatus described in Subsection 3.2.3 can be set up with an arbitrary commutative ring $\mathcal{R}$ with unit taking the role played by $\mathbb{Z}$. In this scenario, $C_n(\Delta; \mathcal{R})$ are free $\mathcal{R}$-modules generated by the simplices of Δ. The boundary operator is still defined by (3.4) and (3.5) and then extended by $\mathcal{R}$-linearity. The $\mathcal{R}$-modules $Z_n(\Delta; \mathcal{R})$ and $B_n(\Delta; \mathcal{R})$ are also defined in the same way, for all $n \geq 0$.

Definition 3.21. *Let Δ be an arbitrary trisp. For all $n \geq 0$, the quotient $H_n(\Delta; \mathcal{R}) := Z_n(\Delta; \mathcal{R})/B_n(\Delta; \mathcal{R})$ is called the nth **homology group of** Δ **with coefficients in** $\mathcal{R}$.*

Again there is some abuse of terminology here, since $H_n(\Delta; \mathcal{R})$ actually has the inherited structure of an $\mathcal{R}$-module.

As a special case, one can take any field, such as $\mathbb{Z}_2$ or $\mathbb{Q}$, to be the field of coefficients. In this situation, the homology groups are vector spaces over the respective field, and the complete homology information is encoded by the Betti numbers over this field, which are defined to be the dimensions of these spaces. It is a comforting thought to know that any field of characteristic 0 will give the same Betti numbers as the ones we got from the homology groups with integer coefficients.

In general, passing between the homology groups with different coefficients requires some work. Just to give an example, let Δ be the real projective plane $\mathbb{RP}^2$, then we have

i	$H_*(\Delta, \mathbb{Z})$	$H_*(\Delta, \mathbb{Z}_2)$
0	$\mathbb{Z}$	$\mathbb{Z}_2$
1	$\mathbb{Z}_2$	$\mathbb{Z}_2$
2	0	$\mathbb{Z}_2$

The description of the exact mechanism of going between various homology groups requires some homological algebra, which lies beyond the intended scope of this book.

3.4.3 Simplicial Cohomology Groups

Simplicial cohomology is a notion dual to that of simplicial homology.

Let Δ be an arbitrary trisp. For $n \geq 0$ we let the *nth cochain group* $C^n(\Delta; \mathbb{Z})$ be the free abelian group of all homomorphisms from $C_n(\Delta; \mathbb{Z})$ to $\mathbb{Z}$. The elements of $C^n(\Delta, \mathbb{Z})$ are called *n-cochains*. One can think of an n-cochain as a way to assign some integer values to all n-dimensional oriented simplices. Algebraically, we set

$$C^n(\Delta; \mathbb{Z}) := \mathrm{Hom}\,(C_n(\Delta), \mathbb{Z}) = \left\{ \sum_{\sigma \in S_n(\Delta)} c_\sigma \sigma^* \,\middle|\, c_\sigma \in \mathbb{Z} \right\},$$

where σ^* is the support function, it takes the value 1 on σ and 0 on the other simplices.

By general linear algebra, for all $n \geq 0$, we can *evaluate* n-cochains on n-chains:

$$C^n(\Delta; \mathbb{Z}) \times C_n(\Delta; \mathbb{Z}) \longrightarrow \mathbb{Z},$$

defined by

$$(c, e) \mapsto \langle c, e \rangle = c(e),$$

which is the same as a bilinear extension of the rule

$$\langle \sigma^*, \tau \rangle = \begin{cases} 1, & \text{if } \sigma = \tau; \\ 0, & \text{otherwise.} \end{cases}$$

Instead of the boundary operator decreasing the dimension, we have the *coboundary operator* $\partial^n : C^n(\Delta; \mathbb{Z}) \longrightarrow C^{n+1}(\Delta; \mathbb{Z})$, which increases it. The operator ∂^n is defined by the rule

$$(\partial^n f)(\sigma) := f(\partial_{n+1}\sigma), \tag{3.11}$$

for all $f \in C^n(\Delta; \mathbb{Z})$, and $\sigma \in C_{n+1}(\Delta; \mathbb{Z})$.

We would now like to make equation (3.11) more explicit. Assume that Δ is an abstract simplicial complex, and replace f by σ^* in (3.11). Then, for an oriented simplex $\tau \in C_{n+1}(\Delta; \mathbb{Z})$, $\tau = [v_0, \ldots, v_{n+1}]$, we have

$$(\partial^n \sigma^*)(\tau) = \sigma^*(\partial_{n+1}\tau) = \sigma^* \left(\sum_{i=0}^{n+1} (-1)^i [v_0, \ldots, \hat{v}_i, \ldots, v_{n+1}] \right)$$

$$= \begin{cases} (-1)^i, & \text{if there exists } i \text{ such that } \sigma = [v_0, \ldots, \hat{v}_i, \ldots, v_{n+1}]; \\ (-1)^{i+1}, & \text{if there exists } i \text{ such that } \sigma = -[v_0, \ldots, \hat{v}_i, \ldots, v_{n+1}]; \\ 0, & \text{otherwise.} \end{cases}$$

$$\tag{3.12}$$

Summarizing we see that (3.12) yields

$$\partial^n \sigma^* = \sum_w \tau_w^*, \tag{3.13}$$

where $\sigma = [v_0, \ldots, v_n]$, $\tau_w = [w, v_0, \ldots, v_n]$, and the sum is taken over all w such that τ_w is a simplex of Δ. It is not difficult to produce a similar explicit description for arbitrary trisps, we leave the details to the reader.

Note that by (3.11) we have $\partial^n \circ \partial^{n-1} = 0$, for all $n \geq 0$. Therefore we can define the cohomology groups as quotients in the same way as we did for homology.

Definition 3.22. *For an arbitrary trisp Δ, and any $n \geq 0$, the nth cohomology group of Δ with integer coefficients is defined by setting* $H^n(\Delta; \mathbb{Z}) := Z^n(\Delta; \mathbb{Z})/B^n(\Delta; \mathbb{Z}) = \text{Ker}(\partial^n)/\text{Im}(\partial^{n-1})$.

Similar to homology, the elements of $Z^n(\Delta)$ are called n-*cocycles*, and the elements of $B^n(\Delta)$ are called n-*coboundaries*.

Example 3.23.
Let us calculate the cohomology groups of a hollow triangle as above, i.e., let Δ be the abstract simplicial complex whose set of vertices is $\{v_0, v_1, v_2\}$, and whose set of edges is $\{\{v_0, v_1\}, \{v_1, v_2\}, \{v_0, v_2\}\}$.

The cochain groups are $C^0(\Delta) = \langle [v_0]^*, [v_1]^*, [v_2]^* \rangle = \mathbb{Z}^3$ and $C^1(\Delta) = \langle [v_0, v_1]^*, [v_1, v_2]^*, [v_0, v_2]^* \rangle = \mathbb{Z}^3$. The values of the coboundary operator are

$$\partial^0([v_0]^*) = [v_1, v_0]^* + [v_2, v_0]^* = -[v_0, v_1]^* - [v_0, v_2]^*,$$
$$\partial^0([v_1]^*) = [v_0, v_1]^* + [v_2, v_1]^* = [v_0, v_1]^* - [v_1, v_2]^*,$$
$$\partial^0([v_2]^*) = [v_0, v_2]^* + [v_1, v_2]^*.$$

We conclude that

$$H^0(\Delta) = Z^0(\Delta)/B^0(\Delta) = Z^0(\Delta) = \langle [v_0]^* + [v_1]^* + [v_2]^* \rangle = \mathbb{Z}$$

and

$$H^1(\Delta) = Z^1(\Delta)/B^1(\Delta) = \langle [v_0, v_1]^* + B^1(\Delta) \rangle = \mathbb{Z}.$$

Clearly, $H^i(\Delta) = 0$ for $i \notin \{0, 1\}$.

In general, the homology and cohomology groups can be different. For example, let Δ again be the real projective plane $\mathbb{RP}^2$. Then we have

i	$H_*(\Delta, \mathbb{Z})$	$H^*(\Delta, \mathbb{Z})$
0	$\mathbb{Z}$	$\mathbb{Z}$
1	$\mathbb{Z}_2$	0
2	0	$\mathbb{Z}_2$

Remark 3.24. The notions of reduced cohomology and of cohomology with coefficients in a field, or, more generally, in a commutative ring with unit, are defined in complete analogy with the homology setting. We leave the details to the reader.

3.4.4 Singular Homology

While the simplicial homology is very elementary to define and is to a certain extent useful for computations, it is not very suitable for theoretical purposes. For example, already proving that it does not depend on the triangulation is quite a hassle. An alternative way to introduce homology is provided by considering the *singular homology groups*. An additional advantage of this approach is that we are no longer bound to trisps, so we can let X be any topological space.

To start with, for any $n \geq 0$, we define a *singular n-simplex* to be any continuous map from the standard n-simplex to our space, $\sigma : \Delta^n \to X$, and let $S_n(X)$ denote the set of all singular n-simplices of X. Here comes the first difference from the simplicial homology: not only is the set of singular n-simplices not finite, in fact it is huge - unless our space X is rather degenerate, it is not countable.

Furthermore, let $\mathcal{R}$ be an arbitrary commutative ring with unit, and set

$$\operatorname{Sing}_n(X;\mathcal{R}) := \left\{ \sum_\sigma c_\sigma \sigma \mid \sigma \in S_n(X),\, c_\sigma \in \mathcal{R} \right\}.$$

At first it does not seem such a good idea to take this extremely large set of singular n-simplices, and then, on top of that, to span a free $\mathcal{R}$-module, taking the elements of this set as a basis. It turns out, however, that most of this module will cancel out, once we pass to homology.

Similar to what we have done before, we can define the *singular boundary operator* $\partial_n : \operatorname{Sing}_n(X;\mathcal{R}) \to \operatorname{Sing}_{n-1}(X;\mathcal{R})$; only this time the boundary is taken on the standard simplex first, before mapping it to X:

$$\partial_n(\sigma) := \sum_{i=0}^n (-1)^n \sigma \circ d_i,$$

where $d_i : \Delta^{n-1} \hookrightarrow \Delta^n$ is the ith boundary $(n-1)$-simplex inclusion to the standard n-simplex. Just as before, we set $Z_n^{\operatorname{Sing}}(X;\mathcal{R}) := \operatorname{Ker}(\partial_n)$ and $B_n^{\operatorname{Sing}}(X;\mathcal{R}) := \operatorname{Im}(\partial_{n+1})$ and arrive at the following definition.

Definition 3.25. *Let X be an arbitrary topological space. For any $n \geq 0$, the quotient $H_n^{\operatorname{Sing}}(X;\mathcal{R}) := Z_n^{\operatorname{Sing}}(X;\mathcal{R})/B_n^{\operatorname{Sing}}(X;\mathcal{R})$ is called the nth* **singular homology group with coefficients in** $\mathcal{R}$.

Some properties of singular homology are immediate; among these we find the following:

- the singular homology groups do not change if the topological space is replaced with a homeomorphic one;
- for any two topological spaces X and Y, for any continuous map $f : X \to Y$, and for any $n \geq 0$, we have the induced $\mathcal{R}$-module homomorphism $f_n : H_n(X;\mathcal{R}) \to H_n(Y;\mathcal{R})$.

To verify the second property note that the composition of f with continuous maps corresponding to singular simplices yields group homomorphisms $\tilde{f}_i : C_i(X;\mathcal{R}) \to C_i(Y;\mathcal{R})$, for all $i \geq 0$. It is easy to see that these homomorphisms commute with the boundary operators, and therefore induce the $\mathcal{R}$-module homomorphisms mentioned above.

We state here without proof the central fact pertaining to singular homology, which will allow us to use it instead of simplicial homology, whenever it appears opportune.

Theorem 3.26. *For an arbitrary trisp Δ, its simplicial and singular homology groups are isomorphic.*

In particular, Example 3.15(1) together with Theorem 3.26 shows that, independently of the triangulation, the homology groups of an n-dimensional sphere are given by $H_i(\mathbb{S}^n; \mathbb{Z}) = 0$ for $i \neq n$, and $H_n(\mathbb{S}^n; \mathbb{Z}) = \mathbb{Z}$.

3.5 Chain Complexes

In this section we introduce a notion from homological algebra that is behind all our definitions of homology and cohomology.

3.5.1 Definition and Homology of Chain Complexes

Definition 3.27. *Let $\mathcal{R}$ be a commutative ring with unit. A* **chain complex** $\mathcal{C}$ *is a sequence*

$$\cdots \xrightarrow{\partial_{n+2}} C_{n+1} \xrightarrow{\partial_{n+1}} C_n \xrightarrow{\partial_n} C_{n-1} \xrightarrow{\partial_{n-1}} \cdots ,$$

where the C_n's are $\mathcal{R}$-modules, and the ∂_n's are $\mathcal{R}$-module homomorphisms, which are called differentials, satisfying the identity $\partial_n \circ \partial_{n+1} = 0$, for all n.

We extend the by now familiar notation $Z_n(\mathcal{C}; \mathcal{R}) := \operatorname{Ker}(\partial_n)$ and $B_n(\mathcal{C}; \mathcal{R}) := \operatorname{Im}(\partial_{n+1})$ to the case of chain complexes. The property $\partial_n \circ \partial_{n+1} = 0$ implies that $Z_n(\mathcal{C}; \mathcal{R}) \supseteq B_n(\mathcal{C}; \mathcal{R})$, for all n, making the following definition possible.

Definition 3.28. *For a chain complex $\mathcal{C}$, the* **homology groups** *are defined by $H_n(\mathcal{C}; \mathcal{R}) := Z_n(\mathcal{C}; \mathcal{R})/B_n(\mathcal{C}; \mathcal{R})$.*

It is also convenient to introduce a *cochain complex*, which is a chain complex with differentials turned around:

$$\mathcal{C} = \cdots \xrightarrow{\partial^{n-2}} C^{n-1} \xrightarrow{\partial^{n-1}} C^n \xrightarrow{\partial^n} C^{n+1} \xrightarrow{\partial^{n+1}} \cdots ,$$

where again the C^n's are $\mathcal{R}$-modules, the ∂^n's are $\mathcal{R}$-module homomorphisms, and we require that $\partial^{n+1} \circ \partial^n = 0$, for all $n \geq 0$. Associated with a cochain complex, one has the *cohomology groups*

$$H^n(\mathcal{C}; \mathcal{R}) := \operatorname{Ker}(\partial^n)/\operatorname{Im}(\partial^{n-1}).$$

In fact, both $H_n(\mathcal{C}; \mathcal{R})$, for a chain complex $\mathcal{C}$, and $H^n(\mathcal{C}; \mathcal{R})$, for a cochain complex $\mathcal{C}$, are $\mathcal{R}$-modules.

3.5.2 Maps Between Chain Complexes and Induced Maps on Homology

Now we define the structure-preserving maps between chain complexes.

Definition 3.29. *Let $\mathcal{C}^1 = (C_*^1, \partial_*^1)$ and $\mathcal{C}^2 = (C_*^2, \partial_*^2)$ be two chain complexes. A collection of $\mathcal{R}$-module homomorphisms $f = \{f_n\}_n$, $f_n : C_n^1 \to C_n^2$, is called a* **chain map**, *written $f : \mathcal{C}^1 \to \mathcal{C}^2$, if the following diagram commutes:*

$$
\begin{array}{ccc}
C_n^1 & \xrightarrow{\ \partial_n^1\ } & C_{n-1}^1 \\
\ \downarrow{\scriptstyle f_n} & & \ \downarrow{\scriptstyle f_{n-1}} \\
C_n^2 & \xrightarrow{\ \partial_n^2\ } & C_{n-1}^2
\end{array}
\tag{3.14}
$$

Turning all the arrows around gives the definition of *cochain maps*. Note furthermore that the identity maps give a chain map, that chain maps can be composed coordinatewise, and that this composition rule is associative.

Probably one of the most used properties of chain maps is that they induce maps on the homology groups. Indeed, let $\mathcal{C}^1 = (C_*^1, \partial_*^1)$ and $\mathcal{C}^2 = (C_*^2, \partial_*^2)$ be two chain complexes, and let $f : \mathcal{C}^1 \to \mathcal{C}^2$ be a chain map. Since the diagram (3.14) commutes, we have:

(1) if $\sigma \in B_n(\mathcal{C}^1)$, then $\sigma = \partial_{n+1}^1(\tau)$, for some $\tau \in C_{n+1}^1$; hence

$$
f_n(\sigma) = f_n(\partial_{n+1}^1(\tau)) = \partial_{n+1}^2(f_{n+1}(\tau)) \in B_n(\mathcal{C}^2),
$$

so we may conclude that

$$
f_n(B_n(\mathcal{C}^1)) \subseteq B_n(\mathcal{C}^2);
\tag{3.15}
$$

(2) on the other hand, if $\sigma \in Z_n(\mathcal{C}^1)$, then $\partial_n^1(\sigma) = 0$; hence

$$
\partial_n^2(f_n(\sigma)) = f_{n-1}(\partial_n^1(\sigma)) = f_{n-1}(0) = 0,
$$

so this time we may conclude that

$$
f_n(Z_n(\mathcal{C}^1)) \subseteq Z_n(\mathcal{C}^2).
\tag{3.16}
$$

Proposition 3.30. *Let $\mathcal{C}^1$ and $\mathcal{C}^2$ be arbitrary chain complexes. Then any chain map $f : \mathcal{C}^1 \to \mathcal{C}^2$ induces homomorphisms on the homology groups $f_* : H_n(\mathcal{C}^1; \mathcal{R}) \to H_n(\mathcal{C}^2; \mathcal{R})$, for all n.*

Proof. For an arbitrary $\sigma \in Z_n(\mathcal{C}^1)$ we set $f_*(\sigma + B_n(\mathcal{C}^1)) := f_*(\sigma) + B_n(\mathcal{C}^2)$. The inclusions (3.15) and (3.16) imply that this map is a well-defined $\mathcal{R}$-module homomorphism. $\square$

The homomorphisms induced on the homology groups enjoy two important properties:

(1) when $\mathcal{C}^1 = \mathcal{C}^2$, and $f = \mathrm{id}$, we have that id_* is the identity map of the homology groups;
(2) when we have two chain maps $f : \mathcal{C}^1 \to \mathcal{C}^2$ and $g : \mathcal{C}^2 \to \mathcal{C}^3$, we can see that $g_* \circ f_* = (g \circ f)_*$.

One phrases these properties by saying that the homology construction is *functorial*. The formal framework for this terminology will be introduced in Section 4.3.

Again, turning all the arrows around, one can verify that the cochain maps induce homomorphisms on the cohomology groups.

3.5.3 Chain Homotopy

An important part of the theory of chain complexes is the observation that the chain maps can be deformed algebraically.

Definition 3.31. *Let $\mathcal{C}^1 = (C_*^1, \partial_*^1)$ and $\mathcal{C}^2 = (C_*^2, \partial_*^2)$ be two chain complexes, and let $f = \{f_n\}_n$ and $g = \{g_n\}_n$ be two chain maps $f, g : \mathcal{C}^1 \to \mathcal{C}^2$. A sequence of homomorphisms $\{\Phi_n\}_n$, where $\Phi_n : C_n^1 \to C_{n+1}^2$,*

$$\cdots \xrightarrow{\partial_{n+2}^1} C_{n+1}^1 \xrightarrow{\partial_{n+1}^1} C_n^1 \xrightarrow{\partial_n^1} C_{n-1}^1 \xrightarrow{\partial_{n-1}^1} \cdots$$
$$\downarrow{\Phi_{n+1}} \qquad \downarrow{\Phi_n} \qquad \downarrow{\Phi_{n-1}}$$
$$\cdots \xrightarrow{\partial_{n+3}^2} C_{n+2}^2 \xrightarrow{\partial_{n+2}^2} C_{n+1}^2 \xrightarrow{\partial_{n+1}^2} C_n^2 \xrightarrow{\partial_n^2} \cdots$$

*is called a **chain homotopy** between f and g if for all n we have*

$$\Phi_{n-1} \circ \partial_n^1 + \partial_{n+1}^2 \circ \Phi_n = f_n - g_n. \qquad (3.17)$$

Clearly, the existence of a chain homotopy between two maps is an equivalence relation: just replace Φ with its negative to show the symmetry, and add two chain homotopy maps to show the transitivity.

Without a doubt, the most important use of chain homotopies is to be able to show indirectly that two chain maps induce the same homology homomorphisms.

Proposition 3.32. *Let $\mathcal{C}^1 = (C_*^1, \partial_*^1)$ and $\mathcal{C}^2 = (C_*^2, \partial_*^2)$ be two chain complexes, and let $f, g : \mathcal{C}^1 \to \mathcal{C}^2$ be two chain maps such that there exists a chain homotopy Φ between f and g. Then the induced maps $f_*, g_* : H_*(\mathcal{C}^1) \to H_*(\mathcal{C}^2)$ are equal.*

Proof. To show that the induced maps f_* and g_* are equal, it is enough to verify that for any $n \geq 0$ and for any $\sigma \in C_n^1$ such that $\sigma \in \mathrm{Ker}(\partial_n^1)$, the homology classes $[f(\sigma)]$ and $[g(\sigma)]$ are equal. Applying (3.17) to this situation, we see that

$$[f(\sigma)] - [g(\sigma)] = [f(\sigma) - g(\sigma)] = [\partial_{n+1}^2(\Phi_n(\upsilon))] - 0,$$

which finishes the proof. $\square$

3.5.4 Simplicial Homology and Cohomology in the Context of Chain Complexes

Let us now return to the simplicial context. For any trisp Δ, the algebraic structure introduced in Subsection 3.2.3, and more generally in Subsection 3.4.2, can be summarized as a sequence of $\mathcal{R}$-modules and $\mathcal{R}$-module homomorphisms

$$0 \longleftarrow C_0(\Delta;\mathcal{R}) \xleftarrow{\partial_1} C_1(\Delta;\mathcal{R}) \xleftarrow{\partial_2} C_2(\Delta;\mathcal{R}) \xleftarrow{\partial_3} \cdots ,$$

which is then called a *simplicial chain complex* of Δ with coefficients in $\mathcal{R}$, and is denoted by $C_*(\Delta;\mathcal{R})$. The homology groups of this chain complex as defined in Subsections 3.4.2 and 3.2.3 coincide with the simplicial homology groups of Δ.

In the dual case, the algebraic structure introduced in Subsection 3.4.3 is best phrased as a *cochain complex* $C^*(\Delta;\mathcal{R})$:

$$0 \longrightarrow C^0(\Delta;\mathcal{R}) \xrightarrow{\partial^0} C^1(\Delta;\mathcal{R}) \xrightarrow{\partial^1} C^2(\Delta;\mathcal{R}) \xrightarrow{\partial^2} \cdots .$$

This one is called a *simplicial cochain complex* of Δ with coefficients in $\mathcal{R}$, and is denoted by $C^*(\Delta;\mathcal{R})$. The cohomology groups of this cochain complex as defined in Subsection 3.4.3 coincide with the simplicial cohomology groups of Δ.

Finally, we mention that we also have the *augmented simplicial chain complex* of Δ with coefficients in $\mathcal{R}$:

$$0 \longleftarrow \mathbb{Z} \xleftarrow{\epsilon} C_0(\Delta;\mathcal{R}) \xleftarrow{\partial_1} C_1(\Delta;\mathcal{R}) \xleftarrow{\partial_2} C_2(\Delta;\mathcal{R}) \xleftarrow{\partial_3} \cdots ,$$

which is denoted by $\widetilde{C}_*(\Delta;\mathcal{R})$ and whose homology groups coincide with the reduced simplicial homology groups of Δ, which were defined in Subsection 3.4.1.

3.5.5 Homomorphisms on Homology Induced by Trisp Maps

We are now ready to tie the previously defined notions together and show that trisp maps induce maps on homology. Handling the general trisps is not difficult, but can be unnecessarily technical. It is much more easily dealt with once one has the machinery of simplicial sets. Therefore we refer the reader to any general text on simplicial sets, such as [May92], for the proof of this fact in full generality. Here we give only an argument for simplicial maps between abstract simplicial complexes.

Proposition 3.33. *Let Δ_1 and Δ_2 be arbitrary abstract simplicial complexes, and let $f : \Delta_1 \to \Delta_2$ be a simplicial map. Then for all $n \geq 0$, the map f induces homomorphisms on the homology groups $f_* : H_*(\Delta_1;\mathcal{R}) \to H_*(\Delta_2;\mathcal{R})$.*

Before the proof, let us introduce the following convention: $[v_0, \ldots, v_n] = 0$ if $v_0 = v_1$. It follows from our orientation discussion that this implies $[v_0, \ldots, v_n] = 0$ if there exist $i \neq j$ such that $v_i = v_j$. This works well with the boundary operator

$$\partial_n([v, v, v_2, \ldots, v_n]) = [v, v_2, \ldots, v_n] - [v, v_2, \ldots, v_n]$$
$$+ \sum_{i=2}^{n} (-1)^i [v, v, v_2, \ldots, \hat{v}_i, \ldots, v_n] = 0.$$

This convention will simplify our arguments considerably, since by using it we can avoid dealing with different cases, and can perform a unified computation.

Proof of Proposition 3.33. By Proposition 3.30 we need to show that a simplicial map $f : \Delta_1 \to \Delta_2$ will induce a chain map between the corresponding simplicial chain complexes $f : C_*(\Delta_1; \mathcal{R}) \to C_*(\Delta_2; \mathcal{R})$.

On the oriented simplices the chain map is simply defined by

$$f_n : [v_0, \ldots, v_n] \mapsto [f(v_0), \ldots, f(v_n)],$$

where we used the above convention that $[w_0, \ldots, w_n] = 0$ if not all the vertices are distinct. By the linear extension, this induces a map of chain groups $f_n : C_n(\Delta_1) \to C_n(\Delta_2)$.

By definition of chain maps, to see that the collection of maps $\{f_n\}_n$ defines a chain map, we need to see that the following diagram commutes:

$$
\begin{array}{ccc}
C_n(\Delta_1) & \xrightarrow{\partial_n} & C_{n-1}(\Delta_1) \\
\downarrow{\scriptstyle f_n} & & \downarrow{\scriptstyle f_n} \\
C_n(\Delta_2) & \xrightarrow{\partial_n} & C_{n-1}(\Delta_2)
\end{array}
$$

By linearity, it is enough to verify this fact for oriented simplices, which is straightforward, as the following diagram shows:

$$
\begin{array}{ccc}
[v_0, \ldots, v_n] & \xrightarrow{\partial_n} & \sum_{i=0}^{n}(-1)^i[v_0, \ldots, \hat{v}_i \ldots, v_n] \\
\downarrow{\scriptstyle f_*} & & \downarrow{\scriptstyle f_*} \\
[f(v_0), \ldots, f(v_n)] & \xrightarrow{\partial_n} & \sum_{i=0}^{n}(-1)^i[f(v_0), \ldots, \widehat{f(v_i)} \ldots, f(v_n)]
\end{array}
$$

We conclude that a simplicial map between abstract simplicial complexes induces a chain map between associated simplicial chain complexes, and hence also the homomorphisms between the corresponding homology groups. $\square$

We would like to make three remarks to expand on Proposition 3.33.

Remark 3.34. Just as for the chain complexes, the construction is functorial in this case as well, that is, the identity map induces an identity map, and a composition of two maps induces the composition of the induced maps.

Remark 3.35. For two trisps Δ_1 and Δ_2, a trisp map $f : \Delta_1 \to \Delta_2$ induces a chain map between associated augmented simplicial chain complexes, and therefore homomorphisms on reduced homology $f^* : \widetilde{H}^*(\Delta_1; \mathcal{R}) \to \widetilde{H}^*(\Delta_2; \mathcal{R})$.

Remark 3.36. For two trisps Δ_1 and Δ_2, a trisp map $f : \Delta_1 \to \Delta_2$ also induces a cochain map between associated simplicial cochain complexes. However, this map goes in the opposite direction $f : C^*(\Delta_2; \mathcal{R}) \to C^*(\Delta_1; \mathcal{R})$, and therefore it induces a map between corresponding cohomology groups going in the opposite direction as well, $f^* : H^*(\Delta_2; \mathcal{R}) \to H^*(\Delta_1; \mathcal{R})$.

3.6 Cellular Homology

3.6.1 An Application of Homology with Integer Coefficients: Winding Number

Let γ be a closed curve in $\mathbb{R}^2$, i.e., $\gamma : \mathbb{S}^1 \to \mathbb{R}^2$, and assume that the origin does not belong to the image of γ. Using homology groups with integer coefficients, it is possible to formalize the intuitive notion of "the number of times the curve γ winds around the origin."

Let X be the unit circle centered at the origin. Define $\tilde{\gamma} : \mathbb{S}^1 \to X$ by mapping each $t \in \mathbb{S}^1$ to $\gamma(t)/\|\gamma(t)\|$. This map is well-defined, since the vector $\gamma(t)$ is always different from 0. By the functoriality, we have the induced map $\tilde{\gamma}_* : H_1(\mathbb{S}^1; \mathbb{Z}) \to H_1(X; \mathbb{Z})$.

On the other hand, we know that $H_1(\mathbb{S}^1; \mathbb{Z}) = H_1(X; \mathbb{Z}) = \mathbb{Z}$. Clearly, the group homomorphisms $\varphi : \mathbb{Z} \to \mathbb{Z}$ are uniquely determined, hence classified, by the value $\varphi(1)$. We fix the unit element in $H_1(\mathbb{S}^1)$ by fixing an orientation of the curve.

Definition 3.37.
*(1) The **winding number**[1] of a closed curve γ is the value $\tilde{\gamma}_*(1)$, where the map $\tilde{\gamma}$ is defined as above.*
*(2) More generally, for an arbitrary continuous $f : \mathbb{S}^n \to \mathbb{S}^n$, we have an induced map $f_* : H_n(\mathbb{S}^n; \mathbb{Z}) \to H_n(\mathbb{S}^n; \mathbb{Z})$. Again, $H_n(\mathbb{S}^n; \mathbb{Z}) = \mathbb{Z}$, and the value $f_*(1)$ is called the **degree** of f, and is denoted by $\deg(f)$.*

The winding number of a closed curve, and more generally, the degree of a continuous map between spheres of the same dimension, are invariants, which are useful in many situations. We shall need these to define the so-called incidence numbers, which in turn allow one to give an explicit description of cellular homology.

---

[1] Sometimes this invariant is called the *signed winding number*.

3.6.2 The Definition of Cellular Homology

The main reason why CW complexes are so handy for concrete computations of homology groups is that it turns out that one can substitute the simplicial chain complex with another, usually much smaller, chain complex, which we now proceed to define.

Given a CW complex Δ, for all $n \geq 0$, the $\mathcal{R}$-module $C_n^{\mathrm{Cell}}(\Delta; \mathcal{R})$, called the *nth cellular chain group*, is generated by all n-cells, rather than by all n-simplices:

$$C_n^{\mathrm{Cell}}(\Delta) := \left\{ \sum_{\sigma \in S_n(\Delta)} c_\sigma \sigma \;\middle|\; c_\sigma \in \mathcal{R} \right\}, \tag{3.18}$$

where $S_n(\Delta)$ denotes the set of all n-cells of Δ. The boundary operator is then defined as follows: for an n-cell σ we have

$$\partial_n^{\mathrm{Cell}} \sigma = \sum_\tau [\tau : \sigma] \tau, \tag{3.19}$$

where the sum is taken over all $(n-1)$-cells of Δ, and the numbers $[\tau : \sigma]$ are the *incidence numbers*. These are defined by

$$[\tau : \sigma] = \deg(p_\tau \circ f_{\partial\sigma}),$$

where $f_{\partial\sigma} : \partial B^n \to \Delta^{(n-1)}$ is the attachment map of the cell σ, and p_τ is the composition $p_\tau : \Delta^{(n-1)} \to \Delta^{(n-1)}/\Delta^{(n-2)} \to \mathbb{S}^{n-1}$, where the first map is the quotient map shrinking the $(n-2)$-skeleton to a point, and the second map is the projection onto the sphere corresponding to τ.

Together, equations (3.18) and (3.19) define the *cellular chain complex* of a given CW complex:

$$\cdots \xrightarrow{\partial_{n+2}^{\mathrm{Cell}}} C_{n+1}^{\mathrm{Cell}} \xrightarrow{\partial_{n+1}^{\mathrm{Cell}}} C_n^{\mathrm{Cell}} \xrightarrow{\partial_n^{\mathrm{Cell}}} C_{n-1}^{\mathrm{Cell}} \xrightarrow{\partial_{n-1}^{\mathrm{Cell}}} \cdots. \tag{3.20}$$

Although in general it may be hard to compute the incidence numbers, in particular situations it is often geometrically clear what they are.

Definition 3.38. *For an arbitrary CW complex Δ, the homology groups of the cellular chain complex defined above are called **cellular homology groups** of Δ, and are denoted by $H_*^{\mathrm{Cell}}(\Delta; \mathcal{R})$.*

The alert reader will notice that we have not proved that the algebraic structure in (3.20) is in fact a chain complex. For that, one would need to verify that the cellular boundary operator satisfies the equation $\partial_n^{\mathrm{Cell}} \circ \partial_{n+1}^{\mathrm{Cell}} = 0$ for all n. To do so directly from definition (3.19) would require a fairly technical analysis of the incidence numbers.

Fortunately, there is another interpretation of the cellular chain groups as certain *relative homology groups*. The above property of the cellular boundary operator would then follow from the exactness of certain parts of the associated *long exact sequence*. We shall therefore assume that the cellular homology is well-defined without proof for now, and will return to it in Chapter 5, where the necessary additional tools will be introduced and studied.

Dually, we can associate a cochain complex $C^*_{\text{Cell}}(\Delta; \mathcal{R})$ to a cell complex Δ in the standard way: $C^n_{\text{Cell}}(\Delta; \mathcal{R})$ is taken to be a free $\mathcal{R}$-module with the generators indexed by the n-dimensional cells (the module of $\mathcal{R}$-valued functionals on cells of Δ), and the differential maps are given by the corresponding coboundary maps. Sometimes this particular cochain complex is called a *cellular cochain complex*.

3.6.3 Cellular Maps and Properties of Cellular Homology

As mentioned above, for CW complexes it does not matter which definition of the homology groups one uses. We give here the precise statement without a proof.

Proposition 3.39. *For any CW complex Δ and any commutative ring $\mathcal{R}$ with unit, the homology groups $H^{\text{Cell}}_*(\Delta; \mathcal{R})$ and $H^{\text{Sing}}_*(\Delta; \mathcal{R})$ are naturally isomorphic.*

Because of this, we shall usually omit designations Sing and Cell and just write $H_*(\Delta)$. As an immediate corollary of this fact we see that if a topological space X has a CW structure that does not have any $q - 1$ or $q + 1$ cells, then the group $H_q(X)$ is free, and $\beta_q(X)$ is equal to the number of q-cells. This, for example, immediately yields the homology groups of spheres $\mathbb{S}^n$, for $n \neq 1$, as well as the homology groups of $\mathbb{CP}^n$.

It is important to note that the cellular homology also enjoys functoriality as long as the maps are taken to be cellular.

Definition 3.40. *Let Δ_1 and Δ_2 be arbitrary CW complexes. A continuous map $f : \Delta_1 \to \Delta_2$ is called* **cellular** *if $f(\Delta_1^{(n)}) \subseteq \Delta_2^{(n)}$, for all n.*

In particular, the identity map is cellular and the composition of two cellular maps is again cellular. We also see that a cellular map f induces a continuous map $\Delta_1^{(n)}/\Delta_1^{(n-1)} \to \Delta_2^{(n)}/\Delta_2^{(n-1)}$ for all $n \geq 0$.

Theorem 3.41. *Let Δ_1 and Δ_2 be CW complexes. A cellular map $f : \Delta_1 \to \Delta_2$ induces group homomorphisms $f_* : H_n(\Delta_1) \to H_n(\Delta_2)$, for every $n \geq 0$.*

Again we see that the identity map induces identity maps, and that the composition of two cellular maps induces the composition of induced maps. Theorem 3.41 is best when one uses the alternative definition of cellular homology using the relative homology groups. It will follow in that context from the naturality of the associated long exact sequence of a pair. We postpone the precise argument until Subsection 5.2.2.

Remark 3.42. More generally, any continuous map $f : \Delta_1 \to \Delta_2$ induces group homomorphisms $f_n : H_n(\Delta_1) \to H_n(\Delta_2)$, for every n.

4

Concepts of Category Theory

Category theory can be viewed as a metalanguage, whose main utility is to create a common framework for seemingly diverse notions. We have already used several of its concepts in disguise and will now proceed with the formal introduction.

4.1 The Notion of a Category

The basic stock-in-trade of the theory is that of a *category*.

4.1.1 Definition of a Category, Isomorphisms

Definition 4.1. *A category C is a pair of classes[1] $(\mathcal{O}, \mathcal{M})$ satisfying certain properties. The class $\mathcal{O}$ is called the class of* **objects**, *and the class $\mathcal{M}$ is called the class of* **morphisms**.
The class $\mathcal{M}$ is actually a disjoint union of sets $\mathcal{M}(a, b)$, for every pair $a, b \in \mathcal{O}$, with a given composition rule

$$\mathcal{M}(a, b) \times \mathcal{M}(b, c) \to \mathcal{M}(a, c), \quad (m_1, m_2) \to m_2 \circ m_1.$$

This composition rule is required to satisfy the following axioms:

- *composition is associative, when defined;*

---

[1] This word is needed here in order to avoid getting into straits of set theory. Intuitively, this means that we assume the existence of a universe U, which consists of all sets, and call the subsets of U *classes*. The main problem is that if we try to consider *all sets with a certain property*, the so-called *comprehension principle*, then we might get something that is not a set, but a class. The famous example of this is considering all sets that do not have themselves as a member. This issue is not central for the kind of mathematics that we consider in this book, so we refer the interested reader to [McL98, Sections I.6, I.7] and to [AHS06] for further details.

- *for each $a \in \mathcal{O}$ there exists a (necessarily unique) identity morphism $1_a \in \mathcal{M}(a,a)$ such that $1_a \circ f = f$, $g \circ 1_a = g$, whenever the compositions are defined.*

A category $\mathcal{C}$ is called **small** *if the class of objects $\mathcal{O}(\mathcal{C})$ is a set.*

We use the following notation: for a morphism $a \xrightarrow{m} b$, we write $a = \operatorname{dom} m$, or $a = \partial^{\bullet} m$, and $b = \operatorname{cod} m$, or $b = \partial_{\bullet} m$. We call $\partial^{\bullet} m$ the *domain* of m, and we call $\partial_{\bullet} m$ the *codomain* of m.

Given a category $\mathcal{C}$, its *opposite category*, denoted by $\mathcal{C}^{\mathrm{op}}$, is the one we get by "inverting all arrows," i.e., $\mathcal{O}(\mathcal{C}^{\mathrm{op}}) = \mathcal{O}(\mathcal{C})$, and for every pair of objects $x, y \in \mathcal{O}(\mathcal{C})$, we have $\mathcal{M}_{\mathcal{C}^{\mathrm{op}}}(x,y) = \mathcal{M}_{\mathcal{C}}(y,x)$, with the same composition rule as in $\mathcal{C}$.

Given categories $\mathcal{C}$ and $\mathcal{D}$, we say that $\mathcal{D}$ is a *subcategory* of $\mathcal{C}$ if $\mathcal{O}(\mathcal{D})$ is a subclass of $\mathcal{O}(\mathcal{C})$, for every $x, y \in \mathcal{O}(\mathcal{D})$ we have $\mathcal{M}_{\mathcal{D}}(x,y) \subseteq \mathcal{M}_{\mathcal{C}}(x,y)$, and the identity morphisms and the composition rule in $\mathcal{D}$ are inherited from $\mathcal{C}$. If, in addition, we have $\mathcal{M}_{\mathcal{D}}(x,y) = \mathcal{M}_{\mathcal{C}}(x,y)$ for all $x, y \in \mathcal{O}(\mathcal{D})$, then we say that $\mathcal{D}$ is a *full subcategory* of $\mathcal{C}$.

The following terminology corresponds to the intuitive notion of equality between objects:

- A morphism m is called an *inverse* of $\tilde{m}$ if both compositions $m \circ \tilde{m}$ and $\tilde{m} \circ m$ exist, which are then both equal to identity morphisms.
- A morphism is called an *isomorphism* if it has an inverse.
- Two objects are called *isomorphic* if there is an isomorphism between them.

Clearly, being isomorphic is an equivalence relation. First, the inverse of the isomorphism has an inverse (the isomorphism itself), hence is an isomorphism too. Second, a composition of two isomorphisms is also an isomorphism, since by associativity its inverse is simply the composition of the inverses of the two isomorphisms, taken in the reverse order.

An *automorphism* is a morphism m, which has an inverse, and whose domain coincides with its codomain. Automorphisms of a fixed object in the category form a group with respect to the composition.

4.1.2 Examples of Categories

The best way to get acquainted with categories is to look at the numerous examples that can be found in almost every branch of mathematics.

Example 4.2. The category of sets, **Sets.**

- The class of objects consists of all sets.
- For two sets A and B, the set $\mathcal{M}(A,B)$ consists of all set maps $A \to B$.

We note that A and B are isomorphic in **Sets** if and only if they have the same cardinality.

Example 4.3. The category of topological spaces, **Top***.*

- The class of objects consists of all topological spaces.
- For two topological spaces A and B, the set $\mathcal{M}(A, B)$ consists of all continuous maps $A \to B$.

We note that A and B are isomorphic in **Top** if and only if A and B are homeomorphic as topological spaces.

Example 4.4. The category of groups, **Grp***.*

- The class of objects consists of all groups.
- For two groups G and H, the set $\mathcal{M}(G, H)$ consists of all group homomorphisms $G \to H$.

We note that G and H are isomorphic in **Grp** if and only if the groups G and H are isomorphic as groups.

Example 4.5. The category of abelian groups, **Ab***.*

- The class of objects consists of all abelian groups.
- For two abelian groups G and H, the set $\mathcal{M}(G, H)$ consists of all group homomorphisms $G \to H$.

We note that H and G are isomorphic in **Ab** if and only if the abelian groups G and H are isomorphic as groups.

Example 4.6. The category of modules over a ring, $\mathcal{R}$**-Mod***.*

Let $\mathcal{R}$ be a ring.

- The class of objects of $\mathcal{R}$**-Mod** consists of all $\mathcal{R}$-modules.
- For two $\mathcal{R}$-modules A and B, the set $\mathcal{M}(A, B)$ consists of all $\mathcal{R}$-module homomorphisms $A \to B$.

We note that A and B are isomorphic in $\mathcal{R}$**-Mod** if and only if they are isomorphic as $\mathcal{R}$-modules.

Example 4.7. The category of vector spaces over the field **k***,* **Vect**$_{\mathbf{k}}$*.*

Let **k** be a field.

- The class of objects consists of all vector spaces over the field **k**.
- For two vector spaces over the field **k**, V and W, the set $\mathcal{M}(V, W)$ consists of all linear transformations $V \to W$.

We note that V and W are isomorphic in **Vect**$_{\mathbf{k}}$ if and only if V and W are isomorphic as vector spaces.

Example 4.8. Free category generated by a directed graph G*,* $\mathcal{FC}_G$*.*

Let G be a directed graph.

- The set of objects consists of all vertices of G.
- For two vertices x and y, the set $\mathcal{M}(x, y)$ consists of all directed paths from x to y (including the empty path in the case $x = y$). The composition is defined by path concatenation.

We note that x and y are isomorphic in $\mathcal{F}C_G$ if and only if $x = y$.

Example 4.9. The category defined by a partially ordered set P.
To an arbitrary partially ordered set P, see Definition 2.18, we can associate a category as follows.

- The set of objects consists of all elements of P.
- For two elements x and y, the set $\mathcal{M}(x, y)$ is empty unless $x \geq y$; otherwise, it consists of a unique element $(x \to y)$. The composition $\mathcal{M}(x, y) \circ \mathcal{M}(y, z)$ is defined only if $x \geq y$ and $y \geq z$, in which case we have

$$(x \to y) \circ (y \to z) = (x \to z).$$

We note that x and y are isomorphic in P if and only if $x = y$.

Example 4.10. The category defined by a group G.

Let G be a group.

- The set of objects consists of one element e.
- The set of morphisms $\mathcal{M}(e, e)$ is indexed by the elements of G, with the composition defined by the group multiplication.

We shall denote this category by $\mathcal{C}G$.

*Example 4.11. The category of graphs, **Graphs**.*

- The class of objects consists of all graphs.
- For two graphs T and G, the set $\mathcal{M}(T, G)$ consists of all graph homomorphisms $T \to G$; see Definition 9.20 and Proposition 9.22.

We note that T and G are isomorphic in **Graphs** if and only if T and G are isomorphic as graphs.

*Example 4.12. The category of abstract simplicial complexes, **ASC**.*

- The class of objects consists of all abstract simplicial complexes.
- For two abstract simplicial complexes $\Delta_1, \Delta_2 \in \mathcal{O}(\textbf{ASC})$, the set of morphisms $\mathcal{M}(\Delta_1, \Delta_2)$ consists of all simplicial maps $\Delta_1 \to \Delta_2$.

We note that Δ_1 and Δ_2 are isomorphic in **ASC** if and only if they are isomorphic as abstract simplicial complexes.

*Example 4.13. The category of trisps, **TriSp**.*

- The class of objects consists of all trisps.

- For two trisps $\Delta_1, \Delta_2 \in \mathcal{O}(\mathbf{TriSp})$, the set of morphisms $\mathcal{M}(\Delta_1, \Delta_2)$ consists of all trisp maps $\Delta_1 \to \Delta_2$.

We have not defined the notion of isomorphisms between trisps. It is natural to take it as a definition that two trisps are isomorphic if and only if they correspond to isomorphic objects in $\mathbf{TriSp}$.

Example 4.14. The category of regular trisps, $\mathbf{RTS}$.

- The class of objects consists of all regular trisps.
- For two regular trisps $\Delta_1, \Delta_2 \in \mathcal{O}(\mathbf{RTS})$, the set of morphisms $\mathcal{M}(\Delta_1, \Delta_2)$ consists of all trisp maps $\Delta_1 \to \Delta_2$.

Example 4.15. The category of chain complexes of $\mathcal{R}$-modules, $\mathcal{R}$-$\mathbf{CCom}$.

Let $\mathcal{R}$ be a ring.

- The class of objects consists of all chain complexes of $\mathcal{R}$-modules.
- For two chain complexes $\mathcal{C}^1, \mathcal{C}^2 \in \mathcal{O}(\mathcal{R}\text{-}\mathbf{CCom})$, the set of morphisms $\mathcal{M}(\mathcal{C}^1, \mathcal{C}^2)$ consists of all chain complex homomorphisms $\mathcal{C}^1 \to \mathcal{C}^2$.

Again, the notion of isomorphisms between chain complexes was not defined. We take it as a definition that two chain complexes are isomorphic if and only if they correspond to isomorphic objects in $\mathcal{R}$-$\mathbf{CCom}$.

4.2 Some Structure Theory of Categories

Let us look at some concepts that can be defined in terms of the notion of a category alone.

4.2.1 Initial and Terminal Objects

Often our category has special objects, distinguished by the structure of morphisms that have this object as their domain or as their codomain.

Definition 4.16.

- *An **initial object** of a category $\mathcal{C}$ is an object* init $\in \mathcal{O}(\mathcal{C})$ *such that for any object $a \in \mathcal{O}(\mathcal{C})$, there exists a unique morphism* init $\to a$.
- *A **terminal object** of a category $\mathcal{C}$ is an object* term $\in \mathcal{O}(\mathcal{C})$ *such that for any object $a \in \mathcal{O}(\mathcal{C})$, there exists a unique morphism $a \to$ term.*

  Following are examples of initial objects:

- in $\mathcal{C} = \mathbf{Sets}$, init is the empty set;
- in $\mathcal{C} = \mathbf{Grp}$, init is the group with one element $G = \{e\}$;
- in $\mathcal{C} = \mathbf{Graphs}$, init is the empty graph;
- in $\mathcal{C} = \mathbf{ASC}$, init is the void abstract simplicial complex;

- when $\mathcal{C}$ is the category defined by a partially ordered set, init is the maximal element, when it exists; it is denoted by $\hat{1}$.

Following are examples of the terminal objects:

- in $\mathcal{C} = \mathbf{Sets}$, any set with one element is a terminal object;
- in $\mathcal{C} = \mathbf{Grp}$, term is the same as init;
- in $\mathcal{C} = \mathbf{Graphs}$, term is the loop graph, i.e., the graph with one vertex and one edge; we call this graph $\mathsf{C}K_1$;
- in $\mathcal{C} = \mathbf{ASC}$, any abstract simplicial complex $\{\emptyset, \{v\}\}$, where v is some vertex, is a terminal object;
- when $\mathcal{C}$ is the category defined by a partially ordered set, term is the minimal element, when it exists; it is denoted by $\hat{0}$.

Initial and terminal objects do not have to exist. In any case, the category can be extended by adding an element with a unique morphism to or from every other object. Even if the initial and terminal objects already exist, they are not necessarily unique. The good news, as the next proposition shows, is that they are unique up to isomorphism.

Proposition 4.17. *In an arbitrary category $\mathcal{C}$, any two initial objects are isomorphic, and any two terminal objects are isomorphic.*

Proof. We give an argument for two initial objects. To obtain the proof for terminal objects, simply reverse all the arrows. Let init_1 and init_2 be initial objects of $\mathcal{C}$. Since init_1 is an initial object, there exists a unique morphism $\alpha : \mathrm{init}_1 \to \mathrm{init}_2$. Since init_2 is an initial object, there exists a unique morphism $\beta : \mathrm{init}_2 \to \mathrm{init}_1$. Since there exists only one morphism from init_1 to itself, we conclude that the composition morphism $\beta \circ \alpha$ must be an identity. In the same way $\alpha \circ \beta$ must be an identity. This proves that α and β are isomorphisms, and hence init_1 and init_2 are isomorphic. $\square$

4.2.2 Products and Coproducts

We start with the notion of a coproduct. Although it is dual to product on the theoretical level, it is easier to construct than the product in most concrete cases in algebra, topology, and combinatorics.

Definition 4.18. *Let $\mathcal{C}$ be a category, and let a and b be two objects of $\mathcal{C}$. The **coproduct** of a and b is an object $c = a \coprod b$ together with morphisms $\alpha : a \to c$, $\beta : b \to c$ (called the structure morphisms) satisfying the following universal property: for every object d and morphisms $\tilde{\alpha} : a \to d$, $\tilde{\beta} : b \to d$, there exists a unique morphism $\gamma : c \to d$ such that the diagram in Figure 4.1 commutes.*

Intuitively, one should think of the coproduct of a and b as the way to combine a and b without any additional constraints. In many situations this

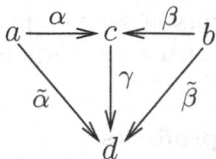

Fig. 4.1. The coproduct universal diagram.

amounts to taking a disjoint union or something induced by a disjoint union. It is also natural to think about the structure morphisms α and β as corresponding inclusion maps.

Example 4.19. **Examples of coproducts:**

- in $\mathcal{C} = \mathbf{Sets}$, the coproduct of two sets is their disjoint union; the structure morphisms are the inclusion maps;
- in $\mathcal{C} = \mathbf{Grp}$, the coproduct of two groups G and H is their *free product*, i.e., the product obtained by forming all strings of the elements from G and H, without any relations additional to those induced by already existing relations in G and in H; the structure morphisms are the inclusion maps;
- in $\mathcal{C} = \mathbf{Graphs}$, the coproduct of two graphs is their disjoint union; the structure morphisms are the inclusion maps;
- in $\mathcal{C} = \mathbf{ASC}$, the coproduct of two abstract simplicial complexes is also their disjoint union, and the structure morphisms are again the inclusion maps.

Reversing all arrows, we obtain the dual notion of a product.

Definition 4.20. *Let $\mathcal{C}$ be a category, and let a and b be two objects of $\mathcal{C}$. The* **product** *of a and b is an object $c = a \prod b$ together with morphisms $\alpha : c \to a$, $\beta : c \to b$, satisfying the following universal property: for every object d and morphisms $\tilde{\alpha} : d \to a$, $\tilde{\beta} : d \to b$, there exists a unique morphism $\gamma : d \to c$ such that the diagram of Figure 4.2 commutes:*

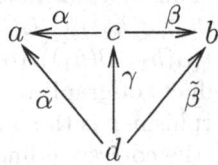

Fig. 4.2. The product universal diagram.

This time around, the intuition for the product of a and b should be that of a direct product. The structure morphisms should be thought of as corresponding projection maps.

Example 4.21. **Examples of products:**

- in $\mathcal{C} = \textbf{Sets}$, the product of two sets is their direct product; the structure morphisms are the projection maps;
- in $\mathcal{C} = \textbf{Grp}$, the product of two groups G and H is their *direct product*, i.e., the set $G \times H$, with the product defined by the rule

$$(g_1, h_1) \cdot (g_2, h_2) = (g_1 \cdot g_2, h_1 \cdot h_2);$$

the structure morphisms are the projection group homomorphisms.

The special case of products and coproducts in a partially ordered set regarded as a category leads to an important notion.

Definition 4.22. *A poset P is called a* **lattice** *if when regarded as a category, it has products and coproducts.*

Let now $\mathcal{C} = \textbf{Graphs}$.

Definition 4.23. *For arbitrary graphs T and G, the* **direct product** $T \times G$ *is defined as follows:* $V(T \times G) = V(T) \times V(G)$, *and* $E(T \times G) = \{((x, y), (x', y')) \mid (x, x') \in E(T), (y, y') \in E(G)\}$.

For example, $K_2 \times K_2$ is a disjoint union of two copies of K_2, whereas $G \times \complement K_1$ is isomorphic to G for an arbitrary graph G.

Proposition 4.24. *For arbitrary graphs T and G, their direct product (together with the projection graph homomorphisms) and their categorical product in* **Graphs** *coincide.*

Proof. Assume that H is a graph, and that $\alpha : H \to T$, $\beta : H \to G$ are graph homomorphisms. Define a set map $\varphi : V(H) \to V(T) \times V(G)$ by $\varphi(h) = (\alpha(h), \beta(h))$, for all $h \in V(H)$.

First, the map φ is a graph homomorphism, since if $(h_1, h_2) \in E(H)$, then $(\alpha(h_1), \alpha(h_2)) \in E(T)$ and $(\beta(h_1), \beta(h_2)) \in E(G)$, since α and β are graph homomorphisms. This implies $((\alpha(h_1), \beta(h_1)), (\alpha(h_2), \beta(h_2))) \in E(T \times G)$ by the definition of the direct products of graphs.

Second, the graph homomorphism φ is the unique one satisfying the commuting relations, since already the corresponding set map is unique. $\square$

Finally, let us consider the case $\mathcal{C} = \textbf{ASC}$. On the one hand, we see already with the example of two intervals that a topological direct product of the geometric realizations of two abstract simplicial complexes does not have an a priori given simplicial structure. On the other hand, it may well be that there is a categorical product that is different from the topological one.

Definition 4.25. *Let Δ_1 and Δ_2 be arbitrary abstract simplicial complexes. We define $\Delta_1 \prod \Delta_2$ to be the following abstract simplicial complex:*

- *for the set of vertices we set $V(\Delta_1 \prod \Delta_2) := V(\Delta_1) \times V(\Delta_2)$, and let $p_1 : V(\Delta_1 \prod \Delta_2) \to V(\Delta_1)$ and $p_2 : V(\Delta_1 \prod \Delta_2) \to V(\Delta_2)$ be the corresponding projections;*
- *the simplices of $\Delta_1 \prod \Delta_2$ are defined by the rule saying that $\sigma \in \Delta_1 \prod \Delta_2$ if and only if $p_1(\sigma) \in \Delta_1$ and $p_2(\sigma) \in \Delta_2$.*

It is clearly seen that set maps p_1 and p_2 from Definition 4.25 actually induce simplicial maps $p_1 : \Delta_1 \prod \Delta_2 \to \Delta_1$ and $p_2 : \Delta_1 \prod \Delta_2 \to \Delta_2$.

Proposition 4.26. *For arbitrary abstract simplicial complexes Δ_1 and Δ_2, the abstract simplicial complex $\Delta_1 \prod \Delta_2$, together with projection maps p_1 and p_2, is the categorical product of Δ_1 and Δ_2 in* **ASC**.

Proof. As we have already mentioned, p_1 and p_2 are simplicial maps. All we need to do is to verify universality. Assume therefore that we have an abstract simplicial complex Δ and two simplicial maps $q_1 : \Delta \to \Delta_1$ and $q_2 : \Delta \to \Delta_2$.

We define $q : V(\Delta) \to V(\Delta_1 \prod \Delta_2)$ by setting $q(v) := (q_1(v), q_2(v))$. Clearly, this is the only possible choice if we want the thus formed triangles to commute, i.e., if we require $q_1 = p_1 \circ q$ and $q_2 = p_2 \circ q$. Hence the uniqueness of the set map follows.

Finally, for $\sigma \in \Delta$, set $\tau := q(\sigma)$. Using the same commuting relations we see that $p_1(\tau) = q_1(\sigma)$ and $p_2(\tau) = q_2(\sigma)$; in particular, both are simplices. By Definition 4.25 we get that $\tau \in \Delta_1 \prod \Delta_2$, and hence the map q is simplicial. □

For example, the categorical product of two simplices Δ^A and Δ^B is the simplex $\Delta^{A \times B}$.

Proposition 4.27. *Let $\mathcal{C}$ be an arbitrary category, and let a, b be objects of $\mathcal{C}$. Any two coproducts of a and b are isomorphic. Also, any two products of a and b are isomorphic.*

Proof. Again, we will prove the statement just for the coproducts. Assume that both c_1 and c_2 are coproducts of a and b. Since c_1 is a coproduct, there exists a unique morphism $\gamma_1 : c_1 \to c_2$ such that $\alpha_2 = \gamma_1 \circ \alpha_1$ and $\beta_2 = \gamma_1 \circ \beta_1$. Since c_2 is a coproduct, there exists a unique morphism $\gamma_2 : c_2 \to c_1$ such that $\alpha_1 = \gamma_2 \circ \alpha_2$ and $\beta_1 = \gamma_2 \circ \beta_2$.

We have $\gamma_1 \circ \gamma_2 \circ \alpha_2 = \gamma_1 \circ \alpha_1 = \alpha_2$, and analogously, $\gamma_1 \circ \gamma_2 \circ \beta_2 = \gamma_1 \circ \beta_1 = \beta_2$. In other words, the morphism $\gamma_1 \circ \gamma_2 : c_2 \to c_2$ commutes with α_2 and with β_2. By the universal property of the coproduct c_2 we have that such a morphism is unique; on the other hand, the identity morphism satisfies these commuting relations as well. Hence $\gamma_1 \circ \gamma_2$ is an identity morphism, and in the same way we can prove that $\gamma_2 \circ \gamma_1$ is an identity morphism as well. This implies that c_1 is isomorphic to c_2. □

Initial and terminal objects, as well as products and coproducts, are special cases of more general constructions of limits and colimits. The latter are defined and investigated in Section 4.4.

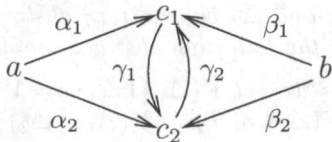

Fig. 4.3. Isomorphism of two coproducts.

4.3 Functors

In this section we introduce structure-preserving maps between categories.

4.3.1 The Category Cat

Definition 4.28. *Given two categories C_1 and C_2, a **functor** $F : C_1 \to C_2$ is a pair of maps $F_\mathcal{O} : \mathcal{O}(C_1) \to \mathcal{O}(C_2)$ and $F_\mathcal{M} : \mathcal{M}(C_1) \to \mathcal{M}(C_2)$ such that*

(1) $F_\mathcal{M}(\mathcal{M}(x,y)) \subseteq \mathcal{M}(F_\mathcal{O}(x), F_\mathcal{O}(y))$, for all $x,y \in \mathcal{O}(C_1)$;
(2) $F_\mathcal{M}(\mathrm{id}_a) = \mathrm{id}_{F_\mathcal{O}(a)}$, for all $a \in \mathcal{O}(C_1)$;
(3) $F_\mathcal{M}(m_1 \circ m_2) = F_\mathcal{M}(m_1) \circ F_\mathcal{M}(m_2)$, for all $m_1, m_2 \in \mathcal{M}(C_1)$.

When no confusion arises we shall just write F instead of $F_\mathcal{O}$ and $F_\mathcal{M}$. Clearly, the identity maps define the identity functor. Also, two functors can be composed, their composition is again a functor, and the rule of the composition is associative.

Example 4.29.

(1) *Forgetful functors.*
    This is a generic name for functors that forget part of a structure. We have forgetful functors **Top** $\to$ **Sets**, **Grp** $\to$ **Sets**, **Graphs** $\to$ **Sets**, **ASC** $\to$ **Sets**, that map topological spaces, groups, graphs to the underlying sets. We also have a forgetful functor **Vect$_k$** $\to$ **Ab**, that maps vector spaces to the underlying abelian group that has vector addition as the group operation.
(2) *Universal functors.*
    These are functors that add some structure in a universal way. For example, we have a functor **Sets** $\to$ **Grp**, which maps each set S to the free group generated by elements of S. Another example is a functor **Sets** $\to$ **Vect$_k$**, which maps each set S to the vector space with a basis indexed by elements of S. Perhaps a more interesting universal functor goes from **Grp** to the category of $\mathbb{C}$-algebras. It takes each group to its group algebra $\mathbb{C}[G]$.
(3) There is a multitude of functors induced by various algebraic structures. For example, any group homomorphism $\varphi : G \to H$ induces a functor $C\varphi : CG \to CH$.

Definition 4.28 allows us now to define a new category.

Definition 4.30. *The category of small categories,* **Cat***, is defined as follows:*

- *the class of objects of* **Cat** *consists of all small categories;*
- *for two small categories C_1 and C_2, the set $\mathcal{M}(C_1, C_2)$ consists of all functors $C_1 \to C_2$.*

We can see all the notions related to a category that we have defined so far in the context of **Cat**.

(1) Two categories $C_1, C_2 \in \mathcal{O}(\textbf{Cat})$ are called *isomorphic* if they are isomorphic as objects in **Cat**.

(2) The category **Cat** has an initial object: it is the empty category, whose sets of objects and of morphisms are both empty.

(3) The category **Cat** also has terminal objects: these are categories with one object, whose only automorphism is the identity.

(4) The coproduct of two categories $C_1, C_2 \in \mathcal{O}(\textbf{Cat})$ is their disjoint union; namely, it is defined by

$$\mathcal{O}(C_1 \coprod C_2) = \mathcal{O}(C_1) \coprod \mathcal{O}(C_2),$$

and for $a, b \in \mathcal{O}(C_1 \coprod C_2)$ we have

$$\mathcal{M}_{C_1 \coprod C_2}(a, b) = \begin{cases} \mathcal{M}_{C_1}(a, b), & \text{if } a, b \in \mathcal{O}(C_1); \\ \mathcal{M}_{C_2}(a, b), & \text{if } a, b \in \mathcal{O}(C_2); \\ \emptyset, & \text{otherwise.} \end{cases}$$

It is easy to see that the corresponding inclusion maps $C_1 \hookrightarrow C_1 \coprod C_2$ and $C_2 \hookrightarrow C_1 \coprod C_2$ are functors.

(5) The product of two categories $C_1, C_2 \in \mathcal{O}(\textbf{Cat})$ is defined by

$$\mathcal{O}(C_1 \times C_2) = \mathcal{O}(C_1) \times \mathcal{O}(C_2),$$

and for $(a_1, a_2), (b_1, b_2) \in \mathcal{O}(C_1 \times C_2)$ we have

$$\mathcal{M}_{C_1 \times C_2}((a_1, b_1), (a_2, b_2)) = \mathcal{M}_{C_1}(a_1, b_1) \times \mathcal{M}_{C_2}(a_2, b_2).$$

Again, it is easy to see that the corresponding projection maps $C_1 \times C_2 \to C_1$ and $C_1 \times C_2 \to C_2$ are functors.

The functors that we have described so far are sometimes called *covariant* functors. This is done in order to distinguish them from the *contravariant* functors. A contravariant functor is defined analogously to the covariant one with the only difference that it "turns the arrows around," i.e., $\mathcal{F}(\mathcal{M}(x, y)) \subseteq \mathcal{M}(\mathcal{F}(y), \mathcal{F}(x))$ Note that a contravariant functor $\mathcal{F} : C_1 \to C_2$ is the same as any of the covariant functors $\mathcal{F} : C_1 \to C_2^{\text{op}}$ and $\mathcal{F} : C_1^{\text{op}} \to C_2$.

Example 4.31.

(1) The dualizing functor $F : \textbf{Vect}_k \to \textbf{Vect}_k$ mapping a vector space V to its dual V^* is a contravariant functor. Indeed, we recall from linear algebra that a linear transformation $f : V \to W$ induces a linear transformation $f : W^* \to V^*$ by mapping $w^* \in W^*$ to $f^*(w^*) \in V^*$ defined by

$$f^*(w^*)(v) := w^*(f(v)), \text{ for all } v \in V.$$

(2) Let G be a group. Then taking an inverse is a contravariant functor from $\mathcal{C}G$ to itself.

4.3.2 Homology and Cohomology Viewed as Functors

One of the reasons why homology groups play such a central role in many contexts is that the whole construction can be viewed as a certain functor.

As we have seen in Chapter 3, a simplicial map $f : \Delta_1 \to \Delta_2$ will induce $\mathcal{R}$-module homomorphisms on the homology groups $f_* : H_d(\Delta_1; \mathcal{R}) \to H_d(\Delta_2; \mathcal{R})$, for any commutative ring $\mathcal{R}$ and any $d \geq 0$. This map satisfies all the properties of Definition 4.28, and hence we have a functor $\textbf{ASC} \to \mathcal{R}\textbf{-Mod}$. This functor is covariant. If we consider homology groups with integer coefficients, or with coefficients in a field k, then we will get covariant functors $\textbf{ASC} \to \textbf{Ab}$ and $\textbf{ASC} \to \textbf{Vect}_k$.

If we consider cohomology groups instead, we will get a classical example of a contravariant functor.

Proposition 4.32. *For any commutative ring $\mathcal{R}$ and any $d \geq 0$, we have a contravariant functor $H^d : \textbf{ASC} \to \textbf{Ab}$ mapping an abstract simplicial complex Δ to $H^d(\Delta; \mathcal{R})$.*

Proof. We just need to see that a simplicial map $f : \Delta_1 \to \Delta_2$ induces an $\mathcal{R}$-module homomorphism $f^* : H^d(\Delta_2; \mathcal{R}) \to H^d(\Delta_1; \mathcal{R})$ which satisfies the properties of the contravariant functor. This can be done by dualizing the proof of Proposition 3.33. $\square$

4.3.3 Group Actions as Functors

Let G be a group, and X an object of the category $\mathcal{C}$. The group action of G on X is by definition a functor from $\mathcal{C}G$ to $\mathcal{C}$ that maps o to X, where we have used o to denote the unique object of G. This is just a formal way of saying that the elements of G give rise to the "structure-preserving maps" (morphisms in our language) from X to itself.

For example, $\mathcal{C}$ could be the category of topological spaces, or abstract simplicial complexes, or posets, in which case the structure-preserving maps would respectively be continuous maps, or simplicial maps, or order-preserving maps (see Definition 10.3).

If we have a functor $\mathcal{F} : \mathcal{C}_1 \to \mathcal{C}_2$ and a G-action on $x \in \mathcal{O}(\mathcal{C}_1)$, then by composing $\mathcal{F}$ with the functor encoding the group action, we get a functor from G to $\mathcal{C}_2$. This is the so-called *induced action*.

For example, an action on an abstract simplicial complex induces an action on the underlying topological space; this is a composition with the geometric realization functor. An action on a poset P induces an action on the order complex $\Delta(P)$, which is defined in Chapter 9; this is a composition with the order complex functor. Furthermore, it induces a G-action on any given homology group $H_i(\Delta(P); \mathcal{R})$, which, in case $\mathcal{R}$ is a field, is the same as a linear representation of G over $\mathcal{R}$.

4.4 Limit Constructions

4.4.1 Definition of Colimit of a Functor

Colimit is a very useful notion in all sorts of combinatorial contexts. We shall give a hands-on definition and refer the reader to standard texts [McL98, Mit65] for further details.

Let $\mathcal{C}_1$ and $\mathcal{C}_2$ be two categories and let $F : \mathcal{C}_1 \to \mathcal{C}_2$ be a functor between them. Intuitively, one can think of this functor as a diagram, a picture in which some objects of $\mathcal{C}_2$ are connected with arrows. These objects are the images of objects of $\mathcal{C}_1$ under F, and the arrows are the images of morphisms of $\mathcal{C}_1$ under F. Accordingly, the arrows "should compose right," as is prescribed by functor axioms.

Let us call a *sink* of F an object L of $\mathcal{C}_2$, together with a collection of morphisms pointing from the objects in the diagram to the object L, i.e., $F(a) \xrightarrow{\lambda_a} L$, for all $a \in \mathcal{O}(\mathcal{C}_1)$, such that these new morphisms commute appropriately with the old morphisms in the diagram. Formally, we require that for all $m \in \mathcal{M}_{\mathcal{C}_1}(a_1, a_2)$ we have $\lambda_{a_2} \circ F(m) = \lambda_{a_1}$.

Definition 4.33. *Given a functor* $F : \mathcal{C}_1 \to \mathcal{C}_2$, *we call a sink of* F, $(L, \{\lambda_a\}_{a \in \mathcal{O}(\mathcal{C}_1)})$, *its* **colimit** *if it is universal in the following sense: for any other sink* $(\widetilde{L}, \{\widetilde{\lambda}_a\}_{a \in \mathcal{O}(\mathcal{C}_1)})$ *there exists a unique morphism* $L \xrightarrow{\varphi} \widetilde{L}$ *such that* $\widetilde{\lambda}_a = \varphi \circ \lambda_a$, *for all* $a \in \mathcal{O}(\mathcal{C}_1)$. *We write* $L = \mathbf{colim}\, X$.

We have already seen several instances of colimits. The easiest, though admittedly somewhat degenerate, case is that in which the category $\mathcal{C}_1$ is empty. In this situation, there are no morphisms λ_a, and the only condition on the object L is that for any other object $\widetilde{L}$ there exist a unique morphism $L \xrightarrow{\varphi} \widetilde{L}$. This is now easily recognized as the definition of the initial object.

Another previously seen case is that in which $\mathcal{C}_1$ consists of two objects and has only identity morphisms. Then, the definition of a colimit translates word for word into the definition of the coproduct.

We invite the reader to take her or his favorite category $\mathcal{C}_1$, choose the functor F, and see what the definition of the colimit translates to.

4.4.2 Colimits and Infinite Unions

One situation in which viewing constructions as colimits brings some clarity is to consider infinite analogues of some familiar objects. Sometimes this can be accomplished by taking an infinite union, and sometimes we need to add some more structure. Let us give here some examples. Consider first the embedding sequence of Euclidean spaces

$$\mathbb{R}^1 \overset{i_1}{\hookrightarrow} \mathbb{R}^2 \overset{i_2}{\hookrightarrow} \dots \overset{i_{n-1}}{\hookrightarrow} \mathbb{R}^n \overset{i_n}{\hookrightarrow} \mathbb{R}^{n+1} \overset{i_{n+1}}{\hookrightarrow} \dots , \tag{4.1}$$

where $i_n : \mathbb{R}^n \hookrightarrow \mathbb{R}^{n+1}$ is the usual initial coordinate embedding, i.e.,

$$i_n : (x_1, \dots, x_n) \mapsto (x_1, \dots, x_n, 0). \tag{4.2}$$

The embedding sequence (4.1) can be viewed as a diagram of spaces and continuous maps, and we can take its colimit. It is a nice exercise to see that in this case the colimit construction actually delivers the infinite union of spaces. Thus we define the *infinite-dimensional Euclidean space* by setting

$$\mathbb{R}^\infty := \bigcup_{n=1}^{\infty} \mathbb{R}^n,$$

where the union is taken with respect to the embedding maps (4.2). The elements of $\mathbb{R}^\infty$ are simply all vectors with countably many coordinates, of which only finitely many are allowed to be different from 0. The usual vector addition and scalar multiplication give the vector space structure on $\mathbb{R}^\infty$.

Let now $\mathbb{S}^n$ denote the unit sphere in $\mathbb{R}^n$. Restrictions of i_n from (4.2) yield an embedding sequence as well:

$$\mathbb{S}^1 \overset{i_1}{\hookrightarrow} \mathbb{S}^2 \overset{i_2}{\hookrightarrow} \dots \overset{i_{n-1}}{\hookrightarrow} \mathbb{S}^n \overset{i_n}{\hookrightarrow} \mathbb{S}^{n+1} \overset{i_{n+1}}{\hookrightarrow} \dots . \tag{4.3}$$

Taking the colimit of this diagram of spaces yields what is known as the *infinite-dimensional sphere*

$$\mathbb{S}^\infty := \bigcup_{n=1}^{\infty} \mathbb{S}^n.$$

This space will be of use once we look at the classifying spaces of finite groups and Stiefel-Whitney characteristic classes in Chapter 8.

It is easy to recognize the category lurking in the background of the two previous examples. Both sequences (4.1) and (4.3) can be viewed as functors $F_1, F_2 : \mathbf{IE} \to \mathbf{Top}$, where $\mathbf{IE}$ denotes the category of *initial embeddings*. It is defined as follows:

- the set of objects of $\mathbf{IE}$ is taken to be the set of natural numbers $\mathbb{N} = \{1, 2, 3, \dots\}$;

- for any $m, n \in \mathbb{N}$, we have

$$|\mathcal{M}_{\mathbf{IE}}(m, n)| = \begin{cases} 1, & \text{if } m \leq n, \\ 0, & \text{otherwise}, \end{cases}$$

which defines the composition rule uniquely.

One possible intuition that one could have when thinking of the category **IE** is to view the elements $m \in \mathbb{N}$ as sets $[m]$, and whenever $m \leq n$, to view the morphism $m \xrightarrow{\alpha} n$ as the embedding of the initial interval $\iota : [m] \hookrightarrow [n]$, taking x to x.

Let us now consider yet another functor with **IE** as the source, though this time having the category of groups as a target, $F : \mathbf{IE} \to \mathbf{Grp}$. Namely, consider the following diagram of symmetric groups:

$$\mathcal{S}_1 \xrightarrow{i_1} \mathcal{S}_2 \xrightarrow{i_2} \cdots \xrightarrow{i_{n-1}} \mathcal{S}_n \xrightarrow{i_n} \mathcal{S}_{n+1} \xrightarrow{i_{n+1}} \cdots , \tag{4.4}$$

where each map $i_n : \mathcal{S}_n \hookrightarrow \mathcal{S}_{n+1}$ is the group homomorphism induced by the initial embedding $[n+1] \hookrightarrow [n]$. The image of i_n is the Young subgroup $\mathcal{S}_n \times \mathcal{S}_1 \subseteq \mathcal{S}_{n+1}$.

Taking the colimit of the functor F, we obtain the *infinite permutation group* $\mathcal{S}_\infty$. It can be explicitly described as follows:

- the elements of $\mathcal{S}_\infty$ are all bijections $\varphi : \mathbb{N} \to \mathbb{N}$ for which there exists a number $N(\varphi)$ such that $\varphi(x) = x$, for all $x \geq N(\varphi)$;
- the multiplication rule is given by the composition of bijections.

Naturally, most of the commonly used colimits can also be defined directly, without using the framework of category theory. However, one usually has many choices for possible definitions. Recognizing that something is a colimit for some functor helps us to make the right choice.

4.4.3 Quotients of Group Actions as Colimits

Our next goal is to interpret the quotients of group actions as colimits. Assume, that X is some mathematical object (e.g., topological space, vector space, abstract simplicial complex, graph), and assume that a group G acts on X. Let us first informally contemplate what a *quotient* X/G should be.

Imagine that X itself consists of some kind of elements (points, vectors, simplices). Then the most natural thing to do would be to glue together two elements whenever one is mapped to the other by the group action. In other words, take orbits of elements as the new elements.

A problem arises when one also tries in a consistent way to impose additional structure to make sure that the collection of orbits of X can be made into the mathematical object of the same nature (read: belonging to the same category) as X. In many situations this will lead to additional identifications, which can be controlled with a varied degree of success.

Colimits provide us with a streamlined formal framework to say precisely what the quotient should be by telling us which universal property this object should satisfy. Namely, assume that X is an object of category $\mathcal{C}$. This is the "hidden" parameter. Choosing the category means that we specify which structure is preserved by the group action, hence in which category we want the quotient to be taken. Most often there are many choices. We can have a topological space X and the group G acting by continuous maps. But we can also just view X as a set and allow all bijections. We will expand on this discussion in Chapter 14.

4.4.4 Limits

An important dual notion of the colimit is that of a limit. It is indeed obtained from the definition of the colimit by reversing all arrows.

Let again $\mathcal{C}_1$ and $\mathcal{C}_2$ be two categories and let $F : \mathcal{C}_1 \to \mathcal{C}_2$ be a functor. Let us call a *source* of F an object L of $\mathcal{C}_2$, together with a collection of morphisms $L \xrightarrow{\lambda_a} F(a)$, for all $a \in \mathcal{O}(\mathcal{C}_1)$, such that for all $m \in \mathcal{M}_{C_1}(a_1, a_2)$ we have $F(m) \circ \lambda_{a_1} = \lambda_{a_2}$.

Definition 4.34. *Given a functor* $F : \mathcal{C}_1 \to \mathcal{C}_2$, *we call a source of* F, $(L, \{\lambda_a\}_{a \in \mathcal{O}(\mathcal{C}_1)})$, *its* **limit** *if it is universal, i.e., if for any other source* $(\widetilde{L}, \{\widetilde{\lambda}_a\}_{a \in \mathcal{O}(\mathcal{C}_1)})$ *there exists a unique morphism* $\widetilde{L} \xrightarrow{\varphi} L$ *such that* $\widetilde{\lambda}_a = \lambda_a \circ \varphi$, *for all* $a \in \mathcal{O}(\mathcal{C}_1)$. *We write* $L = \lim X$.

Examples of limits are provided by terminal objects, as well as by products.

4.5 Comma Categories

4.5.1 Objects Below and Above Other Objects

We start with a standard concept, which is pervasive in combinatorial contexts.

Definition 4.35. *Let* C *be a category, and let* $x \in \mathcal{O}(C)$. *The* **category of objects below** x, *denoted by* $(x \downarrow C)$, *is defined as follows:*

- *the class of objects* $\mathcal{O}(x \downarrow C)$ *consists of all morphisms* $m \in \mathcal{M}(C)$ *such that* $\partial^{\bullet}(m) = x$;
- *the set of morphisms* $\mathcal{M}(x \downarrow C)$ *consists of all composable morphism chains* $x \xrightarrow{m_1} y \xrightarrow{m_2} z$. *Such a chain is seen as a morphism from* $x \xrightarrow{m_1} y$ *to* $x \xrightarrow{m_2 \circ m_1} z$, *and one speaks about a commuting triangle formed by the objects* x, y, z *and the morphisms* m_1, m_2, *and* $m_2 \circ m_1$.

Analogously, one can define $(x \uparrow C)$, the *category of objects above* x, by taking as objects all morphisms $m \in \mathcal{M}(C)$ such that $\partial_{\bullet}(m) = x$, and taking appropriate commuting triangles as morphisms.

4.5.2 The General Construction and Further Examples

The construction from the previous subsection can put into a more general context, which, gratifyingly, will then specialize to further useful and interesting notions.

Definition 4.36. *Let A, B, and C be categories, and let $F : A \to C$ and $G : B \to C$ be arbitrary functors. The* **comma category** *of F and G, denoted by $(F \downarrow G)$, is defined as follows:*

- *the class of objects $\mathcal{O}(F \downarrow G)$ consists of all triples (x, y, m) such that $x \in \mathcal{O}(A)$, $y \in \mathcal{O}(B)$, and $m \in \mathcal{M}(C)$, $m : F(x) \to G(y)$;*
- *let (x_1, y_1, m_1) and (x_2, y_2, m_2) be objects of $(F \downarrow G)$. The set of morphisms between these two objects consists of all pairs of morphisms (α, β) such that $\alpha \in \mathcal{M}(A)$, $\alpha : x_1 \to x_2$, $\beta \in \mathcal{M}(B)$, $\beta : y_1 \to y_2$, and $m_2 \circ F(\alpha) = G(\beta) \circ m_1$, i.e., the diagram (4.5) commutes:*

$$
\begin{array}{ccc}
F(x_1) & \xrightarrow{\ F(\alpha)\ } & F(x_2) \\
{\scriptstyle m_1}\downarrow & & \downarrow{\scriptstyle m_2} \\
G(y_1) & \xrightarrow{\ G(\beta)\ } & G(y_2).
\end{array}
\tag{4.5}
$$

There are several special cases of the comma category construction. First, taking $A = \mathbf{1}$, $B = C$, $G = \mathrm{id}_C$, and taking $F : \mathbf{1} \to C$ to be the functor mapping the unique object of $\mathbf{1}$ to some given object $x \in \mathcal{O}(C)$ gives the category $(x \downarrow C)$ described in Definition 4.35.

Symmetrically, taking $A = C$, $B = \mathbf{1}$, $F = \mathrm{id}_C$, and taking $G : \mathbf{1} \to C$ to be the functor mapping the unique object of $\mathbf{1}$ to some given object $x \in \mathcal{O}(C)$ gives the category $(x \uparrow C)$.

Another example is provided when we take $A = B = C$ and $F = G = \mathrm{id}_C$. This gives the category of all morphisms of C, often called the *category of arrows of C*, with the "new" morphisms being all commuting diagrams of "old" morphisms.

Finally, let us remark that it is possible to take one of the functors F and G in Definition 4.36 contravariant. This will also yield a category, which we denote by $\mathcal{O}(F^{\mathrm{op}} \downarrow G)$ if F is taken to be contravariant, and $\mathcal{O}(F \downarrow G^{\mathrm{op}})$ if G is taken to be contravariant. This construction will come in handy when we define the analogue of intervals for acyclic categories.

5

Exact Sequences

One of the ancient principles for dealing with a larger calculation is to subdivide it into smaller computational tasks. In this chapter we shall investigate one application of this principle, which allows one to break down the calculation of the homology groups of a CW complex into two hopefully simpler computations of homology groups of some chain complexes. The natural question of putting the computed information together to yield the homology groups of the original complex has an elegant algebraic answer due to Eilenberg: the so-called *long exact sequences*.

While the long exact sequences provide us with the first tools for computing the homology groups, they also serve as a good introduction to the more technical subject of Chapter 16: the *spectral sequences*. Intuitively, one could think of the long exact sequences as the gadget that allows one to break the homology group computation into two tasks, whereas the spectral sequences allow us to start with more, hopefully easier, tasks, and then work our way through the deep waters to the final answer.

5.1 Some Structure Theory of Long and Short Exact Sequences

5.1.1 Construction of the Connecting Homomorphism

It is easy to generalize the notions of submodule and quotient module to that of general chain complexes.

Definition 5.1. *Let* $\mathcal{B} = (B_*, \partial_*)$ *be a chain complex of $\mathcal{R}$-modules, and let us choose $\mathcal{R}$-modules A_i such that $A_n \subseteq B_n$, and additionally $\partial(A_n) \subseteq A_{n-1}$, for all n.*

- *The chosen modules form a chain complex $\mathcal{A} = (A_*, \partial_*)$, which is called a* **subcomplex** *of $\mathcal{B}$.*

- *The chain complex $\mathcal{B}/\mathcal{A} = (B_*/A_*, \partial_*)$ is called a* **quotient complex**. *Note that the boundary operator is well-defined on the quotient because of the extra condition $\partial(A_n) \subseteq A_{n-1}$.*

Clearly, the associated collections of inclusion maps $i = \{i_n\}_n$, $i_n : A_n \hookrightarrow B_n$, and projection maps $p = \{p_n\}_n$, $p_n : B_n \longrightarrow B_n/A_n$ both describe chain maps.

Given a subcomplex $\mathcal{A}$ of the chain complex $\mathcal{B}$, there arises a natural question:

What is the relation between the homology groups of $\mathcal{A}$ and those of $\mathcal{B}$?

Do they "sit inside" of $H_*(\mathcal{B})$ just as the chain complex $\mathcal{A}$ sits inside of $\mathcal{B}$? In other words, is the induced map on the homology groups $i_* : H_*(\mathcal{A}) \to H_*(\mathcal{B})$ an injection?

A quick analysis will show that the answer to these questions is negative. The reason is that we may have an element $c \in A_n$ such that there exists an element $b \in B_{n+1}$ satisfying $\partial b = c$, in particular $\partial c = 0$, but there is no $a \in A_{n+1}$ satisfying $\partial a = c$. In other words, the homology class $[c]$ is trivial in $H_*(\mathcal{B})$, but not in $H_*(\mathcal{A})$.

The kernel of the map $i_n : H_n(\mathcal{A}) \to H_n(\mathcal{B})$ consists of those cycles of A_n that are boundaries of elements from B_{n+1}. Intuitively, it may be helpful to think that these homology cycles of $\mathcal{B}$ were considered nontrivial, as long as one looked only inside $\mathcal{A}$; once the scope of attention was widened to all of $\mathcal{B}$, it was revealed that these were actually boundaries, and have to be canceled in the total calculation.

Let us now reformulate our observations so as to produce a map. In the first rough approximation, what we want is a map from B_{n+1} to $H_n(\mathcal{A})$ that is very close to the boundary map. This of course will not work so directly, since ∂B_{n+1} is not contained in A_n. So, to start with, our map will be defined only on the set of those $b \in B_{n+1}$ whose boundary lies in A_n. Second, since we mod out the boundaries ∂A_{n+1} in the target space $H_n(\mathcal{A})$, we may as well define the desired map on the quotient space B_{n+1}/A_{n+1}. The crucial observation now is that the set of elements of B_{n+1}/A_{n+1} whose boundary lies in A_n is by definition precisely the set of $(n+1)$-cycles in the chain complex $\mathcal{B}/\mathcal{A}$, so we have a map $\partial_{n+1} : Z_{n+1}(\mathcal{B}/\mathcal{A}) \to H_n(\mathcal{A})$. Finally, the values of the boundary operator on the boundaries $B_{n+1}(\mathcal{B}/\mathcal{A})$ lie in ∂A_{n+1}; hence we get the induced map

$$\partial_{n+1} : Z_{n+1}(\mathcal{B}/\mathcal{A})/B_{n+1}(\mathcal{B}/\mathcal{A}) = H_{n+1}(\mathcal{B}/\mathcal{A}) \to H_n(\mathcal{A}).$$

By the above discussion, this map is well-defined, and is in fact a homomorphism of $\mathcal{R}$-modules. We also note that since our map is derived from the boundary operator in a very direct way, we find it convenient simply to use the *same* notation.

Definition 5.2. *The maps $\partial_* : H_*(\mathcal{B}/\mathcal{A}) \to H_{*-1}(\mathcal{A})$ defined above are called* **connecting homomorphisms**.

In light of the intuitive picture above, the connecting homomorphisms can be thought of as cancellation maps, which remove the superfluous cycles from the homology groups $H_*(\mathcal{A})$.

The connecting homomorphisms are *natural*, which is just another way of saying that they are functorial in the following sense.

Proposition 5.3. *Assume that* $\mathcal{A}$, $\mathcal{B}$, $\widetilde{\mathcal{A}}$, *and* $\widetilde{\mathcal{B}}$ *are chain complexes such that* $\mathcal{A}$ *is a subcomplex of* $\mathcal{B}$, *and* $\widetilde{\mathcal{A}}$ *is a subcomplex of* $\widetilde{\mathcal{B}}$. *Assume furthermore that we have a chain complex map* $\varphi : \mathcal{B} \to \widetilde{\mathcal{B}}$ *such that* $\varphi(\mathcal{A})$ *is a subcomplex of* $\widetilde{\mathcal{A}}$. *Then the diagram*

$$
\begin{array}{ccc}
H_{n+1}(\mathcal{B}/\mathcal{A}) & \xrightarrow{\ \partial\ } & H_n(\mathcal{A}) \\[4pt]
\varphi \downarrow & & \downarrow \varphi \\[4pt]
H_{n+1}(\widetilde{\mathcal{B}}/\widetilde{\mathcal{A}}) & \xrightarrow{\ \widetilde{\partial}\ } & H_n(\widetilde{\mathcal{A}})
\end{array}
\tag{5.1}
$$

commutes for all n.

Proof. Let us trace the diagram (5.1) in both directions. First,

$$
\varphi(\partial([c] + A_{n+1})) = \varphi([\partial c]) = [\varphi(\partial c)].
$$

On the other hand,

$$
\widetilde{\partial}(\varphi([c] + A_{n+1})) = \widetilde{\partial}([\varphi(c)] + \widetilde{A}_{n+1}) = [\widetilde{\partial}(\varphi(c))].
$$

By our assumptions, φ is a chain complex map; hence $\varphi(\partial c) = \widetilde{\partial}(\varphi(c))$, and the commutativity of the diagram (5.1) follows. $\quad\square$

A natural question is why these maps are called connecting homomorphisms. *What do they connect?* We shall next embark on constructing the algebraic apparatus that will allow us to give an answer to that question.

5.1.2 Exact Sequences

On the level of chain complexes, one has a feeling that one should be able to put $\mathcal{A}$ and $\mathcal{B}/\mathcal{A}$ together to reconstruct $\mathcal{B}$. It is close at hand to ask the same for the homology: how can we put together $H_*(\mathcal{A})$ with $H_*(\mathcal{B}/\mathcal{A})$ to get $H_*(\mathcal{B})$?

To start with, even the prehomology picture is not at all too clear. We certainly do not in general have $B = A \oplus B/A$ when A and B are $\mathcal{R}$-modules, not even when $\mathcal{R} = \mathbb{Z}$, i.e., when A and B are abelian groups. As an example, take $B = \mathbb{Z}$, and take $A = 2\mathbb{Z}$. We have $B/A = \mathbb{Z}_2$, but $\mathbb{Z} \neq \mathbb{Z} \oplus \mathbb{Z}_2$, since $\mathbb{Z}$ does not have nonzero elements of finite order, while $\mathbb{Z} \oplus \mathbb{Z}_2$ does. The exact relationship between B, A, and B/A is best phrased in the language of short exact sequences, which we now proceed to introduce.

Definition 5.4.

*(1) A chain complex of R-modules $C = (C_*, f_*)$ is called an **exact sequence** if for all i, we have $\operatorname{Ker} f_i = \operatorname{Im} f_{i+1}$.*

(2) A sequence of chain complexes and chain maps

$$\cdots \xrightarrow{\varphi^{n+2}} C^{n+1} \xrightarrow{\varphi^{n+1}} C^n \xrightarrow{\varphi^n} C^{n-1} \xrightarrow{\varphi^{n-1}} \cdots$$

*is called **exact** if it is exact in each dimension, i.e., if $C^n = (C_*^n, \varphi_*^n)$, then the sequence*

$$\cdots \xrightarrow{\varphi_d^{n+2}} C_d^{n+1} \xrightarrow{\varphi_d^{n+1}} C_d^n \xrightarrow{\varphi_d^n} C_d^{n-1} \xrightarrow{\varphi_d^{n-1}} \cdots$$

is required to be exact for all d.

*(3) A **map between** two **exact sequences** (of R-modules or chain complexes) (A_*, α_*) and (B_*, β_*) is a collection of maps (R-module homomorphisms or chain complex maps) $\{\varphi_n\}_n$ such that the following diagram commutes:*

$$
\begin{array}{ccccccccc}
\cdots \xrightarrow{\alpha_{n+2}} & A_{n+1} & \xrightarrow{\alpha_{n+1}} & A_n & \xrightarrow{\alpha_n} & A_{n-1} & \xrightarrow{\alpha_{n-1}} & \cdots \\
& \downarrow{\varphi_{n+1}} & & \downarrow{\varphi_n} & & \downarrow{\varphi_{n-1}} & & \\
\cdots \xrightarrow{\beta_{n+2}} & B_{n+1} & \xrightarrow{\beta_{n+1}} & B_n & \xrightarrow{\beta_n} & B_{n-1} & \xrightarrow{\beta_{n-1}} & \cdots
\end{array}
$$

Note that the definition of the chain complex itself already encompasses "half" of the exactness condition: we have $\operatorname{Ker}(f_i) \supseteq \operatorname{Im}(f_{i+1})$. To require the opposite inclusion to hold as well is the same as to ask the chain complex to have trivial homology groups in all dimensions.

The most trivial instance of the exact sequence is the one with all terms equal to 0. Next comes the one with all but two consecutive terms being 0. That special case coincides with the notion of isomorphism. Allowing one more consecutive term to be nontrivial yields what is perhaps the most important special case.

Definition 5.5. *An exact sequence (of R-modules or of chain complexes) is called a **short exact sequence** if all the terms are 0 except possibly three consecutive ones:*

$$0 \longrightarrow A \xhookrightarrow{i} B \xrightarrow{p} C \longrightarrow 0. \tag{5.2}$$

Exact sequences form a subcategory of the category of all chain complexes and chain complex homomorphisms. The same is true for short exact sequences, or in fact for any set of exact sequences with a fixed set of indices where the terms are allowed to be nontrivial.

Let us now return to the situation in which A is a subcomplex of B. We see that even though we may not have $B = A \oplus B/A$, we do have a short exact sequence

$$0 \longrightarrow A \overset{i}{\hookrightarrow} B \overset{p}{\longrightarrow} B/A \longrightarrow 0, \tag{5.3}$$

where i and p are the standard inclusion and projection maps. This is the language in which the relation between A, B, and B/A is phrased.

On the other hand, the fact that the sequence (5.2) is exact means exactly three things:

- the map i is injective;
- the map p is surjective;
- we have $i(A) = \mathrm{Ker}(p)$.

This means that A is isomorphic to $i(A)$, while C is isomorphic to $B/\mathrm{Ker}(p) = B/i(A)$. In other words, all short exact sequences can essentially be written in the form (5.3). To underline this fact, one says that the term B is an *extension of A by C*. Being able to write the middle term as the direct sum of the two others is an additional property that one sometimes has. It is formally defined as follows.

Definition 5.6. *A short exact sequence* (5.2) *is said to* **split** *if there exists an isomorphism $\varphi : B \to A \oplus C$ such that $\iota = \varphi \circ i$ and $p = \pi \circ \varphi$, where $\iota : A \hookrightarrow A \oplus C$ is the injection into the first term, and $\pi : A \oplus C \longrightarrow C$ is the projection onto the second term.*

5.1.3 Deriving Long Exact Sequences from Short Ones

As we have already noted, passing to homology is bound to make things more complex, since, for example, injective maps may cease to be injective. It turns out that instead of one short exact sequence, the relationship between $H_n(A)$, $H_n(B)$, and $H_n(B/A)$ is best described as a *long* exact sequence, which involves the groups $H_*(A)$, $H_*(B)$, and $H_*(B/A)$ in all dimensions.

Theorem 5.7. (Zig-zag lemma).
Assume that

$$0 \longrightarrow \mathcal{A} \overset{i}{\hookrightarrow} \mathcal{B} \overset{p}{\longrightarrow} \mathcal{C} \longrightarrow 0 \tag{5.4}$$

is a short exact sequence of chain complexes. Then we have a long exact sequence of homology groups

$$\cdots \overset{\partial_{n+1}}{\longrightarrow} H_n(\mathcal{A}) \overset{i_*}{\longrightarrow} H_n(\mathcal{B}) \overset{p_*}{\longrightarrow} H_n(\mathcal{C}) \overset{\partial_n}{\longrightarrow} H_{n-1}(\mathcal{A}) \overset{i_*}{\longrightarrow} \cdots, \tag{5.5}$$

where i_* and p_* are the maps between the homology groups induced by the maps i and p, and ∂_n is the connecting homomorphism constructed in Subsection 5.1.1.

Proof. By our discussion above, we may assume that $\mathcal{A}$ is a subcomplex of $\mathcal{B}$, and replace $\mathcal{C}$ with $\mathcal{B}/\mathcal{A}$. We do that to be able to phrase the proof in simple words. Furthermore, we have already checked that all the connecting homomorphisms are well-defined, so it remains only to check the exactness of the sequence (5.5).

(1) *Exactness in $H_n(\mathcal{B})$.* By definition of induced maps, both the kernel of p_* and the image of i_* consist of those homology groups $[b] \in H_*(\mathcal{B})$ that have a representative in $C_*(\mathcal{A})$.
(2) *Exactness in $H_n(\mathcal{B}/\mathcal{A})$.* By definition, the image of p_* consists of those $[b] + A_n$ that have a representative whose boundary is 0. In other words, there must exist $a \in A_n$ such that $\partial(b - a) = 0$, i.e., the condition can be equivalently reformulated as $\partial b \in \partial A_n$. On the other hand, the kernel of the corresponding connecting homomorphism consists of those $[b] + A_n$ for which we have $[\partial([b] + A_n)] = 0$. Since $[\partial([b] + A_n)] = [[\partial b] + \partial A_n] = [\partial b]$, this gives the same condition.
(3) *Exactness in $H_n(\mathcal{A})$.* The image of a connecting homomorphism consists of all those $[a] \in H_*(\mathcal{A})$ for which there exists $b \in B_{n+1}$ such that $\partial b = a$. This is precisely the condition defining the kernel of i_* as well.

Thus the exactness of (5.5) is verified. □

We also note that by Proposition 5.3, any map between short exact sequences

$$0 \longrightarrow \mathcal{A} \overset{i}{\longrightarrow} \mathcal{B} \overset{p}{\longrightarrow} \mathcal{C} \longrightarrow 0$$

$$\alpha \downarrow \qquad \beta \downarrow \qquad \gamma \downarrow$$

$$0 \longrightarrow \widetilde{\mathcal{A}} \overset{\widetilde{i}}{\longrightarrow} \widetilde{\mathcal{B}} \overset{\widetilde{p}}{\longrightarrow} \widetilde{\mathcal{C}} \longrightarrow 0$$

induces a map between the corresponding homology long exact sequences

$$\cdots \overset{\partial_{n+1}}{\longrightarrow} H_n(\mathcal{A}) \overset{i_*}{\longrightarrow} H_n(\mathcal{B}) \overset{p_*}{\longrightarrow} H_n(\mathcal{C}) \overset{\partial_n}{\longrightarrow} H_{n-1}(\mathcal{A}) \overset{i_*}{\longrightarrow} \cdots$$

$$\alpha_* \downarrow \qquad \beta_* \downarrow \qquad \gamma_* \downarrow \qquad \alpha_* \downarrow$$

$$\cdots \overset{\widetilde{\partial}_{n+1}}{\longrightarrow} H_n(\widetilde{\mathcal{A}}) \overset{\widetilde{i}_*}{\longrightarrow} H_n(\widetilde{\mathcal{B}}) \overset{\widetilde{p}_*}{\longrightarrow} H_n(\widetilde{\mathcal{C}}) \overset{\widetilde{\partial}_n}{\longrightarrow} H_{n-1}(\widetilde{\mathcal{A}}) \overset{\widetilde{i}_*}{\longrightarrow} \cdots$$

5.2 The Long Exact Sequence of a Pair and Some Applications

5.2.1 Relative Homology and the Associated Long Exact Sequence

In the next definition we encounter the first and simplest case of a filtration. These will be investigated in greater depth in Chapter 16.

Definition 5.8. *Let X be an arbitrary topological space, and let A be a subspace of X. Then we call the ordered pair (X, A) a **pair of spaces**. When X is a CW complex and A is its CW subcomplex, the pair (X, A) is called a **CW pair**. When X and A are simplicial complexes, the pair (X, A) is called a **simplicial pair**.*

It is easy to see that when (X, A) is a pair of spaces, the singular chain complex of A is a subcomplex of the singular chain complex of X.

Definition 5.9. *Let (X, A) be a pair, and let $\mathcal{R}$ be a commutative ring with unit. The **relative homology** of this pair with coefficients in $\mathcal{R}$ is denoted by $H_*(X, A; \mathcal{R})$, and it is defined to be the homology of the quotient chain complex $H_*(C_*^{\mathrm{Sing}}(X; \mathcal{R})/C_*^{\mathrm{Sing}}(A; \mathcal{R}))$.*

A direct translation of Theorem 5.7 yields the following useful long exact sequence.

Theorem 5.10. (Long exact sequence of a pair).
For any pair of spaces (X, A) we have a long exact sequence of homology groups

$$\cdots \xrightarrow{\partial_{n+1}} H_n(A) \xrightarrow{i_*} H_n(X) \xrightarrow{p_*} H_n(X, A) \xrightarrow{\partial_n} H_{n-1}(A) \xrightarrow{i_*} \cdots, \qquad (5.6)$$

Remark 5.11. There is also a version of the long exact sequence (5.6) using the reduced homology instead:

$$\cdots \xrightarrow{\widetilde{\partial}_{n+1}} \widetilde{H}_n(A) \xrightarrow{\widetilde{i}_*} \widetilde{H}_n(X) \xrightarrow{\widetilde{p}_*} H_n(X, A) \xrightarrow{\widetilde{\partial}_n} \widetilde{H}_{n-1}(A) \xrightarrow{\widetilde{i}_*} \cdots. \qquad (5.7)$$

This fact follows from Theorem 5.7 as well, when one lets $\mathcal{A}$ and $\mathcal{B}$ be the augmented singular chain complexes of A and of X.

We note that, unlike the usual homology, the relative homology does not have a reduced analogue. This is because in taking the relative homology, the augmentations cancel out in the quotient.

Remark 5.12. In full analogy with Definition 5.9 one can define the **relative cohomology** of a pair (X, A) as the cohomology of the cochain complex $C_{\mathrm{Sing}}^*(X, A)$, where $C_{\mathrm{Sing}}^n(X, A)$ consists of all functions on n-cells that have value 0 on the singular simplices inside A. Theorem 5.7 can be applied to the short exact sequence

$$0 \longleftarrow C_{\mathrm{Sing}}^*(A) \longleftarrow C_{\mathrm{Sing}}^*(X) \longleftarrow C_{\mathrm{Sing}}^*(X, A) \longleftarrow 0,$$

yielding the cohomology version of the long exact sequence (5.6):

$$\cdots \xleftarrow{\widetilde{\partial}^{n+1}} \widetilde{H}^n(A) \xleftarrow{\widetilde{i}^*} \widetilde{H}^n(X) \xleftarrow{\widetilde{p}^*} H^n(X, A) \xleftarrow{\widetilde{\partial}^n} \widetilde{H}^{n-1}(A) \xleftarrow{\widetilde{i}^*} \cdots. \qquad (5.8)$$

Definition 5.13. *Given two pairs of spaces (X, A) and $(\widetilde{X}, \widetilde{A})$, a continuous map $\varphi: X \to \widetilde{X}$ is called a **map between pairs** if $\varphi(A) \subseteq \widetilde{A}$.*

The naturality of the connecting homomorphism yields the following fact.

Proposition 5.14. *Any map φ between pairs of spaces (X, A) and $(\widetilde{X}, \widetilde{A})$ induces a chain complex map between the corresponding homology long exact sequences of these pairs.*

5.2.2 Applications

Revising cellular homology

Let X be an arbitrary CW complex. It is easy to see that for every $k \geq 0$, the relative homology group $H_k(X^{(k)}, X^{(k-1)})$ is free abelian, with a basis indexed by the k-cells of X. Hence, we can use the group $H_k(X^{(k)}, X^{(k-1)})$ as an alternative definition for the cellular chain group $C_k^{\text{Cell}}(X)$. The cellular boundary operator $\partial_k^{\text{Cell}} : C_k^{\text{Cell}}(X) \to C_{k-1}^{\text{Cell}}(X)$ can then be defined as the composition of maps

$$H_k(X^{(k)}, X^{(k-1)}) \xrightarrow{\partial} H_{k-1}(X^{(k-1)}) \xrightarrow{i_*} H_{k-1}(X^{(k-1)}, X^{(k-2)}),$$

where the first map is the connecting homomorphism from the long exact sequence of the pair $(X^{(k)}, X^{(k-1)})$, and the second map is the corresponding map from the long exact sequence of the pair $(X^{(k-1)}, X^{(k-2)})$.

Using this approach, it is easy to see that the cellular boundary maps compose to 0. Indeed, by construction we have

$$\partial_{k-1}^{\text{Cell}} \circ \partial_k^{\text{Cell}} = i_*^2 \circ \partial^2 \circ i_*^1 \circ \partial^1,$$

where the maps are the appropriate ones from the corresponding long exact sequences:

$$H_k(X^{(k)}, X^{(k-1)}) \xrightarrow{\partial^1} H_{k-1}(X^{(k-1)}),$$
$$H_{k-1}(X^{(k-1)}, X^{(k-2)}) \xrightarrow{\partial^2} H_{k-2}(X^{(k-2)}),$$
$$H_{k-1}(X^{(k-1)}) \xrightarrow{i_*^1} H_{k-1}(X^{(k-1)}, X^{(k-2)}),$$
$$H_{k-2}(X^{(k-2)}) \xrightarrow{i_*^2} H_{k-2}(X^{(k-2)}, X^{(k-3)}).$$

In particular, the composition $\partial^2 \circ i_*^1$ is a 0-map, since both maps are from the long exact sequence of the pair $(X^{(k-1)}, X^{(k-2)})$. Hence the composition $\partial_{k-1}^{\text{Cell}} \circ \partial_k^{\text{Cell}}$ is also 0.

It is also now easy to prove Theorem 3.41. Indeed, a cellular map $\varphi : X \to Y$ induces a map of pairs $\varphi : (X^{(k)}, X^{(k-1)}) \to (Y^{(k)}, Y^{(k-1)})$, for all k, and hence, by Proposition 5.14, it induces a map between cellular homologies $\varphi_* : H_k^{\text{Cell}}(X) \to H_k^{\text{Cell}}(Y)$.

Geometric model for relative homology

Assume that X is an arbitrary CW complex and A is a subcomplex of X. Construct a new space Q by taking X and adding a cone over A. It is straightforward to extend the cellular structure from X to Q: we will get a new vertex, the apex of the cone, and each cell of A will yield a new additional cell, whose dimension is one higher. The next proposition shows that the space Q is in a sense a geometric model for the relative homology.

Proposition 5.15. *For CW complexes X, A, and Q, as above, we get $\widetilde{H}_n(Q) = H_n(X, A)$, for all n.*

Proof. Let C denote the cone over A, which has been added to X, and consider the reduced version of the long exact sequence of the pair (Q, C):

$$\cdots \xrightarrow{\widetilde{\partial}_{n+1}} \widetilde{H}_n(C) \longrightarrow \widetilde{H}_n(Q) \longrightarrow H_n(Q, C) \xrightarrow{\widetilde{\partial}_n} \widetilde{H}_{n-1}(C) \longrightarrow \cdots .$$

We note that the chain complexes $C_*^{\mathrm{Cell}}(Q)/C_*^{\mathrm{Cell}}(C)$ and $C_*^{\mathrm{Cell}}(X)/C_*^{\mathrm{Cell}}(A)$ are isomorphic; hence $H_n(Q, C) = H_n(X, A)$, for all n. The statement of the proposition now follows from the fact that $\widetilde{H}_n(C) = 0$, for all n. $\square$

Homology of the suspension

Our last application concerns the suspension construction, which was defined in Example 2.32(1). Let X be an arbitrary CW complex. Its suspension is $\operatorname{susp} X = X * \mathbb{S}^1$, and there is a straightforward way to extend the cellular structure of X to the entire $\operatorname{susp} X$. Namely, we add two new vertices, one for each apex, and each cell σ of X gives rise to two new cells σ_+ and σ_-, one inside each cone, such that $\dim \sigma_+ = \dim \sigma_- = \dim \sigma + 1$.

Theorem 5.16. *For an arbitrary CW complex X we have*

$$\widetilde{H}_n(\operatorname{susp} X) = \widetilde{H}_{n+1}(X), \tag{5.9}$$

for all n.

Proof. Let C denote one of the cones over X; this is a "half" of $\operatorname{susp} X$. By Proposition 5.15 we have $\widetilde{H}_n(\operatorname{susp} X) = H_n(C, X)$. On the other hand, the reduced version of the long exact sequence for the pair (C, X) coupled with the fact that $\widetilde{H}_n(C) = 0$ gives the equality $\widetilde{H}_n(X) = H_{n+1}(C, X)$ for all n. This implies (5.9). $\square$

5.3 Mayer–Vietoris Long Exact Sequence

Another standard long exact sequence describes what happens when we glue together two CW complexes. Assume that X is a CW complex that we have represented as a union of two of its subcomplexes $X = A \cup B$. We have four cellular inclusion maps

$$i^1 : A \cap B \hookrightarrow A, \quad i^2 : A \cap B \hookrightarrow B, \quad j^1 : A \hookrightarrow A \cup B, \quad j^2 : B \hookrightarrow A \cup B,$$

which induce corresponding maps between homologies.

Theorem 5.17. (Mayer–Vietoris long exact sequence)
Given A, B, and $X = A \cup B$ as above, we have the following long exact sequence:

$$\cdots \xrightarrow{\partial} H_n(A \cap B) \xrightarrow{\text{diag}} H_n(A) \oplus H_n(B) \xrightarrow{\text{diff}} H_n(A \cup B) \xrightarrow{\partial} H_{n-1}(A \cap B) \xrightarrow{\text{diag}} \cdots, \tag{5.10}$$

where $\text{diag}(x) = (i_*^1(x), i_*^2(x))$ and $\text{diff}(x, y) = j_*^1(x) - j_*^2(y)$.

Proof. Consider the following sequence of chain complexes and chain complex maps:

$$0 \longrightarrow C_*^{\text{Cell}}(A \cap B) \xrightarrow{\text{diag}} C_*^{\text{Cell}}(A) \oplus C_*^{\text{Cell}}(B) \xrightarrow{\text{diff}} C_*^{\text{Cell}}(A \cup B) \longrightarrow 0, \tag{5.11}$$

where $\text{diag}(x) = (i^1(x), i^2(x))$ and $\text{diff}(x, y) = j^1(x) - j^2(y)$. Let us verify that the sequence (5.11) is exact. The injectivity of diag is obvious, as is the exactness in $C_*^{\text{Cell}}(A) \oplus C_*^{\text{Cell}}(B)$. The surjectivity of diff follows from the fact that each cell of $A \cup B$ either lies completely in A or lies completely in B (or both), and therefore every cellular chain of $A \cup B$ can be represented as a sum of a chain of A with a chain of B.

We can therefore apply Theorem 5.7, and using the fact that

$$H_*(C_*^{\text{Cell}}(A) \oplus C_*^{\text{Cell}}(B)) = H_*(C_*^{\text{Cell}}(A)) \oplus H_*(C_*^{\text{Cell}}(B)) = H_*(A) \oplus H_*(B),$$

we derive the long exact sequence (5.10). $\square$

Remark 5.18. We also have a reduced version of the sequence (5.10):

$$\cdots \xrightarrow{\partial} \widetilde{H}_n(A \cap B) \xrightarrow{\text{diag}} \widetilde{H}_n(A) \oplus \widetilde{H}_n(B) \xrightarrow{\text{diff}} \widetilde{H}_n(A \cup B) \xrightarrow{\partial} \widetilde{H}_{n-1}(A \cap B) \xrightarrow{\text{diag}} \cdots. \tag{5.12}$$

Remark 5.19. The cohomology version of the sequence (5.10) is obtained, as expected, by reversing all the arrows:

$$\cdots \xleftarrow{\partial^*} H^n(A \cap B) \xleftarrow{\text{diag}^*} H^n(A) \oplus H^n(B) \xleftarrow{\text{diff}^*} H^n(A \cup B) \xleftarrow{\partial^*} H^{n-1}(A \cap B) \xleftarrow{\text{diag}^*} \cdots. \tag{5.13}$$

In the case of a Mayer–Vietoris sequence, the naturality of connecting homomorphisms yields the following statement.

Proposition 5.20. *Assume that the CW complexes X and $\widetilde{X}$ are represented as unions of CW subcomplexes: $X = A \cup B$ and $\widetilde{X} = \widetilde{A} \cup \widetilde{B}$. Assume furthermore that we have a cellular map $\varphi : X \to \widetilde{X}$ satisfying the additional conditions $\varphi(A) \subseteq \widetilde{A}$ and $\varphi(B) \subseteq \widetilde{B}$. Then φ induces a chain complex map between the corresponding Mayer–Vietoris homology long exact sequences.*

It is interesting to see how Theorem 5.17 can be applied to the two applications of Subsection 5.2.1.

First, when X is a CW complex and A a subcomplex, we have constructed the CW complex Q by attaching a cone C over A. Thus $Q = C \cup X$ and $A = C \cap X$. Since $\widetilde{H}_n(C) = 0$ for all n, it is handy to employ the reduced version of the Mayer–Vietoris long exact sequence. What we then get is the reduced version of the long exact sequence of the pair (X, A), with $H_*(Q)$ instead of $H_*(X, A)$. One can then use this to show the actual isomorphism between the homology groups, using the so-called 5-lemma. We do not need that here, so we refer the reader to any of the standard textbooks in algebraic topology.

Second, in the case of suspension, we note that $\operatorname{susp} X = C_1 \cup C_2$, where C_1 and C_2 are the two cones that we spanned over X when constructing $\operatorname{susp} X$. Furthermore, we have $C_1 \cap C_2 = X$ and $H_n(C_1) = H_n(C_2) = 0$, for all n. Using the reduced version of the Mayer–Vietoris long exact sequence we once again confirm the equalities (5.9).

6

Homotopy

6.1 Homotopy of Maps

Given two topological spaces X and Y, one can introduce an equivalence relation on the space of all continuous maps from X to Y.

Definition 6.1. *Two continuous maps $f, g : X \to Y$ are called* **homotopic** *is there exists a continuous map $F : X \times [0,1] \to Y$ such that $F(-,0) = f$ and $F(-,1) = g$. In that case, we write $f \simeq g$.*

Let us see that Definition 6.1 actually produces an equivalence relation. First, by setting $F(x,t) := f(x)$ independent of t, for all $x \in X$, $t \in [0,1]$, we see that any map f is homotopic to itself. Second, if $f, g : X \to Y$ are homotopic with a homotopy given by $F : X \times [0,1] \to Y$, we can define the reverse homotopy $G : X \times [0,1] \to Y$ by setting $G(x,t) := F(x, 1-t)$. This gives a homotopy from g to f, thus verifying the symmetry. Finally, given three continuous maps $f, g, h : X \to Y$ and two homotopies $F : X \times [0,1] \to Y$ from f to g and $G : X \times [0,1] \to Y$ from g to h, one obtains a concatenated homotopy $H : X \times [0,1] \to Y$ from f to h by setting

$$H(x,t) := \begin{cases} F(x, 2t), & \text{if } t \le 1/2; \\ G(x, 2t - 1), & \text{otherwise.} \end{cases}$$

The equivalence classes of this equivalence relation are called *homotopy classes* of maps, and are denoted by $[f]$.

Example 6.2.
(1) An arbitrary continuous map $f : \mathbb{S}^1 \to \mathbb{S}^2$ is homotopic to any map that takes the entire circle to one point in $\mathbb{S}^2$.
(2) Let T denote the 2-dimensional torus, $T = \mathbb{S}^1 \times \mathbb{S}^1$. The following two maps $f, g : \mathbb{S}^1 \to T$ are homotopic: f wraps $\mathbb{S}^1$ diagonally, i.e., $f(x) := (x, x)$, for $x \in \mathbb{S}^1$, and g traverses the factor circles one after the other, i.e.,

$$g(e^{\varphi i}) := \begin{cases} (e^{2\varphi i}, 0), & \text{if } 0 \leq \varphi \leq \pi; \\ (0, e^{2\varphi i - 2\pi i}), & \text{otherwise.} \end{cases}$$

Proposition 6.3. *The composition of homotopy classes of maps is well-defined if we set*

$$[g] \circ [f] := [g \circ f], \tag{6.1}$$

for any pair of continuous maps $f : X \to Y$ and $g : Y \to Z$.

Proof. Assume first that f is homotopic to some map $\tilde{f} : X \to Y$ and that the homotopy is given by $H : X \times [0,1] \to Y$. The homotopy from $g \circ f$ to $g \circ \tilde{f}$ is given by the composition map $\widetilde{H} : X \times [0,1] \xrightarrow{H} Y \xrightarrow{g} Z$.

Assume now that g is homotopic to some map $\tilde{g} : Y \to Z$ and that the homotopy is given by $H : Y \times [0,1] \to Z$. The homotopy from $g \circ f$ to $\tilde{g} \circ f$ is given by the composition map $\widetilde{H} : X \times [0,1] \xrightarrow{f \times \mathrm{id}} Y \times [0,1] \xrightarrow{H} Z$. $\square$

An important property of the homotopic maps is that they become identical once we pass to the algebraic invariants that we have defined.

Theorem 6.4. *Given two CW complexes X and Y and two homotopic cellular maps $f, g : X \to Y$, the induced maps on homology are the same, i.e., we have $f_* = g_*$.*

The proof of Theorem 6.4 lies outside the scope of this book. It can be found in all standard textbooks in algebraic topology; see, e.g., [Hat02, Theorem 2.10].

6.2 Homotopy Type of Topological Spaces

Theorem 6.4 motivates the introduction of the corresponding equivalence relation on the topological spaces themselves.

Definition 6.5. *Two topological spaces are called* **homotopy equivalent** *if there exist continuous maps $\varphi : X \to Y$ and $\psi : Y \to X$ such that $\varphi \circ \psi \simeq \mathrm{id}_Y$ and $\psi \circ \varphi \simeq \mathrm{id}_X$. In that case, we write $X \simeq Y$.*

Indeed, Theorem 6.4 implies the following fundamental fact, strengthening the homeomorphism invariance that was previously mentioned in Remark 3.16.

Theorem 6.6. *Homology groups are homotopy invariants; in other words, if X and Y are homotopy equivalent CW complexes, then $H_n(X) = H_n(Y)$, for any $n \geq 0$.*

Proof. Let $\varphi : X \to Y$ and $\psi : Y \to X$ be continuous maps such that $\varphi \circ \psi \simeq \mathrm{id}_Y$ and $\psi \circ \varphi \simeq \mathrm{id}_X$. By Theorem 6.4 we have $\varphi_* \circ \psi_* = \mathrm{id}_{H_*(Y)}$ and $\psi_* \circ \varphi_* = \mathrm{id}_{H_*(X)}$, which allows us to conclude that $H_*(X) = H_*(Y)$, and the maps φ_* and ψ_* are inverses of each other. $\square$

The homotopy equivalence is indeed an equivalence relation. The only nontrivial part of this statement is the verification of transitivity. Assume that we have three topological spaces X, Y, and Z, and four maps $\varphi : X \to Y$, $\psi : Y \to X$, $f : Y \to Z$, and $g : Z \to Y$ such that $\varphi \circ \psi \simeq \mathrm{id}_Y$, $\psi \circ \varphi \simeq \mathrm{id}_X$, $g \circ f \simeq \mathrm{id}_Y$, and $f \circ g \simeq \mathrm{id}_Z$. Repeatedly using (6.1), we see that $(\psi \circ g) \circ (f \circ \varphi) = \psi \circ (g \circ f) \circ \varphi \simeq \psi \circ \varphi \simeq \mathrm{id}_X$ and $(f \circ \varphi) \circ (\psi \circ g) = f \circ (\varphi \circ \psi) \circ g \simeq f \circ g \simeq \mathrm{id}_Z$. The equivalence classes of this equivalence relation are called *homotopy types* of topological spaces.

We are now ready to define a new category **HTop**, the so-called *homotopy category*. Its objects are the same as those of **Top**: the topological spaces. However, the morphisms are different: they are all the homotopy classes of continuous maps. Using the previously defined terminology, we say that two objects in **HTop** are isomorphic if and only if the corresponding topological spaces are homotopy equivalent. This is a usual "quotient" situation in category theory: the set of objects is left intact and the set of morphisms is quotiented, producing more isomorphisms.

Example 6.7.

(1) A ball of arbitrary dimension is homotopy equivalent to a point.
(2) The Möbius band is the topological space which we obtain from a rectangle by identifying a pair of opposite sites, changing the directions, e.g., we can take the space $([0,1] \times [0,1])/\sim$, with the identification relation given by $(0,x) \sim (1,1-x)$, for all $x \in [0,1]$. It is easy to see that the Möbius band is homotopy equivalent to a circle. The homotopy equivalence maps are given by embedding $\mathbb{S}^1$ as the middle circle of the Möbius band, respectively by shrinking the Möbius band onto its middle circle.
(3) If an interval is attached to a sphere $\mathbb{S}^2$ by both ends, then the space obtained is homotopy equivalent to the wedge of $\mathbb{S}^2$ and $\mathbb{S}^1$.
(4) Gluing a disk along its boundary to the 2-dimensional hollow torus along one of the torus's generating circles will give a space that is also homotopy equivalent to the wedge of $\mathbb{S}^2$ and $\mathbb{S}^1$.

6.3 Mapping Cone and Mapping Cylinder

Before we proceed, we would like to define two very important constructions for topological spaces, which also behave well if we have cellular structures.

Definition 6.8. *Let* $f : X \to Y$ *be a continuous map between two topological spaces. The* **mapping cylinder** *of f is the quotient space*

$$M(f) = ((X \times I) \cup Y)/\sim,$$

where the equivalence relation $\sim$ is defined by $(x, 1) \sim f(x)$, for all $x \in X$.

Example 6.9. Let $X = \mathbb{S}^1$, $Y = [0, 1]$, and let f map all points in X to the point $1/2$. Then $M(f)$ is the cone over $\mathbb{S}^1$ attached by its apex to the middle point of the interval $[0, 1]$. See Figure 6.1(left).

Definition 6.10. *Let $f : X \to Y$ be a continuous map between two topological spaces. The **mapping cone** of f is the quotient space*

$$\text{Cone}(f) = ((X \times I) \cup Y)/\sim,$$

where the equivalence relation $\sim$ is defined by $(x, 1) \sim f(x)$ and $(x, 0) \sim (y, 0)$, for all $x, y \in X$.

Example 6.11. Let X, Y, and f be as in the previous example. Then $\text{Cone}(f)$ is the sphere $\mathbb{S}^2$ attached to the middle point of the interval $[0, 1]$. See Figure 6.1(right).

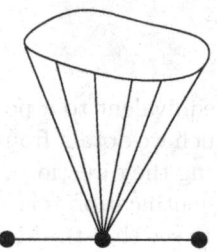 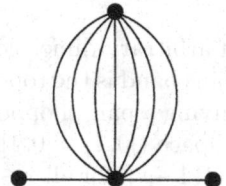

Fig. 6.1. A mapping cylinder and a mapping cone of a map taking everything to a point.

Let us now look to see what happens when X and Y are CW complexes and the map f is cellular. We need three important general facts about CW complexes. We provide intuitive explanations, but no formal proofs.

Fact 1. *If X is a CW complex, then $X \times I$ has a naturally induced product CW structure.*

Consider the CW structure on I consisting of two 0-cells (the endpoints) and one 1-cell. One can take the cells of $X \times I$ simply to be the products of cells in X and in I. In particular, there are two copies of X sitting inside of $X \times I$, as $X \times \{0\}$ and $X \times \{1\}$, and each one is a CW subcomplex.

Fact 2. *If X is a CW complex and Y a CW subcomplex of X, then X/Y has a naturally induced quotient CW structure.*

The cells of X/Y are precisely those cells of X that are not in Y, together with one 0-cell, which denotes the result of shrinking Y to a point. The new attaching maps are best described as compositions of the old ones with the map that shrinks all cells of Y to a point.

Fact 3. *If X and Z are CW complexes, Y is a CW subcomplex of X, and $f : Y \to Z$ is a cellular map, then $X \cup_f Z$ has a naturally induced CW structure.*

The cells of $X \cup_f Z$ are those cells of X that do not belong to Y, together with all cells of Z. For the cells of Z, the attaching maps do not change, whereas for the cells of X, the new attaching maps are again best described as compositions of the old attaching maps with the map that identifies points in Y with their images under the cellular map f.

We are now ready to see, using Facts 1 and 3, that there is a natural induced CW structure on $\mathrm{M}(f)$. In this structure we have the cells of X, the cells of Y, and, additionally, the cells of X multiplied by an interval glued in the appropriate way. This is because $\mathrm{M}(f) = (X \times I) \cup_{\tilde{f}} Y$, where $\tilde{f}$ maps the bottom copy of X inside $X \times I$ to the CW complex Y along the cellular map f.

Observe that it is important in Example 6.9 to choose the CW structure on the interval $[0, 1]$ so that the map f is cellular. For instance, if we take the endpoints to be the vertices and the open interval $(0, 1)$ to be the 1-cell, then the induced cell structure on $\mathrm{M}(f)$ will not be well-defined, since we are not allowed to glue the boundaries of cells onto the interiors of cells of the same dimension or higher.

When both X and Y are CW complexes and the map f is cellular, there is a natural induced CW structure on $\mathrm{Cone}(f)$ as well. This is because $\mathrm{Cone}(f) = \mathrm{M}(f)/X$, where we quotient the top copy of X inside of $X \times I$ to a point, and use Fact 2.

6.4 Deformation Retracts and Collapses

Definition 6.12. *Let X be a topological space, let $A \subseteq X$, and let $i : A \hookrightarrow X$ be the inclusion map. A continuous map $f : X \to A$ is called*

- *a **retraction** if $f|_A = \mathrm{id}_A$;*
- *a **deformation retraction** if $i \circ f : X \to X$ is homotopic to the identity map id_X;*
- *a **strong deformation retraction** if there exists a homotopy $F : X \times I \to X$ between $i \circ f$ and id_X that is constant on A, i.e., $F(a, t) = a$, for all $t \in I$ and $a \in A$.*

Correspondingly, A is called a retract, a deformation retract, or a strong deformation retract of X.

Note that in a *comb space*

$$C = [0,1] \times \{0\} \cup \{0\} \times [0,1] \cup \bigcup_{i=1}^{\infty} \left\{\frac{1}{n}\right\} \times [0,1] \subset \mathbb{R}^2,$$

all points are deformation retracts of C, but not all points are strong deformation retracts of C.

A useful example of a strong deformation retraction is provided by the following.

Definition 6.13. *Let Δ be a generalized simplicial complex. Let $\sigma, \tau \in \Delta$ such that*

(1) $\tau \subset \sigma$, in particular $\dim \tau < \dim \sigma$;
(2) σ is a maximal simplex, and no other maximal simplex contains τ.

*A simplicial **collapse** of Δ is the removal of all simplices γ such that $\tau \subseteq \gamma \subseteq \sigma$. If additionally, we have $\dim \tau = \dim \sigma - 1$, then this is called an **elementary collapse**.*

When Δ_1 and Δ_2 are two generalized simplicial complexes such that there exists a sequence of collapses leading from Δ_1 to Δ_2, we shall use the notation $\Delta_1 \searrow \Delta_2$.

Proposition 6.14. *A sequence of collapses yields a strong deformation retraction, in particular, a homotopy equivalence.*

Another source of strong deformation retracts is provided by mapping cylinders.

Proposition 6.15. *Let $f : X \to Y$ be a continuous map between two topological spaces. Then Y is a strong deformation retract of $M(f)$.*

Proof. Let $i : Y \hookrightarrow M(f)$ be the inclusion map, and let $g : M(f) \to Y$ be defined by $g(x,p) = f(x) \, (= (x,1))$ for $(x,p) \in X \times I$, and $g(y) = y$ for $y \in Y$.

Clearly, $g \circ i : Y \to Y$ is the identity map, so we just need to show that there is a homotopy between $\mathrm{id}_{M(f)}$ and $i \circ g : M(f) \to M(f)$ that keeps Y fixed. Define $G : M(f) \times I \to M(f)$ by

$$G((x,p),t) = (x, 1 - t(1-p))$$

and $G(y,t) = y$, for all $(x,p) \in X \times I$, $y \in Y$, and $t \in I$. Clearly, G is continuous, $G(-,0) = i \circ g$, and $G(-,1) = \mathrm{id}_{M(f)}$. $\square$

Although strong deformation retraction seems like a much stronger operation than homotopy equivalence, it turns out that two topological spaces are homotopy equivalent if and only if there exists a third space that can be strong deformation retracted both onto X and onto Y. One possible choice for this third space is simply the mapping cylinder of the homotopy equivalence map; see Corollary 7.16.

6.5 Simple Homotopy Type

The converse of Proposition 6.14 is not true: there are abstract simplicial complexes that are contractible, yet not collapsible. The situation is somewhat clarified by the following result.

Theorem 6.16. *A generalized simplicial complex Δ is contractible if and only if there exists a sequence of collapses and expansions (operation inverse to the collapse, also called an anticollapse) leading from Δ to a vertex.*

More generally, one benefits from the following definition.

Definition 6.17. *Two generalized simplicial complexes are said to have the same **simple homotopy type** if there exists a sequence of elementary collapses and expansions leading from one to the other. Such a sequence is called a **formal deformation**.*

Theorem 6.16 is a special case of the fundamental theorem that in particular says that two simply connected generalized simplicial complexes are homotopy equivalent if and only if they have the same simple homotopy type.

It is known that a subdivision of any generalized simplicial complex X has the same simple homotopy type as X. Let us show a special case of that.

Proposition 6.18. *Let X be the geometric realization of an arbitrary abstract simplicial complex. Then there exists a formal deformation from X to $\mathrm{Bd}\,X$.*

Proof. To start with, since the barycentric subdivision can be represented as a sequence of stellar subdivisions, see Subsection 2.1.5, it is enough to find a formal deformation leading from X to $\mathrm{sd}(X,\sigma)$ for an arbitrary simplex $\sigma \in X$. One choice of such a deformation is a concatenation of two steps.

Deformation algorithm from X to $\mathrm{sd}(X,\sigma)$.

Step 1. Add a cone over $\mathrm{st}_X(\sigma)$. More precisely, consider a new simplicial complex X' such that $V(X') = V(X) \cup \{v\}$, X is an induced subcomplex of X', and $\mathrm{lk}_{X'}v = \mathrm{st}_X(\sigma)$.

Step 2. Delete from X' all the simplices containing σ.

Since $\mathrm{st}_X(\sigma)$ is a cone, in particular collapsible, Step 1 can be performed as a sequence of elementary expansions. Furthermore, Step 2 can be performed as a sequence of elementary collapses as follows: the set of simplices that are to be deleted can be written as a disjoint union of sets A and B, where B is the set of all simplices that contain both σ and v. Clearly, adding v to a simplex is a bijection $\mu : A \to B$. Let $\{\tau_1 \ldots, \tau_t\}$ be a reverse linear extension order on A then $\{(\tau_1, \mu(\tau_1)), \ldots, (\tau_t, \mu(\tau_t))\}$ is an elementary collapsing sequence.

Finally, we see that performing Steps 1 and 2, in this order, will yield a stellar subdivision of X at σ, and therefore our description is completed. $\square$

6.6 Homotopy Groups

In this section we define yet another family of algebraic invariants of topological spaces. We give just a very brief overview without proofs.

Definition 6.19. *Let X be a topological space, choose $x_0 \in X$, and let $n \geq 0$. The set of homotopy classes of maps $\varphi : \mathbb{S}^n \to X$ mapping the north pole of $\mathbb{S}^n$ to x_0 is denoted by $\pi_n(X, x_0)$.*

When $n \geq 1$ this set can be equipped with a product operation as follows: to compose $\varphi : \mathbb{S}^n \to X$ with $\psi : \mathbb{S}^n \to X$, choose a great circle passing through the north pole, and shrink it to a point (when $n = 1$ we identify the two poles). Now map the obtained wedge of two copies of $\mathbb{S}^n$ to X by taking the map φ on one copy of $\mathbb{S}^n$ and taking the map ψ on the other copy. In both cases, the wedge point plays the role of the north pole on each $\mathbb{S}^n$.

The set $\pi_n(X, x_0)$ with this operation is called the nth **homotopy group** *of X with base point x_0.*

One can show that the product is well-defined, and that one indeed gets a group. For example, the inverse is obtained by precomposing $\varphi : \mathbb{S}^n \to X$ with the reflection in one of the coordinates of $\mathbb{S}^n$, fixing the north pole; here $\mathbb{S}^n$ is viewed as a unit sphere. The set $\pi_0(X, x_0)$ is simply the set of all path-connected components.

As $\pi_1(X, x_0)$ we obtain the so-called *fundamental group* of X with base point x_0. The fundamental group plays an important role in the theory of covering spaces, which we do not cover in this book. It is also closely related to the homology groups; namely, when X is path-connected, the group $H_1(X; \mathbb{Z})$ is the abealinization of the fundamental group.

In can be shown that when the base point is changed within the path-connected component, the new homotopy group is isomorphic to the old one. Therefore, whenever the considered topological space is path-connected, we shall skip the base point from the notation. Again, the proof is not difficult. One basically has to stretch a part of the sphere that is close to the north pole along the path connecting the old base point with the new one.

An important fact is that when $n \geq 2$, the group $\pi_n(X)$ is abelian. This happens because in dimensions 2 and higher it is possible for two balls on the boundary of the sphere to go around each other, whereas on the circle they would have to pass through each other.

Another interesting property, which we state without a proof, is that for any two spaces X and Y and any n, we have

$$\pi_n(X \times Y) = \pi_n(X) \times \pi_n(Y). \tag{6.2}$$

More generally, there is a long exact sequence connecting homotopy groups of the involved spaces in any fibration.

Example 6.20.

(1) We have $\pi_1(\mathbb{S}^1) = \mathbb{Z}$, and $\pi_n(\mathbb{S}^1) = 0$ for $n \geq 2$. The distinct elements of $\pi_1(\mathbb{S}^1)$ can be realized by wrapping $\mathbb{S}^1$ around itself various numbers of times, with direction encoding the sign.

(2) For a two-dimensional torus T we have $\pi_1(T) = \mathbb{Z} \times \mathbb{Z}$, and $\pi_n(T) = 0$ for $n \geq 2$. This follows from the previous example and equation (6.2).

(3) One radical difference between homotopy and homology is illustrated by the case of spheres. Whereas it was very easy to determine the homology groups of spheres, their homotopy groups are in general unknown, and are the subject of current research. Thus our first example of $\mathbb{S}^1$ is an exception rather than the rule.

For $\mathbb{S}^n$, $n \geq 2$, it still remains true that $\pi_n(\mathbb{S}^n) = \mathbb{Z}$, and $\pi_m(\mathbb{S}^n) = 0$, for $m < n$. However, the higher homotopy groups can be nontrivial. The first example is $\pi_3(\mathbb{S}^2) = \mathbb{Z}$. Further examples are $\pi_4(\mathbb{S}^2) = \mathbb{Z}_2$, $\pi_6(\mathbb{S}^2) = \mathbb{Z}_{12}$, and $\pi_{10}(\mathbb{S}^4) = \mathbb{Z}_{24} \times \mathbb{Z}_3$.

An easy but important property of homotopy groups is their functoriality.

Proposition 6.21. *Let X and Y be topological spaces, and let $f : X \to Y$ be a continuous map. Then the composition with f will induce a group homomorphism $f_* : \pi_n(X) \to \pi_n(Y)$, for any n.*

Furthermore, if Z is also a topological space and $g : Y \to Z$ is a continuous map, then we have $(g \circ f)_* = g_* \circ f_*$.

Thus we see that π_n is yet another example of a covariant functor **Top** $\to$ **Ab**. We leave verification of Proposition 6.21 to the reader.

6.7 Connectivity and Hurewicz Theorems

One of the most useful tools for calculating fundamental group is Van Kampen's theorem, which allows us to cover our space by nice subspaces and then reconstruct the total fundamental group from the fundamental groups of the covering spaces and their intersections.

Theorem 6.22. (Van Kampen's theorem).

(1) Assume that X is a topological space with base point x, and assume that A and B are path-connected open subspaces of X such that $A \cap B$ is also path-connected, $x \in A \cap B$, and $X = A \cup B$. Then $\pi_1(X,x)$ is the colimit of the following diagram:

$$
\begin{array}{ccc}
\pi_1(A \cap B, x) & \xrightarrow{\ i_*\ } & \pi_1(A, x) \\
{\scriptstyle j_*}\big\downarrow & & \\
\pi_1(B, x) & &
\end{array}
\qquad (6.3)
$$

where $i_* : \pi_1(A \cap B, x) \to \pi_1(A, x)$ and $j_* : \pi_1(A \cap B, x) \to \pi_1(B, x)$ are induced by the inclusion maps $i : A \cap B \hookrightarrow A$ and $j : A \cap B \hookrightarrow B$.

(2) More generally, assume that X is covered by finitely many path-connected open subspaces A_i, $i = 1, \ldots, n$. Assume that $x \in A_i$, for all $i = 1, \ldots, n$, and that all intersections are again path-connected. Consider a diagram, generalizing (6.3), consisting of the fundamental groups of all possible intersections and all maps between them induced by inclusions of intersections. Then the fundamental group of X is the colimit of this diagram.

The proof of Van Kampen's theorem is elementary, though it requires some topological finesse. We refer the reader to any of the standard textbooks.

Remark 6.23. For combinatorial applications it is useful to note the following variation of Van Kampen's theorem: X is assumed to be a CW complex, and the subspaces A_i are all assumed to be the CW subcomplexes of X.

To derive Remark 6.23 from Theorem 6.22 we can simply replace each subcomplex A_i by an ε-neighborhood of A_i. The ε can be chosen small enough so that the resulting diagram of groups is the same for both the open and the CW covering.

Definition 6.24. *A path-connected topological space X is called* **simply connected** *if its fundamental group is trivial.*

For many calculations, the most frequently used special cases of Theorem 6.22 will suffice.

Corollary 6.25. *Assume that A, B, and $A \cap B$, are path-connected.*

(1) If both A and B are simply connected, then so is $A \cup B$.
(2) If $A \cap B$ is simply connected, then $\pi_1(A \cup B) = \pi_1(A) * \pi_1(B)$.

Proof. These are both corollaries of Van Kampen's Theorem; just substitute the assumptions in diagram (6.3) and take the colimit.   □

Example 6.26. Let us calculate the fundamental group of the projective plane $\mathbb{RP}^2$ using Van Kampen's theorem. Consider a standard representation of $\mathbb{RP}^2$ as a unit disk in the plane with the boundary self-identified by the antipodal map, i.e., $x \sim -x$. Let A be the disk centered at the origin with radius $1/2$, and let B be obtained from $\mathbb{RP}^2$ by removal of the interior of A. We have $A \cup B = \mathbb{RP}^2$, and it is easy to equip $\mathbb{RP}^2$ with a CW structure such that A and B are CW subcomplexes (alternatively, one could thicken A and B a little bit so as to obtain a topologically identical open covering of $\mathbb{RP}^2$).

In either case, Van Kampen's theorem can be applied. The space A is contractible, the intersection $A \cap B$ is homeomorphic to a circle, and the space B deformation retracts to a circle. To see the last of these facts, simply retract the punctured unit disk to its boundary, and note that the antipodal self-identification of the boundary again produces a circle. We have $\pi_1(A) = 0$ and $\pi_1(B) = \pi_1(A \cap B) = \mathbb{Z}$. The fundamental group homomorphisms induced by the inclusion maps $i^A : A \cap B \hookrightarrow A$ and $i^B : A \cap B \hookrightarrow B$ are the following:

- the map $i_*^A : \pi_1(A \cap B) \to \pi_1(A)$ has to be a 0-map;
- the map $i_*^B : \pi_1(A \cap B) \to \pi_1(B)$ is given by $x \mapsto 2x$; this is clear from geometric considerations.

The definition of the colimit translates in this case to saying that the fundamental group $\mathbb{RP}^2$ is the universal group such that the group homomorphism $\varphi : \pi_1(B) \to \pi_1(\mathbb{RP}^2)$ takes the generator of $\pi_1(B)$ to an element of order 2. Hence we conclude that $\pi_1(\mathbb{RP}^2) = \mathbb{Z}_2$.

Definition 6.27. *For $k \geq 1$, a topological space X is called k-connected if its homotopy groups $\pi_1(X), \ldots, \pi_k(X)$ are all trivial.*

In particular, 1-connected is the same as simply connected. For convenience, we also adopt the convention that 0-connected means path-connected, (-1)-connected means nonempty, and any space is (-2)-connected.

We are now ready to formulate an important fact relating homology with homotopy groups.

Theorem 6.28. (Hurewicz theorem)
Let X be an arbitrary CW complex. Assume that X is simply connected, and that $\widetilde{H}_i(X; \mathbb{Z}) = 0$, for $i = 0, \ldots, k$. Then X is k-connected.

Conversely, if X is $(k-1)$-connected, for some $k \geq 2$, then $\widetilde{H}_i(X; \mathbb{Z}) = 0$, for $i = 0, \ldots, k-1$, and $H_k(X; \mathbb{Z}) = \pi_k(X)$.

The proof of Hurewicz theorem is outside the scope of this book. Instead, we formulate some useful corollaries for future reference. For both statements, assume that X is a CW complex, and that A and B are CW subcomplexes, such that $X = A \cup B$.

Corollary 6.29. *If A and B are k-connected, and $A \cap B$ is $(k-1)$-connected, for some $k \geq -1$, then $A \cup B$ is k-connected.*

Proof. The statement is immediate for $k = -1$ and $k = 0$. For $k = 1$ this is just Corollary 6.25(1).

Assume that $k \geq 2$. By Corollary 6.25(1) we see that X is simply connected. Using Hurewicz theorem for spaces $A \cup B$, A, and B, we see that $\widetilde{H}_i(A \cap B; \mathbb{Z}) = \widetilde{H}_i(A; \mathbb{Z}) = \widetilde{H}_i(B; \mathbb{Z}) = 0$, for $i = 0, \ldots, k-1$, and $\widetilde{H}_k(A; \mathbb{Z}) = \widetilde{H}_k(B; \mathbb{Z}) = 0$. Substituting this information into the Mayer-Vietoris long exact sequence for relative homology (5.12) we see that $\widetilde{H}_i(X; \mathbb{Z}) = 0$, for $i = 0, \ldots, k$. The statement follows now if we apply Hurewicz theorem to the space X. $\square$

Corollary 6.30. *If $A \cup B$ and $A \cap B$ are k-connected, for some $k \geq -1$, then both A and B are k-connected.*

Proof. Again the statement is immediate for $k = -1$.

Assume $k \geq 0$. Applying Hurewicz theorem to $A \cup B$, and to $A \cap B$ we see that $\widetilde{H}_i(A \cap B; \mathbb{Z}) = \widetilde{H}_i(A \cup B; \mathbb{Z}) = 0$, for $i = 0, \ldots, k$. By the Mayer-Vietoris long exact sequence for relative homology (5.12), we get $\widetilde{H}_i(A; \mathbb{Z}) = \widetilde{H}_i(B; \mathbb{Z}) = 0$, for $i = 0, \ldots, k$. For $k = 0$ this already proves the statement.

Assume $k \geq 1$. By Corollary 6.25(2) we have $\pi_1(A \cup B) = \pi_1(A) * \pi_1(B)$, implying that both A and B are simply connected. If $k = 1$, this is precisely what we want to prove. If $k \geq 2$, the statement follows by applying Hurewicz theorem to spaces A and B. $\square$

We end this chapter with an important theorem due to Whitehead, which we state here without a proof.

Theorem 6.31. (Whitehead's theorem)
Let X and Y be connected CW complexes, and assume that φ is a cellular map from X to Y that induces isomorphisms on all homotopy groups. Then φ gives a homotopy equivalence between X and Y.

It is important to stress functoriality in Theorem 6.31. Otherwise, the statement is false: there exist connected CW complexes that are not homotopy equivalent but whose homotopy groups are isomorphic. Following is a frequently used corollary of Theorem 6.31.

Corollary 6.32. *A CW complex X whose homotopy groups are all trivial is contractible.*

Proof. A map from X to a one-point space is a cellular map that induces isomorphisms on all homotopy groups; hence by Theorem 6.31, it gives a homotopy equivalence. $\square$

Corollary 6.33. *A simply connected CW complex X whose homology groups $\widetilde{H}_i(X; \mathbb{Z})$ are all trivial is contractible.*

Proof. By Hurewicz Theorem 6.28, all homotopy groups are trivial as well; hence we may use Corollary 6.32. $\square$

Finally, we state without proof some results that are often useful in Combinatorial Algebraic Topology.

Proposition 6.34. *Let $\varphi : X \to Y$ be a map between simply connected CW complexes that induces isomorphism maps $\varphi_* : H_n(X; \mathbb{Z}) \to H_n(Y; \mathbb{Z})$ for all n. Then, the map φ is a homotopy equivalence.*

Proposition 6.35. *Assume that X is a $(k - 1)$-connected CW complex of dimension k, where $k \geq 0$. Then X is homotopy equivalent to a wedge of k-dimensional spheres.*

7

Cofibrations

7.1 Cofibrations and the Homotopy Extension Property

The following notion is one of the most classical and important in homotopy theory.

Definition 7.1. *Let A and X be topological spaces, and let $i : A \to X$ be a continuous map. The map i is said to have the* **homotopy extension property (HEP)** *if for any topological space Y, any continuous map $f : X \to Y$, and any homotopy $H : A \times I \to Y$ satisfying $H(a, 0) = f(i(a))$, for all $a \in A$, there exists a homotopy $\widetilde{H} : X \times I \to Y$ such that $\widetilde{H}(i(a), t) = H(a, t)$, for all $a \in A$ and all $t \in I$, and $\widetilde{H}(x, 0) = f(x)$, for all $x \in X$. A map having homotopy extension property is called a* **cofibration.** *See Figure 7.1.*

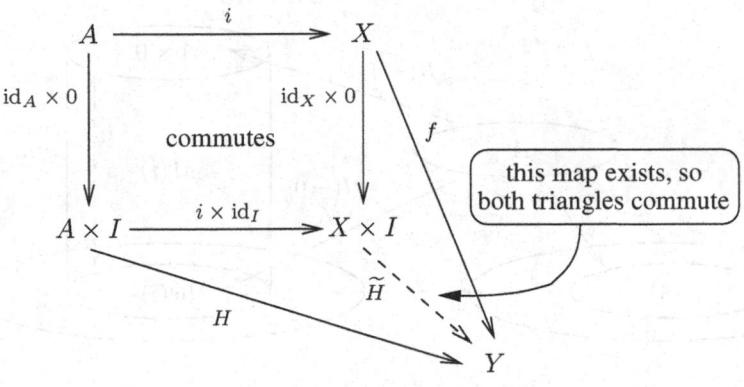

Fig. 7.1. The homotopy extension property.

Let us consider a special case of Definition 7.1, when $i : A \hookrightarrow X$ is an inclusion map. In this case we have a continuous map $f : X \to Y$ and a homotopy

$H : A \times I \to Y$ of the restriction of f to a subspace A. The condition for the inclusion map to be a cofibration says that one should be able to extend the homotopy of A to the homotopy of the entire space X.

Interestingly, this special case captures the whole generality of the notion of a cofibration, as our next proposition shows.

Proposition 7.2. *If a map $i : A \to X$ is a cofibration, then i is a homeomorphism onto its image $i(A)$.*

Proof. Consider the inclusion map $j : A \hookrightarrow M(i)$ taking A to the bottom of the mapping cylinder, i.e., j takes $a \in A$ to $(a,1) \sim i(a)$. See Figure 7.2. Furthermore, consider the homotopy $H : A \times I \to M(i)$, mapping $(a,t) \mapsto (a, 1-t)$. Clearly, the inclusion map $f : X \hookrightarrow M(i)$ (induced by the identity $\mathrm{id}_X : X \to X$) and the homotopy H satisfy the conditions of Definition 7.1; hence there must exist a homotopy $\widetilde{H} : X \times I \to M(i)$ such that $\widetilde{H}|_{X \times 0} = f$ and

$$\widetilde{H}(i(a),t) = H(a,t), \tag{7.1}$$

for all $a \in A$, $t \in I$.

Observe that for $t > 0$, the map $H(-,t) : A \to M(i)$ is a homeomorphism onto its image. Fix $t > 0$, say $t = 1$. The inverse $\widetilde{i}$ of the map $i : A \to i(A)$ is now given by taking the composition of the maps $\widetilde{H}(-,1) : i(A) \to M(i)$ and $H^{-1}(-,1) : A \times 0 \to A$.

Indeed, on the one hand, we have $\widetilde{i} \circ i = H^{-1}(-,1) \circ \widetilde{H}(-,1) \circ i = H^{-1}(-,1) \circ H(-,1) = \mathrm{id}_A$, where we have used (7.1) for the second equality. On the other hand, for every $a \in A$ we have $(i \circ \widetilde{i})(i(a)) = i((\widetilde{i} \circ i)(a)) = i(a)$, and hence $i \circ \widetilde{i} = \mathrm{id}_{i(A)}$. $\square$

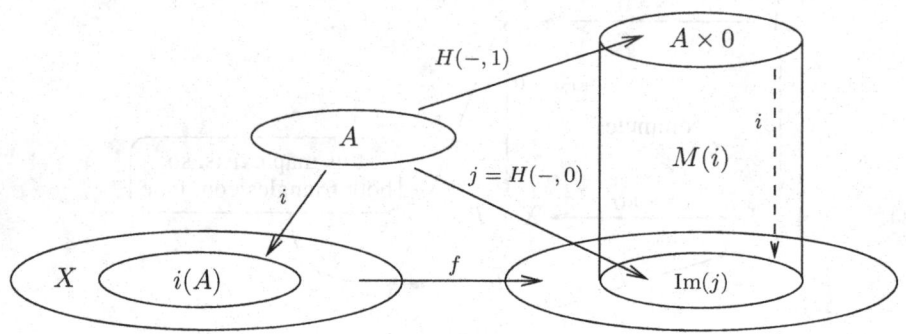

Fig. 7.2. Illustration to the proof of Proposition 7.2.

Let us remark that so far, we have made no implicit assumptions on the topological spaces that we consider. It can be useful to know that when the space X is Hausdorff and the inclusion map $i : A \hookrightarrow X$ is a cofibration, the space A must be closed in X. The proof of this is not difficult and will be extracted from the proof of Proposition 7.7.

Proposition 7.3. *Let A be a subspace of X. Then the following statements are equivalent:*

(1) the inclusion map $i : A \hookrightarrow X$ is a cofibration;
(2) the mapping cylinder $\mathrm{M}(i)$ is a retract of $X \times I$; this means that there exists a map $r : X \times I \to \mathrm{M}(i)$ such that $r \circ j = \mathrm{id}_{\mathrm{M}(i)}$, where j is the standard inclusion map $j : \mathrm{M}(i) \hookrightarrow X \times I$.

Proof. Assume first that the inclusion map $i : A \hookrightarrow X$ is a cofibration. Consider maps $f : X \to \mathrm{M}(i)$ and $H : A \times I \to \mathrm{M}(i)$, where the first one is induced by the identity map $\mathrm{id}_X : X \to X$, and the second one identifies the cylinder $A \times I$ with the corresponding cylinder inside the mapping cylinder $\mathrm{M}(i)$. These maps satisfy the conditions of Definition 7.1; hence there must exist a homotopy $\widetilde{H} : X \times I \to \mathrm{M}(i)$ extending the maps H and f. Clearly, this implies that $\widetilde{H}$ is the desired retract map: $\widetilde{H} \circ j = \mathrm{id}_{\mathrm{M}(i)}$.

Conversely, assume that $r : X \times I \to \mathrm{M}(i)$ is the retract map as specified in (2). Assume furthermore that we are given a map $f : X \to Y$ and a homotopy $H : A \times I \to Y$ such that $H(a,0) = f(a)$, for all $a \in A$. Define $\widetilde{H} : X \times I \to Y$ by taking the composition map $\psi \circ r : X \times I \to \mathrm{M}(i) \to Y$, where $\psi : \mathrm{M}(i) \to Y$ is the map induced by f together with H. $\square$

7.2 NDR-Pairs

Definition 7.4. *A pair of topological spaces (X,A), $A \subseteq X$, is called an* **NDR-pair** *(which is an abbreviation for* neighborhood deformation retract*) if there exist continuous maps $u : X \to I$ (think of it as a separation map) and $h : X \times I \to X$ (think of it as a homotopy) such that*

(1) $A = u^{-1}(0)$;
   "A is the kernel of u"; in particular, A must be closed!
(2) $h(x,0) = x$, for all $x \in X$;
   "homotopy starts with the identity map";
(3) $h(a,t) = a$, for all $t \in I$, $a \in A$;
   "the entire set A stays pointwise fixed through the homotopy process";
(4) $h(x,1) \in A$, for all $x \in X$, such that $u(x) < 1$;
   "all points that are close to A will in the end wander inside A." (See Figure 7.3.)

Intuitively one can think that (X,A) is an NDR-pair if "one can thicken A a little bit without changing its topology."

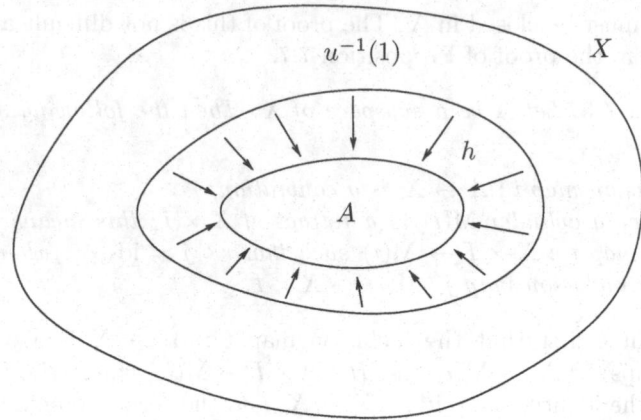

Fig. 7.3. Illustration to Definition 7.4.

Example 7.5.
(0) Let $X = \mathbb{R}^n$, and let A be a point. Then (X, A) is an NDR-pair.
(1) Let $X = [0, 1]$, $A = [0, 1)$. Then (X, A) is not an NDR-pair.
(2) Let X be the whole real line, and let $A = \{0\} \cup \{1/n \mid n \in \mathbb{N}\}$. Then (X, A) is not an NDR-pair.

It is a good exercise to show that the inclusion map $i : A \hookrightarrow X$ in Example 7.5 (1) and (2) is not a cofibration by a direct argument, i.e., without using Proposition 7.7.

Proposition 7.6. *If (X, A) is an NDR-pair, then $(X \times I, A \times I)$ is an NDR-pair as well.*

Proof. Take the appropriate maps u and h for the NDR-pair (X, A), and extend them to $(X \times I, A \times I)$ by simply ignoring the second parameter I, i.e., we set $\widetilde{u}(x, t) := u(x)$ and $\widetilde{h}(x, t, p) := h(x, p)$. $\square$

The next proposition connects the notions of cofibration and NDR-pairs.

Proposition 7.7. *Let A be a closed subspace of X. Then (X, A) is an NDR-pair if and only if the inclusion map $i : A \hookrightarrow X$ is a cofibration.*

Idea of the proof. It might be helpful for the reader to have the following geometric picture in mind. When thinking of NDR-pairs, imagine the open neighborhood of A as made of some flexible material, say of rubber, whereas the outside of the neighborhood is made of some solid material. Then, whenever we have a homotopy (read: movement) of A, we can make the rubber surrounding A move along, stretching or shrinking accordingly, thereby extending the homotopy of A to the homotopy of the entire space X. We shall now make this argument formal.

Proof of Proposition 7.7.
Assume first that (X, A) is an NDR-pair. Let $u : X \to I$ and $h : X \times I \to X$ be appropriate maps. The retraction $r : X \times I \to \mathrm{M}(i)$ is defined by

$$r(x, t) := \begin{cases} (x, t), & \text{if } x \in A \text{ or } t = 0; \\ (h(x, 1), t - u(x)), & \text{if } t \geq u(x) \text{ and } t > 0; \\ (h(x, t/u(x)), 0), & \text{if } u(x) \geq t \text{ and } u(x) > 0. \end{cases}$$

This map is obviously continuous and satisfies all the required properties.

Assume now that $i : A \to X$ is a cofibration, and let $r : X \times I \to \mathrm{M}(i)$ be the retraction whose existence is guaranteed by Proposition 7.3. Furthermore, let $p_1 : X \times I \to X$ and $p_2 : X \times I \to I$ denote the standard projections. Then we define $u : X \to I$ by setting

$$u(x) := \sup_{t \in I} (t - p_2(r(x, t))),$$

for an arbitrary $x \in X$. We also define $h : X \times I \to X$ to be the composition of the retraction with the first projection map

$$h(x, t) := p_1(r(x, t)).$$

Let us check that the maps h and u satisfy the conditions of Definition 7.4:

(1) $u^{-1}(0) = A$, since $u(x) = 0$ means $r(x, t) \in A \times I$, for all $t > 0$. On the other hand, $A \times I$ is closed in $X \times I$; hence $r(x, 0) \in A \times I$ as well, implying $x \in A$;
(2) $h(x, 0) = p_1(r(x, 0)) = p_1(x, 0) = x$, for all $x \in X$;
(3) $h(a, t) = p_1(r(a, t)) = p_1(a, t) = a$, for all $t \in I$, $a \in A$;
(4) if $u(x) < 1$, then $p_2(r(x, 1)) > 0$, hence $p_1(r(x, t)) \in A$. $\quad\square$

7.3 Important Facts Involving Cofibrations

There are several statements that intuitively seem likely to be always true, which in fact require the extra condition that the maps be cofibrations. Here is the first one.

Proposition 7.8. *If (X, A) is an NDR-pair and A is contractible, then the quotient map $q : X \to X/A$ is a homotopy equivalence.*

Proof. Let $a_0 \in A$. By our assumptions, A contracts to a_0. Let us denote the contraction map by $c : A \times I \to A$. Let $H : X \times I \to X$ be the map that extends this contraction. It exists, since $i : A \hookrightarrow X$ is a cofibration. The map $H(-, 1)$ takes all of A to a_0; hence it induces a map $p : X/A \to X$. The map p is a homotopy inverse of the map q. The homotopy $\mathrm{id}_X \simeq p \circ q$ is given by H, and the homotopy $\mathrm{id}_{X/A} \simeq q \circ p$ is induced by H, since $H_{A \times I} \subseteq A$. $\quad\square$

The intuitive statement that "taking a cone is the same as taking a quotient" can also be made precise as follows.

Corollary 7.9. *If (X, A) is an NDR-pair, then the quotient map $q :$ $\mathrm{Cone}(i) \to X/A$ is a homotopy equivalence.*

Proof. It is easy to see that if (X, A) is an NDR-pair, then $(\mathrm{Cone}(i), C)$ is an NDR-pair as well, where C the actual cone inside of $\mathrm{Cone}(i)$, i.e., the image of $A \times I$ under the structural quotient map $(A \times I) \coprod X \to \mathrm{Cone}(i)$. Indeed, if $u : X \to I$ and $h : X \times I \to X$ are the appropriate maps for the NDR-pair (X, A), as in Definition 7.4, the corresponding maps $\widetilde{u} : \mathrm{Cone}(i) \to I$ and $\widetilde{h} : \mathrm{Cone}(i) \times I \to \mathrm{Cone}(i)$ are extensions of u and h. Simply set $\widetilde{u}$ to be 0 and set $\widetilde{h}$ to be the identity map on the interior of the attached cone.

Since we have a homeomorphism $\mathrm{Cone}(i)/C \cong X/A$, which commutes with the quotient maps, and C is contractible inside of $\mathrm{Cone}(i)$, the corollary follows from Proposition 7.8. $\square$

The next proposition states a fundamental property of CW pairs.

Proposition 7.10. *Let (X, A) be a CW pair. Then there exists a strong deformation retraction of $X \times I$ onto $X \times \{0\} \cup A \times I$. In particular, the inclusion map $i : A \hookrightarrow X$ is a cofibration.*

Proof. The topological space $X \times I$ has a natural CW structure with cells labeled $\sigma \times \{0\}$, $\sigma \times \{1\}$, and $\sigma \times (0, 1)$, where σ is an open cell of X. To get $X \times \{0\} \cup A \times I$ from $X \times I$, we need to remove cells of the type $\sigma \times (0, 1)$ and $\sigma \times \{1\}$, where σ is *not* a cell in A. This can be done starting from the top-dimensional cells $\sigma \in X \setminus A$, and moving down in dimension.

At each step, a removal of a pair of cells $(\sigma \times \{1\}, \sigma \times (0, 1))$ is a strong deformation retraction of $\bar{\sigma} \times I$ onto $\sigma \times \{0\} \cup \partial \sigma \times I$, where $\bar{\sigma}$ denotes the closure of the open cell σ. It can be visualized as follows: let B^n be a unit ball, take a radial projection of $B^n \times [0, 1]$ onto $B^n \times \{0\} \cup \partial B^n \times I$, and attach this cylinder onto $\sigma \times \{0\} \cup \partial \sigma \times I$ using the identity map on I and on the interior of B^n, and the characteristic map of σ on ∂B^n. $\square$

A further intuitive statement is that the homotopy type of spaces obtained by gluing over a map should not change if the map is replaced by a homotopic one.

Proposition 7.11. *Let (X, A) be a CW pair. Let Y be an arbitrary topological space, and assume that continuous maps $f, g : A \to Y$ are homotopic. Then the spaces $X \cup_f Y$ and $X \cup_g Y$ are homotopy equivalent.*

Proof. Let $H : A \times I \to Y$ denote a homotopy between the maps f and g, and consider the space $Z = (X \times I) \cup_H Y$. Since (X, A) is a CW pair, Proposition 7.10 implies the existence of a strong deformation retraction of $X \times I$ onto $X \times \{0\} \cup A \times I$. This induces a strong deformation retraction of Z onto $(X \times \{0\} \cup A \times I) \cup_H Y = X \cup_f Y$.

On the other hand, symmetrically, a strong deformation retraction of $X \times I$ onto $X \times \{1\} \cup A \times I$ induces a strong deformation retraction of Z onto

$(X \times \{1\} \cup A \times I) \cup_H Y = X \cup_g Y$. Hence we can conclude that $X \cup_f Y$ is homotopy equivalent to $X \cup_g Y$. $\square$

An important corollary of Proposition 7.11 concerns the homotopy type of CW complexes.

Corollary 7.12. *The homotopy type of a CW complex does not change if the cell attachment maps are replaced by the homotopic maps.*

Proof. This is immediate if we take the constructive definition of CW complexes, and use Proposition 7.11 to deform the attaching maps.

In fact, one can get a very explicit picture. Assume that we are attaching a cell e^n to a CW complex X over the cell's boundary sphere $\mathbb{S}^{n-1}$. Let $H : A \times I \to Y$ denote the homotopy between the maps f and g. We set

$$\varphi(\lambda \mathbf{v}) := \begin{cases} 2\lambda \mathbf{v}, & \text{for } 0 \leq \lambda \leq 1/2, \\ H(\mathbf{v}, 2\lambda - 1), & \text{for } 1/2 \leq \lambda \leq 1, \end{cases} \tag{7.2}$$

where $\mathbf{v}$ denotes any unit vector, and we set $\varphi(x) := x$, for all $x \in X$. Then the map $\varphi : X \cup_g e^n \to X \cup_f e^n$ is a homotopy equivalence. $\square$

7.4 The Relative Homotopy Equivalence

The concept of homotopy equivalence can be made relative.

Definition 7.13. *Let X and Y be topological spaces, and let A be a subspace of both X and of Y. A continuous map $f : X \to Y$ such that $f|_A = \mathrm{id}_A$ is called a **homotopy equivalence rel A** if there exists a continuous map $g : Y \to X$ such that*

(1) $g|_A = \mathrm{id}_A$;
(2) there exists a homotopy from id_X to $g \circ f$ which keeps A fixed, that is, $H(a,t) = a$ for all $a \in A$;
(3) there exists a homotopy from id_Y to $f \circ g$ which keeps A fixed.

The next theorem will be our main technical tool for proving all sorts of gluing statements.

Theorem 7.14. *Suppose that (X, A) and (Y, A) are both NDR-pairs, and that we have a continuous map $f : X \to Y$, such that*

(a) f is a homotopy equivalence;
(b) $f|_A = \mathrm{id}_A$.

Then f is a homotopy equivalence rel A.

Proof. The proof is broken into the following steps: first, we construct a map $g : Y \to X$, which satisfies $g|_A = \mathrm{id}_A$; then we check the conditions (2) and (3) of Definition 7.13.

To start with, let $h : Y \to X$ be some homotopy inverse of f. It exists since we assumed that f is a homotopy equivalence. Since h is a homotopy inverse, there exists a homotopy $H : X \times I \to X$ such that $H(-, 0) = h \circ f$ and $H(-, 1) = \mathrm{id}_X$. When restricted to A, this is a homotopy $H|_{A \times I} : A \times I \to X$ between maps $h|_A$ and id_A (since $f|_A = \mathrm{id}_A$). Set $H_A := H|_{A \times I}$.

We have assumed that (Y, A) is an NDR-pair, so by Proposition 7.7, the inclusion map $i : A \hookrightarrow Y$ has a homotopy extension property. Therefore, there exists a homotopy $G : Y \times I \to X$ that extends h and H_A, i.e., such that $G(-, 0) = h$ and $G|_{A \times I} = H_A$. Set $g := G(-, 1) : Y \to X$. By construction, g is homotopic to h, and $g|_A = \mathrm{id}_A$.

Next, we check condition (2) of Definition 7.13. First, we would like to make it crystal clear why there is anything to be checked at all. The thing is that even though we know that $g \circ f$ is homotopic to id_X and that $g \circ f|_A = \mathrm{id}_A$, we cannot be sure that the homotopy between $g \circ f$ and id_X will fix A along the way. In fact, usually this will not happen. However, fortunately, we need to show only the *existence* of such a homotopy. The idea now is to start with *some* homotopy, and then transform it to another homotopy, one that will fix A along the way. To achieve this we will need a *homotopy of homotopies*.

Consider the homotopies connecting maps

$$g \circ f \rightsquigarrow h \circ f \rightsquigarrow \mathrm{id}_X, \tag{7.3}$$

where the first homotopy is taken to be $G(-, 1 - t) \circ f$, and the second homotopy is taken to be H. When these homotopies are restricted to A, we get a sequence

$$\mathrm{id}_A \rightsquigarrow h \rightsquigarrow \mathrm{id}_A, \tag{7.4}$$

where the first homotopy is $H_A(-, 1 - t)$ and the second one is H_A. A crucial observation now is that the homotopies in (7.4) are time inverses of each other; this follows from the fact that the homotopy G extends the homotopy H_A.

Having noticed this property of the homotopies in (7.4), we shall now describe their deformation. The idea is that instead of "homotoping" the map id_A all the way to h and then the same way back, we can start by stopping short of reaching h, homotoping the same way back, and then waiting the remaining time at id_A. The point where we stop will be at time $1 - t$, so that at time 0 we simply stand at id_A, thus having reached our goal.

Formally, we define $\Omega : (A \times I) \times I \to X$ as follows:

$$\Omega(-, t_1, t_2) = \begin{cases} \mathrm{id}_A, & \text{if } t_1 + t_2 \geq 1; \\ H_A(-, 1 - 2t_1), & \text{if } 2t_1 + t_2 \leq 1; \\ H_A(-, -1 + 2t_1 + 2t_2), & \text{if } 2t_1 + t_2 \geq 1 \geq t_1 + t_2; \end{cases}$$

see Figure 7.4 for a graphic illustration of this deformation.

Now we need the homotopy of homotopies $\widetilde{\Omega} : (X \times I) \times I \to X$ extending the homotopy (7.3) along the homotopy of homotopies Ω. Recall that by one of the previous exercises, $(X \times I, A \times I)$ is an NDR-pair; hence the inclusion map $A \times I \hookrightarrow X \times I$ is a cofibration. This implies the existence of the extension $\widetilde{\Omega}$. To get $\widetilde{H}$ we trace $\widetilde{\Omega}$ along the western, northern, and eastern (following the latter in the opposite direction) sides of the square:

$$\widetilde{H}(x,t) = \begin{cases} \widetilde{\Omega}(x,0,3t), & \text{if } 0 \leq t \leq 1/3; \\ \widetilde{\Omega}(x,3t-1,1), & \text{if } 1/3 \leq t \leq 2/3; \\ \widetilde{\Omega}(x,1,3-3t), & \text{if } 2/3 \leq t \leq 1. \end{cases}$$

This way we obtain a homotopy $X \times I \to X$ that takes $g \circ f$ to id_X, fixing A along the way. In other words, we have verified that $g \circ f \sim \mathrm{id}_X$ rel A.

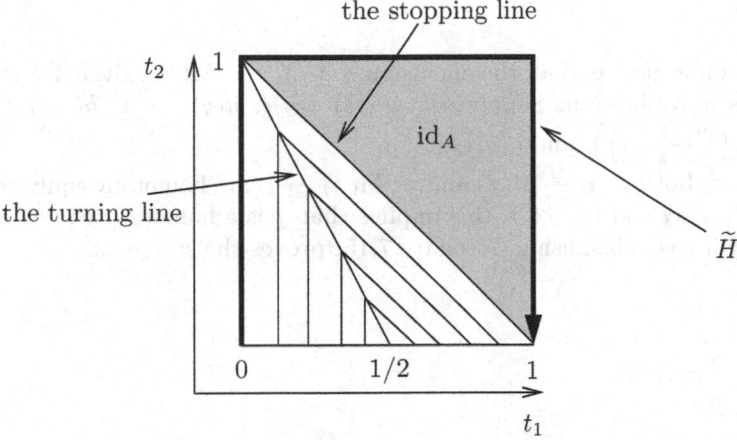

Fig. 7.4. The homotopy of homotopies in the proof of Theorem 7.14.

Finally, we need to check condition (3) of Definition 7.13. The same way as in the previous check of condition (2), we have a homotopy $f \circ g \rightsquigarrow f \circ h \rightsquigarrow \mathrm{id}_Y$. However, this time we cannot be sure that on A the homotopy process breaks down into two symmetric parts (here symmetric means that the two parts are time reverses of each other). Indeed, $f \circ g \rightsquigarrow f \circ h$ by $f \circ G(-, 1-t)$, but $f \circ h \rightsquigarrow \mathrm{id}_Y$ by some other homotopy R, which exists since h is a homotopy inverse of f, and enjoys no a priori known properties.

This means that we must repeat the very first step, i.e., getting a homotopy inverse to satisfy condition (1) as well. This time around we have to do it for the pair (g, f) instead of the pair (f, h). As a result, we obtain a map $\tilde{f} : X \to Y$ such that $\tilde{f}|_A = \mathrm{id}_A$ and $\tilde{f} \circ g \rightsquigarrow \mathrm{id}_Y$ rel A (earlier, we found g in this way). We have $f \sim f \circ (g \circ f) = (f \circ g) \circ f \sim f$ rel A. Hence $f \circ g \sim \tilde{f} \circ g \sim \mathrm{id}_Y$ rel A (since $f \sim \tilde{f}$ rel A and $g|_A = \mathrm{id}_A$). □

We are now ready to derive two important corollaries.

Corollary 7.15. (Subspace retraction).
*When (X, A) is an NDR-pair, the inclusion map $i : A \hookrightarrow X$ is a homotopy
equivalence if and only if A is a strong deformation retract of X.*

Proof. Applying Theorem 7.14 to the map $i : A \hookrightarrow X$ gives that i is actually
a homotopy equivalence relative to A. This means that there exists a map
$r : X \to A$ that is a homotopy inverse of i relative to A. Untangling definitions,
this translates to $r|_A = \mathrm{id}_A$, and $i \circ r \sim \mathrm{id}_X$ rel A, which means precisely that
A is a strong deformation retract of X. $\square$

Corollary 7.16. (Mapping cylinder retraction).
*A continuous map between topological spaces $f : X \to Y$ is a homotopy equiv-
alence if and only if X is a strong deformation retract of the mapping cylinder $\mathrm{M}(f)$.*

Proof. First notice that the inclusion $i : X \hookrightarrow \mathrm{M}(f)$, given by $i(x) =
(x, 0)$, is a cofibration. Simply set $u(x, t) := t$, $u(y) := 1$, $h(x, t_1, t_2) :=
\left(x, \max\left(\frac{2t_1 - t_2}{2 - t_2}, 0\right)\right)$, and $h(y, t_2) := y$.

Second, both $j : Y \hookrightarrow \mathrm{M}(f)$ and $r : \mathrm{M}(f) \to Y$ are homotopy equivalences.
Since $f = r \circ i$ and $i \sim j \circ f$, this implies that f is a homotopy equivalence if
and only if i is. This, using Corollary 7.15, proves the statement. $\square$

8

Principal Γ-Bundles and Stiefel–Whitney Characteristic Classes

This chapter is in no way intended to be a comprehensive introduction to the subject of principal bundles and characteristic classes. Rather, we have tried to assemble a number of relevant results to entice the reader to consult the standard texts. Most of the given proofs are sketchy, though the proofs of the statements that are needed for later applications are complete.

8.1 Locally Trivial Bundles

8.1.1 Bundle Terminology

In many situations the considered topological space has some additional associated structure that is locally trivial, but may contain some important information when considered globally. In such a case, the concept of a *bundle* may turn out to be useful.

Definition 8.1. *Let B and F be topological spaces. A* **locally trivial fiber bundle over** B **with fiber** F *is a topological space E together with a continuous map $p : E \to B$, such that for every $x \in B$ there exists an open neighborhood U of x for which $p^{-1}(U)$ is homeomorphic to $U \times F$, and the restriction of p to $p^{-1}(U)$ is the standard projection onto the first term $p : U \times F \to U$.*

Under the conditions of Definition 8.1, the space B is called the *base space*, the space F is called the *fiber*, and the space E is called the *total space*. The map p is called the *canonical projection* associated to the fiber bundle. Sometimes we denote the bundle by the triple (E, B, p), and sometimes we give it a separate name. We shall always assume that the spaces E, B, and F allow CW structures, and that p is a cellular map.

We shall consider only locally trivial bundles; hence we will just say "fiber bundle" or sometimes just "bundle." In some texts one allows a slightly more general type of bundles, where the fibers may be different over various path-connected components of B. We shall not make use of this generality.

The first, and simplest, example of a bundle is that of a direct product $E = B \times F$, with the canonical projection being the projection onto the first term. Not surprisingly, this bundle is called a *trivial bundle*.

Example 8.2.

(1) The squaring map $\mathbb{S}^1 \to \mathbb{S}^1$, defined by $z \mapsto z^2$, gives a fiber bundle with $E \cong B \cong \mathbb{S}^1$ and the fiber $F \cong \mathbb{S}^0$; this bundle is shown on the left of Figure 8.1.

(2) The projection of a Möbius band onto the middle circle is a fiber bundle with the Möbius band being the total space, the base space $B \cong \mathbb{S}^1$, and the fiber $F \cong [0, 1]$; this bundle shown on the right of Figure 8.1.

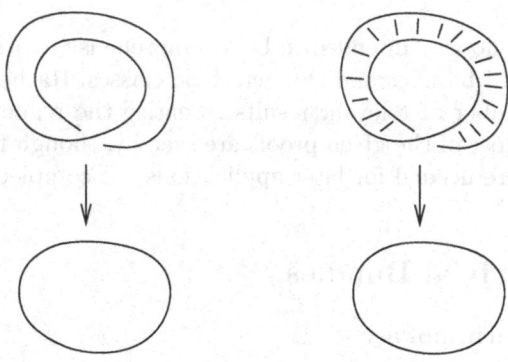

Fig. 8.1. Examples of fiber bundles.

8.1.2 Types of Bundles

Fiber bundles are often distinguished by the allowed type of fiber. For example, one can have *spherical bundles*, where the fiber is a sphere of a certain dimension. Another possibility is to take a discrete set as a fiber. The study of such bundles is the subject of the theory of covering spaces.

A very important and frequently studied type of bundles is that of the *vector bundles*. In this case, the fiber is required to have the structure of a vector space (usually over the real or complex numbers), and the homeomorphism $p^{-1}(U) \cong U \times F$ is required to be a vector space isomorphism for each $x \in U$.

Example 8.3.

(1) The following are two distinct real line bundles (i.e., vector bundles whose fibers are $\mathbb{R}^1$) over $\mathbb{S}^1$: the trivial bundle and the twisted bundle, where the line changes orientation after passing once around the circle; cf. Figure 8.1.

(2) A prominent (and original) instance of vector bundles is that of the so-called *tangent bundles*. Intuitively, for a manifold $M \subseteq \mathbb{R}^t$ of dimension n, one takes $E \subseteq M \times \mathbb{R}^t$ to be the totality of all tangent spaces to points in M. We refer to [MSta74] for a precise and intrinsic (i.e., independent of the particular embedding of M) definition of the tangent bundle.

8.1.3 Bundle Maps

The next definition describes perhaps the single most important operation on fiber bundles.

Definition 8.4. *Let* $\alpha = (E, B, p)$ *be a fiber bundle, and assume that we have a continuous map* $\varphi : \widetilde{B} \to B$. *We define a fiber bundle with* $\widetilde{B}$ *as a base space by taking the total space*

$$\widetilde{E} := \{(\widetilde{b}, e) \mid \widetilde{b} \in \widetilde{B}, \, e \in E, \, \varphi(\widetilde{b}) = p(e)\} \subseteq \widetilde{B} \times E$$

and letting $\widetilde{p} : \widetilde{E} \to \widetilde{B}$ *be the projection onto the first term.*

We call this bundle the **pullback** *of* α *along the map* φ, *and denote it by* $\varphi^*\alpha$. *We also set* $\varphi^*E := \widetilde{E}$.

Another name for the bundle $\varphi^*\alpha$ is the bundle *induced* from α by the map φ.

Definition 8.5. *Assume that we have two fiber bundles* $\alpha_1 = (E_1, B_1, p_1)$ *and* $\alpha_2 = (E_2, B_2, p_2)$. *A* **bundle map** f *from* α_1 *to* α_2 *is a pair of maps* $f = (f_E, f_B)$, $f_E : E_1 \to E_2$ *and* $f_B : B_1 \to B_2$, *such that following diagram commutes:*

$$
\begin{array}{ccc}
E_1 & \xrightarrow{\,f_E\,} & E_2 \\
{\scriptstyle p_1}\downarrow & & \downarrow{\scriptstyle p_2} \\
B_1 & \xrightarrow{\,f_B\,} & B_2
\end{array}
$$

and furthermore, f_E *induces homeomorphisms on fibers, i.e., for any* $b \in B_1$, *the restriction* $f_E : p_1^{-1}(b) \to p_2^{-1}(f_B(b))$ *is a homeomorphism.*

As a whole, we obtain a category of fiber bundles and bundle maps. It is customary to fix the fiber and consider all bundles with this fiber. One can also additionally fix the base space and consider the category of all fiber bundles with a given fiber over this base space.

Definition 8.6. *Given two fiber bundles* $\alpha_1 = (E_1, B_1, p_1)$ *and* $\alpha_2 = (E_2, B_2, p_2)$, *a bundle map* $f = (f_E, f_B)$ *from* α_1 *to* α_2 *is called a* **bundle isomorphism** *if both maps* f_E *and* f_B *are homeomorphisms.*

An important special case of Definition 8.6 arises when $B_1 = B_2$ and $f_B = \mathrm{id}_B$. We can therefore speak of isomorphism classes of bundles over a fixed base space.

Proposition 8.7. *Assume that we have two fiber bundles $\alpha_1 = (E_1, B_1, p_1)$, $\alpha_2 = (E_2, B_2, p_2)$, and a bundle map $f = (f_E, f_B)$ from α_1 to α_2. Then α_1 is isomorphic to the induced map $f_B^* \alpha_2$.*

Proof. The bundle isomorphism is given by the map

$$p_1 \times f_E : E_1 \longrightarrow f_B^* E_2 \subseteq B_1 \times E_2,$$

where $e \mapsto (p_1(e), f_E(e))$. It is well-defined since $f_B \circ p_1 = p_2 \circ f_E$. We leave the verification that this is an isomorphism to the reader. $\square$

Theorem 8.8. *Assume that $\alpha = (E, B, p)$ is a fiber bundle and that we have two continuous maps $f, g : \widetilde{B} \to B$ such that f is homotopic to g. Then the pullback bundles $f^* \alpha$ and $g^* \alpha$ over $\widetilde{B}$ are isomorphic.*

Proof. This is a standard argument in fiber bundle theory. We refer, for example, to [Ste51] for details. $\square$

Theorem 8.9. *Any fiber bundle over a contractible base space is trivial.*

Proof. Assume that $\alpha = (E, B, p)$ is a fiber bundle, and assume that B is contractible. Let $q : B \to B$ be a map that takes the whole space B to some point $b \in B$. By our assumptions, the maps q and id_B are homotopic. It follows by Theorem 8.8 that the pullbacks $q^* \alpha$ and $\mathrm{id}_B^* \alpha$ are isomorphic. On the other hand, we see that $q^* \alpha$ is a trivial bundle, and $\mathrm{id}_B^* \alpha = E$. $\square$

8.2 Elements of the Principal Bundle Theory

8.2.1 Principal Bundles and Spaces with a Free Group Action

The bundles become rigid and have interesting structure theory when the fibers are groups that coherently act on themselves.

Definition 8.10. *Let Γ be a topological group. A **principal Γ-bundle** is a fiber bundle (E, B, p) with Γ-action on E such that*

(1) the group Γ acts on fibers, i.e., $p(g(e)) = p(e)$ for all $e \in E$, $g \in \Gamma$;
(2) the group Γ acts locally trivially, i.e., for all $b \in B$ there exists a neighborhood U of b for which there exists a Γ-homeomorphism $h : p^{-1}(U) \to U \times \Gamma$, where the Γ-action on $U \times \Gamma$ is the multiplication of the second term such that the diagram in Figure 8.2 commutes.

As we shall see later, already the principal $\mathbb{Z}_2$-bundles may contain useful information.

Definition 8.11. *Given two principal Γ-bundles α_1 and α_2, a bundle map $f = (f_E, f_b)$ between α_1 and α_2 is called a **principal Γ-bundle map** if the map f_E is Γ-invariant.*

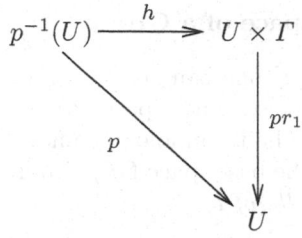

Fig. 8.2. The commuting diagram for Definition 8.10, where pr_1 denotes the projection onto the first term.

The notion of bundle isomorphism generalizes straightforwardly by requiring that the isomorphism map be Γ-invariant. To summarize, we have a category of principal Γ-bundles as objects and principal Γ-bundle maps as morphisms.

An alternative way to describe this category is by considering spaces with a free action of the group Γ. Consider a regular CW complex X with a cellular action of a finite group Γ. As mentioned in Chapter 4, a structural way to visualize such an action is to view it as a functor from the category associated to the group Γ (it has one object and morphisms in bijection with elements of Γ. The composition law corresponds the group multiplication) to the category of regular CW complexes and cellular maps. This functor maps the unique object of the category associated to Γ to X, and it maps the morphisms to the cellular maps of X into itself. When no confusion arises, we shall simply denote these maps by the group element itself. In some cases, for example when the group acts on different spaces and these actions need to be distinguished, we shall write γ_X, indicating that here the specific action on the complex X is considered.

If desired, the Γ-action can be made to be simplicial by passing to the barycentric subdivision (cf. [Bre72, Hat02]). For the interested reader we remark here that sometimes one takes the barycentric subdivision even if the original action already is simplicial. The main point of this is that one can make the action enjoy an additional property: *if a simplex is preserved by one of the group elements, then it must be pointwise fixed by this element.*

Definition 8.12. *We say that a topological space X is a Γ-space if Γ acts on X and this action is free, meaning that for every $x \in X$ and $\gamma \in \Gamma$ such that γ is different from the identity element, we have $\gamma x \neq x$.*

Furthermore, when X and Y are two Γ-spaces and $f : X \to Y$ is a continuous map, we call f a Γ-map if, additionally, it commutes with the corresponding Γ-actions, that is, for every $x \in X$ and $\gamma \in \Gamma$, we have $f(\gamma_X x) = \gamma_Y f(x)$.

It is easily checked that the composition of two Γ-maps is again a Γ map, and that Γ-spaces together with Γ-maps form a category, which we call Γ-**Sp**.

8.2.2 The Classifying Space of a Group

For any topological group Γ one can construct a contractible space $\mathbf{E}\Gamma$ on which the group Γ acts freely. This space $\mathbf{E}\Gamma$ is called the *total space of the universal bundle of Γ*. The terminology comes from the fact that one has a principal Γ-bundle with the base space $\mathbf{E}\Gamma/\Gamma$ and the total space $\mathbf{E}\Gamma$, which one calls the *universal bundle of Γ*.

Definition 8.13. *The quotient space $\mathbf{E}\Gamma/\Gamma$ is called the* **classifying space** *of the group Γ. It is denoted by* $\mathbf{B}\Gamma$.

The classifying space of Γ is also known as the associated Eilenberg–Mac Lane space, or $K(\Gamma, 1)$-space; see, e.g., [AM94, Bre93, GeM96, Hat02, McL95, May99, Wh78].

There are various ways to find the space $\mathbf{E}\Gamma$ with desired properties. A prominent one is the so-called *Milnor construction*:

$$\mathbf{E}\Gamma := \Gamma * \Gamma * \Gamma * \cdots,$$

where the infinite join is to be interpreted as a colimit if the Γ-invariant embedding sequence

$$\Gamma \hookrightarrow \Gamma * \Gamma \hookrightarrow \Gamma * \Gamma * \Gamma \hookrightarrow \cdots, \tag{8.1}$$

with the embedding maps being the identity on the initial terms, and the diagonal Γ-action.

For future reference, we recall the standard way to encode points in joins of topological spaces. Assume that X is a join of topological spaces $\{T_i\}_{i \in I}$, where I is either finite, or countable, in which case the join is defined by ordering elements of I and taking the appropriate colimit of partial joins. The points of X can be encoded as sequences $\{c_i x_i\}_{i \in I}$, where $x_i \in T_i$ for all $i \in I$, and the numbers c_i are the barycentric coordinates, i.e., $0 \le c_i \le 1$, $\sum_{i \in I} c_i = 1$, and only finitely many c_i are different from 0. The topology is naturally reflected in this encoding: small variations of points in X correspond to small variations of points x_i and small changes of coordinates.

Proposition 8.14. *The space $\mathbf{E}\Gamma$ obtained by the Milnor construction is contractible.*

Proof. It is easy to see that a join of a k-connected space with a nonempty one is always $(k + 1)$-connected, and that a join of any two nonempty spaces has to be connected. It follows that an n-fold join of Γ with itself is $(n - 2)$-connected, and therefore all the homotopy groups of $\mathbf{E}\Gamma$ are trivial. Since $\mathbf{E}\Gamma$ has a natural structure of a CW complex, it follows by Corollary 6.32 of Whitehead's theorem that $\mathbf{E}\Gamma$ is contractible.

It is also easy to present a concrete contraction of the space $\mathbf{E}\Gamma$. One can construct a homotopy from the identity map to the map that maps the entire space $\mathbf{E}\Gamma$ to some point in $\mathbf{E}\Gamma$ in two steps:

Step 1. The identity map of $\mathbf{E}\Gamma$ into itself is homotopic to the map φ under which the ith factor is mapped identically to the $(2i)$th factor, i.e.,

$$(c_1x_1, c_2x_2, c_3x_3, \dots) \mapsto (0, c_1x_1, 0, c_2x_2, 0, c_3x_3, 0, \dots).$$

The homotopy can be constructed as an infinite sequence of factor shifts in the same way as in the proof of Theorem 8.17.

Step 2. Let x be any point of Γ. The map φ is homotopic to the one taking all points of $\mathbf{E}\Gamma$ to the point $(x, 0, 0, \dots)$. A concrete homotopy is obtained by simply letting all points of $\mathbf{E}\Gamma$ with 0 as the first coordinate "slide toward" the point $(x, 0, 0, \dots)$; namely, the map $H : [0,1] \times \mathbf{E}\Gamma \to \mathbf{E}\Gamma$ is defined by

$$H(t, (0, c_1x_1, c_2x_2, c_3x_3, \dots)) := (tx, (1-t)c_1x_1, (1-t)c_2x_2, (1-t)c_3x_3, \dots),$$

for all $t \in [0,1]$. $\square$

One alternative to Milnor's construction is to take a trisp whose n-simplices are indexed by all sequences $[g_0, g_1, \dots, g_n]$, $g_i \in \Gamma$, and the boundary map is just skipping elements. The Γ-action can be defined by

$$g \cdot [g_0, \dots, g_n] = [g \cdot g_0, \dots, g \cdot g_n].$$

The quotient of this space $\mathbf{B}\Gamma = \mathbf{E}\Gamma/\Gamma$ coincides with the nerve of the one-object category, which we have previously associated to an arbitrary group.

It seems that there is some choice involved in picking the "right" space $\mathbf{E}\Gamma$. Quite to the contrary, the next theorem, which we give here without a proof, states that all these choices lead to essentially the same object.

Theorem 8.15. *Assume that E_1 and E_2 are both contractible Γ-spaces. Then there exists a Γ-invariant homotopy equivalence between E_1 and E_2; in particular, E_1/Γ and E_2/Γ are homotopy equivalent.*

Perhaps the most important property of the space $\mathbf{E}\Gamma$, and the associated principal Γ-bundle, is its universality. The next proposition states one-half of the universality property: every principal Γ-bundle can be mapped to the universal one.

Proposition 8.16. *Whenever X is a CW complex with a free cellular Γ-action, there exists a Γ-invariant map $w : X \to \mathbf{E}\Gamma$. Moreover, if d is the dimension of X, then the map w can be constructed so that its image is contained in the first $d + 1$ factors of $\mathbf{E}\Gamma = \Gamma * \Gamma * \cdots$.*

Proof. A map w with the required properties can be constructed by induction over the skeletons of X. To start with, pick one vertex from each Γ-orbit in the 0-skeleton of X. Map these vertices to the vertex $(x, 0, 0, \dots) \in \mathbf{E}\Gamma$, where x is some chosen vertex of Γ. Then extend the map to the entire 0-skeleton of X in the unique Γ-invariant way.

Now we describe how to extend the map to the d-skeleton of X, under the assumption that it has already been defined for the $(d-1)$-skeleton. Since the Γ-action on X is cellular, the set of d-cells of X is a disjoint union of Γ-orbits. The extension is now done orbit by orbit.

Assume that O is an orbit to which we would like to extend the map w. Pick an arbitrary d-cell $\sigma \in O$. The map w is defined on the boundary of σ, which is homeomorphic to $\mathbb{S}^{d-1}$ before σ is attached; hence to extend w to σ, we just need to extend this map from the sphere $\mathbb{S}^{d-1}$ to the map from the disk B^d. This can be always done, since the $(d-1)$th homotopy group of $\mathbf{E}\Gamma$ is trivial. There is now a unique way to extend w to the rest of the orbit O by using the Γ-invariance of w.

Since we are using only the triviality of the $(d-1)$th homotopy group to find the extension, the image of the extension of the map w to the d-skeleton can be chosen to lie within the first $d+1$ factors of $\mathbf{E}\Gamma = \Gamma * \Gamma * \cdots$. $\square$

The next theorem covers the other part of universality: the bundle map to the universal bundle is unique up to homotopy.

Theorem 8.17. *For an arbitrary Γ-space X, any two Γ-maps $f, g : X \to \mathbf{E}\Gamma$ are Γ-homotopic. In particular, the induced quotient maps $f/\Gamma, g/\Gamma : X/\Gamma \to \mathbf{E}\Gamma/\Gamma = \mathbf{B}\Gamma$ are homotopic.*

Proof. Let the maps f, g have the form

$$f(x) = (c_1(x)f_1(x), c_2(x)f_2(x), \dots)$$

and

$$g(x) = (d_1(x)g_1(x), d_2(x)g_2(x), \dots),$$

and consider new maps defined by

$$\tilde{f}(x) = (c_1(x)f_1(x), 0, c_2(x)f_2(x), 0, \dots)$$

and

$$\tilde{g}(x) = (0, d_1(x)g_1(x), 0, d_2(x)g_2(x), 0, \dots).$$

We can construct a homotopy from $\tilde{f}$ to f in infinitely many steps. As a first step, take the homotopy

$$H(t, x) := (c_1(x)f_1(x), tc_2(x)f_2(x), (1-t)c_2(x)f_2(x), tc_3(x)f_3(x), \dots),$$

for all $t \in [0, 1]$. This shifts down the terms so as to eliminate the first 0 from the definition of $\tilde{f}$. As further steps we take the homotopies that shift down the terms and eliminate further 0's from left to right. If the image of f was contained in the join of finitely many terms Γ, then only finitely many steps are needed. Otherwise, we observe that each point is kept fixed in all but finitely many homotopies.

Concatenating all these homotopies together, we see that f is homotopic to $\tilde{f}$. Analogously, we can see that g is homotopic to $\tilde{g}$. It is now easy to find a homotopy $\tilde{H}$ between $\tilde{f}$ and $\tilde{g}$. For example, one can take

$$\tilde{H}(t,x) := (tc_1(x)f_1(x), (1-t)d_1(x)g_1(x), tc_2(x)f_2(x), (1-t)d_2(x)g_2(x), \dots),$$

for all $t \in [0,1]$. $\square$

We are now ready to state what is often referred to as the main structure theorem for principal bundles.

Theorem 8.18. (Principal bundle classification theorem)
Let X be a CW complex. The function P from the set of homotopy classes of maps from X to $\mathbf{B}\Gamma$, $[X, \mathbf{B}\Gamma]$, to the set of isomorphism classes of principal Γ-bundles over X that takes each continuous map $f : X \to \mathbf{B}\Gamma$ to the pullback of the universal bundle over $\mathbf{B}\Gamma$ along f is a bijection.

Proof. To start with, P is well-defined, since by Theorem 8.8 it takes homotopic maps $f, g : X \to \mathbf{B}\Gamma$ to isomorphic principal Γ-bundles.

Furthermore, the map P is surjective, since by Proposition 8.16, for any principal Γ-bundle over X we can find a bundle map to the universal bundle, and therefore, by Proposition 8.7, the initial bundle over X can be obtained as the pullback of the universal one.

Finally, a fairly direct application of Theorem 8.17 implies that the map P is injective. $\square$

8.2.3 Special Cohomology Elements

Passing to cohomology, we see that the induced map $(w/\Gamma)^* : H^*(\mathbf{B}\Gamma) \to H^*(X/\Gamma)$ does not depend on the choice of w, and thus the image of $(w/\Gamma)^*$ consists of some canonically distinguished cohomology elements.

Definition 8.19. *For $z \in H^*(\mathbf{B}\Gamma)$, we let $w(z, X)$ denote the element $(w/\Gamma)^*(z)$, which we call the **characteristic class** associated to z.*

Let Y be another regular CW complex with a free action of Γ, and assume that $\varphi : X \to Y$ is a Γ-invariant map. By Proposition 8.16 there exists a Γ-map $v : Y \to \mathbf{E}\Gamma$. Hence, in addition to the map $w : X \to \mathbf{E}\Gamma$, we also have a composition map $v \circ \varphi$. Passing to the quotient map and then to the induced map on cohomology, we get yet another map $((v \circ \varphi)/\Gamma)^* : H^*(\mathbf{B}\Gamma) \to H^*(X/\Gamma)$. However, as we mentioned above, the map on the cohomology algebras does not depend on the choice of the original map to $\mathbf{E}\Gamma$. Thus, since $((v \circ \varphi)/\Gamma)^* = (\varphi/\Gamma)^* \circ (v/\Gamma)^*$, we have commuting diagrams, see Figure 8.3, and therefore we have proved the following proposition.

Proposition 8.20. *Whenever X and Y are Γ-spaces, and whenever $\varphi : X \to Y$ is a Γ-map, we have*

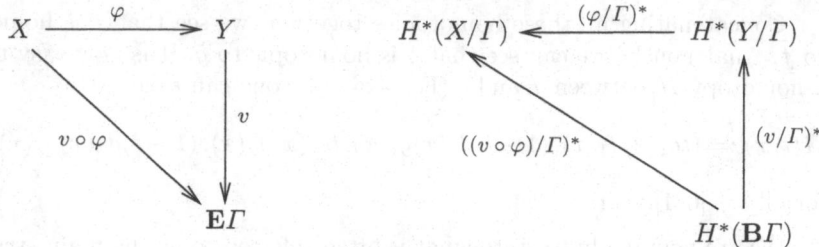

Fig. 8.3. Functoriality of characteristic classes.

$$w(z, X) = (\varphi/\Gamma)^*(w(z, Y)), \qquad (8.2)$$

where z is an arbitrary element of $H^*(\mathbf{B}\Gamma)$.

In other words, the characteristic classes associated to a finite group action are *natural*, or, as one sometimes says, *functorial*. This is a very important and useful property. For example, it allows one to find algebraic obstruction to the existence of Γ-equivariant maps by computing the appropriate characteristic classes.

8.2.4 $\mathbb{Z}_2$-Spaces and the Definition of Stiefel–Whitney Classes

Specifying $\Gamma := \mathbb{Z}_2$ in Definition 8.12, we see that a $\mathbb{Z}_2$-space X is a CW complex equipped with a fixed-point-free involution, that is, with a continuous map $\gamma : X \to X$ such that γ^2 is an identity map. Correspondingly, we have $\mathbb{Z}_2$-maps as continuous maps that commute with these involutions, and a category of $\mathbb{Z}_2$-spaces as objects and $\mathbb{Z}_2$-maps as morphisms, which is called $\mathbb{Z}_2$-**Sp**.

Let us now assume that we would like to construct line bundles over the base space B, which is given as an abstract simplicial complex. We consider the open covering of B by open stars of vertices of B. Since the open stars are contractible, by Theorem 8.9 the line bundle will be trivial when restricted to each star. Therefore, all we need to do is to start with trivial bundles over open stars and then specify how these glue together along the intersections of two open stars. Such a specification is a choice for each intersecting pair, whether the bundles glue with orientations preserved or reversed. Any choice of specifications will define a total bundle as long as it is coherent over the triple intersections.

Clearly, the *same* combinatorial data describes all possible $\mathbb{Z}_2$-principal bundles. Therefore, we can freely switch between these two families of bundles. Starting from the line bundle, the associated $\mathbb{Z}_2$-principal bundle is obtained by replacing each line by the unit sphere $\mathbb{S}^0$. Conversely, starting from a $\mathbb{Z}_2$-principal bundle, the associated line bundle is obtained by "sticking a line" into each fiber, with this line oriented away from the identity element toward the nontrivial element of $\mathbb{Z}_2$.

There are many classical examples of $\mathbb{Z}_2$-spaces. A central one is that of an n-dimensional sphere, with the antipodal map $x \mapsto -x$ playing the role of the structural involution. We denote this one by $\mathbb{S}_a^n$. Note that the embedding sequence (4.3) from Section 4.4 can be viewed as the $\mathbb{Z}_2$-invariant embedding sequence

$$\mathbb{S}_a^1 \overset{i_1}{\hookrightarrow} \mathbb{S}_a^2 \overset{i_2}{\hookrightarrow} \cdots \overset{i_{n-1}}{\hookrightarrow} \mathbb{S}_a^n \overset{i_n}{\hookrightarrow} \mathbb{S}_a^{n+1} \overset{i_{n+1}}{\hookrightarrow} \cdots, \tag{8.3}$$

since the maps i_n, as defined in (4.2), are automatically $\mathbb{Z}_2$-maps. This means that we can extend the antipodal map to the infinite-dimensional sphere as well. We denote the obtained $\mathbb{Z}_2$-space by $\mathbb{S}_a^\infty$.

Specifying $\Gamma = \mathbb{Z}_2$ in the considerations above, we get a $\mathbb{Z}_2$-map $w : X \to \mathbb{S}_a^\infty = \mathbf{E}\mathbb{Z}_2$. Furthermore, we have the induced quotient map

$$w/\mathbb{Z}_2 : X/\mathbb{Z}_2 \to \mathbb{S}_a^\infty/\mathbb{Z}_2 = \mathbb{RP}^\infty = \mathbf{B}\mathbb{Z}_2. \tag{8.4}$$

Let us remark that the *infinite-dimensional projective space* $\mathbb{RP}^\infty$ appearing in (8.4) could also be defined as a colimit of the embedding sequence of the finite-dimensional projective spaces

$$\mathbb{RP}^1 \overset{i_1/\mathbb{Z}_2}{\hookrightarrow} \mathbb{RP}^2 \overset{i_2/\mathbb{Z}_2}{\hookrightarrow} \cdots \overset{i_{n-1}/\mathbb{Z}_2}{\hookrightarrow} \mathbb{RP}^n \overset{i_n/\mathbb{Z}_2}{\hookrightarrow} \mathbb{RP}^{n+1} \overset{i_{n+1}/\mathbb{Z}_2}{\hookrightarrow} \cdots, \tag{8.5}$$

which is obtained from the embedding sequence (8.3) by taking the $\mathbb{Z}_2$-quotient.

The crucial fact now is that in this particular case, the induced $\mathbb{Z}_2$-algebra homomorphism

$$(w/\mathbb{Z}_2)^* : H^*(\mathbb{RP}^\infty; \mathbb{Z}_2) \to H^*(X/\mathbb{Z}_2; \mathbb{Z}_2)$$

is determined by very little data. Namely, let z denote the nontrivial cohomology class in $H^1(\mathbb{RP}^\infty; \mathbb{Z}_2)$. Then $H^*(\mathbb{RP}^\infty; \mathbb{Z}_2) \simeq \mathbb{Z}_2[z]$ as a graded $\mathbb{Z}_2$-algebra, with z having degree 1. We denote the image $(w/\mathbb{Z}_2)^*(z) \in H^1(X/\mathbb{Z}_2; \mathbb{Z}_2)$ by $\varpi_1(X)$. Obviously, the whole map $(w/\mathbb{Z}_2)^*$ is determined by the element $\varpi_1(X)$.

Definition 8.21. *The element* $\varpi_1(X) \in H^1(X/\mathbb{Z}_2; \mathbb{Z}_2)$ *is called the* **Stiefel–Whitney class** *of the $\mathbb{Z}_2$-space X.*

Clearly, $\varpi_1^k(X) = (w/\mathbb{Z}_2)^*(z^k)$. Furthermore, by Proposition 8.20, if Y is another $\mathbb{Z}_2$-space, and $\varphi : X \to Y$ is a $\mathbb{Z}_2$-map, then $(\varphi/\mathbb{Z}_2)^*(\varpi_1(Y)) = \varpi_1(X)$.

As an example we can quickly compute $\varpi_1(\mathbb{S}_a^n)$, for an arbitrary nonnegative integer n. First, for dimensional reasons, $\varpi_1(\mathbb{S}_a^0) = 0$. So we assume $n \geq 1$. Next, we have $\mathbb{S}_a^n/\mathbb{Z}_2 = \mathbb{RP}^n$. The cohomology algebra $H^*(\mathbb{RP}^n; \mathbb{Z}_2)$ is generated by one element $\beta \in H^1(\mathbb{RP}^n; \mathbb{Z}_2)$, with a single relation $\beta^{n+1} = 0$. Finally, the standard inclusion map $\iota : \mathbb{S}_a^n \hookrightarrow \mathbb{S}_a^\infty$ is $\mathbb{Z}_2$-equivariant, and induces another standard inclusion $\iota/\mathbb{Z}_2 : \mathbb{RP}^n \hookrightarrow \mathbb{RP}^\infty$. Identifying $\mathbb{RP}^n$ with the image of $\iota/\mathbb{Z}_2$, we can think of it as the n-skeleton of $\mathbb{RP}^\infty$. Thus the induced $\mathbb{Z}_2$-algebra homomorphism $(\iota/\mathbb{Z}_2)^* : H^*(\mathbb{RP}^\infty; \mathbb{Z}_2) \to H^*(\mathbb{RP}^n; \mathbb{Z}_2)$ maps the canonical generator of $H^1(\mathbb{RP}^\infty; \mathbb{Z}_2)$ to β, and hence we can conclude that $\varpi_1(\mathbb{S}_a^n) = \beta$.

8.3 Properties of Stiefel–Whitney Classes

8.3.1 Borsuk–Ulam Theorem, Index, and Coindex

The Stiefel–Whitney classes can be used to determine the nonexistence of certain $\mathbb{Z}_2$-maps. The following theorem is an example of such a situation.

Theorem 8.22. (Borsuk–Ulam theorem).
Let n and m be nonnegative integers. If there exists a $\mathbb{Z}_2$-map $\varphi : \mathbb{S}_a^n \to \mathbb{S}_a^m$, then $n \leq m$.

Proof. Choose representations for the cohomology algebras $H^*(\mathbb{RP}^n; \mathbb{Z}_2) = \mathbb{Z}_2[\alpha]$ and $H^*(\mathbb{RP}^m; \mathbb{Z}_2) = \mathbb{Z}_2[\beta]$, with the only relations on the generators being $\alpha^{n+1} = 0$ and $\beta^{m+1} = 0$. Since the Stiefel–Whitney classes are functorial, we get $(\varphi/\mathbb{Z}_2)^*(\varpi_1(\mathbb{S}_a^m)) = \varpi_1(\mathbb{S}_a^n)$.

On the other hand, by the computation in Subsection 8.2.4, we have $\varpi_1(\mathbb{S}_a^n) = \alpha$ and $\varpi_1(\mathbb{S}_a^m) = \beta$. So $(\varphi/\mathbb{Z}_2)^*(\beta) = \alpha$, and hence

$$\alpha^{m+1} = (\varphi/\mathbb{Z}_2)^*(\beta)^{m+1} = (\varphi/\mathbb{Z}_2)^*(\beta^{m+1}) = 0.$$

Finally, since $\alpha^{n+1} = 0$ is the only relation on α, this yields the desired inequality $m \geq n$. $\square$

The Borsuk–Ulam theorem makes the following terminology useful for formulating further obstructions to maps between $\mathbb{Z}_2$-spaces.

Definition 8.23. *Let X be a $\mathbb{Z}_2$-space.*
• *The* **index** *of X, denoted by $\mathrm{Ind} X$, is the minimal integer n for which there exists a $\mathbb{Z}_2$-map from X to $\mathbb{S}_a^n$.*
• *The* **coindex** *of X, denoted by $\mathrm{Coind} X$, is the maximal integer n for which there exists a $\mathbb{Z}_2$-map from $\mathbb{S}_a^n$ to X.*

Assume that we have two $\mathbb{Z}_2$-spaces X and Y, and that $\gamma : X \to Y$ is a $\mathbb{Z}_2$-map. Then we have the inequality

$$\mathrm{Coind} X \leq \mathrm{Ind} Y.$$

Indeed, if there exist $\mathbb{Z}_2$-maps $\varphi : \mathbb{S}_a^n \to X$ and $\psi : Y \to \mathbb{S}_a^m$, then the composition

$$\mathbb{S}_a^n \xrightarrow{\varphi} X \xrightarrow{\gamma} Y \xrightarrow{\psi} \mathbb{S}_a^m$$

yields a $\mathbb{Z}_2$-map between two spheres with antipodal actions; hence by the Borsuk–Ulam theorem, we can conclude that $n \leq m$. In particular, taking $Y = X$ and $\varphi = \mathrm{id}$, we get the inequality

$$\mathrm{Coind} X \leq \mathrm{Ind} X,$$

for an arbitrary $\mathbb{Z}_2$-space.

8.3.2 Stiefel–Whitney Height

The following number is a standard benchmark for comparing $\mathbb{Z}_2$-spaces with each other.

Definition 8.24. *Let X be an arbitrary nonempty $\mathbb{Z}_2$-space. The **Stiefel–Whitney height** of X (or simply the height of X), denoted by $\mathrm{h}(X)$, is defined to be the maximal nonnegative integer h such that $\varpi_1^h(X) \neq 0$. If no such h exists, then the space X is said to have infinite height.*

For example, we have $\mathrm{h}(\mathbb{S}_a^n) = n$, for all nonnegative integers n. We remark that for a nonempty $\mathbb{Z}_2$-space X we have $\varpi_1(X) = 0$ if and only if no connected component of X is mapped onto itself by the structural $\mathbb{Z}_2$-action; in other words, the structural involution must swap the connected components of X. In this case, consistent with Definition 8.24, we will say that the height of X is equal to 0.

As mentioned before, if X and Y are two arbitrary $\mathbb{Z}_2$-spaces and $\varphi : X \to Y$ is an arbitrary $\mathbb{Z}_2$-map, then the Stiefel–Whitney characteristic classes are functorial, i.e., we have $(\varphi/\mathbb{Z}_2)^*(\varpi_1(Y)) = \varpi_1(X)$, which in particular implies the inequality

$$\mathrm{h}(X) \leq \mathrm{h}(Y).$$

We note that for an arbitrary $\mathbb{Z}_2$-space X, the existence of a $\mathbb{Z}_2$-equivariant map $\mathbb{S}_a^n \to X$ implies $n = \mathrm{h}(\mathbb{S}_a^n) \leq \mathrm{h}(X)$, whereas the existence of a $\mathbb{Z}_2$-equivariant map $X \to \mathbb{S}_a^m$ implies $m = \mathrm{h}(\mathbb{S}_a^m) \geq \mathrm{h}(X)$. This can be best summarized with the inequality

$$\mathrm{Coind}\,(X) \leq \mathrm{h}(X) \leq \mathrm{Ind}\,(X).$$

8.3.3 Higher Connectivity and Stiefel–Whitney Classes

Many results giving topological obstructions to graph colorings had the k-connectivity of some space as the crucial assumption. We describe here an important connection between this condition and nonnullity of powers of Stiefel–Whitney classes.

First, it is trivial that if X is a nonempty $\mathbb{Z}_2$-space, then one can equivariantly map $\mathbb{S}_a^0$ to X. It is possible to extend this construction inductively to an arbitrary $\mathbb{Z}_2$-space, in a way analogous to our proof of Proposition 8.16.

Proposition 8.25. *Let X and Y be two regular CW complexes with a free $\mathbb{Z}_2$-action such that for some $k \geq 0$, we have $\dim X \leq k$ and Y is $(k-1)$-connected. Assume further that we have a $\mathbb{Z}_2$-map $\psi : X^{(d)} \to Y$, for some $d \geq -1$. Then there exists a $\mathbb{Z}_2$-map $\varphi : X \to Y$ such that φ extends ψ.*

Please note the following convention used in the formulation of Proposition 8.25: $d = -1$ means that we have no map ψ (in other words, $X^{-1} = \emptyset$), hence no additional conditions on the map φ.

Proof. By assumption, the cellular structure on X is $\mathbb{Z}_2$-invariant. We construct φ inductively on the i-skeleton of X, for $i \geq d+1$. If $d = -1$, we start by defining φ on the 0-skeleton as follows: for each orbit $\{a, b\}$ consisting of two vertices of X, simply map a to an arbitrary point $y \in Y$, and then map b to $\gamma(y)$, where γ is the free involution of Y.

Assume now that φ is defined on the $(i-1)$-skeleton of X, and extend the construction to the i-skeleton as follows. Let (σ, τ) be a pair of i-dimensional cells of X such that $\gamma\sigma = \tau$. The boundary $\partial\sigma$ is an $(i-1)$-dimensional sphere. By our assumptions, $i - 1 \leq \dim X - 1 \leq k - 1$; hence the restriction of φ to $\partial\sigma$ extends to σ. Finally, we extend φ to the second cell τ by applying the involution γ: $\varphi|_\tau := (\varphi|_\sigma) \circ \gamma$. $\square$

Corollary 8.26. *Let X be a $\mathbb{Z}_2$-space, and assume that X is $(k-1)$-connected, for some $k \geq 0$. Then there exists a $\mathbb{Z}_2$-map $\varphi : \mathbb{S}_a^k \to X$. In particular, we have $\varpi_1^k(X) \neq 0$.*

Proof. Since $\mathbb{S}_a^k$ is k-dimensional, the statement follows immediately from Proposition 8.25. To see that $\varpi_1^k(X) \neq 0$, recall that the Stiefel–Whitney classes are functorial; therefore we have $(\varphi/\mathbb{Z}_2)^*(\varpi_1^k(X)) = \varpi_1^k(\mathbb{S}_a^k)$, and the latter has been verified to be nontrivial. $\square$

Corollary 8.26 explains the rule of thumb that in dealing with $\mathbb{Z}_2$-spaces, the condition of k-connectivity can be replaced by the weaker condition that the $(k+1)$th power of the appropriate Stiefel–Whitney class is different from 0.

8.3.4 Combinatorial Construction of Stiefel–Whitney Classes

Let us describe how the construction used in the proof of Proposition 8.25 can be employed to obtain an explicit combinatorial description of the Stiefel–Whitney classes.

Let X be a regular CW complex and a $\mathbb{Z}_2$-space, and denote the fixed point free involution on X by γ. As mentioned above, one can choose a simplicial structure on X such that γ is a simplicial map. We define a $\mathbb{Z}_2$-map $\varphi : X \to \mathbb{S}_a^\infty$ following the recipe above.

Take the standard $\mathbb{Z}_2$-equivariant cell decomposition of $\mathbb{S}_a^\infty$ with two antipodal cells in each dimension. Divide $X^{(0)}$, the set of the vertices of X, into two disjoint sets $X^{(0)} = A \cup B$ such that every orbit of the $\mathbb{Z}_2$-action contains exactly one element from A and one element from B. Let $\{a, b\}$ be the 0-skeleton of $\mathbb{S}_a^\infty$, and map all the points in A to a, and all the points in B to b. Call the edges having one vertex in A and one vertex in B *multicolored*, and the edges connecting two vertices in A, resp. two vertices in B, *A-internal*, resp. *B-internal*.

Let $\{e_1, e_2\}$ be the 1-skeleton of $\mathbb{S}_a^\infty$. One can then extend φ to the 1-skeleton as follows. Map the A-internal edges to a; map the B-internal edges to b. Note that the multicolored edges form $\mathbb{Z}_2$-orbits, 2 edges in every orbit.

For each such orbit, map one of the edges to e_1 (there is some arbitrary choice involved here), and map the other one to e_2.

Since the $\mathbb{Z}_2$-action on the space X is free, the generators of the cochain complex $C^*(X/\mathbb{Z}_2; \mathbb{Z}_2)$ can be indexed with the orbits of simplices. For an arbitrary simplex δ we denote by τ_δ the generator corresponding to the orbit of δ; in particular, $\tau_{\gamma(\delta)} = \tau_\delta$.

The induced quotient cell decomposition of $\mathbb{RP}^\infty$ is the standard one, with one cell in each dimension. The cochain z^*, corresponding to the unique edge of $\mathbb{RP}^\infty$, is the generator (and the only nontrivial element) of $H^1(\mathbb{RP}^\infty; \mathbb{Z}_2)$. Its image under $(\varphi/\mathbb{Z}_2)^*$ is simply the sum of all orbits of the multicolored edges:

$$\varpi_1(X) = (\varphi/\mathbb{Z}_2)^*(z^*) = \sum_{\text{multicolored } e} \tau_e, \qquad (8.6)$$

where the sum is taken over representatives of $\mathbb{Z}_2$-orbits of multicolored edges, one representative per orbit.

To describe the powers of the Stiefel–Whitney classes, $\varpi_1^k(X)$, we need to recall how the cohomology multiplication is done simplicially. In fact, to evaluate $\varpi_1^k(X)$ on a k-simplex $(v_0, v_1, \ldots, v_k)$, we need to evaluate $\varpi_1(X)$ on each of the edges (v_i, v_{i+1}), for $i = 0, \ldots, k-1$, and then multiply the results. Thus, the only k-simplices on which the power $\varpi_1^k(X)$ evaluates nontrivially are those whose ordered set of vertices has alternating elements from A and from B. We call these simplices *multicolored*. We summarize with

$$\varpi_1^k(X) = \sum_{\text{multicolored } \sigma} \tau_\sigma, \qquad (8.7)$$

where the sum is taken over representatives of $\mathbb{Z}_2$-orbits of multicolored k-dimensional simplices, one representative per orbit.

8.4 Suggested Reading

We refer the reader to the wonderful book of tom Dieck, [tD87], for further details on equivariant maps and associated bundles. We also recommend the classical book of Milnor and Stasheff, [MSta74], as an excellent source for the theory of characteristic classes of vector bundles. Generalities on bundles, including principal bundles, can be found in [Ste51].

Methods of Combinatorial Algebraic Topology

9

Combinatorial Complexes Melange

A principal constituent of the subject of Combinatorial Algebraic Topology is the fact of existence of a large variety of complexes whose description is purely combinatorial. In this chapter we survey many different situations in which complexes defined by combinatorial data arise. While a certain attempt to structure this set of examples is taken, a complete classification is impossible, due to the nature of the subject.

9.1 Abstract Simplicial Complexes

9.1.1 Simplicial Flag Complexes

Perhaps the simplest situation of an abstract simplicial complex derived from combinatorial data is that of a flag complex. For a graph G and a subset $S \subseteq V(G)$ of its vertices, we let $G[S]$ denote the corresponding induced graph.

Definition 9.1. *Given an arbitrary graph G, we let $\mathrm{Cl}(G)$ denote the abstract simplicial complex whose set of vertices is $V(G)$ and whose simplices are all subsets $S \subseteq V(G)$ such that $G[S]$ is a complete graph.*

The abstract simplicial complex $\mathrm{Cl}(G)$ has various names: it is called a *flag complex* in algebraic topology, while it is called a *clique complex* in combinatorics, prompted by the fact that *clique* is another term used in graph theory for complete subgraphs.

Given a graph G, its *complement* $\overline{G}$ is the graph with the same set of vertices such that (v, w) is an edge of $\overline{G}$ if and only if $v \neq w$ and (v, w) is not an edge of G. A set of vertices $S \subseteq V(G)$ is called *independent* if for all $v, w \in S$ we have $(v, w) \notin E(G)$.

Definition 9.2. *For an arbitrary graph G, the **independence complex** of G, called $\mathrm{Ind}(G)$, is the abstract simplicial complex whose set of vertices is $V(G)$ and whose simplices are all the independent sets (anticliques) of G.*

Since independent sets of G are the same as the cliques of $\overline{G}$, we see that $\text{Ind}\,(G)$ is isomorphic to $\text{Cl}\,(\overline{G})$ as an abstract simplicial complex.

9.1.2 Order Complexes

Definition and examples

Another classical way to describe an abstract simplicial complex by combinatorial data is that of the order complex of a poset. Recall that for a poset P, the set $S \subseteq P$ is called a *chain* if S is totally ordered with respect to the partial order of P. The following is a special case of Definition 15.5.

Definition 9.3. *Let P be a poset. Define $\Delta(P)$ to be the abstract simplicial complex whose vertices are all elements of P and whose simplices are all finite chains of P, including the empty chain. The complex $\Delta(P)$ is called the* **order complex** *of P.*

Example 9.4.

- The order complex of a totally ordered set A with n elements is a simplex Δ^A.

- Let $O_n = \{a_1^1, \ldots, a_n^1, a_1^2, \ldots, a_n^2\}$, with the partial order generated by $a_i^p > a_{i+1}^q$, for all $p, q \in \{1, 2\}$, $i = 1, \ldots, n-1$. Then the order complex of the poset O_n is the join of n copies of $\mathbb{S}^0$. This can be realized as a polytope called a *cross-polytope*; in particular, it is homeomorphic to $\mathbb{S}^{n-1}$. See Figure 9.1(left).

- Let $n \in \mathbb{N}$, and let $\mathcal{B}_n$ be the set of all subsets of $[n]$, partially ordered by inclusion. One can see that the abstract simplicial complex $\Delta(\bar{\mathcal{B}}_n)$ is isomorphic to the barycentric subdivision of the boundary of an $(n-1)$-simplex; in particular, it is homeomorphic to $\mathbb{S}^{n-2}$. See Figure 9.1(right).

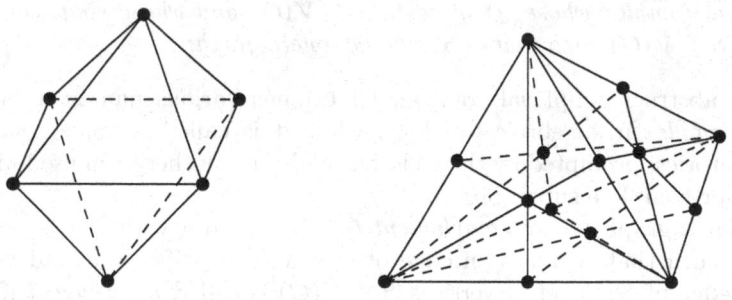

Fig. 9.1. Order complexes of O_3 and $\mathcal{B}_4 \setminus \{\emptyset, [4]\}$.

- Let $n \in \mathbb{N}$. The *partition lattice* Π_n is the partially ordered set whose elements are all set partitions of the set $[n]$, and the partial order is that of refinement. The partition lattice has a minimal element $\{1\}\{2\} \ldots \{n\}$ and a maximal element $[n]$. See Figure 9.2 for the first few examples. It will later be shown that $\Delta(\bar{\Pi}_n)$ is homotopy equivalent to a wedge of $(n-1)!$ copies of $\mathbb{S}^{n-2}$.

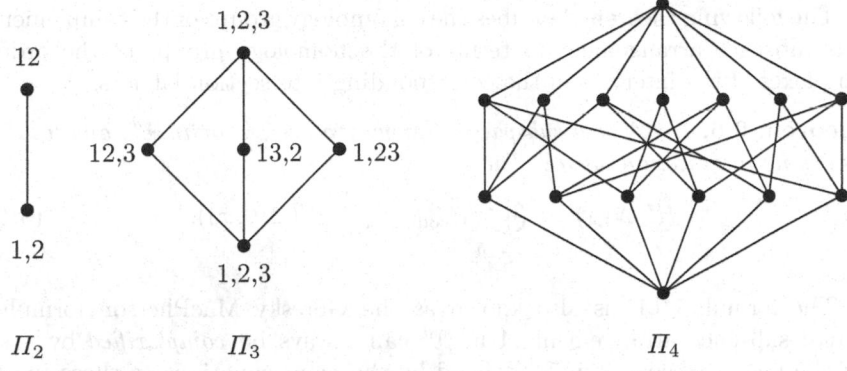

Fig. 9.2. Partition lattices.

Order complexes of posets are special cases of flag complexes, which appear in many contexts. Let us mention two of these here.

Connections to the theory of subspace arrangements

First, the topology of order complexes of certain lattices plays a role in the theory of arrangements. Namely, let $\mathcal{A} = \{A_1, \ldots, A_n\}$ be a collection of finitely many linear subspaces in $\mathbf{k}^d$, where $\mathbf{k} = \mathbb{R}$ or $\mathbf{k} = \mathbb{C}$, such that $A_i \not\supseteq A_j$ for $i \neq j$. Such a collection is called a *subspace arrangement*. Depending on the choice of the field $\mathbf{k}$, one speaks of a real or a complex arrangement. The topological space $\mathcal{M}_{\mathcal{A}} := \mathbf{k}^d \setminus \cup_{i=1}^n A_i$ is called the *complement of arrangement* $\mathcal{A}$, and its algebraic invariants are of interest in various applications.

The intersection data of a subspace arrangement may be represented by a lattice; recall Definition 4.22.

Definition 9.5. *To an arbitrary subspace arrangement* $\mathcal{A} = \{A_1, \ldots, A_n\}$ *in* $\mathbf{k}^d$ *one can associate a partially ordered set* $\mathcal{L}_{\mathcal{A}}$, *called the* **intersection lattice** *of* $\mathcal{A}$. *The set of elements of* $\mathcal{L}_{\mathcal{A}}$ *is*

$$\{K \subseteq \mathbf{k}^d \mid \exists I \subseteq [n], \ such \ that \ \bigcap_{i \in I} A_i = K\} \sqcup \{\mathbf{k}^d\}$$

with the order given by reversing inclusions: $x \leq_{\mathcal{L}_{\mathcal{A}}} y$ if and only if $x \supseteq y$. In particular, the minimal element of $\mathcal{L}_{\mathcal{A}}$ is $\mathbf{k}^d$, and the maximal element is $\bigcap_{K \in \mathcal{A}} K$.

The intersection lattice of an arrangement is in fact a lattice in the order-theoretic sense. For example, if we take the special hyperplane arrangement $\mathcal{A}_{d-1}$, called the *braid arrangement*, consisting of all hyperplanes $x_i = x_j$, for $1 \leq i < j \leq d$, then the intersection lattice is precisely the partition lattice Π_d, independent of the choice of the field $\mathbf{k}$.

The following theorem describes the cohomology groups of the complement of a subspace arrangement in terms of the homology groups of the order complexes of the intervals in the corresponding intersection lattices.

Theorem 9.6. *Let $\mathcal{A}$ be a subspace arrangement in $\mathbb{C}^d$, or in $\mathbb{R}^d$, and let $\mathcal{L}_{\mathcal{A}}$ denote its intersection lattice. Then*

$$\widetilde{H}^i(\mathcal{M}_{\mathcal{A}}) \simeq \bigoplus_{x \in \mathcal{L}_{\mathcal{A}}^{\geq \hat{0}}} \widetilde{H}_{\operatorname{codim}_{\mathbb{R}}(x)-i-2}(\Delta(\hat{0}, x)). \tag{9.1}$$

The formula (9.1) is also known as the Goresky–MacPherson formula. A real subspace arrangement $\mathcal{A}$ in $\mathbb{R}^d$ can always be *complexified* by taking the linear subspaces in $\mathbb{C}^d$ defined by the *same* equations as those in $\mathcal{A}$. The complexified arrangement has the same intersection lattice, and therefore we obtain the following proposition as a direct consequence of the Goresky–MacPherson formula.

Proposition 9.7. *For any real subspace arrangement, the sum of Betti numbers of its complement is equal to the sum of Betti numbers of the complement of its complexification.*

The complement of the real hyperplane arrangement is just a union of contractible spaces: these are the pieces into which the hyperplanes cut the Euclidean space. Proposition 9.7 implies that the sum of the nonreduced Betti numbers of the complement of the complexification of a real hyperplane arrangement is equal to the number of these pieces. For instance, the braid arrangement $\mathcal{A}_{d-1}$ cuts $\mathbb{R}^d$ into $d!$ pieces, indexed by all possible orderings of the coordinates; therefore the sum of the Betti numbers of the complex braid arrangement is equal to $d!$.

Posets of subgroups

The following family of posets appears in the context of homological group theory.

Definition 9.8. *For a group G and a prime p, we denote by $S_p(G)$ the poset of all nontrivial p-subgroups of G, i.e., the subgroups whose cardinality is a power of p. Furthermore, we let $A_p(G)$ denote the poset of all nontrivial elementary abelian p-subgroups of G.*

It turns out that the abstract simplicial complex $\Delta(A_p(G))$ is homotopy equivalent to $\Delta(S_p(G))$. This has led to several investigations of these and other posets of various families of subgroups of a given group G. More specifically, one is trying to understand and to formalize connections between the group-theoretic properties of the considered families of subgroups and the topological properties of the order complexes of the corresponding posets.

9.1.3 Complexes of Combinatorial Properties

Another prominent combinatorial procedure to describe abstract simplicial complexes is the following. Let Ω be a set of combinatorial objects of some kind, equipped with an equivalence relation of isomorphism, and assume that we have a function F that associates to each object from Ω a subset of some universe set V. Any collection C of the isomorphism classes in Ω gives rise to an abstract simplicial complex $\Delta(C)$ as follows:

$\sigma \subseteq V$ is a simplex of $\Delta(C)$ if and only if it is contained in $F(A)$ for some object A whose isomorphism class is in C.

As an example, let Ω be the set of all unoriented graphs on n vertices, where n is fixed. We take V to be the set of all ordered pairs (i, j) such that $1 \leq i < j \leq n$, and we let F map a graph to its set of edges. The isomorphism relation on Ω is the usual graph isomorphism. Frequently one specifies the collection of isomorphism classes C as above in a compact way by simply taking all graphs satisfying some graph property.

Naturally, the examples are endless, and easy to make up. One instance, which has appeared in knot theory, comes from the graph property of being disconnected. It can be shown that the complex of disconnected graphs on n vertices is homotopy equivalent to the order complex of the partition lattice Π_n; see Proposition 13.15.

Another option is to let Ω be the set of oriented graphs on n vertices, again for fixed n. This time we take V to be the set of all ordered pairs (i, j) such that $1 \leq i, j \leq n$, $i \neq j$, and let F map each oriented graph to its set of edges. Again, the isomorphism relation is the usual oriented graph isomorphism, and any property of oriented graphs will yield an abstract simplicial complex along the described scheme. One of the examples that has appeared in the literature is the complex of directed forests.

9.1.4 The Neighborhood and Lovász Complexes

The following construction was one of the first used to export topological ideas to graph theory.

Definition 9.9. *Let G be a graph. The* **neighborhood complex** *of G is the abstract simplicial complex $\mathcal{N}(G)$ defined as follows: its vertices are all nonisolated vertices of G, and its simplices are all the subsets of $V(G)$ that have a common neighbor.*

Let $N(v)$ denote the set of neighbors of v, i.e.,

$$N(v) = \{x \in V(G) \mid (v, x) \in E(G)\}.$$

Then the maximal simplices of $\mathcal{N}(G)$ are precisely the maximal elements $N(v)$, for $v \in V(G)$. Furthermore, for an arbitrary subset $A \subseteq V(G)$, we let $N(A)$ denote the set of common neighbors of A, i.e.,

$$N(A) = \cap_{v \in A} N(v).$$

This gives an order-reversing map $N : \mathcal{F}(\mathcal{N}(G)) \to \mathcal{F}(\mathcal{N}(G))$. It can be seen that $N^3 = N$, cf. Proposition 17.23, and that $N^2(A) \supseteq A$, for any $A \subseteq V(G)$.

Definition 9.10. *The complex* $\Delta(N(\mathcal{F}(\mathcal{N}(G))))$ *is called the* **Lovász complex** *of G and is denoted by* $\mathcal{L}o(G)$.

As we shall see in Subsection 13.2.2, the abstract simplicial complex $\mathrm{Bd}\,(\mathcal{N}(G))$ collapses onto the complex $\mathcal{L}o(G)$.

9.1.5 Complexes Arising from Matroids

In many situations the following concept turns out to be the right formalization.

Definition 9.11. *A* **matroid** *M on a finite ground set V is a nonempty collection I of subsets of V such that every subset of a set in I is in I as well, and such that the following augmentation property is satisfied: if $\sigma, \tau \in I$, and $|\sigma| > |\tau|$, then there exists $x \in \sigma \setminus \tau$ such that $\tau \cup \{x\} \in I$. The elements of I are called* **independent sets**.

By Definition 9.11, the independent sets of a matroid form an abstract simplicial complex. The maximal independent sets, and hence the maximal simplices of the associated complex, are called the *bases* of the matroid. It can be shown that this abstract simplicial complex is always homotopy equivalent to a wedge of spheres of the dimension equal to dimension of the matroid minus one.

9.1.6 Geometric Complexes in Metric Spaces

A number of abstract simplicial complexes used in discrete and computational geometry, as well as in computational topology have gained prominence in the recent years. These complexes are defined starting from some data in a metric space. Strictly speaking, in current applications the input data is not combinatorial, but rather geometric. However, due to a profusion of metric spaces in combinatorial contexts, we include the definitions of these families of abstract simplicial complexes.

Definition 9.12. *Let M be a metric space, and assume that we are given a set S of points in M and a nonnegative real number t. The* **Rips complex** *$R_t(S)$ is the abstract simplicial complex whose set of vertices is S, and $\sigma \subseteq S$ is a simplex if and only if the distance between any two points in σ does not exceed t.*

Clearly, the Rips complex is a special case of a flag complex, with the underlying graph being the graph of all edges that are not longer than t. It is most frequently used to vary the parameter t between 0 and ∞ and then to study the occurring filtrations of the simplex with the set of vertices S, often using the gadget of *persistent homology*.

Sometimes the complex $R_t(S)$ is called *Vietoris-Rips complex*. A possible variation on the Rips complex is to take as simplices those sets of vertices that are contained in a closed ball with radius t.

We shall write a pair (c, r^2) to denote a closed ball in a metric space, where $c \in M$ denotes the center, and r is the radius. Given a point $x \in M$, and a closed ball $B = (c, r^2)$, we set the *weighted square distance* between x and B to be $\mathrm{sd}(x, B) := d(x, c)^2 - r^2$.

Consider now a collection of closed balls $\mathcal{B} = \{B_i\}_{i=1}^n$ in a metric space M, where $B_i = (c_i, r_i^2)$, the c_i's are their centers, and the r_i's are their radii. For each ball B_i we define its *weighted Voronoi region*[1] as

$$V_{B_i} := B_i \cap \{x \in M \mid \mathrm{sd}(x, B_i) \leq \mathrm{sd}(x, B_j), \text{ for all } j \in [n]\}.$$

Clearly, the weighted Voronoi regions give a closed covering of the union of the balls B_i. The nerve of this covering, see Definition 15.14, is denoted by $D(\mathcal{B})$, and is called the *dual complex* of this ball collection. This is an abstract simplicial complex, whose vertices are indexed by the balls $\{B_i\}_{i=1}^n$, and such that σ is a simplex in $D(\mathcal{B})$ if and only if the intersection of the corresponding weighted Voronoi regions is nonempty. In the special case when the radii are taken to be large and equal, the complex $D(\mathcal{B})$ is also called the *Delaunay triangulation*.

In Euclidean space, it can be seen that the weighted Voronoi regions and all their intersections are contractible; therefore it follows by the nerve lemma (Theorem 15.21) that the dual complex of a ball collection is homotopy equivalent to the union of these balls.

The next definition gives an extension of these ideas that has turned out to be very useful in concrete applications.

Definition 9.13. *Let $\mathcal{B} = \{B_i\}_{i=1}^n$, $B_i = (c_i, r_i^2)$, be a collection of balls in a metric space, and let α be a real parameter. The* **alpha complex** *$D_\alpha(\mathcal{B})$ is defined to be the dual complex of the ball collection $\mathcal{B}_\alpha = \{(c_i, r_i^2 + \alpha)\}_{i=1}^n$.*

As the parameter α varies from $-\infty$ to 0 and on to ∞, the corresponding alpha complex goes from the empty set to the dual complex of the ball collection, and on to the Delaunay triangulation associated to the set of the centers

---

[1] Our terminology may differ in minor details from that used in the literature.

of the balls, providing an interesting filtration of the latter, which became the subject of intense study.

Finally, we give a construction that is very recent, yet has already proved itself to be beneficial. Let M be a metric space, and assume that we are given two sets of points $A, B \subseteq M$ such that all distances $d(x, y)$, $x \in A$, $y \in B$, are different; this is just a genericity condition.

Definition 9.14. *The abstract simplicial complex $W(A, B)$, called the **witness complex**, consists of all subsets $\sigma \subseteq B$ such that for every $\tau \subseteq \sigma$ there exists a point w in A that "witnesses" τ in the following sense: every point in τ is closer to w than every point in $B \setminus \tau$.*

The points in A are called *data points*, and the points in B are called *landmark points*. Usually one assumes that there are many more data points than landmark ones.

An alternative way to think of Definition 9.14 is the following. Assume that we have n data points and N landmark points. Every data point induces an order on the landmark points: just sort them with respect to their distances to that point. Every such ordering can be visualized as a path in the Hasse diagram of the Boolean lattice $\mathcal{B}_N$, starting from the point nearest to the chosen data point, then proceeding to the union of the two closest ones, then on to the three closest ones, and so on. Now, the witness complex $W(A, B)$ is the maximal abstract simplicial complex whose face poset is contained in the union of these paths.

In Figure 9.3 we show an example with $N = 4$ landmark points, which are drawn solid black, and $n = 5$ witness points, which are drawn white. On the right of that same figure we see the witness complex of that point configuration. In Figure 9.4 we show the corresponding Boolean algebra with fat paths indicating the five paths corresponding to the witness points.

In concrete successful applications of the witness complex, the crux of the matter is the choice of the landmark points, and many ingenious strategies have been devised.

9.1.7 Combinatorial Presentation by Minimal Nonsimplices

There is a dual presentation of abstract simplicial complexes; namely, instead of describing those sets of vertices that are simplices, one can specify those that are not, with the rule that *if A is not a simplex and $B \supseteq A$, then B is not a simplex either.* This if often handy in the combinatorial context, since the description of the minimal nonsimplices is at times more compact than that of maximal simplices.

For example, the minimal nonsimplices in flag complexes are all of cardinality 2: these are the pairs of vertices that are *not* connected by an edge. As a consequence, for the order complexes of posets, the minimal nonsimplices are pairs of noncomparable elements.

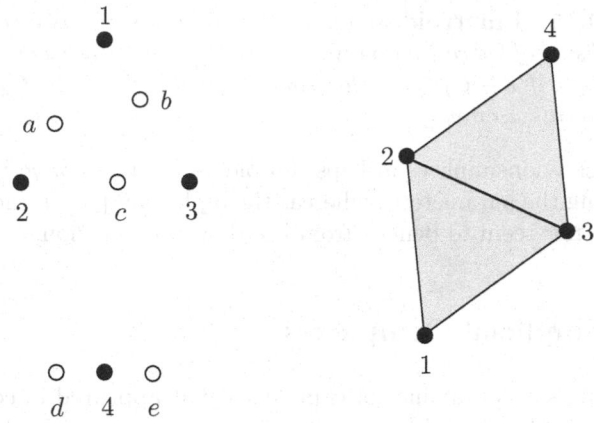

Fig. 9.3. A point configuration and its witness complex.

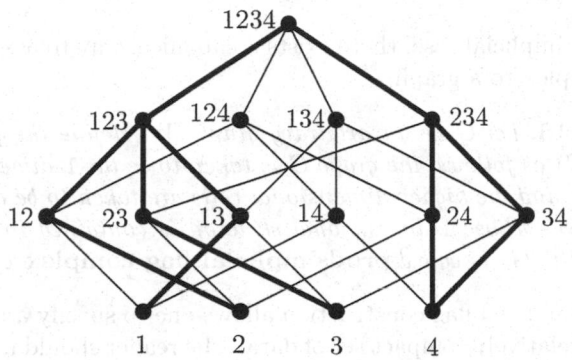

Fig. 9.4. The Boolean algebra with paths corresponding to the point configuration from Figure 9.3.

Also, for the complexes of combinatorial properties, the description can be more succinct when it uses the *forbidden* patterns instead of the allowed ones. For example, for the complex of disconnected graphs, the minimal nonsimplices correspond to spanning trees. For the complex of directed forests, the minimal nonsimplices are all pairs of directed edges that have the same end vertex, together with all directed cycles.

The neighborhood complex of a graph is an example in which this dual description is not useful. On the other hand, the matroid complex is an example in which both descriptions are frequently used, depending on the circumstances. The dual presentation in this case uses the term *circuits*, which play the role of the minimal nonsimplices, and goes as follows.

Definition 9.15. *A* **matroid** *M on a finite ground set V is a collection C of nonempty subsets of V, called* **circuits***, such that no proper subset of a circuit is a circuit, and if $x \in C_1 \cap C_2$, for some $C_1, C_2 \in C$, $C_1 \neq C_2$, then $(C_1 \cup C_2) \setminus \{x\}$ contains a circuit.*

The minimal nonsimplices in Rips complexes are pairs of vertices at a distance exceeding the parameter t, whereas the alpha complexes and the witness complexes do not seem to benefit from the dual presentation.

9.2 Prodsimplicial Complexes

In recent years, several families of complexes that appeared in combinatorics were not simplicial, but rather prodsimplicial.

9.2.1 Prodsimplicial Flag Complexes

Just as in the simplicial case, there exists a canonical way to associate a prodsimplicial complex to a graph.

Definition 9.16. *Let G be an arbitrary graph. We define the prodsimplicial complex $PF(G)$ as follows: the graph G is taken to be the 1-dimensional skeleton of $PF(G)$, and the higher-dimensional cells are taken to be all those products of simplices whose 1-dimensional skeleton is contained in the graph G. The complex $PF(G)$ is called* **prodsimplicial flag complex** *of G.*

The prodsimplicial flag construction allows one to specify a prodsimplicial complex by a relatively compact set of data. The reader should note that while a simplicial complex is always also a prodsimplicial complex, a simplicial flag complex is usually *not* a prodsimplicial flag complex. An example of that is provided by a hollow square.

9.2.2 Complex of Complete Bipartite Subgraphs

First, we define a well-known concept in the generality that we need here.

Definition 9.17. *Let $A, B \subseteq V(G)$, $A, B \neq \emptyset$. We call (A, B) a* **complete bipartite subgraph** *of G if for any $x \in A$, $y \in B$, we have $(x, y) \in E(G)$, i.e., $A \times B \subseteq E(G)$.*

In particular, note that all vertices in $A \cap B$ are required to have loops, and that the edges between the vertices of A (or of B) are allowed; see Figure 9.5.

Let G be a finite graph. Recall that $\Delta^{V(G)}$ is a simplex whose set of vertices is $V(G)$; in particular, the simplices of $\Delta^{V(G)}$ can be identified with the subsets of $V(G)$. Clearly, $\Delta^{V(G)} \times \Delta^{V(G)}$ can be thought of as a polyhedral complex whose cells are direct products of two simplices.

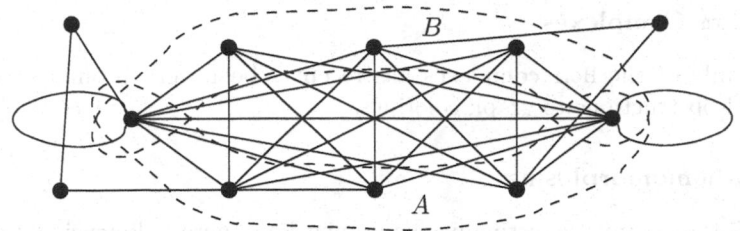

Fig. 9.5. A complete bipartite subgraph.

Definition 9.18. *The complex* $\mathrm{Bip}\,(G)$ *is the subcomplex of* $\Delta^{V(G)} \times \Delta^{V(G)}$ *defined by the following condition:* $\sigma \times \tau \in \mathrm{Bip}\,(G)$ *if and only if* (σ, τ) *is a complete bipartite subgraph of* G.

Note that if (A, B) is a complete bipartite subgraph of G, and $\widetilde{A} \subseteq A$, $\widetilde{B} \subseteq B$, $\widetilde{A}, \widetilde{B} \neq \emptyset$, then $(\widetilde{A}, \widetilde{B})$ is also a complete bipartite subgraph of G. This verifies that $\mathrm{Bip}\,(G)$ is actually a subcomplex.

Example 9.19. Recall that K_n denotes the complete graph on n vertices, and L_n denotes the string graph on n vertices, i.e., we have $V(L_n) = [n]$, and $E(L_n) = \{(i, i+1) \,|\, i = 1, \ldots, n-1\}$:

(1) the complex $\mathrm{Bip}\,(K_1^o)$ is a vertex, here K_1^o is the loop graph;
(2) the complex $\mathrm{Bip}\,(L_3)$ consists of two disjoint 1-simplices;
(3) the complex $\mathrm{Bip}\,(K_3)$ is a hexagon;
(4) the complex $\mathrm{Bip}\,(K_n)$ is isomorphic as a cell complex to the boundary complex of the Minkowski sum $\Delta^n + (-\Delta^n)$. See Figure 9.6.

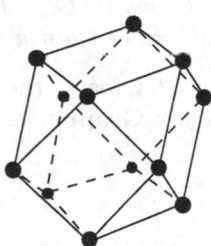

Fig. 9.6. $\mathrm{Bip}\,(K_4)$

The prodsimplicial complex $\mathrm{Bip}\,(G)$ is our first example of the so-called $\mathrm{Hom}\,(-, -)$-construction, namely, as a polyhedral complex it is isomorphic to $\mathrm{Hom}\,(K_2, G)$

9.2.3 Hom Complexes

The family of the Hom complexes has recently been very prominent in the study of obstructions to graph colorings.

Graph homomorphisms

Perhaps the main issue distinguishing graphs from mere 1-dimensional simplicial complexes is that one allows only the rigid maps between them, instead of considering all possible simplicial maps.

Definition 9.20. *For two graphs T and G, a* **graph homomorphism** *from T to G is a map $\varphi : V(T) \to V(G)$ such that if $x, y \in V(T)$ are connected by an edge, then $\varphi(x)$ and $\varphi(y)$ are also connected by an edge.*

In other words, $\varphi : V(T) \to V(G)$ induces $\varphi \times \varphi : V(T) \times V(T) \to V(G) \times V(G)$, and the condition for this map being a graph homomorphism translates into

$$(\varphi \times \varphi)(E(T)) \subseteq E(G).$$

Expressed verbally: *edges map to edges.*

Example 9.21.

(1) The identity map id: $G \to G$;
(2) The unique map $G \to K_1^o$, where K_1^o is the loop graph, i.e., the graph with one vertex and one edge;
(3) A graph homomorphism $\varphi : G \to K_n$ is the same as a vertex coloring of G with n colors. In particular, the *chromatic number* of G, denoted by $\chi(G)$, is the minimal n such that there exists a graph homomorphism $\varphi : G \to K_n$; see Definition 17.2.

Proposition 9.22. *If $\varphi : T \to G$ and $\psi : G \to H$ are graph homomorphisms, then the composition $\psi \circ \varphi : T \to H$ is again a graph homomorphism.*

Therefore, as mentioned in Chapter 4, the graphs together with graph homomorphisms form the category **Graphs**.

Adding topology

We shall now define $\text{Hom}(T, G)$ for an arbitrary pair of graphs T and G. As a model, we take the definition of $\text{Bip}(G)$. Let again $\Delta^{V(G)}$ be a simplex whose set of vertices is $V(G)$. Let $C(T, G)$ denote the direct product $\prod_{x \in V(T)} \Delta^{V(G)}$, i.e., the copies of $\Delta^{V(G)}$ are indexed by vertices of T.

Definition 9.23. *The complex $\text{Hom}(T, G)$ is the subcomplex of $C(T, G)$ defined by the following condition: $\sigma = \prod_{x \in V(T)} \sigma_x \in \text{Hom}(T, G)$ if and only if for any $x, y \in V(T)$, if $(x, y) \in E(T)$, then (σ_x, σ_y) is a complete bipartite subgraph of G.*

Let us make a number of simple but fundamental observations about complexes $\mathrm{Hom}\,(T, G)$.

(1) The topology of $\mathrm{Hom}\,(T, G)$ is inherited from the product topology of $C(T, G)$. By this inheritance, the cells of $\mathrm{Hom}\,(T, G)$ are products of simplices.

(2) The complex $\mathrm{Hom}\,(T, G)$ is a polyhedral complex whose cells are indexed by all functions $\eta : V(T) \to 2^{V(G)} \setminus \{\emptyset\}$ such that if $(x, y) \in E(T)$, then $\eta(x) \times \eta(y) \subseteq E(G)$. The closure of a cell η consists of all cells indexed by $\tilde{\eta} : V(T) \to 2^{V(G)} \setminus \{\emptyset\}$ that satisfy $\tilde{\eta}(v) \subseteq \eta(v)$, for all $v \in V(T)$. We shall make extensive use of the η-notation.

(3) The points of the topological space $\mathrm{Hom}\,(T, G)$ can be given the following explicit description. Let $v_1, \ldots, v_t$ be the vertices of the graph T. Let $S_i \subseteq V(G)$, for all $i = 1, \ldots t$, such that (S_i, S_j) is a complete bipartite graph whenever (i, j) is an edge of T. Then $(S_1, \ldots, S_t)$ indexes a cell whose points are indexed by all tuples $(\lambda(S_1), \ldots, \lambda(S_t))$, where for an arbitrary set S, we let $\lambda(S)$ denote a convex combination of the points in S.

(4) In the literature there are several different notations for the set of all graph homomorphisms from a graph T to the graph G. Since an untangling of the definitions shows that this set is precisely the set of vertices of $\mathrm{Hom}\,(T, G)$, i.e., its 0-skeleton, it feels natural to denote it by $\mathrm{Hom}^{(0)}(T, G)$.

(5) On the intuitive level, one can think of each $\eta : V(T) \to 2^{V(G)} \setminus \{\emptyset\}$ satisfying the conditions of Definition 9.23 as associating nonempty lists of vertices of G to vertices of T with the condition on this collection of lists being that any choice of one vertex from each list will yield a graph homomorphism from T to G.

(6) The standard way to turn a polyhedral complex into a simplicial one is to take the barycentric subdivision. This is readily done by taking the face poset and then taking its nerve (order complex). So here, if we consider the partially ordered set $\mathcal{F}(\mathrm{Hom}\,(T, G))$ of all η as in Definition 9.23, with the partial order defined by $\tilde{\eta} \leq \eta$ if and only if $\tilde{\eta}(v) \subseteq \eta(v)$, for all $v \in V(T)$, then we get that the order complex $\Delta(\mathcal{F}(\mathrm{Hom}\,(T, G)))$ is a barycentric subdivision of $\mathrm{Hom}\,(T, G)$. A cell τ of $\mathrm{Hom}\,(T, G)$ corresponds to the union of all the simplices of $\Delta(\mathcal{F}(\mathrm{Hom}\,(T, G)))$ labeled by the chains with the maximal element τ.

9.2.4 General Complexes of Morphisms

As mentioned above, one way to interpret the definition of the Hom complexes is the following: the cells are indexed by the maps $\eta : V(T) \to 2^{V(G)} \setminus \{\emptyset\}$ such that any choice $\varphi : V(T) \to V(G)$ satisfying $\varphi(x) \in \eta(x)$, for all $x \in V(T)$, defines a graph homomorphism $\varphi \in \mathrm{Hom}^{(0)}(T, G)$. One can generalize this as follows.

Let A and B be two finite sets, and let M be a collection of some set maps $\varphi : A \to B$. Let $C(A, B) = \prod_{x \in A} \Delta^B$, where Δ^B is the simplex having B as a vertex set, and copies in the direct product are indexed by the elements of A.

Definition 9.24. *Let* $\mathrm{Hom}_M(A, B)$ *be the subcomplex of* $C(A, B)$ *consisting of all* $\sigma = \prod_{x \in A} \sigma_x$ *such that any choice* $\varphi : A \to B$ *satisfying* $\varphi(x) \in \sigma_x$, *for all* $x \in A$, *yields a map in* M.

Intuitively one can think of the map φ as the *section* of σ, and the condition can then be verbally stated as: *all sections lie in* M.

Clearly, these complexes are prodsimplicial. It will be shown in Section 18.1.1 that complexes of morphisms $\mathrm{Hom}_-(-, -)$ are in fact prodsimplicial flag complexes.

An example of a complex of morphisms is shown in Figure 9.7. This example comes from the following choice of parameters: $A = \{a, b, c, d\}$, $B = \{1, 2\}$, and a map $\varphi : A \to B$ is in M if and only if $|\varphi^{-1}(2)| \leq 2$. This collection of maps cannot be a set of graph homomorphisms for any graphs T and G such that $V(T) = A$ and $V(G) = B$. One can see this as follows. First, since the map taking all elements in A to 2 is not in M, we conclude that there is no loop on 2, and that the graph G has edges. Let (x, y) be an edge in T, possibly a loop. The map taking both x and y to 2 and the other vertices to 1 is allowed, mapping an edge to a vertex without a loop. This is a contradiction.

The complex in Figure 9.7 can be visualized as follows: take a regular tetrahedron, pick its center of gravity as one of the vertices, and then span six rhombi with one of the vertices in this center, and edges of the tetrahedron as one of the diagonals.

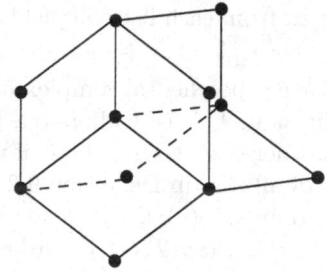

Fig. 9.7. A complex of morphisms.

Here are a few possible specifications of the parameters in the general construction.

(1) As mentioned above, if we take A and B to be the sets of vertices of two graphs T and G, and then take M to be the set of graph homomorphisms from T to G, then $\mathrm{Hom}_M(A, B)$ will coincide with $\mathrm{Hom}(T, G)$.

(2) We think of a *directed* graph G as a pair of sets $(V(G), E(G))$ such that $E(G) \subseteq V(G) \times V(G)$.

Definition 9.25. *For two directed graphs T and G, a **directed graph homomorphism** from T to G is a map $\varphi : V(T) \to V(G)$ such that $(\varphi \times \varphi)(E(T)) \subseteq E(G)$.*

Let A and B be the sets of vertices of two directed graphs T and G, and let M to be the set of directed graph homomorphisms from T to G. Then $\operatorname{Hom}_M(A, B)$ is the analogue of $\operatorname{Hom}(T, G)$ for directed graphs. See Figure 9.8 for an example.

For a directed graph G, let $u(G)$ be the undirected graph obtained from G by forgetting the directions, and identifying the multiple edges. We remark that for any two directed graphs G and H, the complexes $\operatorname{Hom}(G, H)$ and $\operatorname{Hom}(u(G), u(H))$ are isomorphic if $E(H)$ is $\mathbb{Z}_2$-invariant with respect to the $\mathbb{Z}_2$-action that changes the edge orientations.

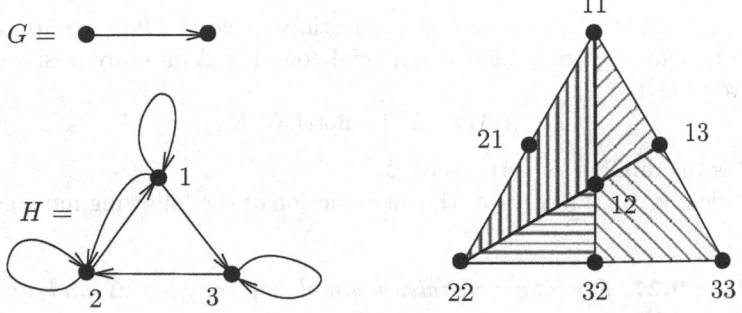

Fig. 9.8. An example of a Hom complex for two directed graphs.

(3) Let A and B be the vertex sets of abstract simplicial complexes Δ_1 and Δ_2, and let M be the set of simplicial maps from Δ_1 to Δ_2. Then $\operatorname{Hom}_M(A, B)$ is the analogue of $\operatorname{Hom}(T, G)$ for the abstract simplicial complexes.

(4) Alternatively, one could also restrict oneself to considering only the "rigid" simplicial maps φ; that is, one could require that $\dim(\varphi(\sigma)) = \dim \sigma$, for all $\sigma \in \Delta_1$.

(5) Recall that a hypergraph with the vertex set V is a subset $\mathcal{H} \subseteq 2^V$. Let A and B be the vertex sets of hypergraphs $\mathcal{H}_1$ and $\mathcal{H}_2$. There are various choices for when to call a map $\varphi : A \to B$ a *hypergraph homomorphism*. Two possibilities that we mention here are that one could require that $\varphi(\mathcal{H}_1) \subseteq \mathcal{H}_2$, or one could ask that for any $H_1 \in \mathcal{H}_1$, there exist $H_2 \in \mathcal{H}_2$ such that $\varphi(H_1) \subseteq H_2$. The example (3) is a special case of both. Either way, the corresponding complex $\operatorname{Hom}_M(A, B)$ provides us with an analogue of $\operatorname{Hom}(T, G)$ for hypergraphs.

(6) Let A and B be the vertex sets of posets P and Q, and let M be the set of order-preserving maps from P to Q (see Definition 10.3) then $\operatorname{Hom}_M(A, B)$ is the analogue of $\operatorname{Hom}(T, G)$ for posets.

9.2.5 Discrete Configuration Spaces of Generalized Simplicial Complexes

A standard concept in algebraic topology is that of a configuration space. Here we consider the discrete version.

Definition 9.26. *Given a generalized simplicial complex Δ and a positive integer n, the* **discrete configuration space** *of Δ is the prodsimplicial subcomplex of $\Delta^{\times n}$, denoted by $DC_n(\Delta)$, defined by the condition: $\sigma_1 \times \cdots \times \sigma_n \in DC_n(\Delta)$ if and only if $\sigma_i \cap \sigma_j = \emptyset$, for all $i, j \in [n]$, $i \neq j$.*

A special case of $DC_n(\Delta)$, when Δ is 1-dimensional, has been a subject of extensive recent study. In this case the complex is actually cubical, a fact that can be used to advantage in applying the methods of geometric group theory.

We can connect the discrete configuration spaces to Hom complexes by noticing that for any generalized simplicial complex Δ and any positive integer n, we have

$$DC_n(\Delta) = \Delta^{\times n} \cap \text{Hom}(K_t, K_n), \tag{9.2}$$

where t is the number of vertices of Δ.

The identity (9.2) motivates the introduction of the following more general object.

Definition 9.27. *For any generalized simplicial complex Δ and arbitrary graphs T and G, we set*

$$\Omega(\Delta, T, G) := \Delta^{\times n} \cap \text{Hom}(T, G), \tag{9.3}$$

where n denotes the number of vertices of T. In other words, $\Omega(\Delta, T, G)$ is the prodsimplicial subcomplex of $\Delta^{\times n}$ defined by the condition: $\sigma_1 \times \cdots \times \sigma_n \in \Omega(\Delta, T, G)$ if and only if whenever $(i, j) \in E(T)$ and $v \in \sigma_i$, $w \in \sigma_j$, we have $(v, w) \in E(G)$.

As remarked earlier, one special case is $\Omega(\Delta, K_t, K_n) = DC_n(\Delta)$, where t is the number of vertices of Δ. Another special case is

$$\Omega(\Delta^{[n]}, T, G) = \text{Hom}(T, G), \tag{9.4}$$

where t is the number of vertices in T, and $\Delta^{[t]}$ is the simplex with vertex set $[t]$.

9.2.6 The Complex of Phylogenetic Trees

In this subsection we describe a cubical complex with combinatorial cell structure that has been suggested as a systematic framework for studying phylogenetic questions in biology.

Let n be a positive integer, $n \geq 3$. We consider all rooted trees, without vertices of degree 2, whose leaves have been labeled 0 to n; call these n-*trees*. The leaf labeled 0 is taken to be the root. We call the vertices that are not leaves *internal vertices*, and we call the edges connecting internal vertices *internal edges*. In such a tree, the number of internal edges varies between 0 and $n - 2$. We fix the lengths of noninternal edges to be 1, but we shall allow the lengths of internal edges to vary.

The trees are considered, as usual, up to isomorphism; that is we do not take into consideration the particular embedding in the plane. For each isomorphism class T we consider an open cube $I(T)$ whose dimension is equal to the number of the internal edges. The coordinates in such a cube are indexed by the internal edges, and we have a bijection between the points of the cube $I(T)$ and the trees of type T, where the lengths of internal edges are taken to be the corresponding coordinates of the point of $I(T)$.

One can think of this as imposing a topology on the space of trees, where a small perturbation of a point corresponds to small change in the lengths of the internal edges. One can go a step further and allow the small perturbations also to create short edges. This corresponds to gluing smaller-dimensional open cubes onto the boundaries of the larger cubes. Finally, we can also allow infinite lengths of edges, and this way close up our cubes by adding cubes at infinity.

Definition 9.28. *For a positive integer $n \geq 3$, the above procedure describes a cubical complex T_n, called the* **complex of phylogenetic trees**.

An example of such complexes is shown in Figure 9.9.

9.3 Regular Trisps

Many of the complexes that arise in combinatorics have the structure of regular trisps, allowing a completely combinatorial description. Let us describe one of the most prominent examples. As we have seen, one way to present an abstract simplicial complex by combinatorial means is to take the order complex of some poset consisting of combinatorial objects. Almost without exception, this poset allows an action of some finite group; in fact, often the acting group is the finite symmetric group.

Assume now that P is a poset and assume that a group G acts on P in an order-preserving way. Since the order complex construction is functorial, this in turn will induce a simplicial action on the complex $\Delta(P)$. In fact, it is easy to see that in this situation the topological quotient has a natural induced cell structure, with new cells being the orbits of the old cells, and that furthermore, one gets a regular trisp. More details on this will be provided in Chapter 14.

It goes without saying that such quotient complexes serve as a bountiful source of combinatorially defined regular trisps. One example is derived

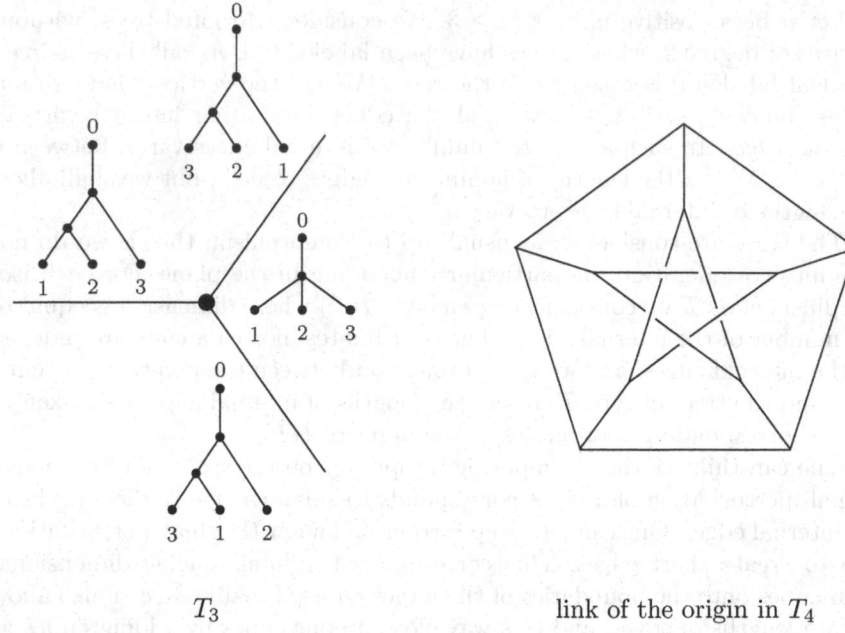

T_3 link of the origin in T_4

Fig. 9.9. Complexes of phylogenetic trees.

from the symmetric group action on the partition lattice; the action is induced by the permutation action on the ground set. The obtained regular trisp $\Delta(\bar{\Pi}_n)/\mathcal{S}_n$ turns out to be contractible, and even stronger, collapsible.

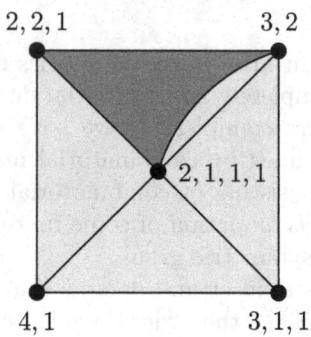

Fig. 9.10. The regular trisp $\Delta(\bar{\Pi}_5)/\mathcal{S}_5$.

In Figure 9.10 we have pictured the regular trisp $\Delta(\bar{\Pi}_5)/\mathcal{S}_5$. This complex consists of five triangles: four in the background, and the fifth one is in front filled with a somewhat darker color. The vertices of this quotient complex are indexed with the number partitions of n; these are precisely the $\mathcal{S}_n$-orbits of the set partitions of n. Several collapsing sequences are readily seen in this special case. We examine the case of general n in greater detail in Section 11.2.

9.4 Chain Complexes

In many pivotal combinatorial applications we do not have a geometric picture. Instead, algebraic objects such as chain complexes arise directly. In this case we are dealing with vector spaces, or more generally, with modules over some ring whose bases have some kind of combinatorial indexing, together with maps between these vector spaces, also these given in a combinatorial way.

One standard situation in which such objects arise occurs when one is proceeding with spectral sequence computations. Even if the original subject of study was a topological space, given by its cell structure, after a suitable filtration is chosen and the first tableau is set up, we are facing the task of computing the next tableaux. Once the appropriate differential is understood, this amounts to computing the homology of a family of chain complexes, each one given combinatorially. One notable larger instance of this framework is given in Chapter 20.

The next definition describes a concrete example that arises in a direct way in the theory of arrangements. Given an arbitrary matroid M, let us describe how to construct an algebra $OS(M)$, called an *Orlik–Solomon algebra*. First, take the exterior algebra $E(M)$ generated by all closed sets in M of rank 1. If M has no loops and no parallel elements, this is the same as taking all elements of M as generators. Assume that we have n generators, which we denote by $e_1, \ldots, e_n$. For any subset $S \subseteq [n]$ we let e_S denote the wedge product $e_{i_1} \wedge \cdots \wedge e_{i_t}$, where $S = \{i_1, \ldots, i_t\}$, and $i_1 < \cdots < i_t$. The exterior algebra $E(M)$ is equipped with the standard boundary operator ∂ defined by

$$\partial(x_1 \wedge \cdots \wedge x_t) = x_2 \wedge \cdots \wedge x_t - x_1 \wedge x_3 \wedge \cdots \wedge x_t + \cdots + (-1)^{t-1} x_1 \wedge \cdots \wedge x_{t-1},$$

for arbitrary generators x_i. Next, consider the ideal I_{OS} of E generated by all the boundaries of circuits. Note that since $x_1 \wedge \partial(x_1 \wedge \cdots \wedge x_t) = x_1 \wedge \cdots \wedge x_t$, for arbitrary generators x_i, this ideal includes e_S for all dependent sets S.

Definition 9.29. *The quotient algebra* $OS(M) := E(M)/I_{OS}$ *is called the* **Orlik–Solomon algebra** *of the matroid* M.

Since the ideal I_{OS} is graded, the quotient $OS(M)$ inherits the natural grading. We denote the graded parts by $OS^i(M)$.

Definition 9.30. *Assume that we are given an arbitrary A and B open in $A \cup B$, a matroid M, and a weight function w, associating a real number to each closed set in M of rank 1. Its Orlik–Solomon algebra $OS(M)$ can be made into a cochain complex $\{OS^i(M), \partial^i\}_{i=0}^\infty$, called the* **Orlik–Solomon cochain**

complex, *by taking multiplication by the element* $w(e_1)e_1 + \cdots + w(e_n)e_n$ *as the coboundary operator.*

Given a hyperplane arrangement $\mathcal{A}$, we know that its intersection lattice is a geometric lattice; hence one can associate to $\mathcal{A}$ a matroid $M(\mathcal{A})$ whose closed sets correspond to the elements of the intersection lattice. It is known that in certain nondegenerate cases, the cohomology of the Orlik–Solomon cochain complex of this matroid coincides with the cohomology of the complement of $\mathcal{A}$, with a local system of coefficients corresponding to the chosen weights.

An additional natural question that arises in the context of the combinatorially defined chain complexes is that of geometric interpretation: construct a geometric framework in which the chain complex at hand can be found.

9.5 Bibliographic Notes

One of the first references on partition lattice is [Ore42]; see also the paper of Folkman, [Fol66], where the homology groups of the order complex of $\bar{\Pi}_n$ were computed.

The classical reference for Theorem 9.6 is the book *Stratified Morse Theory* of Goresky and MacPherson, [GoM88].

The literature on posets of subgroups is rather extensive. A good starting point is still the original paper of Quillen, [Qu78]; see also Shareshian, [Sha02].

For the connection between complexes of disconnected graphs and knot theory, see Vassiliev, [Va93]. Complexes of directed trees first appeared in [Ko99].

Both neighborhood and Lovász complexes were introduced in [Lov78] in connection with the resolution of the Kneser Conjecture.

The subject of geometric complexes in metric spaces is intensively studied in various contexts, in particular in computational topology. A good entry point is the monograph of Zomorodian, [Zo05], as well as papers by Carlsson, Edelsbrunner, Da Silva, Zomorodian, et al; see [CS04b, CZ05, ELZ02].

The term *prodsimplicial complexes* was introduced in the author's survey [Ko05a], where also the prodsimplicial flag construction was emphasized. Complexes of complete bipartite graphs, and more generally Hom complexes associated to graph homomorphisms, were introduced by Lovász end extensively studied by Babson and the author; see [BK03, BK06, BK04]. General complexes of morphisms associated to arbitrary collections of set maps were defined by the author in [Ko05a].

The standard reference for the introduction of the complexes of phylogenetic trees is Billera, Holmes, Vogtmann, [BHV01], where several of their aspects were investigated.

The construction of combinatorial regular trisps in Section 9.3 goes back to the work of Babson and the author on quotient constructions for partial orders; see [BK05]. The case $\Delta(\bar{\Pi}_n)/\mathcal{S}_n$ was specifically the subject of study in [Ko00].

The Orlik–Solomon algebra and Orlik–Solomon cochain complex are cornerstones of a whole direction within arrangement theory. We recommend starting with Orlik & Solomon, [OS80], and Yuzvinsky, [Yuz01], and proceeding with the references in the second paper.

Finally, we mention two general references related to the contents of much of the second part. First, an excellent survey article by Björner, [Bj96]; a portion of the second part is an enhanced and updated version of this article. Second, a textbook by Matoušek, [Ma03], which contains many wonderful applications of topological methods.

10

Acyclic Categories

10.1 Basics

Many results in Combinatorial Algebraic Topology have until now been formulated in the context of posets. In this chapter we would like to emphasize the more general framework of *acyclic categories*. As we shall see in subsequent chapters, no additional difficulties will arise, so there is virtually no penalty to pay for this generality. On the contrary, the situation gets clarified and even simplified in some cases, for example, in dealing with quotients of complexes equipped with group actions in Chapter 14.

Most concepts and constructions involving acyclic categories are introduced in analogy with those used in the context of posets. Often these translate into classical notions of category theory, while in some cases they yield notions that are not standard at all.

10.1.1 The Notion of Acyclic Category

We start by defining the main character of this chapter.

Definition 10.1. *A small category is called* **acyclic** *if only identity morphisms have inverses, and any morphism from an object to itself is an identity.*

We shall always assume that both $\mathcal{O}(C)$ and $\mathcal{M}(C)$ are finite. This makes statements and proofs easier, though many results remain valid in the infinite case as well, either in their original form or with minor alterations.

Recall from Chapter 4 that any poset P can be viewed as a category in the following way: the objects of this category are the elements of P, and for every pair of elements $x, y \in P$, the set of morphisms $\mathcal{M}(x, y)$ has precisely one element if $x \geq y$, and is empty otherwise. Clearly, this determines the composition rule for the morphisms uniquely. When a poset is viewed as a category in this way, it is of course an acyclic category, and intuitively, if posets appear to be more comfortable gadgets, one may think of acyclic categories as

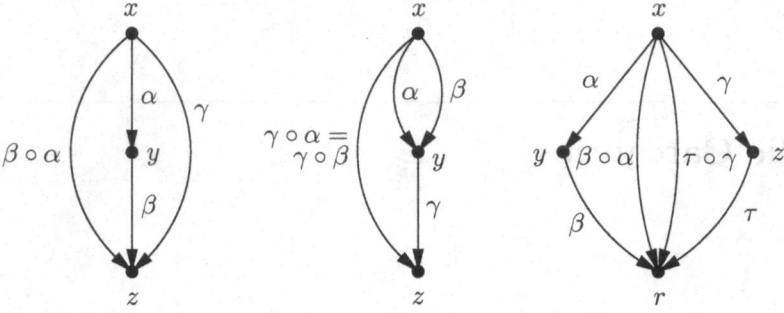

Fig. 10.1. Examples of acyclic categories.

generalizations of posets, with more than one morphism allowed between the elements, and consequently with a more complicated composition rule.

Another way to visualize acyclic categories is to think of them as those that can be drawn on a sheet of paper, with dots indicating the objects, and straight or slightly bent arrows, all pointing down, indicating the nonidentity morphisms; see examples in Figure 10.1.

It can be shown, and is left to the reader, that a category C is acyclic if and only if

- for any pair of distinct objects $x, y \in \mathcal{O}(C)$, at most one of the sets $\mathcal{M}_C(x, y)$ and $\mathcal{M}_C(y, x)$ is nonempty,
- $\mathcal{M}_C(x, x) = \{\mathrm{id}_x\}$, for all $x \in \mathcal{O}(C)$.

One way the acyclic categories generalize posets can be formalized as follows. For any acyclic category C there exists a unique partial order $\geq$ on the set of objects $\mathcal{O}(C)$ such that $\mathcal{M}_C(x, y) \neq \emptyset$ implies $x \geq y$. We denote this poset by $R(C)$. Furthermore, it is immediate that an acyclic category C is a poset if and only if for every pair of objects $x, y \in \mathcal{O}(C)$, the cardinality of the set of morphisms $\mathcal{M}_C(x, y)$ is at most 1.

We call a morphism *indecomposable* if it cannot be represented as a composition of two nonidentity morphisms. An acyclic category C is called *graded* if there is a function $r : \mathcal{O}(C) \to \mathbb{Z}$ such that whenever $m : x \to y$ is a nonidentity indecomposable morphism, we have $r(x) = r(y) + 1$.

10.1.2 Linear Extensions of Acyclic Categories

It is straightforward to generalize the classical notion of linear extension of posets, see Definition 2.20, to the context of acyclic categories. To underline this, we give here the definition using similar wording.

Definition 10.2. *Let C be an acyclic category. A total order $\succ$ on the set of objects of C is called a **linear extension** of C if for any two distinct objects $x, y \in \mathcal{O}(C)$ we have $x \succ y$ whenever $\mathcal{M}_C(x, y) \neq 0$.*

For example, $x \succ y \succ z \succ r$ is a linear extension of the third acyclic category depicted in Figure 10.1. In fact, the set of linear extensions of C coincides with the set of linear extensions of $R(C)$.

It is easy to see, for example using induction, that an acyclic category has at least one linear extension. In a way, a converse of this statement is true as well: if a category C has a linear extension and every morphism of an object into itself is an identity, then C must be acyclic.

10.1.3 Induced Subcategories of Cat

In working with posets, the following notion is standard.

Definition 10.3. *A set map $f : P \to Q$ between sets of elements of two posets is called* **order-preserving** *if $x > y$ in P implies $f(x) \geq f(y)$ in Q.*

In the context of acyclic categories, the role of order-preserving maps is played by *functors*; see Chapter 4 for their definition. It is an important and very useful fact that all acyclic categories together with all possible functors between them form a full subcategory of **Cat**; we call that category **AC**.

We can also consider all posets, and all possible functors between them. This is also a full subcategory of **Cat**, which we shall call **Posets**. The functorial properties provide one explanation for the wide occurrence and accepted usefulness of order-preserving maps, as opposed to, for example, order-reversing ones.

Sometimes it is useful to further restrict our attention to the subcategories consisting only of finite acyclic categories, or posets. The subcategory of finite posets is the one in which most of Combinatorial Algebraic Topology has been done until now.

10.2 The Regular Trisp of Composable Morphism Chains in an Acyclic Category

The appearance of acyclic categories in Combinatorial Algebraic Topology is in large part motivated by the existence of a construction that associates to each acyclic category a geometric object, more precisely, a regular trisp.

10.2.1 Definition and First Examples

The next definition is a special case of a more general construction of the nerve of a category.

Definition 10.4. *Let C be an acyclic category. The* **nerve**[1] *of C is a regular trisp, denoted by $\Delta(C)$, constructed as follows:*

---

[1] Sometimes it is called the **geometric realization** of C.

- *The set of vertices of $\Delta(C)$ is precisely the set of objects of C.*
- *For $k \geq 1$, the set of k-simplices $S_k(\Delta(C))$ is the set of all composable morphism chains consisting of k nonidentity morphisms. The boundary $(k-1)$-simplices of such a simplex $\sigma = (a_0 \xrightarrow{m_1} a_1 \xrightarrow{m_2} \cdots \xrightarrow{m_k} a_k)$ are indexed by composable morphism chains of three types:*

  (1) *the morphism chain $a_0 \xrightarrow{m_1} \cdots \xrightarrow{m_{t-1}} a_{t-1} \xrightarrow{m_{t+1} \circ m_t} a_{t+1} \xrightarrow{m_{t+2}} \cdots \xrightarrow{m_k} a_k$ obtained by skipping the object a_t, for $t = 1, \ldots, k-1$, and composing the adjacent morphisms;*

  (2) *the morphism chain $a_1 \xrightarrow{m_2} a_2 \xrightarrow{m_3} \cdots \xrightarrow{m_k} a_k$ obtained by skipping the first object a_0 and the adjacent morphism;*

  (3) *the morphism chain $a_0 \xrightarrow{m_1} \cdots \xrightarrow{m_{k-2}} a_{k-2} \xrightarrow{m_{k-1}} a_{k-1}$ obtained by skipping the last object a_k, along with its adjacent morphism.*

  The $(k-1)$-simplex indexed by the composable morphism chain skipping the object a_i, for $i = 0, \ldots, k$, is glued to σ by the map $B_{f_i}(\sigma)$, where $f_i : [k] \hookrightarrow [k+1]$ is the order-preserving injection skipping the index i.

It is easy to see that this description yields a well-defined regular trisp, according to Definition 2.47. Indeed, a k-simplex indexed by a composable morphism chain $a_0 \xrightarrow{m_1} a_1 \xrightarrow{m_2} \cdots \xrightarrow{m_k} a_k$ is attached without any identifications along its boundary, since the category C is assumed to be acyclic.

When it is more appropriate, e.g., in Chapter 12, we will forget about the orientations of the simplices of $\Delta(C)$ and just view $\Delta(C)$ as a generalized simplicial complex.

Example 10.5. Figure 10.2 shows the regular trisps that realize the nerves of previously considered acyclic categories. Note that in these examples, the nerves are *not* (geometric realizations of) abstract simplicial complexes.

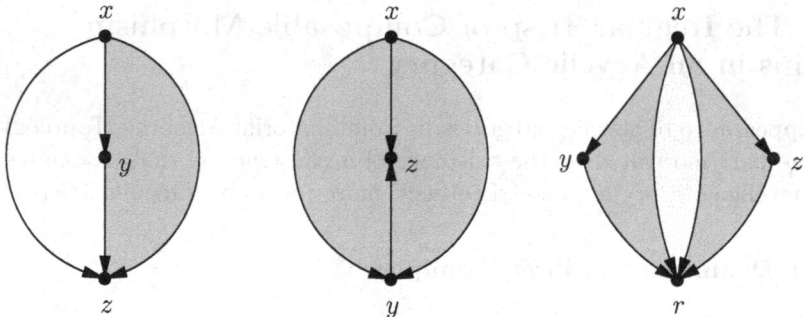

Fig. 10.2. Nerves of acyclic categories from Figure 10.1.

For posets, Definition 10.4 simplifies somewhat; in fact, we recover the definition of the order complex; see Definition 9.3. Thus, in this case, the regular trisps $\Delta(P)$ are always abstract simplicial complexes. On the other hand, when P is a poset, the abstract simplicial complex $\Delta(P)$ is always a flag complex; see Section 9.1.1; hence, for example, a hollow triangle cannot be realized as an order complex of a poset, or even more generally as a nerve of an acyclic category.

10.2.2 Functoriality

The following proposition is the basis for many applications of the construction Δ.

Proposition 10.6. *The nerve construction above gives a functor* $\Delta : \mathbf{AC} \to \mathbf{RTS}$, *where* **RTS** *denotes the category of regular trisps.*

Proof. Given two acyclic categories C and D, and a functor $\mathcal{F} : C \to D$, each composable morphism chain $a_0 \xrightarrow{m_1} a_1 \xrightarrow{m_2} \cdots \xrightarrow{m_k} a_k$ maps to the composable morphism chain $\mathcal{F}(a_0) \xrightarrow{\mathcal{F}(m_1)} \mathcal{F}(a_1) \xrightarrow{\mathcal{F}(m_2)} \cdots \xrightarrow{\mathcal{F}(m_k)} \mathcal{F}(a_k)$, where some of the morphisms may actually end up being identities. By the discussion in Section 2.3 we see that this induces a trisp map $\Delta(\mathcal{F}) : \Delta(C) \to \Delta(D)$.

Furthermore, a trivial functor $\mathcal{F} : C \to C$ induces a trivial trisp map $\Delta(\mathcal{F}) : \Delta(C) \to \Delta(C)$, and Δ commutes with taking a composition: $\Delta(\mathcal{F}_2 \circ \mathcal{F}_1) = \Delta(\mathcal{F}_2) \circ \Delta(\mathcal{F}_1)$. Therefore we may conclude that Δ yields a functor as claimed. $\square$

An example of a functor between acyclic categories and an associated trisp map between their nerves is shown in Figures 10.3 and 10.4. For posets, Proposition 10.6 translates into saying that an order-preserving map between posets P and Q (these are the morphisms in **Posets**) induces a simplicial map between the corresponding order complexes $\Delta(P)$ and $\Delta(Q)$, which additionally preserves edge orientations.

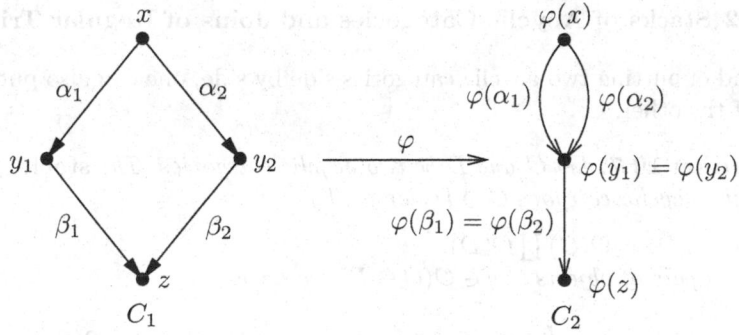

Fig. 10.3. A functor between acyclic categories.

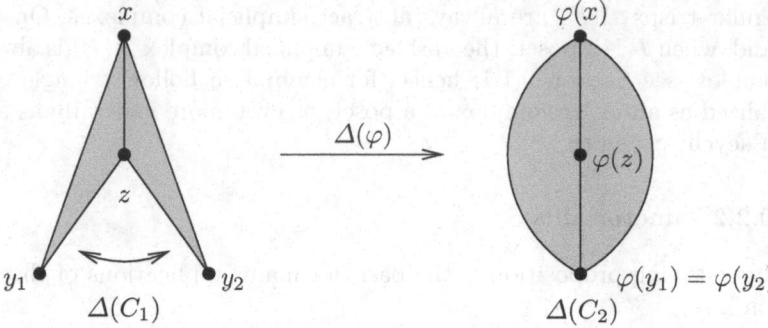

Fig. 10.4. The trisp map between nerves of acyclic categories induced by the functor in Figure 10.3.

10.3 Constructions

10.3.1 Disjoint Union as a Coproduct

For two arbitrary acyclic categories C and D, we let $C \coprod D$ denote the disjoint union of C and D. As a matter of fact, this is the coproduct of C and D in **AC**; see Definition 4.18.

It is obvious that Δ commutes with coproducts, in other words, we have $\Delta(C \coprod D) = \Delta(C) \coprod \Delta(D)$. At the same time, it is important to note at this point that Δ *does not commute* with colimits in general. The question of commutativity will be investigated in much more detail in Chapter 14.

For future use, let **n** denote the poset with n elements and no order relations, i.e., $\mathbf{n} = \underbrace{\mathbf{1} \coprod \cdots \coprod \mathbf{1}}_{n}$.

10.3.2 Stacks of Acyclic Categories and Joins of Regular Trisps

Instead of putting two acyclic categories side by side, one can also put one on top of the other.

Definition 10.7. *Let C and D be two acyclic categories. The **stack** of C and D is the acyclic category $C \oplus D$ defined by*

- $\mathcal{O}(C \oplus D) = \mathcal{O}(C) \coprod \mathcal{O}(D)$;
- *for a pair of objects $x, y \in \mathcal{O}(C \oplus D)$ we have*

$$
\mathcal{M}_{C \oplus D}(x, y) = \begin{cases} \mathcal{M}_C(x, y), & \text{if } x, y \in \mathcal{O}(C); \\ \mathcal{M}_D(x, y), & \text{if } x, y \in \mathcal{O}(D); \\ \text{a single morphism } x \to y, & \text{if } x \in \mathcal{O}(C), y \in \mathcal{O}(D). \end{cases}
$$

The composition rule for morphisms of $C \oplus D$ is induced by the composition rules for $\mathcal{M}(C)$ and $\mathcal{M}(D)$.

Clearly, the composition rule for morphisms of $C \oplus D$ is uniquely determined by the condition that the images of the inclusion maps $C \hookrightarrow C \oplus D$ and $D \hookrightarrow C \oplus D$ are full subcategories, and the fact that $\mathcal{M}(x, y)$ has cardinality 1 when $x \in \mathcal{O}(C), y \in \mathcal{O}(D)$.

The second acyclic category in Figure 10.1 can be seen as a stack of the acyclic category with two objects and two nonidentity morphisms and the acyclic category with one object. Another example of a stack is shown in Figure 10.5.

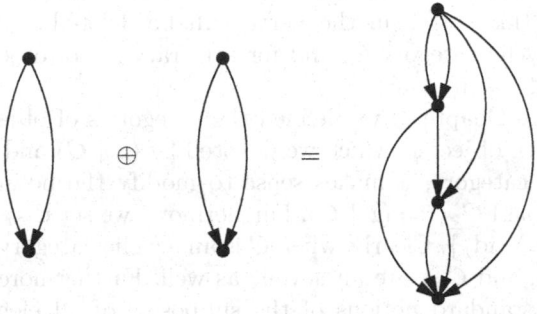

Fig. 10.5. An example of a stack of two acyclic categories.

We note that for an arbitrary acyclic category C, the acyclic category $C \oplus 1$ is obtained from C by adding a terminal object, while the category $1 \oplus C$ is obtained from C by adding an initial one. In the special case of posets we recover the following classical concept.

Definition 10.8. (Stacks of posets). *Let P and Q be two posets. The* **stack**[2] *of P and Q is the poset $P \oplus Q$ whose set of vertices is $P \coprod Q$ and whose order relation is given by*

$$x \leq y \text{ if and only if } \begin{cases} either & x, y \in P, \quad x \leq y; \\ or & x, y \in Q, \quad x \leq y; \\ or & x \in P, y \in Q. \end{cases}$$

For arbitrary acyclic categories C and D we have an isomorphism of regular trisps:

$$\Delta(C \oplus D) = \Delta(C) * \Delta(D), \tag{10.1}$$

since the simplices of the regular trisp $\Delta(C \oplus D)$ are precisely all ordered pairs of simplices of $\Delta(C)$ and $\Delta(D)$ glued in the right way, see Subsection 2.3.2.

---

[2] Also called **ordinal sum**.

Recall the poset $O_n = \{a_1^1, \ldots, a_n^1, a_1^2, \ldots, a_n^2\}$, with the partial order generated by $a_i^p > a_{i+1}^q$, for all $p, q \in \{1, 2\}$, $i = 1, \ldots, n-1$, which was introduced in Subsection 9.1.2. In the new notations, we have $O_n = \mathbf{2} \oplus \cdots \oplus \mathbf{2}$, where we sum up n copies. Then, we have

$$\Delta(O_n) = \Delta(\underbrace{\mathbf{2} \oplus \cdots \oplus \mathbf{2}}_{n}) = \underbrace{\Delta(\mathbf{2}) * \cdots * \Delta(\mathbf{2})}_{n} \cong \underbrace{\mathbb{S}^0 * \cdots * \mathbb{S}^0}_{n} \cong \mathbb{S}^{n-1}.$$

10.3.3 Links, Stars, and Deletions

One of the simplest operations one can do on an acyclic category is that of a deletion of a set of vertices. Once the vertex is deleted, all the simplices that contained this vertex in the nerve will be deleted as well; hence for an arbitrary acyclic category C, and for arbitrary set of objects $S \subseteq \mathcal{O}(C)$, we have $\Delta(C \setminus S) = \mathrm{dl}_{\Delta(C)} S$.

Recall that in Chapter 4 we defined the categories of objects in C below and above a given object x, which we denoted by $(x \downarrow C)$ and $(x \uparrow C)$. When C is an acyclic category, it makes sense to modify the notation as follows: $C_{\leq x} := (x \downarrow C)$ and $C_{\geq x} := (x \uparrow C)$. Furthermore, we set $C_{<x} := C_{\leq x} \setminus \{\mathrm{id}_x\}$ and $C_{>x} := C_{\geq x} \setminus \{\mathrm{id}_x\}$. Clearly, when C is an acyclic category, the categories $C_{\leq x}$, $C_{<x}$, $C_{\geq x}$, and $C_{>x}$ are all acyclic as well. Furthermore, for posets we simply get the standard notions of the subposets of all elements below or above x.

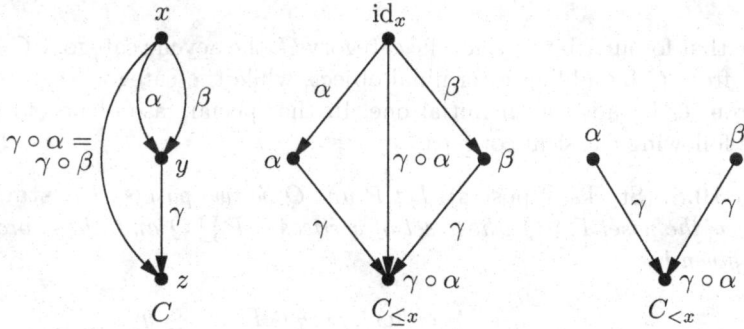

Fig. 10.6. Acyclic categories of objects below x and strictly below x.

For any acyclic category C, and $x \in \mathcal{O}(C)$, we have following formulas:

$$\mathrm{lk}_{\Delta(C)} x = \Delta(C_{<x} \oplus C_{>x}) = \Delta(C_{<x}) * \Delta(C_{>x}), \qquad (10.2)$$

$$\mathrm{star}_{\Delta(C)} x = \mathrm{Cone}(\mathrm{lk}_{\Delta(C)} x) = \Delta(C_{<x} \oplus \{\mathrm{id}_x\} \oplus C_{>x})$$
$$= \Delta(C_{<x}) * \{x\} * \Delta(C_{>x}). \qquad (10.3)$$

We can see how these formulas are satisfied for the example in Figure 10.6. To prove formula (10.2), let us describe a bijection between the sets of simplices on the left- and on the right-hand sides. By definition, the k-simplices of the regular trisp $\mathrm{lk}_{\Delta(C)}x$ are indexed by those $(k+1)$-simplices of the regular trisp $\Delta(C)$ that have x as a vertex. This is the same as the composable morphism chains

$$a_0 \xrightarrow{m_1} a_1 \xrightarrow{m_2} \cdots \xrightarrow{m_p} a_p \xrightarrow{m} x \xrightarrow{n} b_0 \xrightarrow{n_1} b_1 \xrightarrow{n_2} \cdots \xrightarrow{n_q} b_q,$$

where $p+q+1 = k$. On the other hand, k-simplices of the join $\Delta(C_{<x}) * \Delta(C_{>x})$ are indexed by all pairs of simplices (σ, τ) such that σ is a p-simplex in $\Delta(C_{<x})$, τ is a q-simplex in $\Delta(C_{>x})$, and $p+q+1 = k$. In turn, the p-simplices of $\Delta(C_{<x})$ are indexed precisely by all composable morphism chains $a_0 \xrightarrow{m_1} a_1 \xrightarrow{m_2} \cdots \xrightarrow{m_p} a_p \xrightarrow{m} x$, whereas the q-simplices of $\Delta(C_{>x})$ are indexed precisely by all composable morphism chains $x \xrightarrow{n} b_0 \xrightarrow{n_1} b_1 \xrightarrow{n_2} \cdots \xrightarrow{n_q} b_q$. Thus we have the promised bijection.

It is easy to see that this bijection induces a trisp isomorphism, and hence formula (10.2) is proved. Formula (10.3) is proved in just the same way, with the element x inserted everywhere.

10.3.4 Lattices and Acyclic Categories

The natural generalization of the notion of a lattice would be to require that the acyclic category have all finite products and coproducts. Unfortunately, this does not bring anything new, as the next proposition shows.

Proposition 10.9. *Let C be an acyclic category that has all products of two elements. Then C is a poset.*

Proof. Assume that there exist two objects $x, y \in \mathcal{O}(C)$ such that $|\mathcal{M}(x,y)| \geq 2$. Choose $m_1, m_2 \in \mathcal{M}(x,y)$ such that $m_1 \neq m_2$. Let (z, p_1, p_2) be the product of x and y, which we have assumed must exist. Recall that in this notation, $z \in \mathcal{O}(C)$, and $p_1 : z \to x$ and $p_2 : z \to y$ are the projection morphisms.

Consider the morphisms $\mathrm{id}_x : x \to x$ and $m_1 : x \to y$. By the universality property of the product construction, these two morphisms should factor through (z, p_1, p_2), that is, there exists $\alpha : x \to z$ such that $p_1 \circ \alpha = \mathrm{id}_x$ and $p_2 \circ \alpha = m_1$. Since the category C is acyclic, we can conclude that $\alpha = p_1 = \mathrm{id}_x$, and hence $z = x$. It follows that $p_2 = m_1$.

Repeating the same argument with a different initial choice of morphisms, namely with $\mathrm{id}_x : x \to x$ and $m_2 : x \to y$, we get $p_2 = m_2$. This contradicts our assumption that the morphisms m_1 and m_2 are different. $\square$

Also, by symmetry, any acyclic category C that has all finite coproducts of two elements must be a poset.

10.3.5 Barycentric Subdivision and Δ-Functor

The following definition allows us to turn an acyclic category into a poset, as far as the topology of the associated space is concerned.

Definition 10.10. *For an arbitrary acyclic category C, let $\mathrm{Bd}\,C$ denote the poset whose minimal elements are objects of C, and whose other elements are all composable morphism chains consisting of nonidentity morphisms of C. The elementary order relations are given by composing morphisms in the chain, and by removing the first or the last morphism. The poset $\mathrm{Bd}\,C$ is called the **barycentric subdivision** of C.*

In Figure 10.7 we show an acyclic category and its barycentric subdivision, whereas in Figure 10.8 we show corresponding nerves.

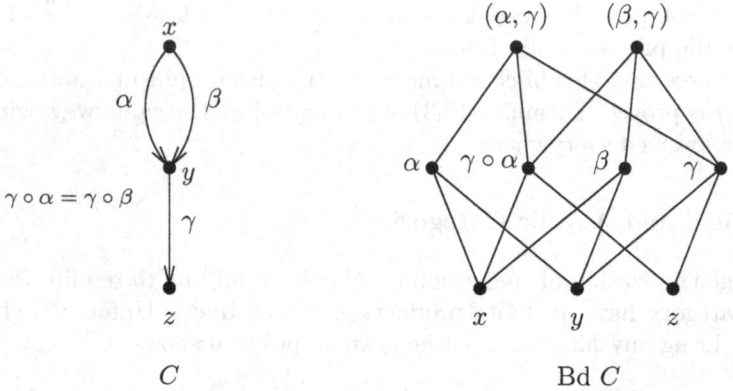

Fig. 10.7. An acyclic category and its barycentric subdivision.

In Section 10.2 we have saw to construct a cell complex out of an acyclic category. There is a standard way to go in the opposite direction as well.

Definition 10.11. *For a regular CW complex K, let $\mathcal{F}(K)$ denote the poset of all closures of nonempty cells of K ordered by inclusion.*

Just like Δ, this is a functor from the category of regular CW complexes to the category of posets, since a cellular map between regular CW complexes will induce an order-preserving map between their face posets.

Given a regular CW complex K, we recall that the simplices of $\mathrm{Bd}\,K$ are exactly those sets $\{b_{F_1}, \ldots, b_{F_t}\}$ that correspond to sequences of closures of cells including each other $F_1 \subset F_2 \subset \cdots \subset F_t$. So we have

$$\mathrm{Bd}\,K = \Delta(\mathcal{F}(K)),\tag{10.4}$$

which geometrically means

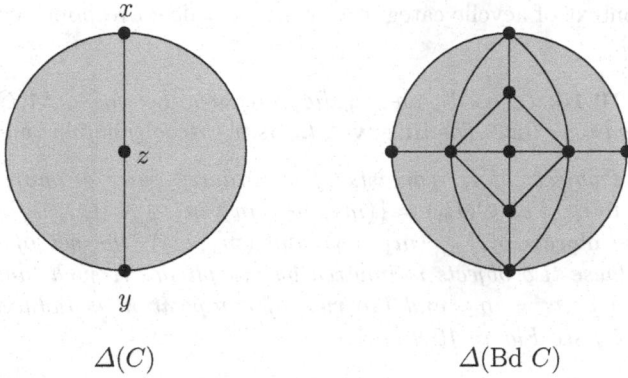

$\Delta(C)$ $\Delta(\mathrm{Bd}\, C)$

Fig. 10.8. Nerves of the categories in Figure 10.7.

$$|\mathrm{Bd}\, K| \cong |\Delta(\mathcal{F}(K))| \cong |K|. \qquad (10.5)$$

In particular, as an immediate corollary of (10.5) we see that *an arbitrary regular CW complex is homeomorphic to the order complex of some poset.* We also obtain an alternative way to define the barycentric subdivision of an acyclic category C:

$$\mathrm{Bd}\, C = \mathcal{F}(\Delta(C)). \qquad (10.6)$$

The combination of equations (10.4) and (10.6) implies

$$\Delta(\mathrm{Bd}\, C) = \Delta(\mathcal{F}(\Delta(C))) = \mathrm{Bd}\,(\Delta(C)). \qquad (10.7)$$

Example 10.12. Let P be a totally ordered set with n elements. Clearly, its barycentric subdivision is the Boolean algebra with the minimal element removed: $\mathrm{Bd}\, P = \mathcal{B}_n \setminus \{\hat{0}\}$. Equation (10.7) implies that

$$\Delta(\mathcal{B}_n \setminus \{\hat{0}\}) = \mathrm{Bd}\,(\Delta(P)) = \mathrm{Bd}\, \Delta^{[n]}.$$

Furthermore, $\mathcal{B}_n \setminus \{\hat{0}, \hat{1}\}$ is the face poset of the boundary of $\Delta^{[n]}$. Applying (10.4), we get

$$\Delta(\mathcal{B}_n \setminus \{\hat{0}, \hat{1}\}) = \Delta(\mathcal{F}(\partial \Delta^{[n]}) = \mathrm{Bd}\,(\partial \Delta^{[n]})) \cong \mathbb{S}^{n-2},$$

which was already observed at an earlier point.

10.4 Intervals in Acyclic Categories

10.4.1 Definition and First Properties

Given a poset $(P, >)$ and two elements x, y of P such that $x \geq y$, the (closed) interval $[y, x]$ is the set of all elements z of P satisfying $x \geq z \geq y$, equipped with the partial order induced from the partial order of P.

In the context of acyclic categories, the intervals correspond to the following objects.

Definition 10.13. *Let C be an acyclic category. Let $m \in \mathcal{M}(C)$, and set $x := \partial^{\bullet} m$ and $y := \partial_{\bullet} n$. The* **interval** I_m *is a category defined by:*

- *the set of objects of I_m consists of all ordered pairs or morphisms that compose to m, i.e., $\mathcal{O}(I_m) = \{(m_1, m_2) \mid m_1, m_2 \in \mathcal{M}(C), m_2 \circ m_1 = m\}$;*
- *given two objects of I_m, (m_1, m_2) and $(\widetilde{m}_1, \widetilde{m}_2)$, the set of morphisms between these two objects is indexed by morphisms α such that $\alpha \circ m_1 = \widetilde{m}_1$ and $\widetilde{m}_2 \circ \alpha = m_2$, and the rule of composition is induced from the category C; see Figure 10.9.*

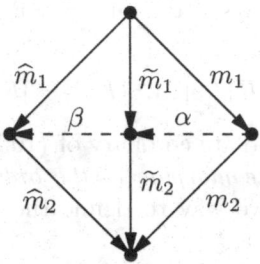

Fig. 10.9. The composition rule in the interval.

An example is shown in Figure 10.10. We remark that when we say that, for example, $\alpha \circ m_1 = \widetilde{m}_1$, we do implicitly require that in particular the composition be well-defined, i.e., in this case, we require that $\partial_{\bullet} m_1 = \partial^{\bullet} \alpha$.

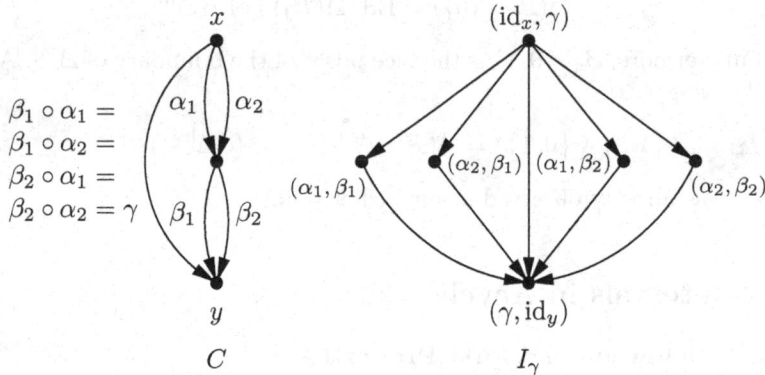

Fig. 10.10. Example of an interval.

Furthermore, we say that the set of morphisms between (m_1, m_2) and $(\tilde{m}_1, \tilde{m}_2)$ is *indexed* by the morphisms α, rather than just taking the morphisms themselves, because the same morphism α may give rise to different morphisms in I_m: this happens, for example, in Figure 10.10. Alternatively, formally we could have said that the set of all morphisms between (m_1, m_2) and $(\tilde{m}_1, \tilde{m}_2)$ in the category I_m consists of all quintuples $(m_1, m_2, \tilde{m}_1, \tilde{m}_2, \alpha)$ of morphisms of C such that $m_2 \circ m_1 = \tilde{m}_2 \circ \tilde{m}_1 = m$, $\alpha \circ m_1 = \tilde{m}_1$, and $\tilde{m}_2 \circ \alpha = m_2$. The composition would then be defined by the rule

$$(\tilde{m}_1, \tilde{m}_2, \hat{m}_1, \hat{m}_2, \beta) \circ (m_1, m_2, \tilde{m}_1, \tilde{m}_2, \alpha) = (m_1, m_2, \hat{m}_1, \hat{m}_2, \beta \circ \alpha), \quad (10.8)$$

where again we refer the reader to Figure 10.9. With the formal definition (10.8) it is easy to verify that I_m is well-defined as a category, i.e., that it has identity morphisms and that the composition is associative.

The interval category has both an initial and a terminal object. Indeed, given an acyclic category C and its morphism m, the initial object of I_m is (id_x, m), whereas the terminal one is (m, id_y), where $x = \partial^{\bullet} m$ and $y = \partial_{\bullet} m$. When (m_1, m_2) is any object of I_m, it is easy to check that $(\mathrm{id}_x, m, m_1, m_2, m_1)$ is the unique morphism from (id_x, m) to (m_1, m_2), and that $(m_1, m_2, m, \mathrm{id}_y, m_2)$ is the unique morphism from (m_1, m_2) to (m, id_y).

Proposition 10.14. *For any acyclic category C, and any morphism of C, Definition 10.13 produces a well-defined acyclic category.*

Proof. Let m be the chosen morphism of C. We verify that I_m is a well-defined category. To start with, the rule of the composition is obviously well-defined and associative. This is easily checked, using the fact that the composition is associative in the category C itself. Furthermore, for any object (m_1, m_2) of I_m, there is the corresponding identity morphism $(m_1, m_2, m_1, m_2, \mathrm{id}_x)$, where $x = \partial_{\bullet} m_1 = \partial^{\bullet} m_2$; the latter follows directly from the definition.

Let us now verify that I_m is acyclic. Given a morphism $(m_1, m_2, \tilde{m}_1, \tilde{m}_2, \alpha)$ of I_m, its inverse (if one exists) must be of the form $(\tilde{m}_1, \tilde{m}_2, m_1, m_2, \beta)$, such that $\beta \circ \alpha = \mathrm{id}_x$, where $x = \partial_{\bullet} m_1 = \partial^{\bullet} m_2$. Since C is acyclic, only identity morphisms have inverses in C; hence $\alpha = \beta = \mathrm{id}_x$. This implies that $m_1 = \tilde{m}_1$ and $m_2 = \tilde{m}_2$. Therefore, also in I_m only identity morphisms have inverses.

Furthermore, a morphism from an object (m_1, m_2) of I_m to itself is of the form $(m_1, m_2, m_1, m_2, \alpha)$, where α is a morphism from $x = \partial_{\bullet} m_1 = \partial^{\bullet} m_2$ to itself. Since C is acyclic, this implies that $\alpha = \mathrm{id}_x$, hence $(m_1, m_2, m_1, m_2, \alpha)$ itself is an identity morphism. $\square$

Clearly, the interval construction is symmetric, i.e., if C is an acyclic category, m its morphism, and m^{op} is the morphism corresponding to m in C^{op}, then

$$I_{m^{\mathrm{op}}} = (I_m)^{\mathrm{op}}. \quad (10.9)$$

The next proposition is an important observation connecting intervals in acyclic categories to the arrow categories.

Proposition 10.15. *Let C be an acyclic category with a terminal object t. Let m be any morphism of C satisfying $t = \partial_\bullet m$, and set $x := \partial^\bullet m$. Then we have $I_m = C_{\leq x}$.*

Proof. We observe that since t is a terminal object, for any $m_1 \in \mathcal{M}(C)$ satisfying $\partial^\bullet m = \partial^\bullet m_1$, there exists a unique morphism m_2 such that $m_1 \circ m_2 = m$. In particular, we have an object (m_1, m_2) of I_m, and the second morphism in this pair is uniquely determined by the first. We recall Definition 4.35 and see that this fact gives a bijection between objects of I_m and objects of $C_{\leq x}$.

Furthermore, we see that in the definition of the morphisms of I_m the commutativity requirement for the lower triangle is always trivially satisfied. This leaves us with requiring just the commutativity of the upper triangle, which under the bijection above translates exactly to the commutativity condition in Definition 4.35. This shows that the above bijection extends to yield an isomorphism of categories. $\square$

By symmetry we also see that if a category has an initial object s, and m is a morphism satisfying $\partial^\bullet m = s$, then $I_m = C_{\geq \partial_\bullet m}$.

10.4.2 Acyclic Category of Intervals and Its Structural Functor

The entirety of intervals of the category C itself can be equipped with the structure of the category.

Definition 10.16. *Let C be an acyclic category. The **category of intervals** of C, which we denote by $I(C)$ is defined as follows:*

- *the set of objects of $I(C)$ is given by the set of morphisms of C;*
- *for two objects m_1 and m_2 of $I(C)$, the set of morphisms from m_1 to m_2 in $I(C)$ is indexed by all pairs (α, β) such that $\alpha, \beta \in \mathcal{M}(C)$ and $m_2 = \beta \circ m_1 \circ \alpha$.*

The composition rule in $I(C)$ is given by the trapezoidal combination rule shown in Figure 10.11.

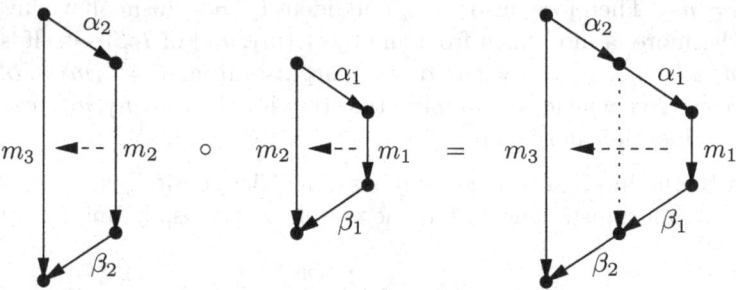

Fig. 10.11. Composition rule in the category of intervals.

Formally, if (α_1, β_1) is a morphism from m_1 to m_2, and (α_2, β_2) is a morphism from m_2 to m_3, then the composition morphism from m_1 to m_3 is given by $(\alpha_1 \circ \alpha_2, \beta_2 \circ \beta_1)$.

Some examples of categories of intervals are shown in Figure 10.12. The first one in this figure is the category of intervals of the acyclic category with two objects and two nonidentity morphisms. The second, resp. the third one is the category of intervals of the first, resp. the second category shown in Figure 10.1. Please note that the first two categories in Figure 10.12 are posets. The third category is also drawn in a simplified way: we skipped the arrows and did not draw those compositions of nonidentity morphisms that are unique in this case.

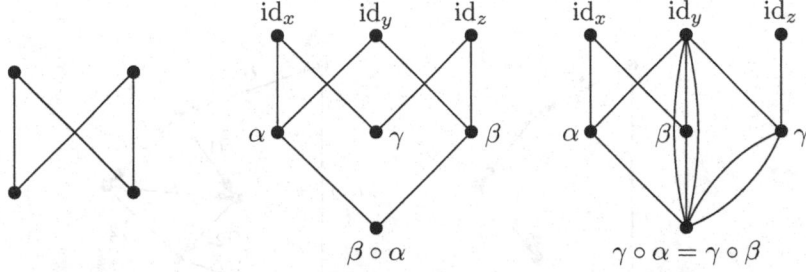

Fig. 10.12. Examples of categories of intervals of acyclic categories.

When C is a poset, the category $I(C)$ is simply the poset of all closed intervals of C ordered by reverse inclusion.

Proposition 10.17. *For an arbitrary acyclic category C, the category of intervals $I(C)$ is acyclic as well.*

Proof. Indeed, an inverse of a morphism (α, β) would be of the form $(\widetilde{\alpha}, \widetilde{\beta})$, such that $\partial^\bullet \alpha = \partial_\bullet \widetilde{\alpha}$, $\partial_\bullet \alpha = \partial^\bullet \widetilde{\alpha}$, $\partial^\bullet \beta = \partial_\bullet \widetilde{\beta}$, and $\partial_\bullet \beta = \partial^\bullet \widetilde{\beta}$. Since C is acyclic, this implies that α and β are identity morphisms in C, hence (α, β) is an identity morphism in $I(C)$.

Furthermore, a morphism of an object m to itself must have the form (α, β) such that α is a morphism of $\partial^\bullet m$ to itself, and β is a morphism of $\partial_\bullet m$ to itself. Again, this implies that α, β, and hence also (α, β), are identity morphisms in their respective categories. $\square$

The fact that objects of $I(C)$ actually index intervals of C is best expressed by phrasing it in terms of a functor.

Definition 10.18. *For an arbitrary acyclic category C, we define a functor* $\text{Int} : I(C) \to \mathbf{AC}$ *as follows. First, it takes every morphism of C to the associated interval, i.e., for $m \in \mathcal{M}(C)$, we set* $\text{Int}(m) := I_m$. *Second, given* *a morphism (α, β) in $I(C)$ going from m_1 to m_2, for $m_1, m_2 \in \mathcal{M}(C)$, we set* Int (α, β) *(we skip the double brackets for clarity) to be the functor from I_{m_1} to I_{m_2} defined by (10.10) and (10.11):*

- *for $(\gamma_1, \gamma_2) \in \mathcal{O}(I_{m_1})$ we set*

$$\text{Int} (\alpha, \beta)(\gamma_1, \gamma_2) := (\gamma_1 \circ \alpha, \beta \circ \gamma_2) \in \mathcal{O}(I_{m_2}), \tag{10.10}$$

see the left part of Figure 10.13;
- *for $(\gamma_1, \gamma_2, \widetilde{\gamma}_1, \widetilde{\gamma}_2, \tau) \in \mathcal{M}(I_{m_1})$ we set*

$$\text{Int} (\alpha, \beta)(\gamma_1, \gamma_2, \widetilde{\gamma}_1, \widetilde{\gamma}_2, \tau) := (\gamma_1 \circ \alpha, \beta \circ \gamma_2, \widetilde{\gamma}_1 \circ \alpha, \beta \circ \widetilde{\gamma}_2, \tau); \tag{10.11}$$

see the right part of Figure 10.13.

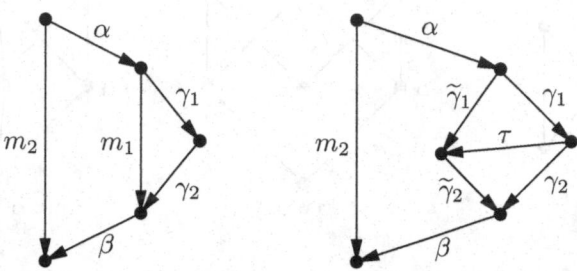

Fig. 10.13. The definition of functor Int (α, β).

We think of Int as a *structural functor* associated to $I(C)$. Let us see that it is well-defined. First, we check that Int (α, β) is a functor for any $(\alpha, \beta) \in \mathcal{M}(I(C))$. We see that Int (α, β) maps identity morphisms to identity morphisms, since this simply means that τ in (10.11) is an identity morphism. It is also easy to check that Int (α, β) maps composition of morphisms to composition of morphisms:

$$\text{Int} (\alpha, \beta)((\widetilde{\gamma}_1, \widetilde{\gamma}_2, \widehat{\gamma}_1, \widehat{\gamma}_2, \tau_2) \circ (\gamma_1, \gamma_2, \widetilde{\gamma}_1, \widetilde{\gamma}_2, \tau_1))$$
$$= \text{Int} (\alpha, \beta)(\gamma_1, \gamma_2, \widehat{\gamma}_1, \widehat{\gamma}_2, \tau_2 \circ \tau_1) = (\gamma_1 \circ \alpha, \beta \circ \gamma_2, \widehat{\gamma}_1 \circ \alpha, \beta \circ \widehat{\gamma}_2, \tau_2 \circ \tau_1)$$
$$= (\widetilde{\gamma}_1 \circ \alpha, \beta \circ \widetilde{\gamma}_2, \widehat{\gamma}_1 \circ \alpha, \beta \circ \widehat{\gamma}_2, \tau_2) \circ (\gamma_1 \circ \alpha, \beta \circ \gamma_2, \widetilde{\gamma}_1 \circ \alpha, \beta \circ \widetilde{\gamma}_2, \tau_1)$$
$$= \text{Int} (\alpha, \beta)(\widetilde{\gamma}_1, \widetilde{\gamma}_2, \widehat{\gamma}_1, \widehat{\gamma}_2, \tau_2) \circ \text{Int} (\alpha, \beta)(\gamma_1, \gamma_2, \widetilde{\gamma}_1, \widetilde{\gamma}_2, \tau_1).$$

Second, we can check that Int itself is a well-defined functor. An identity morphism in $I(C)$ is of the form $(\text{id}_{\partial \bullet m}, \text{id}_{\partial \bullet m})$; this is mapped to the identity by (10.10) and (10.11). Furthermore, Int $(\alpha_2 \circ \alpha_1, \beta_2 \circ \beta_1) = $ Int $(\alpha_2, \beta_2) \circ$ Int (α_1, β_1), as one again can verify from (10.10) and (10.11).

10.4.3 Topology of the Category of Intervals

As it turns out, the topology of the category of intervals of an acyclic category is the same as that of the original category itself.

Theorem 10.19. *For any acyclic category C, the regular trisp $\Delta(I(C))$ is a subdivision of the regular trisp $\Delta(C)$, i.e., we have*

$$\Delta(I(C)) \rightsquigarrow \Delta(C). \tag{10.12}$$

Proof. We define continuous maps $f : \Delta(C) \to \Delta(I(C))$ and $g : \Delta(I(C)) \to \Delta(C)$, which are inverses of each other.

First, take $x \in \Delta(C)$. By definition, x is encoded by a composable morphism chain $a_0 \xrightarrow{m_1} a_1 \xrightarrow{m_2} \cdots \xrightarrow{m_k} a_k$, consisting of nonidentity morphisms and coefficients $t_0, \ldots, t_k$ such that $t_0 + \cdots + t_k = 1$ and $0 \leq t_i \leq 1$, for all $i = 0, \ldots, k$. If $k = 0$, then f maps x to the vertex $\mathrm{id}_{a_0} \in \Delta(I(C))$. Otherwise, assume $k \geq 1$. To describe $f(x)$, we need to give a composable morphism chain in $I(C)$ and the corresponding set of coefficients. Let $t = \min(t_0, t_k)$. We take $m_k \circ m_{k-1} \circ \cdots \circ m_1$ as the last object in this composable morphism chain, with the corresponding coefficient $2t$. After that, we construct the rest of the composable morphism chain, and the coefficients recursively according to the following rule:

- if $t_0 > t_k$, then proceed with $a_0 \xrightarrow{m_1} a_1 \xrightarrow{m_2} \cdots \xrightarrow{m_{k-1}} a_{k-1}$, using the connecting morphism (id_{a_0}, m_k);
- if $t_0 < t_k$, then proceed with $a_1 \xrightarrow{m_2} a_2 \xrightarrow{m_3} \cdots \xrightarrow{m_k} a_k$, using the connecting morphism (m_0, id_{a_k});
- otherwise proceed with $a_1 \xrightarrow{m_2} a_2 \xrightarrow{m_3} \cdots \xrightarrow{m_{k-1}} a_{k-1}$, if it is nonempty, using the connecting morphism (m_0, m_k),

where in all three cases we hand over the coefficients t_i intact. In other words, t is subtracted both from t_0 and from t_k, after which the objects with zero coefficients get deleted.

In the end, we either have an empty composable morphism chain, which means that our process has terminated, or we have one object a_i, with some coefficient t, left. In the latter case, we finish our construction by putting the object a_i with coefficient t as the first object in the composable morphism chain that we are constructing.

For example, if $k = 4$, and $t_0 = 0.1$, $t_1 = 0.3$, $t_2 = 0.2$, $t_3 = 0.2$, $t_4 = 0.2$, then $f(x)$ is encoded by the composable morphism chain

$$\mathrm{id}_{a_2} \xrightarrow{(m_2, m_3)} m_3 \circ m_2 \xrightarrow{(\mathrm{id}_{a_1}, m_4)} m_4 \circ m_3 \circ m_2 \xrightarrow{(m_1, \mathrm{id}_{a_4})} m_4 \circ m_3 \circ m_2 \circ m_1,$$

with the corresponding coefficients 0.2, 0.4, 0.2, 0.2; see the left part of Figure 10.14.

Let now $y \in \Delta(I(C))$. It is encoded by a composable morphism chain $\gamma_0 \xrightarrow{(\alpha_1, \beta_1)} \cdots \xrightarrow{(\alpha_k, \beta_k)} \gamma_k$, consisting of nonidentity morphisms and coefficients

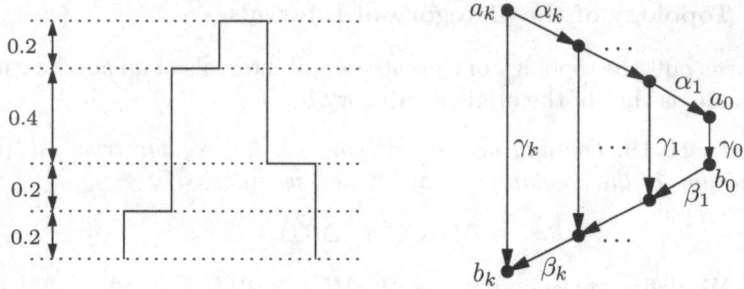

Fig. 10.14. Illustrating definitions of f and of g.

$t_0, \ldots, t_k$ such that $t_0 + \cdots + t_k = 1$ and $0 \leq t_i \leq 1$, for all $i = 0, \ldots, k$. Set $a_i := \partial^\bullet \gamma_i$ and $b_i := \partial_\bullet \gamma_i$, for $i = 0, \ldots, k$. An example of such a chain is shown in the right part of Figure 10.14. Consider the following composable morphism chain in C going along the perimeter of the diagram in Figure 10.14:

$$a_k \xrightarrow{\alpha_k} \cdots \xrightarrow{\alpha_1} a_0 \xrightarrow{\gamma_0} b_0 \xrightarrow{\beta_1} \cdots \xrightarrow{\beta_k} b_k.$$

We set $g(y)$ to be the point that is encoded by this chain with the coefficients of a_i and of b_i being each equal to $t_i/2$, for all $i = 0, \ldots, k$. Notice that this composable chain may contain identity morphisms. As is customary in such a case, we then skip this morphism, identify its endpoints, and add their coefficients.

To check that the maps f and g are continuous, we just need to see that this is the case on closed simplices. This is straightforward, since all our manipulations with coefficients give continuous functions, including the case in which the coefficients are set to 0. Also, clearly these maps are inverses of each other. This shows that the spaces $\Delta(C)$ and $\Delta(I(C))$ are homeomorphic. Finally, one can see that the image of any closed simplex of $\Delta(I(C))$ under g is contained in some closed simplex of $\Delta(C)$. Therefore $\Delta(I(C))$ is actually a subdivision of the regular trisp $\Delta(C)$. $\square$

The interested reader is invited to see how the statement of Theorem 10.19 works out for the categories of intervals depicted in Figure 10.12.

10.5 Homeomorphisms Associated with the Direct Product Construction

10.5.1 Simplicial Subdivision of the Direct Product

The product of two arbitrary objects, see Definition 4.20, exists in the category **Cat**. The next definition describes an explicit construction.

Definition 10.20. *Let C and D be arbitrary categories. The* **direct product** *$C \times D$ is the category defined by the following:*

- *the class of objects is given by $\mathcal{O}(C \times D) := \mathcal{O}(C) \times \mathcal{O}(D)$;*
- *for objects $(x_1, y_1), (x_2, y_2) \in \mathcal{O}(C \times D)$, the set of morphisms is given by $\mathcal{M}_{C \times D}((x_1, y_1), (x_2, y_2)) := \mathcal{M}_C(x_1, x_2) \times \mathcal{M}_D(y_1, y_2)$.*

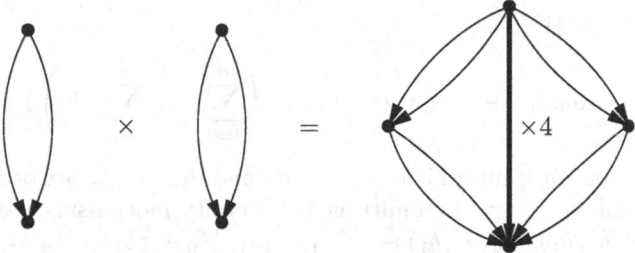

Fig. 10.15. A direct product of two acyclic categories. There are actually four morphisms denoted by the fat middle morphism on the right-hand side. They correspond to the four possible compositions of morphisms to the left of it, or, symmetrically, to the right of it.

When C and D are acyclic categories, their product $C \times D$ will be an acyclic category as well, see Figure 10.15 for an example. Furthermore, when P and Q are posets, we recover the classical notion of a direct product of posets, namely the poset whose set of elements is the set of all pairs (p, q), $p \in P$, $q \in Q$, equipped with the partial order given by $(p_1, q_1) \leq (p_2, q_2)$ if and only if $p_1 \leq p_2$ and $q_1 \leq q_2$.

The next theorem is a classical statement about acyclic categories.

Theorem 10.21. *For arbitrary acyclic categories C and D, the topological space $\Delta(C \times D)$ is homeomorphic to $\Delta(C) \times \Delta(D)$. In fact, a stronger statement is true:*

$$\Delta(C \times D) \rightsquigarrow \Delta(C) \times \Delta(D). \tag{10.13}$$

Proof. Let us compare the point descriptions of the spaces in (10.13).

On the right-hand side (RHS), the points can be described as pairs of linear combinations $(s_0 a_0 + \cdots + s_k a_k, t_0 b_0 + \cdots + t_m b_m)$, where $a_0 \xrightarrow{\alpha_1} a_1 \xrightarrow{\alpha_2} \cdots \xrightarrow{\alpha_k} a_k$ is a composable morphism chain in C, $b_0 \xrightarrow{\beta_1} b_1 \xrightarrow{\beta_2} \cdots \xrightarrow{\beta_m} b_m$ is a composable morphism chain in D, $s_i, t_j > 0$, $\sum_{i=0}^{k} s_i = 1$, and $\sum_{j=0}^{m} t_j = 1$, where both chains consist of nonidentity morphisms. It is important to remark here that even though the morphisms $\alpha_1, \ldots, \alpha_k$ and $\beta_1, \ldots, \beta_m$ do not enter the description of the points explicitly, implicitly they remain part of the data, and a different choice of morphisms between the same objects will yield different points in $\Delta(C) \times \Delta(D)$.

On the left-hand side (LHS), the points can be described as linear combinations $r_0(\tilde{a}_0, \tilde{b}_0) + \cdots + r_n(\tilde{a}_n, \tilde{b}_n)$, where $\tilde{a}_0 \xrightarrow{\tilde{\alpha}_1} \tilde{a}_1 \xrightarrow{\tilde{\alpha}_2} \cdots \xrightarrow{\tilde{\alpha}_n} \tilde{a}_n$ is a composable morphism chain in C, $\tilde{b}_0 \xrightarrow{\tilde{\beta}_1} \tilde{b}_1 \xrightarrow{\tilde{\beta}_2} \cdots \xrightarrow{\tilde{\beta}_n} \tilde{b}_n$ is a composable morphism chain in D, $\sum_{i=0}^n r_i = 1$, $r_i \geq 0$, and for every $i = 1, \ldots, n$, at least one of the morphisms $\tilde{\alpha}_i$ and $\tilde{\beta}_i$ is a nonidentity morphism.

Let us now describe the bijection between the points of $\Delta(C \times D)$ and $\Delta(C) \times \Delta(D)$.

(1) (LHS) $\longrightarrow$ (RHS).

$$r_0(\tilde{a}_0, \tilde{b}_0) + \cdots + r_n(\tilde{a}_n, \tilde{b}_n) \mapsto \left(\sum_{i=0}^n r_i \tilde{a}_i, \sum_{i=0}^n r_i \tilde{b}_i \right),$$

and the corresponding morphisms $\alpha_1, \ldots, \alpha_k$ and $\beta_1, \ldots, \beta_m$ are obtained from $\tilde{\alpha}_1, \ldots, \tilde{\alpha}_n$ and $\tilde{\beta}_1, \ldots, \tilde{\beta}_n$ by omitting the identity morphisms. For example, $\frac{1}{3}(a_0, b_0) + \frac{1}{3}(a_0, b_1) + \frac{1}{3}(a_1, b_1) \mapsto \left(\frac{2}{3}a_0 + \frac{1}{3}a_1, \frac{1}{3}b_0 + \frac{2}{3}b_1 \right)$.

(2) (RHS) $\longrightarrow$ (LHS).

Take the collection of partial sums $0, s_1, s_1 + s_2, \ldots, s_1 + \cdots + s_{k-1}, t_1, t_1 + t_2, \ldots, t_1 + \cdots + t_{m-1}, 1$ and order them in increasing order, eliminating the duplicates. Denote the obtained sequence by $\Sigma_0 = 0, \Sigma_1, \ldots, \Sigma_{n-1}, \Sigma_n = 1$, and set $r_i = \Sigma_i - \Sigma_{i-1}$, for $i = 1, \ldots, n$. Furthermore, set $(\tilde{a}_i, \tilde{b}_i) = (a_{f(i)}, b_{g(i)})$, where $f(i)$ is the minimal index such that $s_1 + \cdots + s_{f(i)} \geq \Sigma_i$, and $g(i)$ is the minimal index such that $t_1 + \cdots + t_{g(i)} \geq \Sigma_i$. The morphism from $(\tilde{a}_i, \tilde{b}_i)$ to $(\tilde{a}_{i+1}, \tilde{b}_{i+1})$ is given by the input data morphisms $\tilde{a}_i \to \tilde{a}_{i+1}$ and $\tilde{b}_i \to \tilde{b}_{i+1}$, one of which (but not both) might be an identity morphism.

Our two constructions yield continuous maps, which are inverses of each other; in particular, they must be bijections. Furthermore, since every simplex of $\Delta(C \times D)$ gets mapped inside some cell of $\Delta(C) \times \Delta(D)$, we see that $\Delta(C \times D)$ is a simplicial subdivision of $\Delta(C) \times \Delta(D)$, which incidentally has the same set of vertices. $\square$

Example 10.22.
(1) Recall that $\mathcal{B}_2$ consists of two ordered elements. It is not difficult to check that $\mathcal{B}_2 \times \cdots \times \mathcal{B}_2 = \mathcal{B}_n$, where there are n factors. On the other hand, applying (10.13), we obtain

$$\Delta(\mathcal{B}_n) = \Delta(\mathcal{B}_2) \times \cdots \times \Delta(\mathcal{B}_2) = I^n.$$

Thus we see that $\Delta(\mathcal{B}_n)$ is a triangulation of the n-dimensional unit cube. It is perhaps worth noting that this triangulation does not introduce new vertices, and that the number of maximal simplices is equal to $n!$. In fact, they are naturally indexed by the permutations of n, as are the maximal chains of $\mathcal{B}_n$. See the left and the middle parts of Figure 10.16.

(2) Taking P to be a chain with m elements, and Q to be a chain with n elements, we see that $\Delta(P \times Q)$ gives a simplicial subdivision of the direct

product of two simplices $\Delta^{[m]} \times \Delta^{[n]}$, without introducing new vertices. See the right part of Figure 10.16.

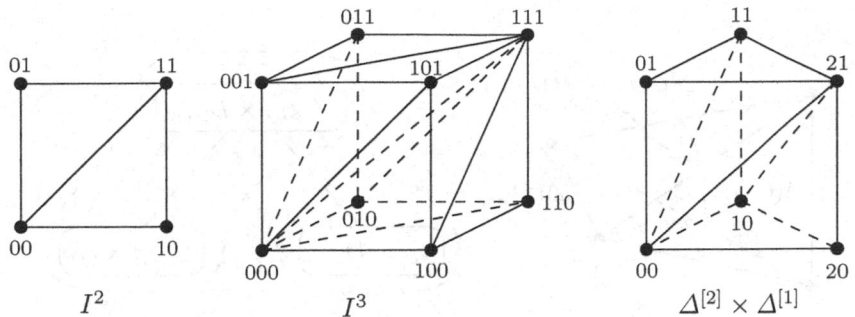

Fig. 10.16. Subdivisions in Example 10.22.

10.5.2 Further Subdivisions

When the considered acyclic categories have terminal and/or initial objects, there exist further subdivision results that are sometimes useful. Both results are corollaries of the main Theorem 10.21.

Theorem 10.23. *Assume that C and D are acyclic categories.*

(1) If both C and D have terminal objects t_C and t_D, then

$$\Delta((C \times D)_{>(t_C, t_D)}) \rightsquigarrow \Delta(C_{>t_C}) * \Delta(D_{>t_D}), \qquad (10.14)$$

where the simplicial structure on the right-hand side is the standard one of the join of two regular trisps.

(2) If both C and D have initial objects s_C and s_D, as well as terminal objects t_C and t_D, then

$$\Delta(\overline{C \times D}) \rightsquigarrow susp\,(\Delta(\bar{C}) * \Delta(\bar{D})), \qquad (10.15)$$

where $\bar{C}$ and $\bar{D}$ denote the acyclic categories obtained from C and D by removing the initial and the terminal elements, and the simplicial structure on the right-hand side is again the standard one.

Proof. To see (10.14), we represent the space $\Delta(C_{>t_C}) * \Delta(D_{>t_D})$ as a union of spaces $\Delta(C_{>t_C}) \times \Delta(D)$ and $\Delta(C) \times \Delta(D_{>t_D})$, which intersect over a copy of $\Delta(C_{>t_C}) \times \Delta(D_{>t_D})$; here we recall that, according to (10.1), we have $\Delta(C) = \Delta(C_{>t_C}) * \{t_C\}$ and $\Delta(D) = \Delta(D_{>t_D}) * \{t_D\}$. Geometrically, one could think of the join $\Delta(C_{>t_C}) * \Delta(D_{>t_D})$ being cut along the "middle" copy of $\Delta(C_{>t_C}) \times \Delta(D_{>t_D})$; see the left part of Figure 10.17. Correspondingly, we can decompose $\Delta((C \times D)_{>(t_C, t_D)})$ as a union of $\Delta(C_{>t_C} \times D)$ and $\Delta(C \times D_{>t_D})$, which in turn intersect in $\Delta(C_{>t_C} \times D_{>t_D})$; see the right side of Figure 10.17.

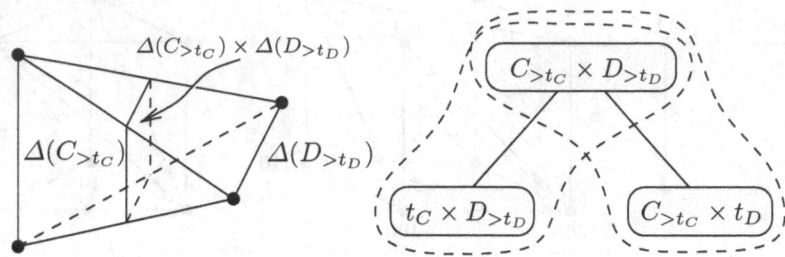

Fig. 10.17. Decompositions proving (10.14).

By Theorem 10.21 we know that we have the following subdivisions: $\Delta(C_{>t_C} \times D) \rightsquigarrow \Delta(C_{>t_C}) \times \Delta(D)$, $\Delta(C \times D_{>t_D}) \rightsquigarrow \Delta(C) \times \Delta(D_{>t_D})$, and $\Delta(C_{>t_C} \times D_{>t_D}) \rightsquigarrow \Delta(C_{>t_C}) \times \Delta(D_{>t_D})$. Furthermore, these subdivisions are natural, in the sense that this subdivision of $\Delta(C_{>t_C}) \times \Delta(D_{>t_D})$ coincides with the restrictions of these subdivisions of $\Delta(C_{>t_C}) \times \Delta(D)$ and $\Delta(C) \times \Delta(D_{>t_D})$. The statement (10.14) follows from this.

The subdivision statement (10.15) is slightly more complicated, but follows by a similar argument. This time, we represent the space susp $(\Delta(\bar{C}) * \Delta(\bar{D}))$ as a union of spaces $\Delta(C_{<s_C}) * \Delta(\bar{D})$, which is the same as $\{t_C\} * \Delta(\bar{C}) * \Delta(\bar{D})$, and $\Delta(\bar{C}) * \Delta(D_{<s_D})$, which is the same as $\{t_D\} * \Delta(\bar{C}) * \Delta(\bar{D})$. These intersect over a copy of $\Delta(\bar{C}) * \Delta(\bar{D})$. The decomposition of $\Delta(\overline{C \times D})$ that we get here is a little more interesting: we represent it as a union of $\Delta((C \times D_{>t_D})_{<(s_C, s_D)})$ and $\Delta((C_{>t_C} \times D)_{<(s_C, s_D)})$, which intersect in $\Delta((C_{>t_C} \times D_{>t_D})_{<(s_C, s_D)})$; see Figure 10.18.

By the proved statement (10.14), we have

$$\Delta((C \times D_{>t_D})_{<(s_C, s_D)}) \rightsquigarrow \Delta(C_{<s_C}) * \Delta(\bar{D}) = \{t_C\} * \Delta(\bar{C}) * \Delta(\bar{D}),$$
$$\Delta((C_{>t_C} \times D)_{<(s_C, s_D)}) \rightsquigarrow \Delta(\bar{C}) * \Delta(D_{<s_C}) = \{t_D\} * \Delta(\bar{C}) * \Delta(\bar{D}),$$
$$\Delta((C_{>t_C} \times D_{>t_D})_{<(s_C, s_D)}) \rightsquigarrow \Delta(\bar{C}) * \Delta(\bar{D}).$$

Again, this subdivision of $\Delta(\bar{C}) * \Delta(\bar{D})$ coincides with the restrictions of these subdivisions of $\{t_C\} * \Delta(\bar{C}) * \Delta(\bar{D})$ and $\{t_D\} * \Delta(\bar{C}) * \Delta(\bar{D})$. The statement (10.15) follows from this. □

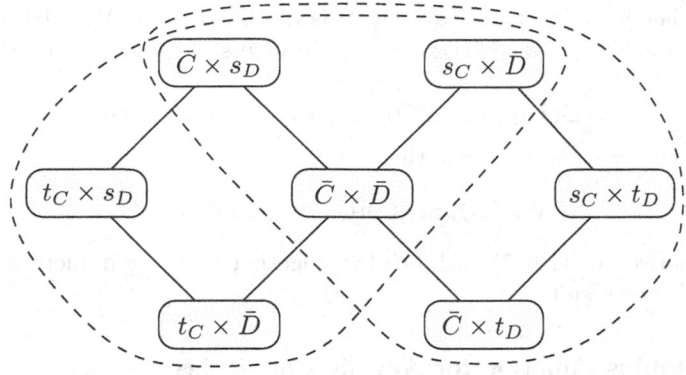

Fig. 10.18. Decomposition proving (10.15).

10.6 The Möbius Function

10.6.1 Möbius Function for Posets

First, we recall the classical facts about the Möbius function for posets. As mentioned in Section 10.4, for an arbitrary poset P, let $I(P)$ denote the set of the finite intervals of P. Define the function $\mu : I(P) \to \mathbb{C}$ as follows:

- $\mu(x,x) := 1$, for all $x \in P$;
- $\mu(x,y) := -\sum_{x \le z < y} \mu(x,z)$, for all $x < y$, $x, y \in P$.

The function μ is called the **Möbius function**. An alternative way to define μ is via the *incidence algebra* of P. All functions $f : I(P) \to \mathbb{C}$ form a $\mathbb{C}$-algebra, with pointwise addition, pointwise multiplication by complex numbers, and the following convolution product: for $f, g : I(P) \to \mathbb{C}$ we define

$$(f \circ g)(x,y) := \sum_{x \le z \le y} f(x,z)g(z,y).$$

Define special functions $e, \xi : I(P) \to \mathbb{C}$, by $\xi(x,y) := 1$, for all $x \le y$, and $e(x,x) = 1$, $e(x,y) = 0$, for $x \ne y$. The Möbius function can now be defined as the solution of the identity $\mu \circ \xi = e$.

Theorem 10.24. *For any finite poset P with a maximal and minimal elements, we have*

$$\mu(\hat{0}, \hat{1}) = \tilde{\chi}(\Delta(\bar{P})). \tag{10.16}$$

Proof. Clearly, (10.16) is true for $|P| \le 3$. We proceed by induction on $|P|$.

Let x be a maximal element of $\bar{P}$. Set $Q = P \setminus \{x\}$. By the definition of the Möbius function we have

$$\mu_P(\hat{0}, \hat{1}) = -\mu_P(\hat{0}, x) + \mu_Q(\hat{0}, \hat{1}). \tag{10.17}$$

On the other hand, topologically $\Delta(\bar{P})$ is obtained from $\Delta(\bar{Q})$ by attaching a cone with apex in x and base $\Delta(\hat{0}, x)$. In terms of Euler characteristics we obtain

$$\chi(\Delta(\bar{P})) = \chi(\Delta(\bar{Q})) + \chi(cone) - \chi(\Delta(\hat{0}, x)).$$

Since $\chi(cone) = 1$, we conclude that

$$\tilde{\chi}(\Delta(\bar{P})) = \tilde{\chi}(\Delta(\bar{Q})) - \tilde{\chi}(\Delta(\hat{0}, x)). \tag{10.18}$$

The comparison of (10.17) with (10.18), together with the induction assumption, yields the result. $\square$

10.6.2 Möbius Function for Acyclic Categories

The Möbius function for posets is in fact a special case of the notion of a Möbius function for acyclic categories.

Definition 10.25. *Let C be an acyclic category with a terminal object t. A function $\mu : \mathcal{O}(C) \to \mathbb{Z}$ is defined as follows:*

- *$\mu(t) = 1$;*
- *for $x \in \mathcal{O}(C)$, $x \neq t$, we set*

$$\mu(x) := -\sum_{m} \mu(\partial_{\bullet} m), \tag{10.19}$$

where the sum is taken over all nonidentity morphisms $m \in \mathcal{M}(C)$ such that $\partial^{\bullet} m = x$.

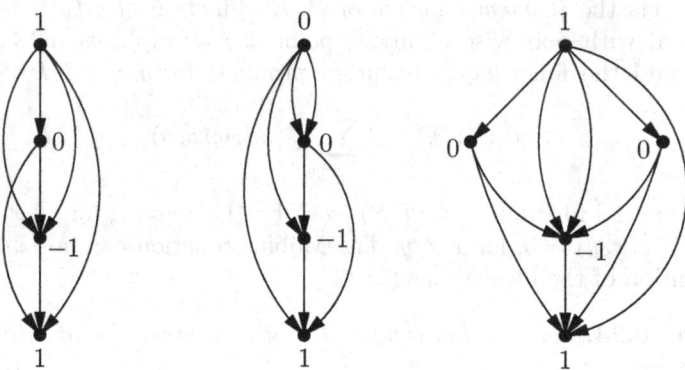

Fig. 10.19. Möbius functions for acyclic categories from Figure 10.1 with added terminal objects.

For future reference, we mention that moving all summands in (10.19) over to the left-hand side, we get

$$\sum_{m:\partial^\bullet m=x} \mu(\partial_\bullet m) = 0. \tag{10.20}$$

In analogy with Hall's theorem we have the following statement.

Theorem 10.26. *For any finite acyclic category C with an initial object s and a terminal object t, we have*

$$\mu(s) = \tilde{\chi}(\Delta(\bar{C})). \tag{10.21}$$

Proof. Choose a linear extension of C. We can see by induction on the position in this linear extension that for any $x \in \mathcal{O}(C)$ we have

$$\mu(x) = \sum_c (-1)^{l(c)}, \tag{10.22}$$

where the sum is taken over all composable morphism chains c of the type

$$x \xrightarrow{m_1} a_1 \xrightarrow{m_2} \cdots \xrightarrow{m_{l(c)-1}} a_{l(c)-1} \xrightarrow{m_{l(c)}} t,$$

where all morphisms are nonidentity, and with $l(c)$, the length of c, being the number of morphisms.

Indeed, equation (10.22) is clearly true for $x = t$, where we use the convention that an object of C is a composable morphism chain of length 0. The induction step follows immediately from equation (10.19), since the term $\mu(\partial_\bullet m)$ counts those composable morphism chains that start with m, and the minus sign is there to account for the fact that we have just added another morphism.

Since t is an initial object and s is a terminal one, substituting $x = s$ in equation (10.22), and recalling the definition of the nerve of an acyclic category, we obtain (10.21). $\square$

The notion of incidence algebra has its analogue for acyclic categories as well. As is to be expected, the role of intervals in a poset is taken over by morphisms of a category.

Definition 10.27. *Let C be an arbitrary finite acyclic category. The **incidence algebra** of C is the $\mathbb{C}$-algebra of all functions $\alpha : \mathcal{M}(C) \to \mathbb{C}$, with pointwise addition, pointwise multiplication by complex numbers, and with the following convolution product $\circ$: for $\alpha, \beta : \mathcal{M}(C) \to \mathbb{C}$, $m \in \mathcal{M}(C)$, we define*

$$(\alpha \circ \beta)(m) := \sum_{m_1, m_2} \alpha(m_1)\beta(m_2),$$

where the sum is taken over all pairs of morphisms $m_1, m_2 \in \mathcal{M}(C)$ such that $m_2 \circ m_1 = m$.

This convolution product is associative, since we have

$$((\alpha \circ \beta) \circ \gamma)(m) = (\alpha \circ (\beta \circ \gamma))(m) = \sum_{m_3 \circ m_2 \circ m_1 = m} \alpha(m_1)\beta(m_2)\gamma(m_3).$$

Define special functions $e, \xi : \mathcal{M}(C) \to \mathbb{C}$ by $\xi(m) := 1$, for all $m \in \mathcal{M}(C)$, $e(\mathrm{id}_x) = 1$, for all $x \in \mathcal{O}(C)$, and $e(m) = 0$, for all nonidentity morphisms $m \in \mathcal{M}(C)$. It is readily seen that e is a two-sided unit for this convolution, that is, $\alpha \circ e = e \circ \alpha = \alpha$, for any $\alpha : \mathcal{M}(C) \to \mathbb{C}$.

Proposition 10.28. *Let C be an arbitrary finite acyclic category. The equation $\mu \circ \xi = e$ has a unique solution, which is also called the **Möbius function**. This solution will automatically satisfy the equation $\xi \circ \mu = e$ as well.*

Proof. Used on identity morphisms, the equation $\mu \circ \xi = e$ implies that $\mu(\mathrm{id}_x) = 1$, for all $x \in \mathcal{O}(C)$. For nonidentity morphisms the equation $\mu \circ \xi = e$ translates to

$$\mu(m) = - \sum_{\substack{m_2 \circ m_1 = m \\ m_2 \neq \mathrm{id}}} \mu(m_1). \tag{10.23}$$

Let $n = |\mathcal{O}(C)|$, and let $L : \mathcal{O}(C) \to [n]$ be a linear extension bijective on elements; such an L exists by the discussion in Subsection 10.1.2. Order all morphisms of C in any way so that the difference $L(\partial^{\bullet} m) - L(\partial_{\bullet} m)$ does not decrease. Then (10.23) expresses $\mu(m)$ through the values of μ on morphisms that precede m in this order. This proves, by induction on the order L, that $\mu \circ \xi = e$ has a solution, and that it must be unique.

Analogously, the equation $\xi \circ \mu = e$ has a unique solution. If these two solutions are different, then denote them by μ_1 and μ_2, and consider the product $\mu_1 \circ \xi \circ \mu_2$. The following is a standard computation from, e.g., group theory:

$$\mu_2 = e \circ \mu_2 = (\mu_1 \circ \xi) \circ \mu_2 = \mu_1 \circ (\xi \circ \mu_2) = \mu_1 \circ e = \mu_1,$$

and it shows that $\mu_1 = \mu_2$. $\square$

Some examples of this Möbius function are shown in Figure 10.20. When x is an object our previously used shorthand notation from Definition 10.25 translates to $\mu(x) = \mu(m)$, where m is the unique morphism $m : x \to t$.

The set of all functions $f : \mathcal{O}(C) \to \mathbb{C}$ is the dual of the vector space whose basis elements are indexed by the set $\mathcal{O}(C)$. It is important to notice that the following equation describes the algebra representation of $I(C)$ on this vector space:

$$(\alpha f)(x) = \sum_{\partial^{\bullet} m = x} \alpha(m) f(\partial_{\bullet} m), \tag{10.24}$$

for all $x \in \mathcal{O}(C)$, $f : \mathcal{O}(C) \to \mathbb{C}$, and $\alpha : \mathcal{M}(C) \to \mathbb{C}$. To see that (10.24) actually defines an algebra representation, we need to prove that

$$ef = f, \tag{10.25}$$

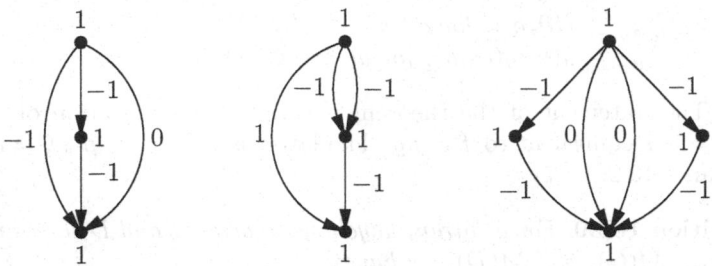

Fig. 10.20. Möbius functions on morphisms of acyclic categories from Figure 10.1.

for all $f : \mathcal{O}(C) \to \mathbb{C}$, and

$$\alpha(\beta f) = (\alpha \circ \beta)f, \tag{10.26}$$

for all $f : \mathcal{O}(C) \to \mathbb{C}$ and $\alpha, \beta : \mathcal{M}(C) \to \mathbb{C}$. Let $x \in \mathcal{O}(C)$. The following computation verifies (10.25):

$$(ef)(x) = \sum_{\partial^\bullet m = x} e(m)f(\partial_\bullet m) = f(\partial_\bullet \mathrm{id}_x) = f(x).$$

To see (10.26), we expand both of its sides. We have

$$(\alpha(\beta f))(x) = \sum_{\partial^\bullet m_1 = x} \alpha(m_1)(\beta f)(\partial_\bullet m_1)$$

$$= \sum_{\partial^\bullet m_1 = x} \left(\alpha(m_1) \sum_{\partial^\bullet m_2 = \partial_\bullet m_1} \beta(m_2)f(\partial_\bullet m_2) \right)$$

$$= \sum_{\substack{\partial^\bullet m_1 = x \\ \partial^\bullet m_2 = \partial_\bullet m_1}} \alpha(m_1)\beta(m_2)f(\partial_\bullet m_2),$$

and on the other hand,

$$((\alpha \circ \beta)f)(x) = \sum_{\partial^\bullet m = x} (\alpha \circ \beta)(m)f(\partial_\bullet m)$$

$$= \sum_{\partial^\bullet m = x} \left(\sum_{m_2 \circ m_1 = m} \alpha(m_1)\beta(m_2) \right) f(\partial_\bullet m)$$

$$= \sum_{\substack{\partial^\bullet m = x \\ m_2 \circ m_1 = m}} \alpha(m_1)\beta(m_2)f(\partial_\bullet m).$$

Comparing the two outcomes, we see that they are actually equal; hence (10.26) is verified.

Theorem 10.29. (Möbius inversion for acyclic categories).
Let $f, g : \mathcal{O}(C) \to \mathbb{C}$. *Then the following two statements are equivalent:*

- $g(x) = \sum_{\partial^\bullet m = x} f(\partial_\bullet m)$, for all $x \in \mathcal{O}(C)$;
- $f(x) = \sum_{\partial^\bullet m = x} \mu(m)g(\partial_\bullet m)$, for all $x \in \mathcal{O}(C)$.

Proof. The statement of the theorem is simply the translation of the fact that $\xi f = g$ is equivalent to $f = \mu g$. The latter is true since $\mu \circ \xi = e$, using (10.25) and (10.26). $\square$

Proposition 10.30. *For arbitrary acyclic categories C and D, and any morphisms $\alpha \in \mathcal{M}(C)$, $\beta \in \mathcal{M}(D)$, we have*

$$\mu_{C \times D}(\alpha, \beta) = \mu_C(\alpha)\mu_D(\beta). \tag{10.27}$$

Proof. We have

$$\sum_{(\alpha_2,\beta_2)\circ(\alpha_1,\beta_1)=(\alpha,\beta)} \mu_C(\alpha_1)\mu_D(\beta_1) = \sum_{\alpha_2\circ\alpha_1=\alpha} \mu_C(\alpha_1) \cdot \sum_{\beta_2\circ\beta_1=\beta} \mu_C(\beta_1)$$

$$= \delta_{\alpha=\mathrm{id}} \cdot \delta_{\beta=\mathrm{id}} = \delta_{(\alpha,\beta)=\mathrm{id}},$$

where δ_S is the Kronecker delta function: it is 1 if S is true, and 0 if S is false; for example, $\delta_{\alpha=\mathrm{id}}$ is 1 if and only if α is an identity morphism. $\square$

10.7 Bibliographic Notes

Most of the material of this chapter is new and has not appeared elsewhere. The interested reader may benefit from a standard text in category theory, such as [McL98], as an additional source of information. In the special case of posets instead of acyclic categories, the standard references are [Sta97] and [Bj96].

Originally, the Möbius function was introduced for integers. The textbook [Sta97] contains a great deal of material about the Möbius function for posets, including Theorem 10.24, which is also called Hall's Theorem.

11

Discrete Morse Theory

11.1 Discrete Morse Theory for Posets

When the set of cells of a CW complex is given by means of a combinatorial enumeration, and the cell attachment maps are not too complicated, for instance if the CW complex in question is regular, it is natural to attempt to use the standard notion of *cellular collapse* to simplify the considered topological space, while preserving its homotopy type.

Since the presentation of the cell complex is combinatorial, once this course of action is taken, it becomes imperative to have a language as well as an appropriate combinatorial machinery for dealing with allowed sequences of collapses. Accordingly, we shall first investigate what happens on the purely combinatorial level of posets, before proceeding to drawing topological conclusions and looking at applications.

11.1.1 Acyclic Matchings in Hasse Diagrams of Posets

Recall from Definition 6.13 that for a generalized simplicial complex Δ, *a simplicial collapse* is simply a removal of interiors of two simplices σ and τ such that

- $\dim \sigma = \dim \tau + 1$;
- the only simplex containing σ is σ itself;
- the only simplices containing τ are σ and τ.

Sometimes such a collapse is called an *elementary collapse*. Note that a simplicial collapse is possible if and only if there exists a simplex τ whose link in Δ consists of a single vertex; the simplex σ is then given by the span of τ and v. For a general CW complex one has to take care of some additional technicalities; see Definition 11.12.

In any case, we see that the combinatorial encoding of a set of collapses is best provided by a matching consisting of a collection of pairs of cells (τ, σ)

such that σ contains τ, and $\dim \sigma = \dim \tau + 1$. Clearly, not every matching of this type can be turned into a *sequence of collapses*. For instance, no allowed sequence of collapses for the simplicial complex in Figure 11.1 can be found in the matching depicted on the right of that figure.

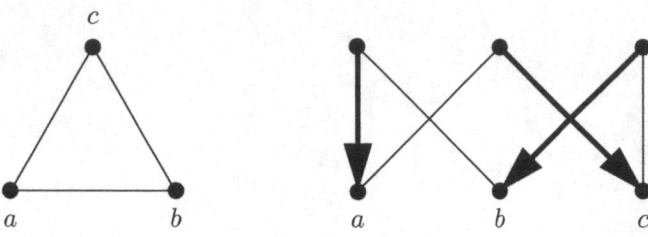

Fig. 11.1. A hollow triangle and a matching in its face poset.

It is easy to see what goes wrong in this example: the prospective collapses are all "hooked up" with each other in a cyclic pattern, which we are unable to break by doing only single collapses. This simple observation leads to the following formalization.

Definition 11.1.
*(1) A **partial matching**[1] in a poset P is a partial matching in the underlying graph of the Hasse diagram of P, i.e., it is a subset $M \subseteq P \times P$ such that*

- $(a, b) \in M$ *implies* $b \succ a$;
- *each $a \in P$ belongs to at most one element in M.*

When $(a, b) \in M$, we write $a = d(b)$ and $b = u(a)$.
*(2) A partial matching on P is called **acyclic** if there does not exist a cycle*

$$b_1 \succ d(b_1) \prec b_2 \succ d(b_2) \prec \cdots \prec b_n \succ d(b_n) \prec b_1, \qquad (11.1)$$

with $n \geq 2$, and all $b_i \in P$ being distinct.

A popular way to reformulate condition (2) of Definition 11.1 is the following. Given a poset P, we can orient all edges in the Hasse diagram of P so that they point from the larger element to the smaller one. After that, given a partial matching M, change the orientation of the edges in M to the opposite one. The condition in question now says that the oriented graph obtained in this fashion has no cycles.

We see that Definition 11.1 allows a more general situation than just the collapses that we described above. This makes our situation quite different from the simple homotopy theory considered in Section 6.5. For example, a partial matching consisting of a single pair of simplices $b \succ a$ is always

---

[1] Also called *discrete vector field*.

acyclic. The reader is invited to intuitively think about such pairs as *internal collapses*. The idea is to remove all the matched elements in some appropriate order, so that the homotopy type of the underlying space is kept intact. We call the unmatched elements, i.e., the elements that will remain, *critical*, and denote the set of critical elements by $C(P, M)$.

The next theorem is the crucial combinatorial fact pertaining to matchings in Hasse diagrams of posets. It characterizes acyclic matchings by means of linear extensions.

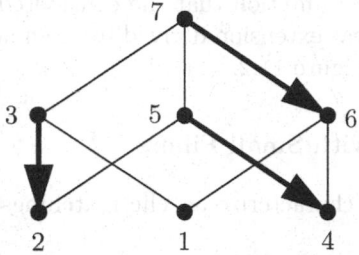

Fig. 11.2. Acyclic matching and the corresponding linear extension.

Theorem 11.2. (Acyclic matchings via linear extensions)
A partial matching on P is acyclic if and only if there exists a linear extension L of P such that the elements a and $u(a)$ follow consecutively in L.

Proof. Assume first that we have a linear extension L satisfying this property, and that at the same time, we have a cycle as in (11.1). Set $a_i = d(b_i)$, for $i = 1, \ldots, n$. Then

$$b_{i+1} \succ a_i \Rightarrow a_i <_L b_{i+1} \Rightarrow a_i <_L a_{i+1}$$

(since a_{i+1}, b_{i+1} follow consecutively in L). Thus $a_n >_L a_{n-1} >_L \cdots >_L a_1 >_L a_0 = a_n$, yielding a contradiction.

Assume now that we are given an acyclic matching, and let us define L inductively. Let Q denote the set of elements that are already ordered in L. We start with $Q = \emptyset$. Let W denote the set of minimal elements in $P \setminus Q$. At each step we have one of the following cases.

Case 1. *One of the elements c in W is critical.*

In this case, we simply add c to the order L as the largest element, and proceed with $Q \cup \{c\}$.

Case 2. *All elements in W are matched.*

Consider the subgraph of the underlying graph of the Hasse diagram of $P \setminus Q$ induced by $W \cup u(W)$. Orient its edges as described above, i.e., they should point from the larger element to the smaller one in all cases, except when these two elements are matched, in which case the edge should point from the smaller element to the larger one. Call this oriented graph G.

If there exists an element $a \in W$ such that the only element in $W \cup u(W)$ that is smaller than $u(a)$ is a itself, then we can add elements a and $u(a)$ on top of L and proceed with $Q \cup \{a, u(a)\}$. Otherwise, we see that the outdegree of $u(a)$ in G is positive, for each $a \in W$. On the other hand, the outdegrees in G of all $a \in W$ are equal to 1. Since therefore outdegrees of all vertices in the oriented graph G are positive, we conclude that G must have a cycle, which clearly contradicts the assumption that the considered matching is acyclic.

An example of a linear extension derived from an acyclic matching by this procedure is shown in Figure 11.2.   □

11.1.2 Poset Maps with Small Fibers

Next, we would like to characterize acyclic matchings by means of a special class of poset maps.

Definition 11.3. *Given two posets P and Q, a poset map $\varphi : P \to Q$ is said to have* **small fibers** *if for any $q \in Q$, the fiber $\varphi^{-1}(q)$ is either empty or consists of a single element or consists of two comparable elements.*

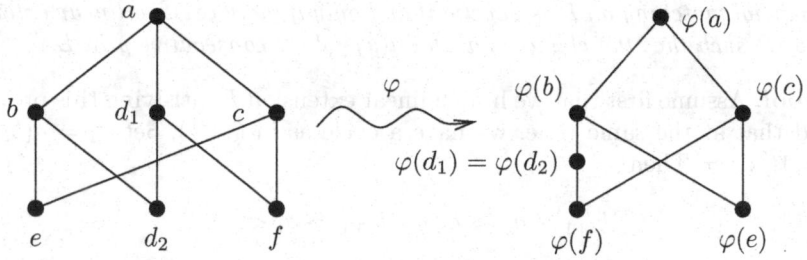

Fig. 11.3. A poset map with small fibers.

An example is shown in Figure 11.3. We remark that since φ is a poset map, if for some $q \in Q$ the fiber $\varphi^{-1}(q)$ consists of two comparable elements, then one of these two elements must actually cover the other one. Therefore, to any given poset map with small fibers $\varphi : P \to Q$ we can associate a partial matching $M(\varphi)$ consisting of all fibers of cardinality 2.

Theorem 11.4. *(Acyclic matchings via poset maps with small fibers) For any poset map with small fibers $\varphi : P \to Q$, the partial matching $M(\varphi)$ is acyclic. Conversely, any acyclic matching on P can be represented as $M(\varphi)$ for some poset map with small fibers φ.*

Proof. The fact that $\varphi : P \to Q$ is a poset map implies that the induced matching $M(\varphi)$ is acyclic: for if it were not, there would exist a cycle as in (11.1), and φ would be mapping this cycle to a set of distinct elements $q_1 > q_2 > \cdots > q_t > q_1$ of Q, for some t, yielding a contradiction.

On the other hand, by Theorem 11.2, for any acyclic matching on P there exists a linear extension L of P such that the elements a and $u(a)$ follow consecutively in L. Gluing a with $u(a)$ in this order yields a poset map with small fibers from P to a chain. $\square$

In the proof of Theorem 11.4 we have actually constructed a poset map with small fibers into a chain. These maps are especially important, and we give them a separate name.

Definition 11.5. *A poset map with small fibers $\varphi : P \to Q$ is called a* **collapsing order** *if φ is surjective as a set map, and Q is a chain.*

Given an acyclic matching M, we say that a collapsing order φ is a *collapsing order for M* if it satisfies $M(\varphi) = M$. The etymology of this terminology is fairly clear: the chain Q gives us the order in which it is allowed to perform the prescribed collapses.

11.1.3 Universal Object Associated to an Acyclic Matching

It turns out that for any poset P and any acyclic matching on P, there exists a universal object: a poset whose linear extensions enumerate all allowed collapsing orders.

Definition 11.6. *Let P be a poset, and let M be an acyclic matching on P. We define $U(P, M)$ to be the poset whose set of elements is $M \cup C(P, M)$, and whose partial order is the transitive closure of the elementary relations given by $S_1 \leq_U S_2$, for $S_1, S_2 \in U(P, M)$ if and only if $x \leq y$, for some $x \in S_1$, $y \in S_2$.*

Note that in the formulation of Definition 11.6 we think of elements of M as subsets of P of cardinality 2, while we think of elements of $C(P, M)$ as subsets of P of cardinality 1. One can loosely say that Definition 11.6 states that $U(P, M)$ is obtained from P by gluing each matched pair together to form a single element, with the new partial order induced by the partial order of P in a natural way. See Figure 11.4 for an example.

Of course, the first natural question is whether this new order is actually well-defined. The next proposition answers that question and also explains in what sense $U(P, M)$ is a universal object.

Theorem 11.7. (Universality of $U(P, M)$)
For any poset P and for any acyclic matching M on P, we have:

(1) the partial order on $U(P, M)$ is well-defined;

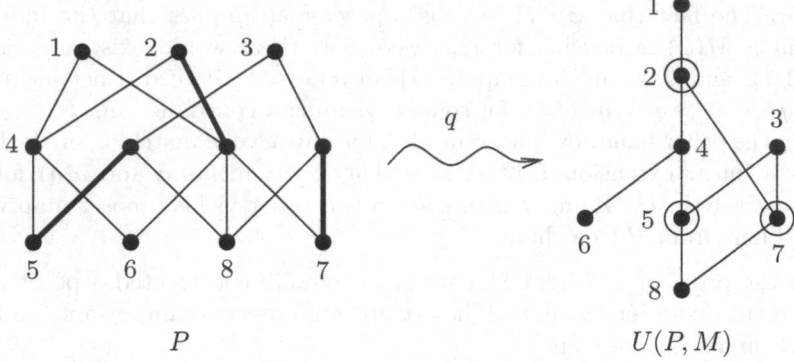

$$P \qquad\qquad\qquad U(P,M)$$

Fig. 11.4. A universal poset associated with an acyclic matching.

(2) *the induced quotient map* $q : P \to U(P,M)$ *is a poset map with small fibers;*

(3) *the linear extensions of* $U(P,M)$ *are in 1-to-1 correspondence with collapsing orders for* M; *this correspondence is given by the composition of the quotient map* q *with a linear extension map.*

Proof. To prove (1) we need to check the three axioms of partial orders. The reflexivity is obvious, and the transitivity is automatic, since we have taken the transitive closure. The only property that needs to be proved is antisymmetry. So assume that it does not hold, and take $X, Y \in U(P,M)$ such that $X \leq_U Y$, $Y \leq_U X$, and $X \neq Y$. Choose a sequence

$$X <_U S_1 <_U \cdots <_U S_p <_U Y <_U T_1 <_U \cdots <_U T_q <_U X, \qquad (11.2)$$

with the minimal possible p and q. Since p and q are chosen to be minimal, all the sets $S_1, \ldots, S_p$ and $T_1, \ldots, T_q$ must have cardinality 2.

Let us first deal with the case $p = q = 0$ separately. If $|X| = |Y| = 1$, say $X = \{x\}$, $Y = \{y\}$, then we have $x \leq y$ and $y \leq x$, hence $x = y$, since P itself is a poset. If $|X| = 1$ and $|Y| = 2$, say $X = \{x\}$, $Y = (a, b)$, then $b > x$ and $x > a$, since $x \neq b$, $x \neq a$. This gives $b > x > a$, yielding a contradiction to the assumption that b covers a. By symmetry of (11.2) this argument covers the case $|X| = 2$, $|Y| = 1$ as well, so we can assume that $|X| = |Y| = 2$. In this case $X \leq_U Y \leq_U X$ is a cycle, contradicting the assumption that our matching is acyclic.

From now on, we have $p + q \geq 1$. Assume first that $|X| = |Y| = 1$, say $X = \{x\}$, $Y = \{y\}$. If $p = 0$ and $q = 1$, let $T_1 = (a, b)$, with $b \succ a$. On the one hand, we have $x \leq y$; on the other, $b \geq y$, $x \geq a$. Combining, we get $b \geq y \geq x \geq a$, implying $x = y$, since b covers a. Again by symmetry this takes care of the case $p = 1$ and $q = 0$ as well.

Without loss of generality we may now assume that either $p + q \geq 2$, or $|Y| = 2$ and $p + q \geq 1$. In the first case,

$$S_1 <_U \cdots <_U S_p <_U T_1 <_U \cdots <_U T_q$$

yields a cycle, contradicting the assumption that our matching was acyclic; in the second case such a cycle is given by

$$S_1 <_U \cdots <_U S_p <_U Y <_U T_1 <_U \cdots <_U T_q.$$

Part (2) is straightforward. If $x < y$ in P and $x \in X$, $y \in Y$, for $X, Y \in U(P, M)$, then $X \leq Y$ in $U(P, M)$ by the definition of the partial order on $U(P, M)$, though we may actually get equality. So q is a poset map, and the fibers are small, since we have just proved that $X \leq_U Y$ together with $Y \leq_U X$ implies $X = Y$.

Let us now prove (3). Given a linear extension $l : U(P, M) \to Q$, the composition $l \circ q : P \to Q$ is of course a poset map with small fibers, and it is surjective since both l and q are surjective.

Conversely, assume that $\varphi : P \to Q$ is a collapsing order for M. Since φ is surjective, $\varphi^{-1}(x)$ is nonempty for every $x \in Q$; in fact, we have a bijection between sets $\varphi^{-1}(x)$, for $x \in Q$, and elements of $U(P, M)$. To factor φ through $U(P, M)$, we set $l(q(\varphi^{-1}(x))) := x$, for each $x \in Q$. We have $l \circ q = \varphi$ as set maps. To see that the map l is order-preserving, notice that an elementary relation $S \geq T$, for $S, T \in U(P, M)$, implies that there exist $x \in S$, $y \in T$ such that $x \geq y$, which in turn implies $\varphi(x) \geq \varphi(y)$, since φ is order-preserving, and notice furthermore that all relations $S \geq T$ are just the transitive closures of the elementary ones.

Thus, we get the desired 1-to-1 correspondence between linear extensions of $U(P, M)$, and collapsing orders for M. $\square$

11.1.4 Poset Fibrations and the Patchwork Theorem

Beyond the encoding of all allowed collapsing orders as the set of linear extensions of the universal object $U(P, M)$, viewing the posets with small fibers as the central notion of the combinatorial part of discrete Morse theory is also invaluable for the structural explanation of a standard way to construct acyclic matchings as unions of acyclic matchings on fibers of a poset map.

The following construction generalizes Definition 10.7 of the stack of acyclic categories. Since we will need this only for posets, we satisfy ourselves here with formulating the special case. The generalization to acyclic categories is straightforward.

Definition 11.8. *A **poset fibration** is a pair $(B, \mathcal{F})$, where*

- *B is a poset, thought of as the **base** of the fibration;*
- *$\mathcal{F} = \{F_x\}_{x \in B}$ is a collection of posets, indexed by the elements of B, thought of as individual **fibers**.*

Associated to such a fibration we have a poset $E(B, \mathcal{F})$, defined as the union $\cup_{x \in B} F_x$, with the order relation given by $\alpha \geq \beta$ if either $\alpha, \beta \in F_x$, and $\alpha \geq \beta$ in F_x, for some $x \in B$, or $\alpha \in F_x$, $\beta \in F_y$, and $x > y$ in B. This is the *total space*.

Furthermore, we have a poset map $p : E(B, \mathcal{F}) \to B$ defined by $p(\alpha) := x$ if $\alpha \in F_x$. In particular, we have $p^{-1}(x) = F_x$, for all $x \in B$. This is the structural *projection map* of the total space to the base space, whose preimages are the fibers.

The notion of poset fibrations satisfies the following universality property.

Theorem 11.9. (Decomposition theorem)
For an arbitrary poset fibration $(B, \mathcal{F})$, where $\mathcal{F} = \{F_x\}_{x \in B}$, and an arbitrary poset P, there is a 1-to-1 correspondence between

- *poset maps $\varphi : P \to E(B, \mathcal{F})$;*
- *pairs $(\psi, \{g_x\}_{x \in B})$, where ψ and each g_x's are poset maps $\psi : P \to B$ and $g_x : \psi^{-1}(x) \to F_x$, for each $x \in B$.*

Under this bijection, the fibers of φ are the same as the fibers of the maps g_x.

Proof. One direction of this bijection is trivial: given a poset map $\varphi : P \to E(B, \mathcal{F})$, we obtain the poset map $\psi : P \to B$ by composing φ with the structural projection map $p : E(B, \mathcal{F}) \to B$, and we obtain the poset maps g_x by taking the appropriate restrictions of the map φ.

In the opposite direction, assume that we have a poset map $\psi : P \to B$ and a collection of poset maps $g_x : \psi^{-1}(x) \to F_x$, for all $x \in B$. Define $\varphi : P \to E(B, \mathcal{F})$ by taking the value of the appropriate fiber map:

$$\psi(\alpha) := g_{\varphi(\alpha)}(\alpha),$$

for all $\alpha \in P$. Let us see that this defines a poset map. For $\alpha > \beta$, $\alpha, \beta \in P$, we have $\varphi(\alpha) \geq \varphi(\beta)$, since φ is a poset map. If $\varphi(\alpha) = \varphi(\beta)$, then $g_{\varphi(\alpha)}(\alpha) \geq g_{\varphi(\beta)}(\beta)$, since $g_{\varphi(\alpha)} (= g_{\varphi(\beta)})$ is a poset map. Otherwise, we have $\varphi(\alpha) > \varphi(\beta)$, and hence $g_{\varphi(\alpha)}(\alpha) > g_{\varphi(\beta)}(\beta)$, by the definition of the partial order on the total space $E(B, \mathcal{F})$. $\square$

The decomposition theorem 11.9, is often used as a rationale to construct an acyclic matching on a poset P in several steps: first map P to some other poset Q, then construct acyclic matchings on the fibers of this map. By the observation above, these acyclic matchings will "patch together" to form an acyclic matching for the whole poset. See Figure 11.5 for an example. For future reference, we summarize this observation in the next theorem.

Theorem 11.10. (Patchwork theorem)
Assume that $\varphi : P \to Q$ is an order-preserving map, and assume that we have acyclic matchings on subposets $\varphi^{-1}(q)$, for all $q \in Q$. Then the union of these matchings is itself an acyclic matching on P.

Proof. The role of the base space here is played by the poset Q, and the fiber maps g_q are given by the acyclic matchings on the subposets $\varphi^{-1}(q)$. The decomposition theorem tells us that there exists a poset map from P to the total space of the corresponding poset fibration, and that the fibers of this map are the same as the fibers of the fiber maps g_q. Since the latter are given by acyclic matchings, we conclude that we have a poset map from P with small fibers that corresponds precisely to the patching of acyclic matchings on the subposets $\varphi^{-1}(q)$, for $q \in Q$. $\quad\square$

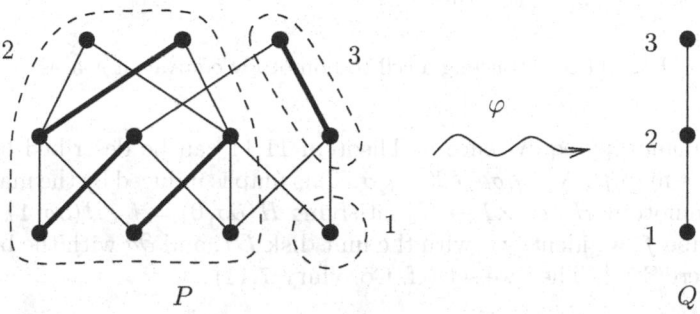

Fig. 11.5. Acyclic matching composed of acyclic matchings on fibers.

We conclude our discussion of poset maps with small fibers by mentioning that this point of view yields a rich class of generalizations. Indeed, any choice of the set of allowed fibers will yield a combinatorial theory that could be interesting to study. One could, for instance, allow any Boolean algebra as a fiber. This would correspond to the theory of all collapses, not just the elementary ones, which we get when considering the small fibers. One can take any other infinite family of posets. One prominent family is that of partition lattices $\{\varPi_n\}_{n=1}^{\infty}$. What happens if we consider all poset maps with partition lattices as fibers?

11.2 Discrete Morse Theory for CW Complexes

11.2.1 Attaching Cells to Homotopy Equivalent Spaces

We shall use the following standard fact of algebraic topology, which we state here with only a sketch of a proof.

Theorem 11.11. *Assume that X_1 and X_1 are two homotopy equivalent topological spaces, and let $h : X_1 \rightarrow X_2$ be some homotopy equivalence. Let σ be a cell with attachment maps $f_1 : \partial\sigma \rightarrow X_1$ and $f_2 : \partial\sigma \rightarrow X_2$ such that $h \circ f_1$ is homotopic to f_2; see Figure 11.6.*

Under these conditions, the space $X_1 \cup_{f_1} \sigma$ is homotopy equivalent to the space $X_2 \cup_{f_2} \sigma$.

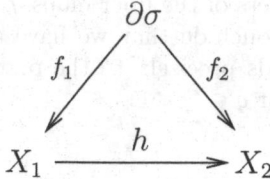

Fig. 11.6. Attaching a cell to homotopy equivalent spaces.

The homotopy equivalence in Theorem 11.11 can be described by giving an explicit map $f : X_1 \cup_{f_1} \sigma \rightarrow X_2 \cup_{f_2} \sigma$. This map is induced by the map h, and by the homotopy $H : \partial\sigma \times I \rightarrow X_2$ satisfying $H(\partial\sigma, 0) = f_2$, $H(\partial\sigma, 1) = h \circ f_1$. To describe f, we identify σ with the unit disk D^n, and $\partial\sigma$ with the bounding unit sphere $\mathbb{S}^{n-1}$. Then we set (cf. Corollary 7.12)

$$f(x) := h(x), \quad \text{for } x \in X_1,$$

$$f(t\mathbf{v}) := \begin{cases} 2t\mathbf{v}, & \text{for } 0 \leq t \leq 1/2, \quad \mathbf{v} \in \mathbb{S}^{n-1}, \\ H(\mathbf{v}, 2t-1), & \text{for } 1/2 \leq t \leq 1, \quad \mathbf{v} \in \mathbb{S}^{n-1}. \end{cases}$$

The following two special cases of Theorem 11.11 are often distinguished as being of particular importance:

Case 1. $X_1 = X_2$, and $h = \text{id}_{X_1}$.

This is a special case of Proposition 7.11, which is used, for example, in justifying the fact that the homotopy type of a CW complex is uniquely determined even if the cell attachment maps are given only up to homotopy; see Figure 11.7.

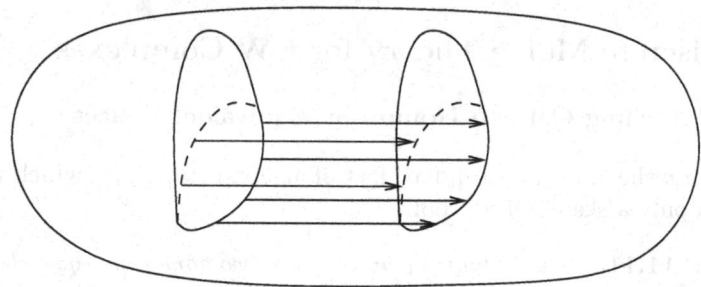

Fig. 11.7. Changing the attachment map by a homotopy.

Case 2. $h \circ f_1 = f_2$.

In fact, if $h \circ f_1 = f_2$, then it is much simpler to describe the homotopy equivalence map $f : X_1 \cup_{f_1} \sigma \to X_2 \cup_{f_2} \sigma$. We may simply set

$$f(x) := \begin{cases} h(x), & \text{for } x \in X_1; \\ x, & \text{for } x \in \text{Int } \sigma. \end{cases} \tag{11.3}$$

11.2.2 The Main Theorem of Discrete Morse Theory for CW Complexes

Intuitively, a cellular collapse is a strong deformation retract that pushes the interior of a maximal cell in, using one of its free boundary cells as the starting point, much like compressing a body made of clay. The cellular collapses can be defined for arbitrary CW complexes.

Definition 11.12. *Let X be a topological space and let Y be a subspace of X. We say that Y is obtained from X by an **elementary collapse** if X can be represented as a result of attaching a ball B^n to Y along one of the hemispheres. In other words, if there exists a map $\varphi : B_-^{n-1} \to Y$ such that $X = Y \cup_\varphi B^n$, where B_-^{n-1} denotes one of the closed hemispheres on the boundary of B^n.*

*Such a collapse is called **cellular** if additionally X is a CW complex, and X is a CW complex obtained from Y by attaching two cells: B_+^{n-1} (this is the opposite hemisphere of B_-^{n-1}) and B^n, with φ inducing the necessary attaching maps as above.*

The simplicial collapse defined in Section 6.4 is a special case of Definition 11.12. We are now ready to formulate the central result of this section. For technical convenience, we restrict ourselves to considering cellular collapses in the setting of polyhedral complexes only.

Theorem 11.13.
(Main theorem of discrete Morse theory for CW complexes)
Let Δ be a polyhedral complex, and let M be an acyclic matching on $\mathcal{F}(\Delta) \setminus \{\hat{0}\}$. Let c_i denote the number of critical i-dimensional cells of Δ.

(a) If the critical cells form a subcomplex Δ_c of Δ, then there exists a sequence of cellular collapses leading from Δ to Δ_c.

(b) In general, the space Δ is homotopy equivalent to Δ_c, where Δ_c is a CW complex with c_i cells in dimension i.

(c) There is a natural indexing of cells of Δ_c with the critical cells of Δ such that for any two cells σ and τ of Δ_c satisfying $\dim \sigma = \dim \tau + 1$, the incidence number $[\tau : \sigma]$ is given by

$$[\tau : \sigma] = \sum_c w(c). \tag{11.4}$$

Here the sum is taken over all alternating paths c connecting σ with τ, i.e., over all sequences $c = (\sigma, a_1, u(a_1), \ldots, a_t, u(a_t), \tau)$ such that $\sigma \succ a_1$, $u(a_t) \succ \tau$, and $u(a_i) \succ a_{i+1}$, for $i = 1, \ldots, a_{t-1}$. For such an alternating path, the quantity $w(c)$ is defined by

$$w(c) := (-1)^t [a_1 : \sigma][\tau : u(a_t)] \prod_{i=1}^{t} [a_i : u(a_i)] \prod_{i=1}^{t-1} [a_{i+1} : u(a_i)], \quad (11.5)$$

where the incidence numbers in the right-hand side are taken in the complex Δ.

Remark 11.14. The converse of Theorem 11.13(a) is clearly true in the following sense: if Δ_c is a subcomplex of Δ and if there exists a sequence of collapses from Δ to Δ_c, then the matching on the cells of $\Delta \setminus \Delta_c$ induced by this sequence of collapses is acyclic. In particular, a polyhedral complex Δ is collapsible if and only if the poset $\mathcal{F}(\Delta) \setminus \{\hat{0}\}$ allows a complete acyclic matching.

Proof of Theorem 11.13. Take the linear extension L satisfying the conditions of Theorem 11.2.

Proof of (a). Clearly, the linear extension can be chosen so that all the critical cells come first. Hence, if read in decreasing order, L gives a sequence of cellular collapses leading from Δ to Δ_c.

Proof of (b). We perform induction on the cardinality of $\mathcal{F}(\Delta)$. If $|\mathcal{F}(\Delta)| = 1$, the statement is clear. For the induction step, let σ be the last cell in L.

Case 1. *The cell σ is critical.*

Let $\widetilde{\Delta} = \Delta \setminus \mathrm{Int}\,\sigma$, and let $\varphi : \partial\sigma \to \widetilde{\Delta}$ be the attaching map of σ in Δ. The matching M restricted to $\widetilde{\Delta}$ is again acyclic, and the critical cells are the same, with σ missing. Hence by induction, there exist a CW complex $\widetilde{\Delta}_c$ and a homotopy equivalence $h : \widetilde{\Delta} \to \widetilde{\Delta}_c$.

Consider the composition attaching map $h \circ \varphi : \partial\sigma \to \widetilde{\Delta}_c$; see Figure 11.8. By Theorem 11.11, we conclude that $\widetilde{\Delta} \cup_\varphi \sigma \simeq \widetilde{\Delta}_c \cup_{h\circ\varphi} \sigma$. Note that $\Delta = \widetilde{\Delta} \cup_\varphi \sigma$. The theorem follows by induction if we set $\Delta_c := \widetilde{\Delta}_c \cup_{h\circ\varphi} \sigma$. The new homotopy equivalence map is given by equation (11.3).

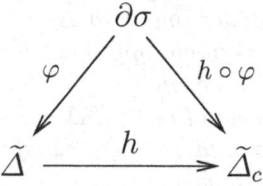

Fig. 11.8. Attaching a critical cell.

Case 2. *The cell σ is not critical.*

In this case we must have $(d(\sigma), \sigma) \in M$. Note that $d(\sigma)$ is maximal in $\mathcal{F}(\Delta) \setminus \{\sigma\}$, and let $\widetilde{\Delta} = \Delta \setminus (\text{Int } \sigma \cup \text{Int } d(\sigma))$.

Clearly, removing the pair $(d(\sigma), \sigma)$ is a cellular collapse; in particular, there exists a homotopy equivalence $f : \Delta \to \widetilde{\Delta}$. On the other hand, by the induction assumption, there exist a CW complex $\widetilde{\Delta}_c$ with c_i i-dimensional cells and a homotopy equivalence $\widetilde{f} : \widetilde{\Delta} \to \widetilde{\Delta}_c$. Hence, setting $\Delta_c := \widetilde{\Delta}_c$, we have obtained the desired homotopy equivalence $\widetilde{f} \circ f : \Delta \to \Delta_c$.

Proof of (c). We would like to give an elementary geometric argument. Let σ be some critical cell of Δ of dimension n. Initially, σ was attached along its boundary sphere, but after all the internal collapses, the attachment became more intricate. We would like to envision the attaching map as "a map of the world" drawn on the boundary sphere $\partial\sigma$. When the attaching map changes, we "redraw" this map, usually only locally.

Recall that the collapses can be performed so that the dimension of the collapsing pairs does not increase (the dimension is measured by the one of the two cells that has higher dimension). This means that first all collapses of dimension n can be performed, then all collapses of dimension $n - 1$.

We think of this collapsing as a dynamic procedure, and we start by tracing the changes of the attachment map of σ when the collapses of dimension n are executed. Let $(a, u(a))$ be such a collapse. If a was not in the image of the attaching map of σ at this point, then this collapse does not alter the attaching map. If a is in the image of the attaching map of σ at this point, then this collapse alters the attaching map (the map of the world on $\partial\sigma$) as follows: a gets replaced with $\partial u(a) \setminus \text{Int } a$. In a polyhedral complex this says that a gets replaced with the Schlegel diagram of $u(a)$ with respect to a; see Figure 11.9 for an example. This process will continue until all the collapses of dimension n are done.

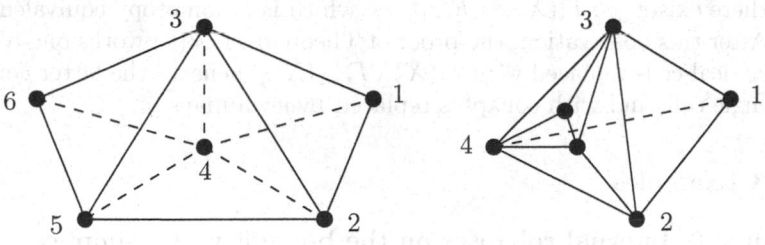

Fig. 11.9. Internal collapses as boundary subdivisions: the result of collapsing $(234, 2345)$ and $(345, 3456)$.

Once it is finished, the only cells of dimension $n-1$ that appear in the image of the attaching map of σ are the critical ones and those that are matched to the cells of dimension $n - 2$. The execution of collapses of dimension $n - 1$, which follows after that, has a simple effect on the latter ones: they are being internally collapsed, leaving no contribution to the incidence numbers.

This means that the only thing that we need to understand is how often and with which orientations the critical cells of dimension $n - 1$ will appear on the boundary sphere $\partial\sigma$. It follows from our iterative procedure above that appearances of a given critical cell τ are in one-to-one correspondence with the alternating paths. Indeed, every replacement of a with $\partial u(a) \setminus \text{Int}\, a$ corresponds to prolonging the alternating path with the edge up $(a, u(a))$, and then extending it with all possible edges down $(b, u(a))$.

The correctness of (11.4) follows from the following observation, which allows us to trace the evolution of incidence numbers of the cells on $\partial\sigma$. When the cell a with the incidence number ϵ gets replaced by the cells $\partial u(a) \setminus \text{Int}\, a$, each such cell b gets the incidence number $-\epsilon[a : u(a)][b : u(a)]$. $\square$

It is easy to see that the proof of Theorem 11.13(b) actually works in greater generality: one can take arbitrary CW complexes, at the same time replacing cellular collapses by arbitrary homotopy equivalences. More precisely, we have the following result.

Theorem 11.15. *Let X be a CW complex, and let*

$$F_0(X) \subset F_1(X) \subset \cdots \subset F_t(X) = X$$

be a CW filtration of X such that the subcomplex $F_0(X)$ is just a single vertex, and such that for all $i = 1, \ldots, t$, either $F_i(X) \setminus F_{i-1}(X)$ consists of a single cell, or the inclusion map $f_i : F_{i-1}(X) \hookrightarrow F_i(X)$ is a homotopy equivalence.

Then X is homotopy equivalent to a CW complex whose cells are in dimension-preserving bijection with the cells of X, which appear as $F_i(X) \setminus F_{i-1}(X)$.

Proof. If the inclusion map $f_i : F_{i-1}(X) \hookrightarrow F_i(X)$ is a homotopy equivalence, then there exists $g_i : F_i(X) \to F_{i-1}(X)$, which is a homotopy equivalence as well. After this observation, the proof of Theorem 11.13(b) works one-to-one, with critical cells replaced with $F_i(X) \setminus F_{i-1}(X)$ whenever the latter consists of a single cell, and with collapses replaced by such maps g_i. $\square$

11.2.3 Examples

Example 0: Internal collapses on the boundary of a simplex

Our first example is rather simple. Let Δ be the boundary of an n-dimensional simplex. We see that $\mathcal{F}(\Delta) \setminus \{\hat{0}\} = \bar{\mathcal{B}}_{n+1}$. Consider the following matching M on $\bar{\mathcal{B}}_{n+1}$: $(S, S \cup \{1\}) \in M$ for all $S \subset \{2, \ldots, n+1\}$. Clearly, this is an acyclic matching, and the only critical simplices are $\sigma = \{1\}$ and $\sigma = \{2, \ldots, n+1\}$. Therefore $c_0 = c_{n-1} = 1$, whereas $c_1 = \cdots = c_{n-2} = 0$. It follows that $\Delta \simeq \mathbb{S}^{n-1}$, as is of course expected.

Example 1: The independence complexes of strings and cycles

Our first real application is concerned with the independence complexes of graphs, which were defined in Subsection 9.1.1. Recall that for an arbitrary integer $n \geq 1$, we let L_n denote the graph consisting of n vertices and $n - 1$ edges that connect these vertices so as to form a string.

Proposition 11.16. *For any $n \geq 1$, we have*

$$\mathrm{Ind}(L_n) \simeq \begin{cases} \mathbb{S}^{k-1}, & \text{if } n = 3k; \\ \text{pt}, & \text{if } n = 3k + 1; \\ \mathbb{S}^k, & \text{if } n = 3k + 2. \end{cases}$$

Proof. Assume that the vertices of L_n are labeled 1 through n in the same sequence as they occur along the string. Let k denote the maximal integer such that $3k \leq n$. Furthermore, let C be a chain with $k + 1$ elements labeled as follows:

$$c_3 > c_6 > \cdots > c_{3k} > c_r.$$

We define a map $\varphi : \mathcal{F}(\mathrm{Ind}\,(L_n)) \to C$ by the following rule. The simplices that contain the vertex labeled 3 get mapped to c_3; the simplices that do not contain the vertex labeled 3, but contain the vertex labeled 6 get mapped to c_6; the simplices that do not contain the vertices labeled 3 and 6, but contain the vertex labeled 9 get mapped to c_9; and so on. Finally, the simplices that contain none of the vertices labeled 3, 6, ..., $3k$ all get mapped to c_r (the index r stands here for *the rest*).

Clearly, the map φ is order-preserving, since if one takes a larger simplex, it will have more vertices, and the only way its image may change is to go up when a new element from the set $\{3, 6, \ldots, 3k\}$ is added and is smaller than the previously smallest one.

Let us now define acyclic matchings on the preimages of C under the map φ. We split our argument into three cases.

Case 1. First we consider the preimage $\varphi^{-1}(c_3)$. For any simplex $\sigma \in \varphi^{-1}(c_3)$ we have $3 \in \sigma$ and hence $2 \notin \sigma$. It follows that pairing $\sigma \leftrightarrow \sigma \oplus \{1\}$ provides a matching that is well-defined and is obviously acyclic.

Case 2. Next, we consider the preimages $\varphi^{-1}(c_6)$ through $\varphi^{-1}(c_{3k})$. Let t be an integer such that $2 \leq t \leq k$. The preimage $\varphi^{-1}(c_{3t})$ consists of all simplices σ such that $3, 6, \ldots, 3t - 3 \notin \sigma$, while $3t \in \sigma$. In particular, we have $3t - 1 \notin \sigma$. This means that the pairing $\sigma \leftrightarrow \sigma \oplus \{3t-2\}$ provides a well-defined matching, which is acyclic.

Case 3. Finally, we consider the preimage $\varphi^{-1}(c_r)$. We consider three subcases.

If $n = 3k + 1$, then this preimage is a face poset with a cone with apex in n, in particular, the pairing $\sigma \leftrightarrow \sigma \oplus \{n\}$ provides an acyclic matching with one critical cell $\sigma = \{n\}$, which has dimension 0. By Theorem 11.10 we can conclude that $\mathrm{Ind}\,(L_{3k+1})$ is collapsible.

If $n = 3k$, we see that $X = \varphi^{-1}(c_r)$ is a face poset of the boundary of a k-dimensional cross-polytope, which is the same as the k-fold join of $\mathbb{S}^0$ with itself. By Theorem 11.10 the matching constructed up to now is acyclic, and it gives us a collapsing sequence leading to X. In particular, this shows that $\mathrm{Ind}\,(L_{3k})$ is homotopy equivalent to $\mathbb{S}^{k-1}$.

If $n = 3k + 2$, we see that $X = \varphi^{-1}(c_r)$ is a face poset of the boundary of a $(k + 1)$-dimensional cross-polytope, since this time around we have the k-fold join of $\mathbb{S}^0$ with itself. The rest is just the same, and we conclude that $\mathrm{Ind}\,(L_{3k+2})$ is homotopy equivalent to $\mathbb{S}^k$. $\square$

Note that the proof of Proposition 11.16 actually yields a stronger statement: instead of contractibility, we actually get a collapsibility, and instead of a mere homotopy equivalence to a sphere of some dimension, we get a sequence of collapses, leading to an explicit sphere, sitting inside $\mathrm{Ind}\,(L_n)$ as a subcomplex.

For an arbitrary integer $n \geq 2$, we let C_n denote the cycle with n vertices. These vertices are labeled $1, \ldots, n$, with arithmetic operations on labels performed modulo n. The homotopy type of the independence complexes of cycles allows an easy description as well.

Proposition 11.17. *For any $n \geq 2$, we have*

$$\mathrm{Ind}(C_n) \simeq \begin{cases} \mathbb{S}^{k-1} \vee \mathbb{S}^{k-1}, & \text{if } n = 3k; \\ \mathbb{S}^{k-1}, & \text{if } n = 3k \pm 1. \end{cases}$$

Proof. Let k denote the maximal integer such that $3k \leq n + 1$, and let the chain C be defined in the same way as in Proposition 11.16. Furthermore, let $\varphi : \mathcal{F}(\mathrm{Ind}\,(C_n)) \to C$ be the order-preserving map also described by the same rule as the one in Proposition 11.16. Again, we are looking for acyclic matchings on the preimages.

To start with, the matchings on the preimages $\varphi^{-1}(c_6)$ through $\varphi^{-1}(c_{3k})$ defined identically to Proposition 11.16 are again well-defined and acyclic, without any critical cells. The cases of the remaining two preimages are slightly different.

The preimage $\varphi^{-1}(c_3)$ is the same as the face poset of $\mathrm{Ind}\,(L_{n-3})$ with a minimal element added. Taking the acyclic matching for $\mathrm{Ind}\,(L_{n-3})$ and augmenting it by matching the critical 0-cell with the minimal element yields a new acyclic matching. If $n = 3k+1$, this matching has no critical cells at all. Otherwise, i.e., if $n = 3k$, or $n = 3k - 1$, it has one critical cell in dimension $k - 1$.

Finally, we describe an acyclic matching on the preimage $\varphi^{-1}(c_r)$ by considering three cases.

If $n = 3k - 1$, then we know that $3, 6, \ldots, 3k - 3, 3k \notin \sigma$, where we recall that with our conventions $3k = 1$. Therefore, we are dealing with a face poset of a cone with apex in 2, and hence pairing $\sigma \leftrightarrow \sigma \oplus \{2\}$ gives a well-defined acyclic matching with one critical cell $\{2\}$ in dimension 0.

If $n = 3k$, then we again have a face poset of the join of k copies of $\mathbb{S}^0$. Denote the sets of vertices of these k copies of $\mathbb{S}^0$ by $\{x_1, y_1\}$, ..., $\{x_k, y_k\}$. Consider the pairing $\sigma \leftrightarrow \sigma \oplus \{x_i\}$, where i is the minimal index such that $y_i \notin \sigma$. This is a well-defined acyclic matching with critical cells $\{x_1\}$ of dimension 0, and $\{y_1, \ldots, y_k\}$ of dimension $k - 1$.

If $n = 3k+1$, then we have a face poset of $k-1$ copies of $\mathbb{S}^0$ and one copy of $\mathrm{Ind}\,(L_3)$. Denote the sets of vertices of these $k-1$ copies of $\mathbb{S}^0$ by $\{x_1, y_1\}$, ..., $\{x_{k-1}, y_{k-1}\}$, and let $\{x_k, y_k, z_k\}$ be the vertices of $\mathrm{Ind}\,(L_3)$, with y_k being the middle vertex. Consider the pairing with the same rule: $\sigma \leftrightarrow \sigma \oplus \{x_i\}$, where i is the minimal index such that $y_i \notin \sigma$. This is a well-defined acyclic matching with critical cells $\{x_1\}$ of dimension 0, and $\{y_1, \ldots, y_k\}$ of dimension $k - 1$. $\square$

For future reference, let us remark that the proofs of Proposition 11.16 and Proposition 11.17 imply that the inclusion graph homomorphism $i : L_{3k} \hookrightarrow C_{3k+1}$ induces an isomorphism on the homology groups $i_* : H_*(\mathrm{Ind}\,(L_{3k})) \hookrightarrow H_*(\mathrm{Ind}\,(C_{3k+1}))$.

Example 2: The simplicial complex $\Delta(\bar{\Pi}_n)$ is homotopy equivalent to a wedge of $(n - 1)!$ spheres of dimension $n - 3$

Recall the partition lattice Π_n introduced in Chapter 9.

Theorem 11.18. *For $n \geq 3$, the simplicial complex $\Delta(\bar{\Pi}_n)$ is homotopy equivalent to a wedge of spheres of dimension $n - 3$.*

Proof. The statement is obviously true for $n = 3$, so we assume that $n \geq 4$ and proceed by induction. Set $\alpha := (1)(2, 3, \ldots, n)$, and let Q to be the interval consisting of all partitions having a singleton block (1) with reversed order, i.e., $Q := [\hat{0}, \alpha]_{\bar{\Pi}_n}^{\mathrm{op}}$. We define an order-preserving map $\varphi : \mathcal{F}(\Delta(\bar{\Pi}_n)) \to Q$ by the following rule:

a chain c is taken to the minimal element of Q that can be added to c.

One example is shown in Figure 11.10. To analyze this rule, take $c \in \mathcal{F}(\Delta(\bar{\Pi}_n))$, assume $c = (\pi_1 < \pi_2 < \cdots < \pi_t)$, and consider various cases.

Case 1. If $\alpha \geq \pi_t$, then $\varphi(c) = \alpha$.

Case 2. If $\alpha \not\geq \pi_k$ and either $\alpha \geq \pi_{k-1}$ or $k = 1$, then $\varphi(c) = \pi_k \wedge \alpha$.

In words: find the smallest partition π_k in c, where 1 is a part of a nonsingleton block B, and then partition B into (1) and $B \setminus \{1\}$. This also shows that the minimal element in this rule is unique; hence the map φ is well-defined.

To see that φ is order-preserving, it is enough to notice that if the chain is increased, then the minimal possible element of Q, i.e., the maximal possible element of $[\hat{0}, \alpha]_{\bar{\Pi}_n}$ that can be added to this chain will either remain the same or increase in Q (resp. decrease in $[\hat{0}, \alpha]_{\bar{\Pi}_n}$).

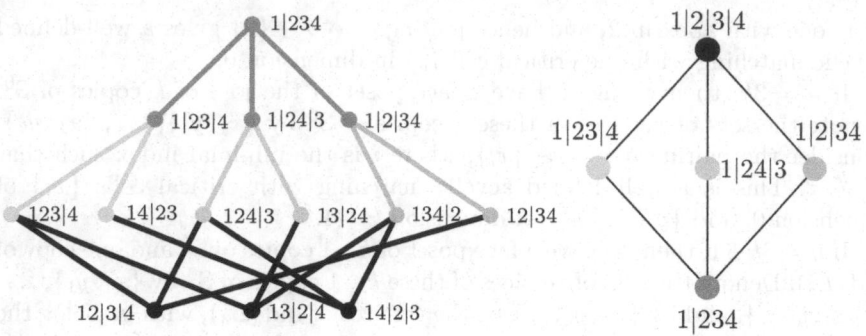

Fig. 11.10. The map φ for $n = 4$.

By Theorem 11.10 it is now sufficient to construct acyclic matchings on the fibers $\varphi^{-1}(x)$. We do this again with case-by-case analysis.

Case 1. Let $S = \varphi^{-1}((1)(2)\ldots(n))$. Clearly, the poset S is in fact a disjoint union $S = S_2 \cup \cdots \cup S_n$, where S_i is the subposet consisting of all chains containing the element $(1i)(2)\ldots(i-1)(i+1)\ldots(n)$, for $i = 2,\ldots,n$. Each poset S_i is actually a copy of $\mathcal{F}(\Delta(\bar{\Pi}_{n-1})) \cup \{\hat{0}\}$. By induction, there exists an acyclic matching on $\mathcal{F}(\Delta(\bar{\Pi}_{n-1}))$ that has one critical cell in dimension 0 and $(n-2)!$ critical cells in dimension $n-4$.

In the poset $\mathcal{F}(\Delta(\bar{\Pi}_{n-1})) \cup \{\hat{0}\}$ this acyclic matching can be extended to have only the top-dimensional critical element, since the other one is matched with $\hat{0}$. When considered in S_i, these maximal chains consist of $n-2$ elements; hence they correspond to critical simplices of dimension $n-3$ in $\Delta(\bar{\Pi}_n)$.

Case 2. Let $S = \varphi^{-1}(\pi)$, for $\pi \neq (1)(2)\ldots(n)$. The matching rule in this case is the following: add π to the chain if it is not there already; otherwise, remove it. Obviously this gives an acyclic matching. The only element that is not matched is the chain consisting only of π: this one would have to be matched with an empty chain, which, by our assumptions, is not there. This corresponds to one critical cell of dimension 0.

Summarizing our considerations, we get $(n-1) \times (n-2)! = (n-1)!$ critical cells of dimension $n-3$ and one critical cell of dimension 0. We may therefore conclude that $\Delta(\bar{\Pi}_n)$ is homotopy equivalent to a wedge of $(n-1)!$ spheres of dimension $n-3$.

The spheres are enumerated by these critical cells of dimension $n-3$. If desired, the recurrence above can be avoided and the chains corresponding to these critical cells can be listed explicitly. These are indexed by permutations of $\{2,\ldots,n\}$, where for every such permutation $(i_1,\ldots,i_{n-1})$ the corresponding chain c is given by

$$c = (1,i_1)(2)\ldots(n) < (1,i_1,i_2)(2)\ldots(n) < \cdots < (1,i_1,\ldots,i_{n-2})(i_{n-1}).$$

The dual cochains of these simplices can also be taken as a basis for the cohomology group $\widetilde{H}^{n-3}(\Delta(\bar{\varPi}_n); \mathbb{Z})$. □

In the next two examples we illustrate how one can check the acyclicity of a partial matching directly, bypassing the patchwork theorem.

Example 3: The generalized simplicial complex $\Delta(\bar{\varPi}_n)/\mathcal{S}_n$ is collapsible

The indexing of the simplices of $\Delta(\bar{\varPi}_n)/\mathcal{S}_n$.
The simplices of $\Delta(\bar{\varPi}_n)$ can be indexed with sequences of the set partitions of $[n]$, where the partitions refine each other. One can think of such a sequence as a forest, where vertices are ordered into levels, each level correspond to a set partition of $[n]$, and each vertex on that level corresponds to a block of this partition.

We can therefore index the simplices of $\Delta(\bar{\varPi}_n)$ with such "leveled forests," where the vertices carry subsets of $[n]$ as labels, and the label of each vertex is equal to the union of the labels of its children; see Figure 11.11 for an example. If one desires, one can also add two artificial levels: one on top, consisting of one vertex having the label $[n]$, and one on the bottom, consisting of n leaves having labels 1 through n; this way we obtain labeled trees. Clearly, the labels of the bottom level determine all other labels.

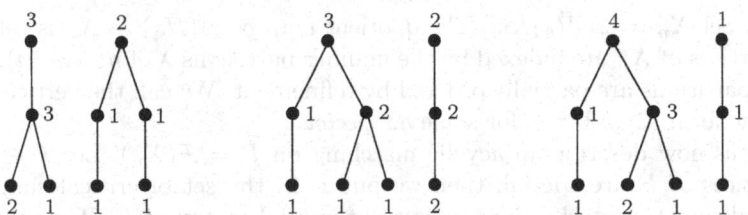

Fig. 11.11. Examples of labeled forests indexing 2-simplices of the generalized simplicial complex $\Delta(\bar{\varPi}_n)/\mathcal{S}_n$.

The symmetric group $\mathcal{S}_n$ acts by permuting the ground set $[n]$. We leave it as an exercise to see that an $\mathcal{S}_n$-orbit of a labeled tree as above consists of all labeled trees with the same cardinalities of the labels on the vertices. As a result of this observation, we can index the simplices of $\Delta(\bar{\varPi}_n)/\mathcal{S}_n$ with the labeled trees as above, with the difference that the labels are positive integers, and labels on each level form a number partition of n (instead of the set partition of $[n]$). For example, the vertices of $\Delta(\bar{\varPi}_n)/\mathcal{S}_n$ are indexed with number partitions of n, edges of $\Delta(\bar{\varPi}_n)/\mathcal{S}_n$ are indexed with the ways two such number partitions can refine each other, and so on.

We also see that, both in the case of $\Delta(\bar{\Pi}_n)$ and in that of $\Delta(\bar{\Pi}_n)/\mathcal{S}_n$, the boundary operator is obtained by deleting entire levels from trees and reconnecting vertices transitively through the deleted level.

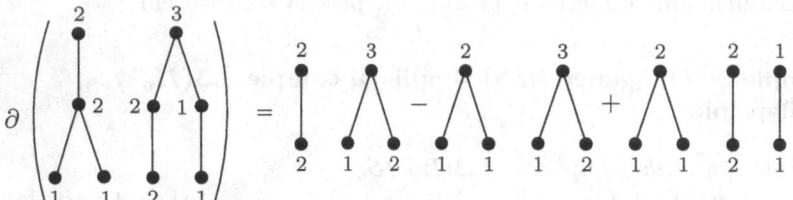

Fig. 11.12. An example of the boundary operator in the generalized simplicial complex $\Delta(\bar{\Pi}_n)/\mathcal{S}_n$.

When λ is a number partition of n, $\lambda = (\lambda_1, \ldots, \lambda_t)$, we define $s_2(\lambda) = \sum_{i=1}^{t} \lfloor \frac{\lambda_i}{2} \rfloor$ and $\mu_2(\lambda) = (2^{s_2(\lambda)}, 1^{n-2 \cdot s_2(\lambda)}) = (\underbrace{2, \ldots, 2}_{s_2(\lambda)}, \underbrace{1, \ldots, 1}_{n-2 \cdot s_2(\lambda)})$. Clearly $\mu_2(\lambda)$ refines λ.

Theorem 11.19. *The generalized simplicial complex $\Delta(\bar{\Pi}_n)/\mathcal{S}_n$ is collapsible, for all $n \geq 3$.*

Proof. Set $X_n := \Delta(\bar{\Pi}_n)/\mathcal{S}_n$. The quotient map $p : \Delta(\bar{\Pi}_n) \to X_n$ is cellular. The vertices of X_n are indexed by the number partitions λ of n, $\lambda \neq (n), (1^n)$. These partitions are partially ordered by refinement. We call the vertices that have the form $(2^a, 1^{n-2a})$, for some a, *special*.

Let us now describe an acyclic matching on $P = \mathcal{F}(X_n)$. Let $F \in P$. If all vertices of F are special, then we put F in the set of critical simplices. If not, define $\lambda(F)$ to be the smallest not special vertex of F. If $\mu_2(\lambda(F))$ is a vertex of F, then we match F with $F \setminus \{\mu_2(\lambda(F))\}$; otherwise, we match F with $F \cup \{\mu_2(\lambda(F))\}$.

To see that this is a valid matching, note that

$$\lambda(F) = \lambda(F \cup \mu_2(\lambda(F))).$$

Next, we show that the obtained matching M is acyclic. Assume that there exists a sequence $\sigma_0, \ldots, \sigma_t \in P$ such that all σ_i are different, with the exception $\sigma_0 = \sigma_t$, and such that $u(\sigma_i) \succ \sigma_{i+1}$ for $0 \leq i \leq t - 1$. Assume that $u(\sigma_0) = (a_1, \ldots, a_\alpha, b_1, \ldots, b_\beta)$, where the a_i's are special and b_1 is not. Then $\sigma_0 = (a_1, \ldots, a_{\alpha-1}, b_1, \ldots, b_\beta)$. Since σ_1 is matched upward, and $u(\sigma_0) \neq u(\sigma_1)$, we have $\sigma_1 = (a_1, \ldots, a_\alpha, b_2, \ldots, b_\beta)$. We see that the number of special vertices in σ_1 is larger by 1 than in σ_0. Repeating the argument, we see that σ_t has t special vertices more than σ_0; therefore $\sigma_0 \neq \sigma_t$. This leads to the conclusion that M is an acyclic matching.

The critical simplices form a subcomplex of X_n, which we call X_n^C. By Theorem 11.13, there exists a sequence of elementary collapses leading from X_n to X_n^C. Observe that if $A = (a_1, \ldots, a_t)$ and $B = (b_1, \ldots, b_t)$ are two simplices of $\Delta(\bar{\Pi}_n)$ such that for $i = 1, \ldots, t$, both a_i and b_i are of the type $(2^{\alpha_i}, 1^{n-2\alpha_i})$ for some α_i, then there exists $g \in S_n$ such that $gA = B$, i.e., $p(A) = p(B)$. This implies that X_n^C is a simplex, so we can conclude that X_n is collapsible.

Note that this matching can also be found in a functorial way as follows. Let Q be a chain with $\lfloor n/2 \rfloor$ elements labeled with the numbers $1, \ldots, \lfloor n/2 \rfloor$ in reverse order; i.e., 1 labels the maximal element. Define $\varphi : P \to Q$ by mapping the cell $F \in P$ to the maximal number k such that $(2^k, 1^{n-2k})$ is either a vertex of F or can be added to F to form a new cell. This is an order-preserving map, since taking a bigger cell will either keep this number k the same or decrease it.

An acyclic matching on the fiber $\varphi^{-1}(k)$, for $k \in Q$, is simply obtained by adding the vertex $(2^k, 1^{n-2k})$ to cells that do not have it, or removing it from those that do. The only critical cell has dimension 0, and it can be found in the fiber $\varphi^{-1}(k)$, for $k = \lfloor n/2 \rfloor$: it is the vertex $(2^k, 1^{n-2k})$, which cannot be removed, since the empty cell is not being matched. $\square$

Example 4: Bounded sets in a lattice

As we have seen in Chapter 9, there are a number of constructions associating a simplicial complex to a poset (or more generally, to a category); here is yet another one that works for lattices.

Definition 11.20. *Let $\mathcal{L}$ be an arbitrary finite lattice. We define $\mathcal{J}(\mathcal{L})$ to be the simplicial complex whose set of vertices is equal to the set of elements of $\bar{\mathcal{L}}$, and whose simplices are all subsets $S \subseteq \bar{\mathcal{L}}$ that have a nontrivial lower bound, i.e., such that $\bigwedge S \neq \hat{0}$.*

Clearly, the simplicial complex $\mathcal{J}(\mathcal{L})$ contains $\Delta(\bar{\mathcal{L}})$ as a subcomplex. It turns out that much more is true.

Theorem 11.21. *Let $\mathcal{L}$ be an arbitrary finite lattice. Then $\mathcal{J}(\mathcal{L}) \searrow \Delta(\bar{\mathcal{L}})$.*

Proof. As the centerpiece of the argument we define the following partial acyclic matching on $\mathcal{F}(\mathcal{J}(\mathcal{L}))$. Let S be an arbitrary simplex of $\mathcal{J}(\mathcal{L})$. Assume that $\mathcal{F}(\mathcal{J}(\mathcal{L}))[S]$ is not a chain. Set $t := |S|$, and let $\{a_1, a_2, \ldots, a_t\}$ be a linear extension of $\mathcal{F}(\mathcal{J}(\mathcal{L}))[S]$, i.e., if $1 \le i < j \le t$, then $a_i \not\geq a_j$.

Let $k(S)$ be the maximal index, $1 \le k(S) \le t$, such that $a_1 < a_2 < \cdots < a_{k(S)}$, and $a_{k(S)} < a_i$, for all $k(S) + 1 \le i \le t$; see Figure 11.13. If S has no minimal element, then we set $k(S) := 0$. Set $a(S) := a_{k(S)+1} \wedge \cdots \wedge a_t$. Since $\mathcal{F}(,\mathcal{J}(\mathcal{L}))[S]$ is not a chain, we have $k(S) \le t - 2$, and hence $a(S)$ is well-defined.

Let Σ be the set of all subsets $S \subseteq \bar{\mathcal{L}}$ such that $\mathcal{F}(\mathcal{J}(\mathcal{L}))[S]$ is not a chain and such that $a(S) \notin S$. For $S \in \Sigma$ define $\mu(S) := S \cup \{a(S)\}$; again see Figure 11.13. Clearly, μ defines a partial matching, and, since for any $S \in \Sigma$ we have $a(\mu(S)) = a(S)$, we see that the set $\mu(\Sigma) \cup \Sigma$ consists of all subsets $S \subseteq \bar{\mathcal{L}}$ such that $\mathcal{F}(\mathcal{J}(\mathcal{L}))[S]$ is not a chain. Consequently, the set of critical elements $\mathcal{C}(\mathcal{F}(\mathcal{J}(\mathcal{L})), \mu)$ consists of all chains $S \in \mathcal{F}(\Delta(\bar{\mathcal{L}}))$.

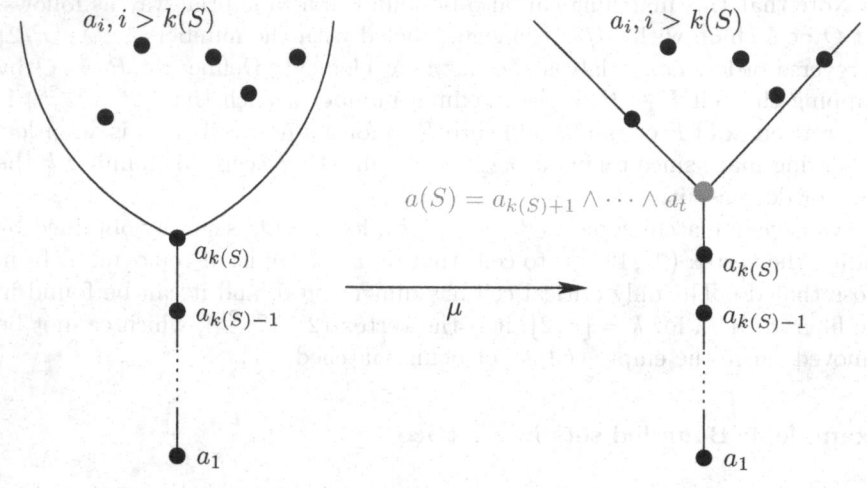

Fig. 11.13. The partial matching μ.

Let us see that the partial matching μ is acyclic. Assume that there exists a sequence $S_1, \ldots, S_t \in \Sigma$, where $t \geq 2$, such that $\mu(S_1) \succ S_2, \mu(S_2) \succ S_3, \ldots,$ $\mu(S_t) \succ S_1$. Let again $\{a_1, a_2, \ldots, a_t\}$ be a linear extension of $\mathcal{F}(\mathcal{J}(\mathcal{L}))[S_1]$, as above. By the definition of covering relations, and since $S_2 \neq S_1$, we have $S_2 = \mu(S_1) \setminus \{a_i\}$, for some $1 \leq i \leq t$. If $1 \leq i \leq k(S_1)$, then $a(S_2) = a(S_1)$, which, together with $S_1 = \mu(S_1) \setminus \{a(S_1)\}$, implies $a(S_2) \in S_2$, and hence $S_2 \in \mu(\Sigma)$, giving a contradiction.

Finally, the only option left is that $k(S_1) + 1 \leq i \leq t$, in which case $a(S_2) \geq a(S_1)$, since the join is taken over a set in which each element is larger than $a(S_1)$. If the equality $a(S_2) = a(S_1)$ holds, then $S_2 \in \mu(\Sigma)$, again giving a contradiction. Thus we have shown that a strict inequality must hold: $a(S_2) > a(S_1)$.

Analogously, we can prove that $a(S_{i+1}) > a(S_i)$, for all $1 \leq i \leq t - 1$, and that $a(S_1) > a(S_t)$, which, when combined together, yields a contradiction to the assumption that the matching is not acyclic. By Theorem 11.13 we see that the acyclic matching μ provides a sequence of elementary collapses leading from $\mathcal{J}(\mathcal{L})$ to $\Delta(\bar{\mathcal{L}})$. $\square$

More applications will appear in the subsequent sections.

11.3 Algebraic Morse Theory

In this section we give a version of discrete Morse theory that is adapted to the setting of arbitrary free chain complexes.

11.3.1 Acyclic Matchings on Free Chain Complexes and the Morse Complex

Let $\mathcal{R}$ be an arbitrary commutative ring with unit. Recall that a chain complex C_* consisting of $\mathcal{R}$-modules,

$$C_* = \cdots \xrightarrow{\partial_{n+2}} C_{n+1} \xrightarrow{\partial_{n+1}} C_n \xrightarrow{\partial_n} C_{n-1} \xrightarrow{\partial_{n-1}} \cdots,$$

is called *free* if C_n is a finitely generated free $\mathcal{R}$-module for all n. When no confusion can occur, we simply write ∂ instead of ∂_n. We also always require that C_* be bounded on the right.

In order to introduce a combinatorial element into this setting, we need to choose *a basis* (i.e., a set of free generators) Ω_n for each C_n. When this is done, we say that we have chosen a basis $\Omega = \bigcup_n \Omega_n$ for the entire chain complex C_*. We write (C_*, Ω) to denote a chain complex with a basis. A *free chain complex with a basis* is the main object of study of algebraic Morse theory.

Given a free chain complex with a basis (C_*, Ω) and two elements $\alpha \in C_n$ and $b \in \Omega_n$, we denote the coefficient of b in the representation of α as a linear combination of the elements of Ω_n by $\mathbf{k}_\Omega(\alpha, b)$, or, if the basis is clear, simply by $\mathbf{k}(\alpha, b)$. For $x \in C_n$ we write $\dim x = n$. By convention, we set $\mathbf{k}_\Omega(\alpha, b) = 0$ if the dimensions do not match, i.e., if $\dim \alpha \neq \dim b$.

Note that a free chain complex with a basis (C_*, Ω) can be represented as a ranked poset $P(C_*, \Omega)$, with $\mathcal{R}$-weights on the order relations. The elements of rank n correspond to the elements of Ω_n, and the weight of the covering relation $b \succ a$, for $b \in \Omega_n, a \in \Omega_{n-1}$, is simply defined by $w_\Omega(b \succ a) := \mathbf{k}_\Omega(\partial b, a)$. In other words,

$$\partial b = \sum_{b \succ a} w_\Omega(b \succ a) a,$$

for each $b \in \Omega_n$. Again, if the basis is clear, we simply write $w(b \succ a)$.

Definition 11.22. *Let (C_*, Ω) be a free chain complex with a basis. A **partial matching** $\mathcal{M} \subseteq \Omega \times \Omega$ on (C_*, Ω) is a partial matching on the covering graph of $P(C_*, \Omega)$ such that if $b \succ a$, and b and a are matched, i.e., if $(a, b) \in M$, then $w(b \succ a)$ is invertible.*

It is important to note that Definition 11.22 is different from the topological one, which was used in Theorem 11.13. In the algebraic setting, the condition that the matched cells form a regular pair (in the CW sense) is replaced by requiring that the covering weights in matched pairs be invertible. However, the notion of *acyclic matching*, which is purely combinatorial, since it is defined on the level of posets, remains the same.

Given such a partial matching $\mathcal{M}$, we denote by $\mathcal{U}_n(\Omega)$ the set of all $b \in \Omega_n$ such that b is matched with some $a \in \Omega_{n-1}$, and analogously, we denote by $\mathcal{D}_n(\Omega)$ the set of all $a \in \Omega_n$ that are matched with some $b \in \Omega_{n+1}$. We let $\mathcal{C}_n(\Omega) := \Omega_n \setminus \{\mathcal{U}_n(\Omega) \cup \mathcal{D}_n(\Omega)\}$ denote the set of critical basis elements of dimension n. Finally, we set $\mathcal{U}(\Omega) := \bigcup_n \mathcal{U}_n(\Omega)$, $\mathcal{D}(\Omega) := \bigcup_n \mathcal{D}_n(\Omega)$, and $\mathcal{C}(\Omega) := \bigcup_n \mathcal{C}_n(\Omega)$.

Given two basis elements $s \in \Omega_n$ and $t \in \Omega_{n-1}$, the *weight* of an alternating path

$$p = (s \succ d(b_1) \prec b_1 \succ d(b_2) \prec b_2 \succ \cdots \succ d(b_n) \prec b_n \succ t), \qquad (11.6)$$

where $n \geq 0$ and all $b_i \in \mathcal{U}(\Omega)$ are distinct, is defined to be the quotient

$$w(p) := (-1)^n \frac{w(s \succ d(b_1)) \cdot w(b_1 \succ d(b_2)) \cdots w(b_n \succ t)}{w(b_1 \succ d(b_1)) \cdot w(b_2 \succ d(b_2)) \cdots w(b_n \succ d(b_n))}. \qquad (11.7)$$

The reader is invited to compare (11.7) with formula (11.5). Additionally, we shall use the notation $p^\bullet = s$ and $p_\bullet = t$.

Definition 11.23. *Let* (C_*, Ω) *be a free chain complex with a basis, and let* $\mathcal{M}$ *be an acyclic matching. The* **Morse complex**

$$\cdots \xrightarrow{\partial_{n+2}^{\mathcal{M}}} C_{n+1}^{\mathcal{M}} \xrightarrow{\partial_{n+1}^{\mathcal{M}}} C_n^{\mathcal{M}} \xrightarrow{\partial_n^{\mathcal{M}}} C_{n-1}^{\mathcal{M}} \xrightarrow{\partial_{n-1}^{\mathcal{M}}} \cdots,$$

is defined as follows. The $\mathcal{R}$-*module* $C_n^{\mathcal{M}}$ *is freely generated by the elements of* $\mathcal{C}_n(\Omega)$. *The boundary operator is defined by*

$$\partial_n^{\mathcal{M}}(s) = \sum_p w(p) \cdot p_\bullet,$$

for all $s \in \mathcal{C}_n(\Omega)$, *where the sum is taken over all alternating paths* p *satisfying* $p^\bullet = s$. *Again, if the indexing is clear, we simply write* $\partial^{\mathcal{M}}$ *instead of* $\partial_n^{\mathcal{M}}$.

Given a free chain complex with a basis (C_*, Ω), we can choose a different basis $\widetilde{\Omega}$ by replacing each $a \in \mathcal{D}_n(\Omega)$ by $\tilde{a} = w(u(a) \succ a) \cdot a$, because $w(u(a) \succ a)$ is required to be invertible. Since

$$\mathbf{k}_{\widetilde{\Omega}}(x, \tilde{a}) = \mathbf{k}_\Omega(x, a) / w(u(a) \succ a), \qquad (11.8)$$

for any $x \in \Omega_n$, we see that the weights of those alternating paths that do not begin with or end in an element from $\mathcal{D}_n(\Omega)$ remain unaltered, since the quotient $w(x \succ z) / w(y \succ z)$ stays constant as long as $x, y \neq a$. In particular, the Morse complex will not change. On the other hand, by (11.8), $w_{\widetilde{\Omega}}(u(a) \succ$

$a) = 1$, for all $a \in \mathcal{D}(\widetilde{\Omega})$, so the total weight of the alternating path in (11.6) will simply become

$$w_{\widetilde{\Omega}}(p) = (-1)^n w_{\widetilde{\Omega}}(s \succ d(b_1)) \cdot w_{\widetilde{\Omega}}(b_1 \succ d(b_2)) \cdots w_{\widetilde{\Omega}}(b_n \succ t).$$

Because of these observations, we may always replace any given basis of C_* with the basis $\widetilde{\Omega}$ satisfying $w_{\widetilde{\Omega}}(u(a) \succ a) = 1$, for all $a \in \mathcal{D}(\widetilde{\Omega})$.

11.3.2 The Main Theorem of Algebraic Morse Theory

The chain complex $\cdots \longrightarrow 0 \longrightarrow \mathcal{R} \xrightarrow{\text{id}} \mathcal{R} \longrightarrow 0 \longrightarrow \cdots$, where the only nontrivial modules are in the dimensions d and $d-1$, is called an *atom chain complex*, and is denoted by $\texttt{Atom}(d)$.

The main theorem of algebraic Morse theory brings to light a certain structure in C_*. Namely, by choosing a different basis, one can represent C_* as a direct sum of two chain complexes, of which one is a direct sum of atom chain complexes, in particular acyclic, and the other one is isomorphic to $C_*^{\mathcal{M}}$. For convenience, the choice of basis can be performed in several steps, one step for each matched pair of the basis elements.

Theorem 11.24.
(Main theorem of discrete Morse theory for free chain complexes)
Assume that we have a free chain complex with a basis (C_*, Ω), and an acyclic matching $\mathcal{M}$. Then C_* decomposes as a direct sum of chain complexes $C_*^{\mathcal{M}} \oplus T_*$, where $T_* \simeq \bigoplus_{(a,b) \in \mathcal{M}} \texttt{Atom}(\dim b)$.

It can be advisable to use the example in Subsection 11.3.3 as an illustration for the following proof.

Proof. To start with, let us choose a linear extension L of the partially ordered set $P(C_*, \Omega)$ satisfying the conditions of Theorem 11.2, and let $<_L$ denote the corresponding total order.

Assume first that C_* is bounded; without loss of generality, we can assume that $C_i = 0$ for $i < 0$, and $i > N$. Let $m = |M|$ denote the size of the matching, and let $l = |\Omega| - 2m$ denote the number of critical cells.

We shall now inductively construct a sequence of bases $\Omega^0, \Omega^1, \ldots, \Omega^m$ of C_*. More specifically, each basis will be divided into three parts: $\mathcal{C}(\Omega^k) = \{c_1^k, \ldots, c_l^k\}$, $\mathcal{D}(\Omega^k) = \{a_1^k, \ldots, a_m^k\}$, and $\mathcal{U}(\Omega^k) = \{b_1^k, \ldots, b_m^k\}$, such that $a_i^k = d(b_i^k)$, for all $i \in [m]$.

We start with $\Omega^0 = \Omega$ and the initial condition $b_i^0 <_L b_{i+1}^0$, for all $i \in [m-1]$. Since the lower index of $\mathbf{k}_-(-,-)$ and $w_-(- \succ -)$ will be clear from the arguments, we shall omit it to make the formulas more compact.

When constructing the bases, we shall simultaneously prove by induction the following statements:

(i) $C_* = C_*[k] \oplus \mathcal{A}_1^k \oplus \cdots \oplus \mathcal{A}_k^k$, where $C_*[k]$ is the subcomplex of C_* generated by $\Omega^k \setminus \{a_1^k, \ldots, a_k^k, b_1^k, \ldots, b_k^k\}$, and $\mathcal{A}_i^k$ is isomorphic to $\texttt{Atom}(\dim b_i^k)$, for $i \in [k]$;

(ii) for every $x^k \in \{b_{k+1}^k, \ldots, b_m^k\} \cup \mathcal{C}(\Omega^k)$, $y^k \in \mathcal{C}(\Omega^k)$, we have $w(x^k \succ y^k) = \sum_p w(p)$, where the sum is restricted to those alternating paths from x^0 to y^0 that use only the pairs (a_i^0, b_i^0), for $i \in [k]$.

Clearly, all of the statements are true for $k = 0$. Assume $k \geq 1$.

Transformation of the basis Ω^{k-1} into the basis Ω^k:
we set

- $a_k^k := \partial b_k^{k-1};$
- $b_k^k := b_k^{k-1};$
- $x^k := x^{k-1} - w(x^{k-1} \succ a_k^{k-1}) \cdot b_k^{k-1},$ for all $x^{k-1} \in \Omega^{k-1}$, $x \neq a_k, b_k$.

First, we see that Ω^k is a basis. Indeed, assume $b_k^{k-1} \in C_n$. For $i \neq n, n-1$, we have $\Omega_i^k = \Omega_i^{k-1}$; hence by induction, it is a basis. Then Ω_{n-1}^k is obtained from Ω_{n-1}^{k-1} by adding a linear combination of other basis elements to the basis element a_k^{k-1}; hence Ω_{n-1}^k is again a basis. Finally, Ω_n^k is obtained from Ω_n^{k-1} by subtracting multiples of the basis element b_k^{k-1} from the other basis elements; hence it is also a basis.

Next, we investigate how the poset $P(C_*, \Omega^k)$ differs from $P(C_*, \Omega^{k-1})$. If $x \neq b_k$, we have $w(x^k \succ a_k^k) = \mathbf{k}(\partial x^k, a_k^k) = \mathbf{k}(\partial x^k, a_k^{k-1}) = \mathbf{k}(\partial x^{k-1}, a_k^{k-1}) - w(x^{k-1} \succ a_k^{k-1}) \cdot \mathbf{k}(\partial b_k^{k-1}, a_k^{k-1}) = 0$, where the second equality follows from the fact that Ω_{n-1}^k is obtained from Ω_{n-1}^{k-1} by adding a linear combination of other basis elements to the basis element a_k^{k-1}, and the last equality follows from $\mathbf{k}(\partial b_k^{k-1}, a_k^{k-1}) = 1$.

Furthermore, since Ω_n^k is obtained from Ω_n^{k-1} by subtracting multiples of the basis element b_k^{k-1} from the other basis elements, we see that for $x \in \Omega_{n+1}^k$, $y \in \Omega_n^k$, $y \neq b_k$, we have $w(x^k \succ y^k) = w(x^{k-1} \succ y^{k-1})$. Additionally, since the differential of the chain complex squares to 0, we have $0 = \sum_{z^k \in \Omega_n^k} w(x^k \succ z^k) \cdot w(z^k \succ a_k^k) = w(x^k \succ b_k^k) \cdot w(b_k^k \succ a_k^k) = w(x^k \succ b_k^k)$, where the second equality follows from $w(z^k \succ a_k^k) = 0$, for $z \neq b_k$.

We can summarize our findings as follows: all weights in the poset $P(C_*, \Omega^k)$ are the same as in $P(C_*, \Omega^{k-1})$, with the following exceptions:

(1) $w(x^k \succ b_k^k) = 0$, and $w(b_k^k \succ x^k) = 0$, for $x \neq a_k$;

(2) $w(a_k^k \succ x^k) = 0$, and $w(x^k \succ a_k^k) = 0$, for $x \neq b_k$;

(3) $w(x^k \succ y^k) = w(x^{k-1} \succ y^{k-1}) - w(x^{k-1} \succ a_k^{k-1}) \cdot w(b_k^{k-1} \succ y^{k-1})$, for $x \in \Omega_n^k$, $y \in \Omega_{n-1}^k$, $x \neq b_k$, $y \neq a_k$.

In particular, the statement *(i)* is proved. Furthermore, the following fact can be seen by induction, using (1), (2), and (3):

Fact $(*)$. *If $w(x^k \succ y^k) \neq w(x^{k-1} \succ y^{k-1})$, then $b_k^0 \geq_L y^0$.*

Indeed, either $y \in \{a_k, b_k\}$ or y is critical or $y = a_{\tilde{k}}$, for $\tilde{k} > k$ such that $w(b_k^{k-1} \succ y^{k-1}) \neq 0$. In the first two cases, $b_k^0 \geq_L y^0$ by the construction of L, and the last case is impossible by induction, and again, by the construction of L.

We have $w(b_j^k \succ a_j^k) = w(b_j^{k-1} \succ a_j^{k-1})$, for all j, k. Indeed, this is clear for $j = k$. The case $j < k$ follows by induction, and the case $j > k$ is a consequence of Fact $(*)$.

Next, we see that the partial matching $\mathcal{M}^k := \{(a_i^k, b_i^k) \mid i \in [m]\}$ is acyclic. For $j \leq k$, the poset elements b_j^k, a_j^k are incomparable with the rest; hence they cannot be a part of a cycle. For $i > k$, we have $w(b_j^k \succ a_i^k) = w(b_j^{k-1} \succ a_i^{k-1})$, by Fact $(*)$. Hence by induction, no cycle can be formed by these elements either.

Finally, we trace the boundary operator. Let $x^k \in \{b_{k+1}^k, \ldots, b_m^k\} \cup \mathcal{C}(\Omega^k)$, $y^k \in \mathcal{C}(\Omega^k)$. We have $w(x^k \succ y^k) = w(x^{k-1} \succ y^{k-1}) - w(x^{k-1} \succ a_k^{k-1})w(b_k^{k-1} \succ y^{k-1})$. By induction, the first term counts the contribution of all the alternating paths from x^0 to y^0 that do not use the edges $b_l^0 \succ a_l^0$, for $l \geq k$. The second term contains the additional contribution of the alternating paths from x^0 to y^0 that use the edge $b_k^0 \succ a_k^0$. Observe that if this edge occurs, then by the construction of L, it must be the second edge of the path (counting from x^0), and by Fact $(*)$, we have $w(x^{k-1} \succ a_k^{k-1}) = w(x^0 \succ a_k^0)$. This proves the statement (ii), and therefore concludes the proof of the finite case.

It is now easy to deal with the infinite case, since the basis stabilizes as we proceed through the dimensions, so we may take the union of the stable parts as the new basis for C_*. $\square$

We remark that even if the chain complex C^* is infinite in both directions, one can still define the notions of the acyclic matching and of the Morse complex. Since each particular homology group is determined by a finite excerpt from C^*, we may still conclude that $H_*(C_*) = H_*(C_*^{\mathcal{M}})$.

11.3.3 An Example

Note that the proof of Theorem 11.24 is actually an algorithm. In this subsection we illustrate the workings of this algorithm on a concrete chain complex, namely one associated to some chosen triangulation of the projective plane; see Figure 11.14. For the sake of clarity, we restrict ourselves to $\mathbb{Z}_2$-coefficients. In our figures, a solid line directed from a basis element x down to the basis element y means that ∂x contains y with coefficient 1 when ∂x is decomposed in the *current* basis of the chain group in dimension $\dim y$, that is, the basis consisting of elements that are depicted in the figure in question on the same level as y.

The algorithm starts by picking an extension L satisfying the conditions of Theorem 11.2; this is done in Figure 11.15.

This linear order yields the initial basis Ω^0; see Figure 11.16.

Applying the basis transformation rule to Ω^0 is especially easy, since we are dealing with $\mathbb{Z}_2$-coefficients: all we need to do is to replace all basis elements

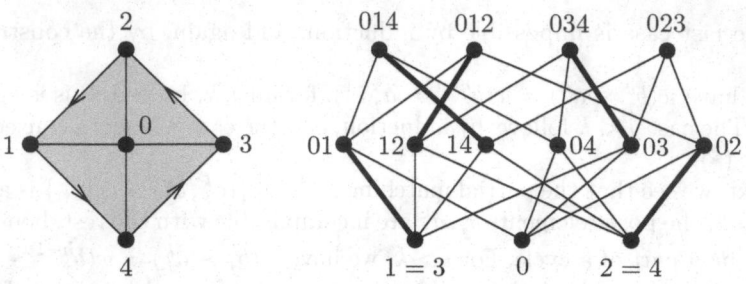

Fig. 11.14. A triangulation of $\mathbb{RP}^2$ and the corresponding face poset.

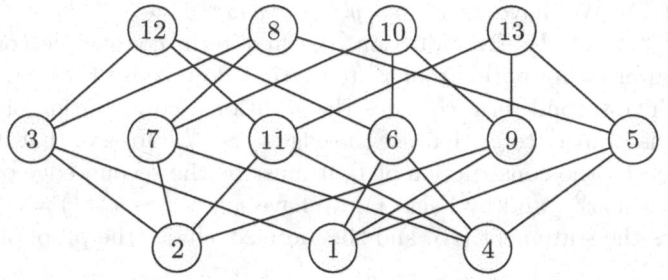

Fig. 11.15. The linear extension L.

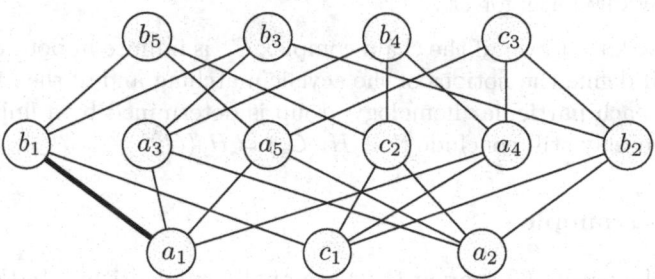

Fig. 11.16. The basis Ω^0.

x containing a_1 in their boundary by $x + b_1$, and recompute the boundaries in the new basis. Again, since we are working over $\mathbb{Z}_2$, the new boundaries of the basis elements in the same dimension as b_1 are simply obtained by taking the symmetric sum (exclusive **or**) of the sets covered by x and by b_1. The analogy with Gaussian elimination is apparent. The resulting basis with corresponding boundaries is shown in Figure 11.17.

We continue in the same way to obtain the bases Ω^2 through Ω^5, as shown in Figures 11.18–11.21. In each figure, the thick line denotes the next collapse. The important thing that one should keep in mind is that at each step, the

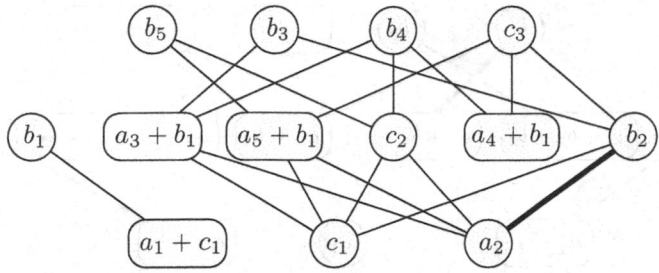

Fig. 11.17. The basis Ω^1.

poset in the figure is given with respect to the new basis, i.e., the boundaries all have to be recalculated accordingly.

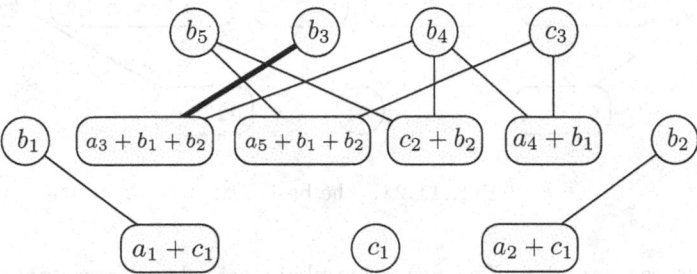

Fig. 11.18. The basis Ω^2.

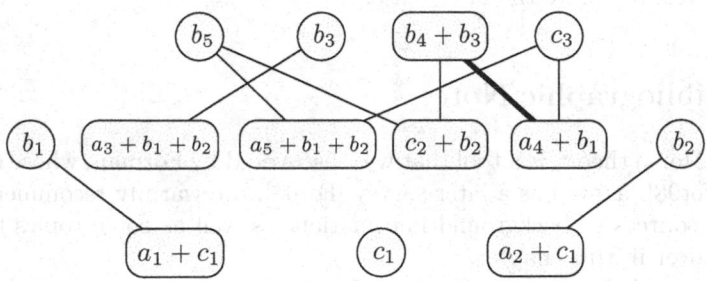

Fig. 11.19. The basis Ω^3.

The final answer is presented in Figure 11.21. It is a good illustration of Theorem 11.24 in this special case. In the new basis, shown in circles and rounded rectangles, our chain complex splits into five atom chain complexes

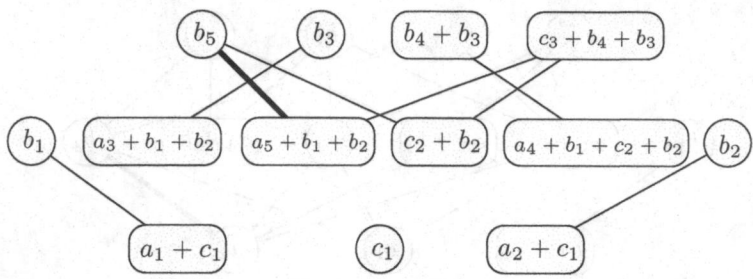

Fig. 11.20. The basis Ω^4.

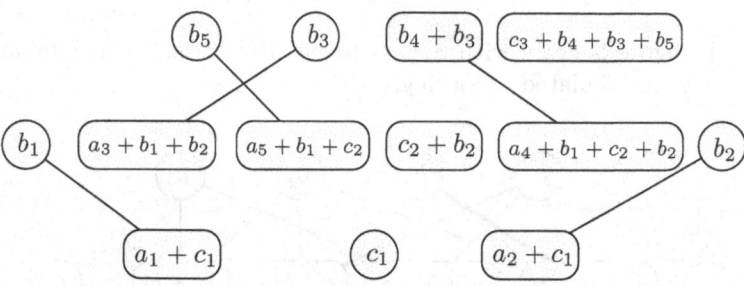

Fig. 11.21. The basis Ω^5.

and the Morse complex. The five atom chain complexes correspond to the initially matched pairs, and have no influence on the homology of the chain complex. The Morse complex is especially simple in this case, since all the differentials ∂_*^M come out to be trivial. This, of course, would have been different if we had worked with integer coefficients. The interested reader is invited to contemplate the latter case.

11.4 Bibliographic Notes

Discrete Morse theory is a tool that was discovered by Forman, whose original article [For98], as well as a later survey [For03], are warmly recommended as excellent sources of background information, as well as some topics that we did not cover in this chapter.

Our main innovation in Section 11.1 is the equivalent reformulation of acyclic matchings in terms of poset maps with small fibers, as well as the introduction of the universal object connected to each acyclic matching. The patchwork theorem 11.10 is a standard tool, used previously by several authors. We think that the terminology of poset fibrations together with the decomposition theorem 11.9 give the patchworking particular clarity.

The material of Sections 11.2.1 and 11.2.2 is quite standard, though our proof of Theorem 11.13(c) is a new ad hoc argument.

We have taken our examples from various sources. Proposition 11.16 and Proposition 11.17 were both originally proved in [Ko99], though in a different way. Theorem 11.18 has been proved in [Bj80], whereas Theorem 11.19 can be found in [Ko00]. Finally, Theorem 11.21 is taken from [Ko06c].

Algebraic Morse theory was discovered independently by several sets of authors. We invite the reader to consult the original sources [JW05, Ko05c, Sk06], providing excellent insight into various aspects of the subject, as well as supplying further applications. Our treatment here follows the algorithmic presentation in [Ko05c].

12

Lexicographic Shellability

Lexicographic shellability is an important tool for studying the topological properties of the order complexes of partially ordered sets. Although, as we shall see in Remark 12.4, discrete Morse theory is more powerful as a method, shellability may still be useful in concrete applications. We take a detailed look at this concept in this section.

Our presentation centers on the lexshellable posets, as the most general form of lexicographic shellability. We start with the classical situation of order complexes of posets, and then proceed to describe how this generalizes to nerves of acyclic categories.

12.1 Shellability

12.1.1 The Basics

The topological fact that gets the whole theory going is that for an n-dimensional simplex σ, any union of $(n-1)$-dimensional simplices $\tau_1 \cup \cdots \cup \tau_k$ from the boundary of σ is either homeomorphic to an $(n-1)$-dimensional sphere, if we take the entire boundary, or is contractible. In the latter case, it is in fact a cone, in which any vertex from $\tau_1 \cap \cdots \cap \tau_k$ can be taken as an apex.

Definition 12.1. *A generalized simplicial complex Δ is called **shellable** if its maximal simplices can be arranged in linear order $F_1, F_2, \ldots, F_t$ in such a way that the subcomplex $(\bigcup_{i=1}^{k-1} F_i) \cap F_k$ is pure and $(\dim F_k - 1)$-dimensional for all $k = 2, \ldots, t$.*

Note that in Definition 12.1 we use the notion of a generalized simplicial complex. In particular, multiple simplices on the same set of vertices are explicitly allowed. Also, we shall call a generalized simplicial complex *pure* if all of its maximal simplices have the same dimension.

An ordering of maximal simplices satisfying the conditions of Definition 12.1 is called a *shelling order*. For the right intuition here, one should think of the whole complex as being decomposed (shelled) by removing the maximal simplices (along with all those subsimplices that do not belong to the remaining maximal simplices) in order of *decreasing* index. The condition in Definition 12.1 then says that the intersection of the simplex that is being removed with the remainder of the complex is pure, having dimension one less than this simplex. Alternatively, one could also think that our generalized simplicial complex is being glued together from the maximal simplices, in order of *increasing* index. In this light, the term *gluing order* would probably be more precise in describing what happens.

A maximal simplex σ is called *spanning* with respect to the given shelling order if it is glued along its entire boundary. We denote the set of the spanning simplices by Σ. As a matter of fact, this set depends on the shelling order. However, for reasons of brevity, we shall omit its mention from our notation.

We remark for future use that the conditions of Definition 12.1 can equivalently be formulated as follows.

Proposition 12.2. *Assume that Δ is a generalized simplicial complex and $F_1, \ldots, F_t$ are its maximal simplices. The complex Δ is shellable if and only if the following holds: whenever σ is a simplex in $F_i \cap F_j$, for some $1 \leq i < j \leq t$, such that $\dim \sigma \leq \dim F_j - 2$, there must exist $1 \leq k < j$ and a simplex τ in $F_k \cap F_j$ such that τ contains σ as a proper subsimplex.*

By repeated application of the condition in Proposition 12.2, one can always choose the simplex τ so that $\dim \tau = \dim F_j - 1$, i.e., $\tau = F_j \setminus \{x\}$, for some vertex x of F_j.

Note that in the case of an abstract simplicial complex, the condition of Proposition 12.2 becomes somewhat simpler; namely, it says that for any $1 \leq i < j \leq t$, there exists $1 \leq k < j$ such that $F_k \cap F_j = F_j \setminus \{x\}$, for some $x \notin F_i$; in other words, $|F_j \setminus F_k| = 1$ and $F_k \cap F_j \supseteq F_i \cap F_j$. In the actual proofs, it is often more convenient to check this condition instead.

Another obvious but useful fact is that if we are given a shelling order, then those maximal simplices that are glued along the entire boundary may be glued at any later point, and this will give a shelling order as well. We can therefore always assume that these simplices occur as the tail part of the shelling order. The order in which they appear in the tail is again of no importance.

The next theorem summarizes the most important properties of a shellable generalized simplicial complex.

Theorem 12.3. *Assume that Δ is a shellable generalized simplicial complex, with $F_1, F_2, \ldots, F_t$ being the corresponding shelling order of the maximal simplices, and Σ being the set of spanning simplices. Then the following facts hold:*

(1) The generalized simplicial complex obtained by the removal of the interiors of the spanning simplices, that is, the complex $\widetilde{\Delta} := \Delta \setminus \bigcup_{\sigma \in \Sigma} \mathrm{Int}\,\sigma$, is collapsible. Even stronger, this complex can be obtained from a simplex by a sequence of gluing simplices not merely over collapsible subcomplexes, but in fact over cones.

(2) The generalized simplicial complex Δ is homotopy equivalent to a wedge of spheres that are indexed by the spanning simplices and have corresponding dimensions. More precisely, we have

$$\Delta \simeq \bigvee_{\sigma \in \Sigma} \mathbb{S}^{\dim \sigma}. \tag{12.1}$$

(3) The cohomology groups of Δ with integer coefficients are free, and the set of elementary cochains $\{\sigma^*\}_{\sigma \in \Sigma}$ can be taken as a basis.

Proof. We think of Δ as being glued from the maximal simplices, adding them one by one, with the index increasing in the shelling order. As mentioned in the beginning of this section, the condition in Definition 12.1 implies that each new simplex is glued either over a contractible subcomplex, which is in fact a cone, or over its entire boundary. The statement (1) follows immediately.

To see statement (2) note that by the previous remark, we can first glue all the maximal simplices that are not spanning, and then proceed with the spanning ones. After the first part we have a contractible complex $\widetilde{\Delta}$ in hand. By Proposition 7.8 we see that Δ is homotopy equivalent to $\Delta/\widetilde{\Delta}$, with the quotient map giving a homotopy equivalence. The latter is in effect obtained by gluing the spanning simplices over their entire boundaries to a point, which is thus a wedge point. Both statements (2) and (3) now follow. $\square$

When additionally the complex Δ is pure, it follows from Theorem 12.3(2) that the reduced Betti number is nonzero only in the top dimension. Therefore, by the Euler–Poincaré formula, in this case the cohomology groups can be computed simply by computing the Euler characteristic. In the even more special case that Δ is an order complex of a poset $\Delta = \Delta(P)$, by Hall's theorem, it suffices to compute the value of the Möbius function $\mu_P(\hat{0}, \hat{1})$.

Remark 12.4. As mentioned above discrete Morse theory is more powerful as a method than shellability. The rationale for this fact is provided by Theorem 12.3(1), saying that the complex $\widetilde{\Delta} := \Delta \setminus \bigcup_{\sigma \in \Sigma} \mathrm{Int}\,\sigma$ is collapsible, coupled with the fact that a generalized simplicial complex is collapsible if and only if there exists an acyclic matching on the set of its simplices; see Theorem 11.13(a) and Remark 11.14.

In fact, any shelling order gives several collapsing sequences on the complex $\widetilde{\Delta}$. To find one, notice that adding a new simplex along the shelling order adds a Boolean lattice to the face poset; hence the acyclic matching can be extended by any acyclic matching on this Boolean lattice.

12.1.2 Shelling Induced Subcomplexes

The next theorem provides a handy criterion for being able to conclude that an *induced* subcomplex (see Definition 2.40) of a shellable complex is shellable as well.

Theorem 12.5. *Assume that Δ is a shellable generalized simplicial complex, and let S be some subset of the set of vertices of Δ. Assume furthermore that the induced generalized simplicial complex $\Delta[S]$ satisfies the following condition:*

if σ is a maximal simplex in Δ, then $\sigma \cap \Delta[S] = \sigma[S]$ is a maximal simplex in $\Delta[S]$.

In this case, the generalized simplicial complex $\Delta[S]$ is shellable as well, and a shelling order on its maximal simplices is induced by any shelling order on Δ.

Proof. First we notice that any maximal simplex τ of $\Delta[S]$ can be obtained as an intersection of $\Delta[S]$ with some maximal simplex of Δ: just take any one that contains τ, and use the fact that we are dealing with an induced subcomplex. Since the converse of that statement is the assumption of the theorem, we see that the list of intersections of maximal simplices of Δ with $\Delta[S]$ consists of all maximal simplices of $\Delta[S]$, though each maximal simplex will typically appear several times on that list.

Let now $F_1, \ldots, F_t$ be the maximal simplices of Δ arranged in a shelling order. Let $D_1, \ldots, D_m$ be the list of the maximal simplices of $\Delta[S]$ obtained from the list $F_1 \cap \Delta[S], \ldots, F_t \cap \Delta[S]$ by going left to right and deleting a simplex if it has already occurred on the list. We let i_j denote the index of the source of the maximal simplex D_j, i.e., we have $D_j = F_{i_j} \cap \Delta[S]$, and furthermore, $D_j = F_i \cap \Delta[S]$ implies $i_j \leq i$. We claim that $D_1, \ldots, D_m$ is a shelling order for $\Delta[S]$.

We shall verify the equivalent shelling condition from Proposition 12.2. Let $1 \leq a < b \leq m$, and consider a simplex σ in the intersection $D_a \cap D_b$ such that $\dim \sigma \leq \dim D_b - 2$. By our construction, we have $i_a < i_b$, and of course, σ is a simplex of $F_{i_a} \cap F_{i_b}$, with $\dim \sigma \leq \dim F_{i_b} - 2$. Therefore, since $F_1, \ldots, F_t$ is a shelling order for the generalized simplicial complex Δ, there must exist $j < i_b$ such that $F_j \cap F_{i_b}$ contains a simplex τ, which in turn contains σ as a proper subsimplex. As remarked earlier, this simplex τ can always be chosen so that $\dim \tau = \dim F_{i_b} - 1$; in other words, $\tau = F_{i_b} \setminus \{x\}$, with $x \notin \sigma$. We now consider two different cases.

Case 1. Assume that $x \notin S$. Then, since $F_j \supseteq \tau$, we have

$$F_j \cap \Delta[S] \supseteq \tau \cap \Delta[S] = F_{i_b} \cap \Delta[S] = D_b.$$

On the other hand, $F_j \cap \Delta[S]$ is a maximal simplex of $\Delta[S]$; hence $F_j \cap \Delta[S] = D_b$. This, together with $j < i_b$, contradicts our construction of the list $D_1, \ldots, D_m$.

Case 2. Assume that $x \in S$. Then $\tau \cap \Delta[S] = D_b \setminus \{x\}$. Set $D_k := F_j \cap \Delta[S]$. Clearly, by the way the list $D_1, \ldots, D_m$ was constructed, the inequality $j < i_b$ implies $k < b$. On the other hand, we have

$$D_k \cap D_b = (F_j \cap \Delta[S]) \cap (F_{i_b} \cap \Delta[S]) = (F_j \cap F_{i_b}) \cap \Delta[S] \supseteq \tau \cap \Delta[S] = D_b \setminus \{x\}.$$

Since $x \notin \sigma$ and $\dim \sigma \leq \dim D_b - 2$, we conclude that the simplex $D_b \setminus \{x\}$ contains σ as a proper subsimplex.

These two cases verify that $D_1, \ldots, D_m$ is a shelling order for $\Delta[S]$. $\square$

The conditions of Theorem 12.5 are reasonably restrictive. They have to be, since most subcomplexes of the shellable complexes are not shellable. However, it turns that in combinatorial situations these conditions are often satisfied in a natural way.

12.1.3 Shelling Nerves of Acyclic Categories

One standard situation in which shellability has often been used is the study of the order complexes of partially ordered sets. Classically, posets whose order complexes are shellable are themselves called shellable. There is, however, no difficulty whatsoever to extend the framework of shellability to encompass the case of nerves of acyclic categories, which are generalized simplicial complexes as well.

One standard convention in the literature is always to assume that the poset P has the maximal and the minimal elements, and that we are shelling the order complex $\Delta(\bar{P})$. For acyclic categories this would correspond to the assumption of existence of initial and terminal elements. We shall be explicit in our statements and not use this assumption implicitly.

When an acyclic category is shown to be shellable, and all of its maximal chains have the same number of morphisms, the number of spheres can be found by computing the Möbius function. The latter can often be done by some explicit combinatorial counting procedure.

Theorem 12.5 can frequently be applied in this situation. When the acyclic category C is graded, one can take any rank selection R. Clearly, the nerve of R is an induced subcomplex of the nerve of C. Furthermore, one can see that any maximal chain of C will intersect R in a maximal chain; hence the conditions of Theorem 12.5 are satisfied. We can therefore conclude the following proposition.

Proposition 12.6. *If under the conditions above, the nerve of C is shellable, then so is the nerve of R. The explicit shelling order for $\Delta(R)$ can be obtained from any shelling order for $\Delta(C)$ by taking the restriction and then omitting repetitions.*

One can also consider questions of what happens if one performs other constructions involving shellable acyclic categories. However, since our foremost interest in shellability is its topological consequences, we are not interested in constructions for which it is clear what happens with the topology. Therefore, we have limited our considerations to the case above.

12.2 Lexicographic Shellability

12.2.1 Labeling Edges as a Way to Order Chains

One possible procedure for ordering maximal simplices of Δ is to associate to each simplex a string of numbers, and then take the lexicographic order on these.

More specifically, let P be a poset, and let $\widehat{P}$ denote the poset obtained from P by augmenting it with a minimal and a maximal element. Let us do as follows:

(1) Label with integers all the covering edges in the poset $\widehat{P}$, including the edges $(\hat{0}, x)$ and $(y, \hat{1})$.

(2) The maximal simplices of the order complex $\Delta(P)$ correspond to maximal chains of $\widehat{P}$. Associate to each such chain a string of integers by reading off the labels from the covering edges, starting from below, and then order the maximal simplices of $\Delta(P)$ following the lexicographic order on these strings.

Naturally, in order to be able to decide uniquely which of the simplices we should take first, we require that no two maximal simplices receive the same string of labels. Observe that this actually implies that in every interval the maximal chains can be lexicographically ordered as well, and no two chains will receive the same string of labels.

The natural question that arises now is, *what conditions should we put on the edge labeling so that the order of the maximal simplices that is obtained in the way described above will actually be a shelling order?*

There is almost no difference between considering pure posets (i.e., posets in which maximal chains all have the same length) and the nonpure ones, so we will not make any distinction. However, there is one additional condition, which we always require to be satisfied, whenever we are labeling a nonpure poset.

Prefix condition. *For any interval $[x, y]$ of $\widehat{P}$ and for any two maximal chains m_1 and m_2 in $[x, y]$, the label sequence of m_1 is not a prefix of the label sequence of m_2.*

Clearly, the prefix condition subsumes the requirement that in any interval no two chains may have the same string of labels.

Remark 12.7. The additional condition for the nonpure poset is needed in order to make sure that we avoid the following peculiar situation. It may happen that there are two maximal chains c and d differing from each other

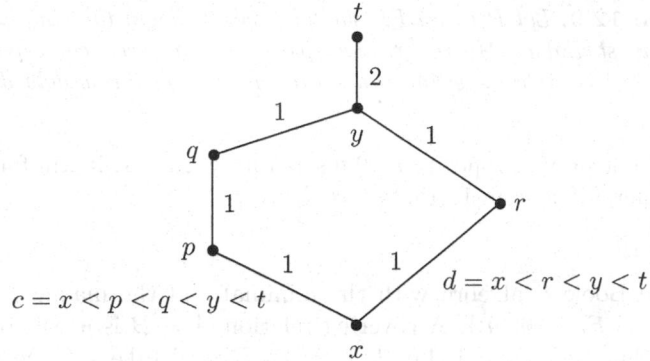

Fig. 12.1. What can go wrong in the nonpure case.

only in the interval $[x, y]$ such that $c \prec d$, but $c|_{[x,y]} \succ d|_{[x,y]}$. See Figure 12.1 for an example.

12.2.2 EL-Labeling

The following condition is sufficient to guarantee shellability of the order complex.

Definition 12.8. *A poset P is said to be **EL-shellable** if one can label covering edges of $\widehat{P}$ with elements from a poset Λ so that for every interval $[x, y]$ in $\widehat{P}$, the following **EL-conditions** are satisfied:*

(i) there is a unique increasing maximal chain c in $[x, y]$ (increasing means that the associated labels form a strictly increasing sequence);
(ii) $c \prec c'$ for all other maximal chains c' in $[x, y]$.

*The labeling satisfying these conditions is called an **EL-labeling**.*

In the formulation of Definition 12.8 we have used the symbol $\prec$ to mean "lexicographically preceding." We will often say "lexicographically less" or just "less." We recall here that a sequence of poset elements $(\lambda_1, \ldots, \lambda_t)$ is said to lexicographically precede another sequence of elements $(\mu_1, \ldots, \mu_q)$ from the same poset if there exists $k \leq \min(t, q)$ such that $\lambda_i = \mu_i$, for all $1 \leq i \leq k - 1$ and $\lambda_k < \mu_k$, $\lambda_k \neq \mu_k$.

Given an edge labeling, we call a maximal chain *weakly decreasing* if the associated string of labels is weakly decreasing. Here, a sequence of poset elements $(\lambda_1, \ldots, \lambda_t)$ is said to be weakly decreasing if for any $1 \leq i \leq t - 1$ we do not have $\lambda_{i+1} > \lambda_i$, i.e., either the elements λ_{i+1} and λ_i are incomparable, or $\lambda_{i+1} \leq \lambda_i$

Proposition 12.9. *Let P be an EL-shellable poset. Then the simplicial complex $\Delta(P)$ is shellable. Moreover, the spanning simplices corresponding to the induced lexicographic shelling order are indexed by the weakly decreasing chains.*

We shall not prove Proposition 12.9 separately; rather, it will follow from the more general Theorem 12.15.

Example 12.10.

(1) Take the Boolean algebra with the minimal and the maximal elements removed, $P := \mathcal{B}_n \setminus \{\emptyset, [n]\}$. A covering relation $A \succ B$ is a pair of subsets of $[n]$ such that $|A \setminus B| = 1$. Let $\{x\} = A \setminus B$, and take x to be the label of $A \succ B$. One can check that this is an EL-labeling. Furthermore, there is exactly one weakly decreasing chain, obtained by arranging the elements of the set $[n]$ in decreasing order.

By Proposition 12.9 we can therefore conclude that $\Delta(P)$ is homotopy equivalent to a sphere of dimension $n - 2$. Of course, we know that $\Delta(P)$ is actually homeomorphic to a sphere of dimension $n-2$, but we cannot conclude this from the existence of the EL-labeling alone.

(2) Let k, m be positive integers such that $1 \leq k \leq m \leq n$. Take P to be the following rank selection of $\mathcal{B}_n$: for $S \in \mathcal{B}_n$ we have $S \in P$ if and only if $k \leq |S| \leq m$. It follows from the previous example and Proposition 12.6 that $\Delta(P)$ is shellable.

Let us now see that the poset P is EL-shellable. We label the edges of $\widehat{P}$ as follows:

- the edges $S \prec T$, for $S \neq \hat{0}$, $T \neq \hat{1}$, are labeled with the unique element of $T \setminus S$, as in (1);
- the edges $\hat{0} \prec S$ are labeled with $\max S$;
- finally, the edges $(S, \hat{1})$ are labeled with $\min([n] \setminus S)$.

Let us check that this yields an EL-labeling.

First, the intervals $[S, T]$, with $S \neq \hat{0}$, $T \neq \hat{1}$, are the same as in (1); therefore the conditions for the EL-labeling are satisfied.

Second, consider the interval $[\hat{0}, S]$, for $S \neq [n]$. The lexicographically least chain is obviously the one that starts with the set K, consisting of the k smallest elements of S, and then proceeds toward S by adding the elements of $S \setminus K$ in increasing order. It is also the unique increasing chain, since if we start our chain with some k-subset K' that is different from K, then $\max K' > \max K$, and somewhere along the chain we will have a label that is less than $\max K'$.

Third, consider the interval $[S, \hat{1}]$, for $S \neq \emptyset$. Let T be the subset consisting of the $n - m$ largest elements of $[n] \setminus S$; we have $S \subseteq [n] \setminus T \subseteq [n]$. The lexicographically least chain in $[S, \hat{1}]$ proceeds from S to $[n] \setminus T$ by adding the elements of $[n] \setminus (S \cup T)$ in increasing order, and then as the last step the whole subset T is added. This is the unique increasing chain, since otherwise, the chain would have to end with adding some subset T' different from T, because $\min T' < \min T$ implies that we would somewhere before have to have a label that is larger than $\min T'$.

Finally, consider the interval $[\hat{0}, \hat{1}]$. By essentially amalgamating the arguments of the two previous cases, we see that the unique increasing chain is given by $\emptyset < \{1, \ldots, k\} < \{1, \ldots, k, k+1\} < \cdots < \{1, \ldots, n-m\} < [n]$.

The poset is pure, so all the weakly decreasing chains have the same length, and we just have to count how many there are. Let c be a weakly decreasing chain. Assume that c starts with S and ends with $[n] \setminus T$, and let $[n] \setminus (T \cup S) = \{a_1, \ldots, a_{m-k}\}$ such that $a_1 > \cdots > a_{m-k}$. Then the part of c between S and $[n] \setminus T$ is uniquely determined, and the string of labels assigned to c is $(\max S, a_1, \ldots, a_{m-k}, \min T)$. We see that c is weakly decreasing if and only if $\max S > \max([n] \setminus (S \cup T))$ and $\min T < \min([n] \setminus (S \cup T))$, and that the number of weakly decreasing chains is equal to the number of ways to choose the k-subset S and the $(n-m)$-subset T satisfying these inequalities.

12.2.3 General Lexicographic Shellability

Definition 12.11. *We say that a poset P has a **LEX-labeling** if we can label edges of $\widehat{P}$ with elements of a poset Λ so that the following condition is satisfied:*

LEX-condition. *For any interval $[x, t]$, any maximal chain c in $[x, t]$, and any $y, z \in c$ such that $x < y < z < t$, if $c|_{[x,z]}$ is lexicographically least in $[x, z]$ and $c|_{[y,t]}$ is lexicographically least in $[y, t]$, then c is lexicographically least in $[x, t]$.*

*A poset is called **lexshellable** if it possesses a LEX-labeling.*

We shall see that, in a sense that will be made precise by Theorem 12.15(1), this is the most general condition possible. Yet it is not more difficult to check it than the EL-conditions, and additionally, this condition allows labelings that are often more natural for the examples considered.

The LEX-condition is illustrated in Figure 12.2. It is immediate that the EL-condition implies the LEX-condition. Indeed, the chains $c|_{[x,z]}$ and $c|_{[y,t]}$ would be increasing, so since they overlap, the chain $c = c|_{[x,z]} \cup c|_{[y,t]}$ would be increasing too.

Remark 12.12. It is not difficult to see that taking integers as labels in Definition 12.11 does not make it less general, since taking a linear extension of Λ will give us a LEX-labeling again. However, it is often more natural to have the elements of some poset as labels.

The corresponding question for EL-shellability is still open. However, since EL-shellability implies lexshellability, it loses its attractiveness.

The LEX-condition has several equivalent formulations.

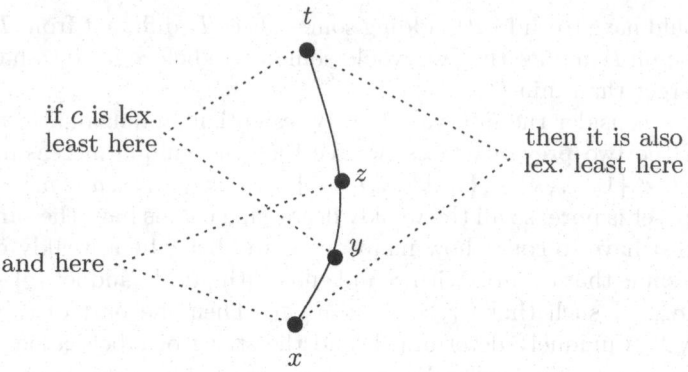

Fig. 12.2. The condition for the lexicographic shelling.

Proposition 12.13. *For any poset P and for any edge labeling of $\widehat{P}$, the following conditions are equivalent to the LEX-condition:*

(1) **SLEX-condition ("short lexicographic condition").**
For any interval $[x,t]$, any maximal chain c in $[x,t]$, and any $y, z \in c$ such that $x \prec y \prec z$, we know that if $c|_{[x,z]}$ is lexicographically least in $[x,z]$ and $c|_{[y,t]}$ is lexicographically least in $[y,t]$, then c is lexicographically least in $[x,t]$.

(2) **BS-condition ("bad subchain condition").**
For any interval $[x,t]$, any maximal chain c in $[x,t]$ such that c is not lexicographically least in $[x,t]$, and $c|_{[x,t]}$ contains at least two elements excluding x and t, there exist elements $y, z \in c$ such that $c|_{[y,z]}$ is a proper subchain of c and $c|_{[y,z]}$ is not lexicographically least in $[y,z]$.

(3) **SBS-condition ("short bad subchain condition").**
For any interval $[x,t]$, any maximal chain c in $[x,t]$ such that c is not lexicographically least in $[x,t]$, there exist elements $y, q, z \in c$ such that $y \prec q \prec z$ and $c|_{[y,z]}$ is not lexicographically least in $[y,z]$.

Proof. It is obvious that (LEX) $\Rightarrow$ (SLEX) and (BS) $\Leftrightarrow$ (SBS).
(SLEX) $\Rightarrow$ (SBS).
Condition (SLEX) can be reformulated in the following way: if c is not lexicographically least in $[x,t]$, then either $c|_{[x,z]}$ is not lexicographically least in $[x,z]$ or $c|[y,t]$ is not lexicographically least in $[y,t]$, which proves the SBS-condition.
(SBS) $\Rightarrow$ (LEX).
Consider an interval $[x,t]$, c a maximal chain in $[x,t]$, $y, z \in c$, $x < y < z < t$, such that c is not lexicographically least in $[x,t]$, but $c|_{[x,z]}$ is lexicographically least in $[x,z]$ and $c|_{[y,t]}$ is lexicographically least in $[y,t]$. Then there exist $p, q, r \in c$ such that $p \prec q \prec r$ and $c|_{[p,r]}$ is not lexicographically least in $[p,r]$. Obviously either $y \leq p$ or $r \leq z$. Assume $y \leq p$ (the other case goes along

the same lines). Then $p, q, r \in c|_{[y,t]}$. Since $c|_{[p,r]}$ is not lexicographically least in $[p, r]$, we conclude that $c|_{[y,t]}$ is not lexicographically least in $[y, t]$, which gives a contradiction. $\square$

Definition 12.14. *Given an edge labeling of a poset, we say that a saturated chain c is **mediocre** if for any $x, y, z \in c$ such that $x \prec y \prec z$, the chain $c|_{[x,z]}$ is not lexicographically least in $[x, z]$.*

So, a mediocre chain c does not have to be the lexicographically worst (maximal) one, but it is also never the lexicographically best (minimal) when restricted to any given interval. Clearly, in the special case of the EL-labeling, the mediocre chains are precisely the weakly decreasing ones.

Theorem 12.15. *Let P be a poset.*

(1) The following two statements are equivalent:

(a) the simplicial complex $\Delta(P)$ is shellable and it is possible to label the edges of $\widehat{P}$ with elements of some poset so that the induced lexicographic ordering of the maximal chains of $\widehat{P}$ gives a shelling order;
(b) the poset P is lexshellable.

(2) In the shelling order induced by a LEX-labeling, the mediocre maximal chains correspond to the spanning simplices, and can therefore be used to deliver a basis for the cohomology $H^*(\Delta(P); \mathbb{Z})$.

Proof. We start by proving (1).
(b) $\Rightarrow$ (a).
Let P be a lexshellable poset, and let λ be a LEX-labeling of edges of $\widehat{P}$. We will prove that the lexicographic ordering of the maximal chains in $\widehat{P}$ gives a shelling order on the maximal simplices of $\Delta(P)$.

Let c_1, c_2 be two maximal chains such that $c_1 \prec c_2$, and let $[a, b]$ be the first interval on which c_1 and c_2 differ; see Figure 12.3. Consider the restrictions $\alpha = c_1|_{[a,b]}$ and $\beta = c_1|_{[a,b]}$. Since $c_1 \prec c_2$, by Remark 12.7 we must have $\alpha \prec \beta$. In particular, by the SBS-condition, there must exist $x, y, z \in \beta$ such that $x \prec y \prec z$, and the chain $c_2|_{[x,z]}$ is not lexicographically least in $[x, z]$.

This implies that there must exist a chain γ in $[x, z]$ that lexicographically precedes $c_2|_{[x,z]}$. Consider the following concatenation: $d := c_2|_{[\hat{0},x]} \circ \gamma \circ c_2|_{[z,\hat{1}]}$. Clearly, $d \prec c_2$ and $d \cap c_2 = c_2 \setminus \{y\} \supseteq c_1 \cap c_2$, which verifies the shelling condition of Definition 12.1.
(a) $\Rightarrow$ (b).
Reading the proof of the other direction backward, we see that the shelling condition implies the SBS-condition, which by Proposition 12.13 implies that the poset is lexshellable.

Finally, we verify (2). The spanning simplices are by definition those that are glued along the entire boundary in the shelling process. Clearly, on the level of chains, this means those chains c that can be replaced by a lexicographically

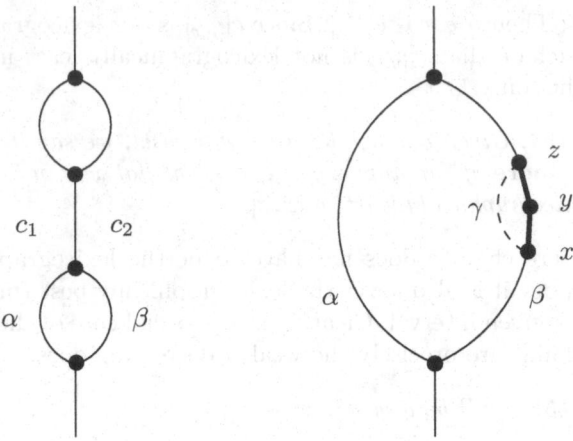

Fig. 12.3. Two maximal chains.

preceding chain on any subchain of length 2. This, in turn, is precisely the definition of the mediocre chains. □

Let us return to example (2) above, that of the rank selection of a Boolean algebra. Consider the following labeling:

- the edges $S \prec T$, for $S \neq \hat{0}$, $T \neq \hat{1}$, are labeled with the unique element of $T \setminus S$, as in (1);
- the edges $\hat{0} \prec S$ are labeled with the set S (these sets are ordered lexicographically);
- finally, the edges $(S, \hat{1})$ get arbitrary labels (one may also simply abstain from labeling these).

Let us verify that this is a LEX-labeling. Consider an interval $[x, t]$, a maximal chain c in $[x, t]$, and $y, z \in c$ such that $x < y < z < t$, and such that $c|_{[x,z]}$ is lexicographically least in $[x, z]$ and $c|_{[y,t]}$ is lexicographically least in $[y, t]$.

The fact that $c|_{[x,z]}$ is lexicographically least in $[x, z]$ means that either the labels of the chain are increasing, or, if $x = \emptyset$, that we start with the minimal subset of z and then proceed with an increasing sequence of labels. Symmetrically, the fact that $c|_{[y,t]}$ is lexicographically least in $[y, t]$ means that we start with an increasing sequence of labels, which additionally, if $t = [n]$, constitute the minimal elements of $[n] \setminus y$.

This implies that c is lexicographically least in $[x, t]$: if $x = \emptyset$, then c must start with the set consisting of the k smallest elements in t; if $t = [n]$, then c must end by adding the $n - m$ largest elements of $[n] \setminus t$, and along the way the elements are added in increasing order.

We remark that this example is rather characteristic for what happens in the lexshellable posets. The labels that one uses are often more natural and suitable for the combinatorial situation at hand than those in EL-labelings.

Importantly, one has to know only how to compare labels on the edges on the same level. In fact, if additionally, all edges leaving a vertex upward happen to have different labels, then we need to know only how to compare these.

A large class of posets for which it appears inopportune to try to compare edges on different levels is that of the intersection lattices of subspace arrangements. This is because the edges $(\hat{0}, x)$ encode the subspaces themselves, whereas other edges encode the combinatorial nature of the subspace intersections. On the other hand, knowing the homology groups of the nerve complexes of the intersection lattices (and intervals therein) is very important, since according to the Goresky–MacPherson formula, these encode the cohomology groups of the complements of subspace arrangements.

Finally, we note that a rank selection of a lexshellable ranked poset is again lexshellable. In fact, the labels are easy to make out of the labels for the original poset: just label each new covering relation $x \prec y$ with the lexicographically least label sequence for a maximal chain connecting x with y in the original poset. Once again, we benefit from the fact that we do not have to compare labels of edges on different levels.

12.2.4 Lexicographic Shellability and Nerves of Acyclic Categories

For the sake of simplicity we have so far discussed lexicographic shellability in the context of order complexes of posets. It turns out that working in the generality of nerves of acyclic categories does not cause any substantial problems, and essentially everything can be extended to this context.

So let us assume that C is a finite acyclic category, and let $\widehat{C}$ denote the category obtained from C by augmenting it with an initial and a terminal object. We would like to shell the nerve $\Delta(C)$. This time around, we shall label not the edges in the Hasse diagram of a poset, but rather the morphisms that cannot be represented as a composition of two morphisms, none of which is an identity. Recall that we called such morphisms indecomposable.

The main difference in the acyclic category case is that the notion of the interval is replaced with the notion of the morphism. Accordingly, a maximal chain in the interval is replaced with the composable sequence of indecomposable morphisms, which together compose to yield the corresponding morphism. To abbreviate our language, we say that a maximal composable sequence of morphisms $(m_1, \ldots, m_k)$ is "in m" if it composes to m.

The translation of the prefix condition is then straightforward. The translation of the LEX-condition is very easy as well. However, we write it out explicitly for future reference.

LEX-condition for acyclic categories.
For any morphism m, and any composable sequence of indecomposable morphisms $c = (m_1, \ldots, m_k)$ in m, let $i, j \in [k]$ be some indices such that $1 \leq i < j < k$. If $(m_1, \ldots, m_j)$ is lexicographically least in $m_j \circ \cdots \circ m_1$, and $(m_{i+1}, \ldots, m_k)$ is lexicographically least in $m_k \circ \cdots \circ m_{i+1}$, then c is lexicographically least in m.

Definition 12.16. *We call an acyclic category C* **lexshellable** *if the augmented category $\widehat{C}$ possesses a labeling of indecomposable morphisms satisfying the LEX-condition.*

The notion of mediocre chains gets replaced by the composable sequences of indecomposable morphisms such that each subsequence is not lexicographically least in the corresponding composition morphism.

Theorem 12.17. *Let C be an acyclic category.*

(1) *The following two statements are equivalent:*

(a) *the generalized simplicial complex $\Delta(C)$ is shellable and it is possible to label the edges of $\widehat{C}$ with integers so that the induced lexicographic ordering of the maximal chains of C gives a shelling order;*
(b) *the acyclic category C is lexshellable.*

(2) *In the shelling order induced by a LEX-labeling, the mediocre composable sequences of indecomposable morphisms correspond to the spanning simplices, and can therefore be used to deliver a basis for the cohomology $H^*(\Delta(C); \mathbb{Z})$.*

Proof. The proofs of both Proposition 12.13 and Theorem 12.15 hold mutatis mutandis with the notion of interval replaced with that of a morphism, and the notion of covering replaced with the notion of indecomposable morphism. It is also handy to use the equivalent condition for shellability given in Proposition 12.2. We leave the details to the reader.   □

12.3 Bibliographic Notes

Results on shellability date back at least to [BM71]. The notion of EL-shellability for pure posets was first introduced in [Bj80, Chapter 2]. The general lexicographic shellability (lexshellability) described in Subsection 12.2.3 was first introduced in the author's Ph.D. thesis; see also [Ko97]. It is curious that this concept was not discovered earlier, taking into account the equivalence given by Theorem 12.15.

Lexicographic shellability for the nerves of acyclic categories appears in Subsection 12.2.4 for the first time.

We also refer the interested reader to the wonderful book [GoM88] and to [Ko97] for further comments on connections between topology of posets and arrangements.

13

Evasiveness and Closure Operators

One of the classical applications of topological methods in combinatorics is the proof of the so-called Evasiveness Conjecture for graphs whose number of vertices is a prime power. In this chapter we describe the framework of the problem, sketch the original argument, and prove some important facts about nonevasiveness. One of the important tools is the so-called closure operators, which are also useful in other contexts.

13.1 Evasiveness

13.1.1 Evasiveness of Graph Properties

Let us fix a natural number n and consider the set of all graphs with n vertices. If the vertices are labeled, then we have a permutation action of $\mathcal{S}_n$, which induces an action on the set of labeled graphs. This action maps graphs to isomorphic graphs, and in fact by definition, the orbits of this action are precisely the isomorphism classes of graphs. Recall that a *graph property* is simply a set of unlabeled graphs, which is the same as the set of labeled graphs closed under the above-mentioned permutation action. A graph from this set is then said to satisfy this property.

Definition 13.1. *A graph property is called* **monotone** *if the set of graphs that satisfy this property is closed under removal of edges.*

Examples of monotone properties include planarity and the property of being disconnected. The condition in Definition 13.1 is basically the same condition as the one defining abstract simplicial complexes. In fact, the following is the standard construction associating an abstract simplicial complex to a monotone graph property.

Definition 13.2. *Given a monotone graph property $\mathcal{G}$ of graphs with n vertices, the abstract simplicial complex $\Delta(\mathcal{G})$ has $\binom{n}{2}$ vertices labeled by ordered* *pairs (i,j), for $1 \leq i < j \leq n$, which correspond to potential edges in the graph. A set of such pairs forms a simplex in $\Delta(\mathcal{G})$ if and only if the corresponding graph has the property $\mathcal{G}$.*

A graph property is called *trivial* if either no graphs with n vertices satisfy it, in which case $\Delta(\mathcal{G})$ is void, or all graphs with n vertices satisfy it, in which case $\Delta(\mathcal{G})$ is an $(n-1)$-dimensional simplex.

Let us now consider the following algorithmic situation. We are given a certain graph property $\mathcal{G}$, which we know, and a certain graph G, which we do not know. However, we can ask an oracle questions of the type

is the edge (i,j) in G?

Our task is to decide whether graph G satisfies the property $\mathcal{G}$. We would like to ask as few questions as possible. More precisely, we are interested in knowing how many questions we have to ask in the worst-case scenario. For example, if the graph property is trivial, then we do not need to ask any questions at all. On the other hand, if the property is being a complete graph, then we might be forced into asking $\binom{n}{2}$ questions. This would happen if the answers to the first $\binom{n}{2} - 1$ questions were all positive.

Definition 13.3. *A graph property $\mathcal{G}$ of graphs on n vertices is called* **evasive** *if in the worst case we need to ask $\binom{n}{2}$ questions in the course of the algorithm described above. Otherwise, the graph property is called* **nonevasive**.

By a simple induction on the number of vertices, it will follow that if $\mathcal{G}$ is a nonevasive graph property, then the abstract simplicial complex $\Delta(\mathcal{G})$ is collapsible; see Proposition 13.7 and Proposition 13.9.

It is not all too trivial to construct nonevasive graph properties. One example of such a property is being a so-called *scorpion graph*. Here a graph G on n vertices is called scorpion if it has 3 vertices s, t, and b such that

- vertex s, called the sting, is connected only to vertex t;
- vertex t, called the tail, is connected only to vertices s and b;
- vertex b, called the body, is connected to all vertices but s.

Let us call vertices s, t, and b as above *special*. See Figure 13.1 for an example of a scorpion graph.

To see that being a scorpion graph is nonevasive, note first that if we have a scorpion graph, then the special vertices are determined uniquely: the body is the unique vertex of degree $n-2$, the sting is the unique vertex that is not connected to the body, and the tail is the unique vertex that is connected to the sting. Moreover, given any one of them, the other two can be found; alternatively, it can be checked in linear time whether they exist. This can be done by asking all the edges connected to the vertex suspected to be special and following the found edges. For example, if we suspect that x is a sting, we can first test all the adjacent edges. We must find precisely one edge, say (x, y). Then we test all the edges adjacent to y. We must find precisely one

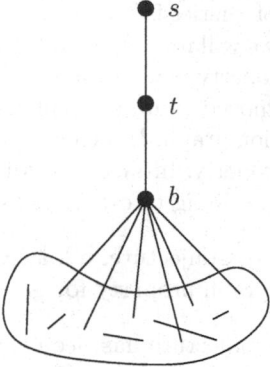

Fig. 13.1. An example of a scorpion graph.

more edge, say (y, z). Finally, test all the edges adjacent to z and verify that z has valency $n - 2$; otherwise, the graph is not a scorpion graph. We proceed similarly if we suspect x to be a tail or a body.

Equipped with these observations we can now perform the following algorithm. Take any vertex x and test all edges adjacent to x. If the valency of x, denoted by val(x), is 0 or $n - 1$, then the graph is not a scorpion graph. If its valency is 1, 2, or $n - 2$, then either x itself has to be special, or, when val$(x) = 1$ or 2, one of its neighbors has to be special. Each one can be checked in linear time. Assume therefore that $3 \leq$ val$(x) \leq n - 3$. In this case the only option for x is to be a nonspecial vertex. Let B denote the set of vertices adjacent to x, and let S denote the set of vertices not adjacent to x. The sting and the tail, if they exist, must lie in S, and the body, if it exists, must lie in B.

The algorithm now runs as follows. Take a vertex $y \in B$ and a vertex $z \in S$. Test the edge (y, z). If the oracle says that this is an edge, then delete z, choose a new vertex $\tilde{z}$ instead, and proceed. In this case we know that the deleted vertex z is not a sting, since it is connected to something that cannot be a tail. If, on the other hand, the oracle says that (y, z) is not an edge, then delete y, choose a new vertex $\tilde{y}$ instead, and proceed. In this case we know that either the deleted vertex is not a body, or the remaining vertex z is a sting. The algorithm terminates when either B or S becomes empty. If S is empty at termination, then our graph is not a scorpion graph, since we could not have deleted the sting in the process, and yet we did not find it. If, on the other hand, the set B is empty at termination, then for our graph to be a scorpion graph, the last tested vertex in S must have been the sting, since otherwise we did not delete the body from B and we also did not find it. Thus we finish by testing this last vertex for being a sting.

Since the total number of queries is linear in the number of vertices, we see that for sufficiently large n we will need substantially fewer than $\binom{n}{2}$ questions. In particular, this graph property is nonevasive.

Clearly, if an edge is removed from a scorpion graph, the obtained graph no longer has to be a scorpion graph. In other words, being a scorpion graph is not a monotone graph property. In fact, the following is perhaps the most important open question pertaining to evasiveness of graphs.

Conjecture 13.4. (Evasiveness Conjecture, a.k.a. Karp Conjecture)
Every nontrivial monotone graph property for graphs on n vertices is evasive.

So far, the Evasiveness Conjecture has been verified in the case when n is a prime power, and additionally, when $n = 6$. Beyond being an important fact, the proof has also acquired quite a bit of resonance due to its nontrivial use of topological techniques. We would like to sketch the argument here. First we need the following result.

Theorem 13.5. *Let Γ be a finite group, and assume that there exist a prime p and a normal p-subgroup H such that*

(1) the quotient group Γ/H is cyclic,
(2) the group Γ acts on a $\mathbb{Z}_p$-acyclic simplicial complex Δ.

Then we have $\chi(\Delta_\Gamma) = 1$, where Δ_Γ denotes the fixed-point set of the Γ-action on Δ.

Here we shall assume Theorem 13.5 without proof. Let us instead prove the Evasiveness Conjecture for prime powers, using some simple statements, which will be proved in the next subsection.

Proof of the Evasiveness Conjecture for prime powers.
Assume now that $n = p^t$, for some prime number p. Assume that the Evasiveness Conjecture is false for that value of n, and let $\mathcal{G}$ denote a monotone, but nonevasive, graph property. Furthermore, let $\mathbf{GF}(n)$ denote a field with n elements, which exists because n is a prime power, and let $\mathbf{GF}(n)^*$ denote the multiplicative group of that field. Let Γ be a subgroup of $\mathcal{S}_n$ consisting of all affine maps in $\mathbf{GF}(n)$,

$$\Gamma := \{x \mapsto ax + b \,|\, a \in \mathbf{GF}(n)^*, b \in \mathbf{GF}(n)\},$$

and let H be the subgroup of Γ consisting of all parallel translations,

$$H := \{x \mapsto x + b \,|\, b \in \mathbf{GF}(n)\}.$$

We see that H is a p-group; more precisely, $|H| = p^t$. Furthermore, H is a normal subgroup of Γ, since it is the kernel of the group homomorphism $c : \Gamma \to \mathbf{GF}(n)^*$ that takes the affine map $(x \mapsto ax + b)$ to a. The quotient group Γ/H is isomorphic to $\mathbf{GF}(n)^*$, with the isomorphism induced by the already mentioned group homomorphism c. A general fact from field theory tells us now that the group Γ/H is cyclic.

Since the graph property $\mathcal{G}$ is nonevasive, we know that the abstract simplicial complex $\Delta(\mathcal{G})$ is collapsible; see Proposition 13.7(2). In particular, it has to be $\mathbb{Z}_p$-acyclic. We can now use Theorem 13.5 to conclude that $\chi(\Delta(\mathcal{G})_\Gamma) = 1$. On the other hand, the group Γ acts doubly transitively on the set $[n]$, i.e., any ordered pair of points can mapped to any other ordered pair of points by a transformation from Γ. This means that the group Γ acts transitively on the set of vertices of the abstract simplicial complex $\Delta(\mathcal{G})$. Therefore, the only point of $\Delta(\mathcal{G})$ that can possibly be fixed by every element in Γ is the barycenter of the simplex on all $\binom{n}{2}$ vertices, which of course can be the case only when $\Delta(\mathcal{G})$ is a full simplex. This contradicts our assumption that the graph property $\mathcal{G}$ is nontrivial. $\square$

13.1.2 Evasiveness of Abstract Simplicial Complexes

As we have seen in the previous section, some of the most important results concerning evasiveness were obtained by topological methods. As a matter of fact, the whole concept can be extended to a purely simplicial context, as the next definition shows.

Definition 13.6.
(1) A finite nonempty abstract simplicial complex X is called **nonevasive** *if either X is a point, or, inductively, there exists a vertex v of X such that both $X \setminus \{v\}$ and $\mathrm{lk}_X v$ are nonevasive. Otherwise, the complex X is called* **evasive**.
(2) For two nonempty abstract simplicial complexes X and Y we write $X \searrow_{NE} Y$ (or, equivalently, $Y \nearrow_{NE} X$) if there exists a sequence $X = A_1 \supset A_2 \supset \cdots \supset A_t = Y$ such that for all $i \in \{1, \dots, t-1\}$ we can write $A_i \setminus \{x_i\} = A_{i+1}$, and the abstract simplicial complex $\mathrm{lk}_{A_i} x_i$ is nonevasive.

In the situation described in Definition 13.6(2), we say that the abstract simplicial complex X NE-*reduces* to its subcomplex Y. The following facts about NE-reduction are useful for our arguments

Proposition 13.7.

(1) If X_1 and X_2 are abstract simplicial complexes such that $X_1 \searrow_{NE} X_2$ and Y is an arbitrary abstract simplicial complex, then $X_1 * Y \searrow_{NE} X_2 * Y$. A cone over any abstract simplicial complex is nonevasive.
(2) The reduction $X \searrow_{NE} Y$ implies the existence of a collapsing sequence $X \searrow Y$, which, in turn implies that, viewed as a topological space, Y is a strong deformation retract of X.
(3) On the numerical side, if $X \searrow_{NE} Y$, then the Euler characteristics of X and Y are the same. In particular, a nonevasive abstract simplicial complex has reduced Euler characteristic equal to 0.

Proof. Statement (1) follows from the fact that if v is any vertex of an abstract simplicial complex X, then we have the equalities $\mathrm{lk}_{X*Y}v = (\mathrm{lk}_X v) * Y$ and $(X * Y) \setminus \{v\} = (X \setminus \{v\}) * Y$.

To see statement (2) we can first prove that a nonevasive abstract simplicial complex is collapsible. Perhaps the simplest argument is to use induction on the number of vertices. Indeed, if our abstract simplicial complex has one vertex, then it must be nonempty and is therefore collapsible. For the induction step we notice that for any abstract simplicial complex X and any vertex v of X, if both $\mathrm{lk}_X(v)$ and $\mathrm{dl}_X(v)$ are collapsible, then so is X. To find a collapsing sequence for X, start by collapsing away all the simplices that contain v, following some collapsing sequence for $\mathrm{lk}_X(v)$. This will remove the open star of v, and we can finish off by continuing with any collapsing sequence for $\mathrm{dl}_X(v)$.

The full generality of statement (2) follows now from the definition of NE-reduction and the fact that the open star of v can be collapsed away as long as the link of v is collapsible.

Statement (3) follows immediately from (2). $\square$

The following is a very important conjecture about evasive abstract simplicial complexes, which has now been open for quite some time.

Conjecture 13.8. (Evasiveness Conjecture for abstract simplicial complexes) Let X be a nonempty abstract simplicial complex with the vertex set $[n]$. Assume furthermore that Γ is a subgroup of $\mathcal{S}_n$ such that the permutation action of Γ on $[n]$ is transitive. Then either X is evasive or it is isomorphic to the full $(n-1)$-dimensional simplex.

The Evasiveness Conjecture for abstract simplicial complexes is still open for general n. It has been verified for the case when n is a prime power, as well as for the special cases $n = 6$, 10, and 12.

Definition 13.6(1) can be reinterpreted algorithmically as follows: we know a certain abstract simplicial complex X, and there is some chosen subset σ of the set of vertices of X that we do not know. We need to decide whether σ is a simplex of X by asking the oracle questions of the type *"is v in σ?"* where v is some vertex of X. The abstract simplicial complex X is then called evasive if in the worst case scenario, we may have to ask this question for all vertices of X.

An example of a nonevasive complex is a simplex: we need to ask no questions at all. An example of an evasive complex is given by the boundary of a simplex: if the answer to all our questions hitherto is "yes," then we will have to keep querying until the very last vertex, since the question whether σ is a simplex will hinge on the question whether it contains all vertices.

The algorithmic reformulation above makes the next proposition immediate.

Proposition 13.9. *A monotone graph property $\mathcal{G}$ is evasive if and only if the associated abstract simplicial complex $\Delta(\mathcal{G})$ is evasive.*

Proof. To decide whether a graph G has property $\mathcal{G}$ is the same as to decide whether its set of edges is a simplex of $\Delta(\mathcal{G})$. The queries "is this edge in G?" get translated precisely into the queries "is this vertex in your presumed simplex?" □

Clearly, when interpreted this way, the notion of evasiveness can be extended to an arbitrary set system. Having done that, one can as well attempt to drop the monotonicity in Conjecture 13.8, which will yield the following conjecture.

Conjecture 13.10. (Generalized Aanderaa–Rosenberg Conjecture)
Let Σ be a collection of subsets of the set $[n]$ such that $\emptyset \in \Sigma$ and $[n] \notin \Sigma$. Assume furthermore that Γ is a subgroup of $\mathcal{S}_n$ such that the permutation action of Γ on $[n]$ is transitive. Then Σ is evasive.

It is easy to prove the Generalized Aanderaa–Rosenberg Conjecture in the case when n is a prime power. First, we note that the notion of Euler characteristic can be defined for any set system Σ by setting $\chi(\Sigma) := \sum_{S \in \Sigma}(-1)^{|S|}$. Just as in the simplicial case we see that the reduced Euler characteristic of a nonevasive set system is equal to 0.

Proof of the Generalized Aanderaa–Rosenberg Conjecture in the case when n is a prime power.

Assume that $n = p^t$, for some prime number p, and assume that Σ is a nonevasive set system. Assume furthermore that the group Γ is a subgroup of $\mathcal{S}_n$ satisfying the conditions of Conjecture 13.10. Let O be an arbitrary orbit of the Γ-action on Σ. Let k denote the cardinality of the sets in O. Since the permutation action of Γ on $[n]$ is transitive, every element of $[n]$ is contained in the *same* number of sets in O, which we denote by m.

Double counting the number of pairs (A, x), where $x \in [n]$, $A \in O$, and $x \in A$, yields the identity $|O| \cdot k = p^t \cdot m$. Since $[n] \notin \Sigma$, we have $1 \leq k \leq p^t - 1$, and therefore we see that p must divide $|O|$. Clearly, the entire collection of subsets Σ is a disjoint union of Γ-orbits, and since we have shown that the cardinality of each orbit is divisible by p, we can conclude that the reduced Euler characteristic of Σ is congruent to -1 modulo p. Thus, the set system Σ cannot be nonevasive. □

The Generalized Aanderaa–Rosenberg Conjecture in its full generality is wrong. We describe here a counterexample for $n = 12$; cf. Figure 13.2. Let Γ be the cyclic subgroup of $\mathcal{S}_{12}$ generated by the cycle $(1\,2\,3\,4\,5\,6\,7\,8\,9\,10\,11\,12)$. Clearly, Γ acts transitively on the set of vertices. Let $\mathcal{A}$ be the family of all subsets of the sets of the Γ-orbit of $\{1, 4, 7, 10\}$, let $\mathcal{B}$ be the family of all subsets of the sets of the Γ-orbit of $\{1, 7\}$, and let $\mathcal{C}$ be the family of all subsets of the sets of the Γ-orbit of $\{1, 5, 9\}$. Consider the Γ-invariant set system $\Sigma := (\mathcal{A} \setminus \mathcal{B}) \cup \mathcal{C}$.

A concrete algorithm proving that the set system Σ the so-called Illies system, is nonevasive is shown in Figure 13.3. In this figure, the following

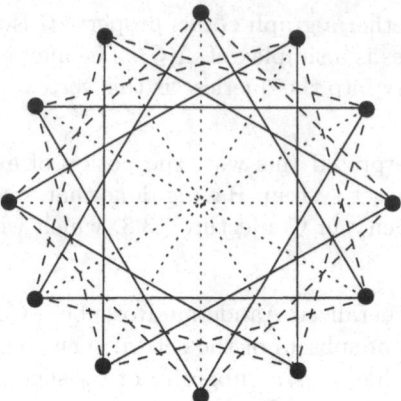

Fig. 13.2. A counterexample to the Generalized Aanderaa–Rosenberg Conjecture.

conventions are used: the rectangular box with straight corners with one or more numbers in it means "ask whether these numbers are in our set"; the rectangular box with rounded corners with a number a in it means "ask about all remaining elements *except for* a whether they are in our set" (the point here being that we will not have to ask about a, verifying that this set system is in fact nonevasive).

13.2 Closure Operators

13.2.1 Collapsing Sequences Induced by Closure Operators

The following concept is of fundamental importance in studying the topological properties of order complexes of posets.

Definition 13.11. *An order-preserving map φ from a poset P to itself is called   a **descending closure operator** if $\varphi^2 = \varphi$ and $\varphi(x) \leq x$, for any $x \in P$; analogously, φ is called an **ascending closure operator** if $\varphi^2 = \varphi$ and $\varphi(x) \geq x$, for any $x \in P$.*

Ascending and descending closure operators induce strong deformation retractions of $\Delta(P)$ onto $\Delta(\varphi(P))$. Here we give a short and self-contained inductive proof of the following stronger fact.

Theorem 13.12. *Let P be a poset, and let φ be a descending closure operator. Then $\Delta(P)$ collapses onto $\Delta(\varphi(P))$. By symmetry, the same is true for an ascending closure operator.*

Proof. We use induction on $|P| - |\varphi(P)|$. If $|P| = |\varphi(P)|$, then φ is the identity map and the statement is obvious. Assume that $P \setminus \varphi(P) \neq \emptyset$ and let $x \in P$ be one of the minimal elements of $P \setminus \varphi(P)$.

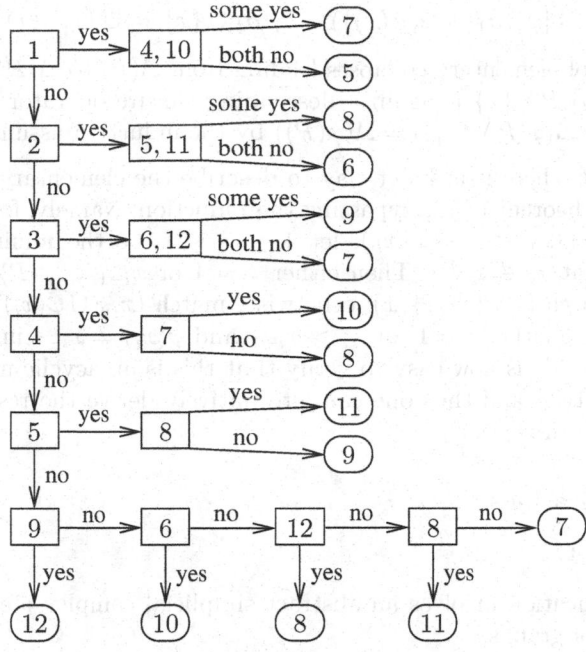

Fig. 13.3. An algorithm for the Illies set system.

Since φ fixes each element in $P_{<x}$, $\varphi(x) < x$, and φ is order-preserving, we see that $P_{<x}$ has $\varphi(x)$ as a maximal element; see Figure 13.4. Thus the link of x in $\Delta(P)$ is $\Delta(P_{>x}) * \Delta(P_{<x}) = \Delta(P_{>x}) * \Delta(P_{<\varphi(x)}) * \varphi(x)$; in particular, it is a cone with apex $\varphi(x)$.

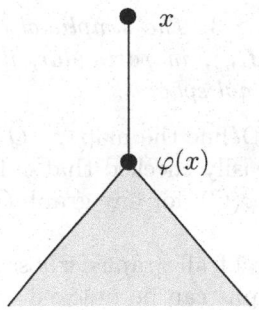

Fig. 13.4. $P_{<x} = P_{\leq\varphi(x)}$.

Let $\sigma_1, \ldots, \sigma_t$ be the simplices of $\Delta(P_{>x}) * \Delta(P_{<\varphi(x)})$ ordered so that the dimension is weakly decreasing. Then

$$(\sigma_1 \cup \{x\}, \sigma_1 \cup \{x, \varphi(x)\}), \ldots, (\sigma_t \cup \{x\}, \sigma_t \cup \{x, \varphi(x)\})$$

is a sequence of elementary collapses leading from $\Delta(P)$ to $\Delta(P \setminus \{x\})$. Since φ restricted to $P \setminus \{x\}$ is again a descending closure operator, $\Delta(P \setminus \{x\})$ collapses onto $\Delta(\varphi(P \setminus \{x\})) = \Delta(\varphi(P))$ by the induction assumption. $\quad\square$

Remark 13.13. There is a direct way to describe the elementary collapses in the proof of Theorem 13.12, bypassing the induction. Namely, for $x \in \mathrm{Bd}\, P \setminus \mathrm{Bd}\,\varphi(P)$, $x = (x_1 < \cdots < x_k)$, let $1 \le i \le k$ be the minimal possible index such that $x_i \notin \varphi(P)$. Then either $i = 1$ or $x_{i-1} \in \varphi(P)$. If $\varphi(x_i) = x_{i-1}$, then match $(x \setminus \{x_{i-1}\}, x)$; otherwise, match $(x, x \cup \varphi(x_i))$. The latter is possible since either $i = 1$, or $x_i > x_{i-1}$ and $\varphi(x_i) \ne x_{i-1}$ imply $\varphi(x_i) > \varphi(x_{i-1}) = x_{i-1}$. It is now easy to verify that this is an acyclic matching, see Definition 11.1(2), and thus one can alternatively derive the result by using discrete Morse theory.

13.2.2 Applications

Application 1.

Our first application involves an abstract simplicial complex stemming from a certain set of graphs.

Definition 13.14. *Let* $\mathrm{DG_n}$ *be the abstract simplicial complex of all discon-nected graphs on n labeled vertices. In other words, the vertices of* $\mathrm{DG_n}$ *are all pairs* (i, j), *with* $i < j$, $i, j \in [n]$, *i.e., all possible edges of a graph on n labeled vertices; and simplices of* $\mathrm{DG_n}$ *are all collections of edges that form a graph with at least two connected components.*

It turns out that topologically, the simplicial complex of all disconnected graphs is essentially equivalent to the already considered order complex of the partition lattice.

Proposition 13.15. *Let* $n \ge 3$. *The simplicial complex* $\mathrm{Bd}\,(\mathrm{DG_n})$ *collapses onto the order complex* $\Delta(\bar{\bar{\Pi}}_n)$; *in particular, it is homotopy equivalent to a wedge of* $(n-3)$-*dimensional spheres.*

Proof. Set $Q := \mathcal{F}(\mathrm{DG_n})$. Define the map $\varphi : Q \to Q$, taking each graph to its transitive closure. It is easily checked that φ is an order-preserving map, that $\varphi^2 = \varphi$, and that $G \le \varphi(G)$, for any graph G. We conclude that φ is an ascending closure operator.

The image of φ consists of all graphs whose connected components are complete graphs. These graphs can be indexed with the set partitions of $[n]$ by associating to each graph the set partition given by its connected compo-nents. Since the graphs are nonempty and disconnected, we get exactly all set partitions of $[n]$ except for (1^n) and (n). These partitions are ordered by refinement, and we conclude that $\varphi(Q) = \bar{\bar{\Pi}}_n$.

Clearly, $\Delta(Q) = \mathrm{Bd}\,(\mathrm{DG_n})$, so by the Theorem 13.12 we conclude that $\mathrm{Bd}\,(\mathrm{DG_n})$ collapses onto $\Delta(\bar{\bar{\Pi}}_n)$. $\quad\square$

Application 2.

Let G be an arbitrary graph. Recall from the Subsection 9.1.4 that we have an order-reversing map $N : \mathcal{F}(\mathcal{N}(G)) \to \mathcal{F}(\mathcal{N}(G))$, which maps every set of vertices to their common neighbors. Recall also that the complex $\Delta(N(\mathcal{F}(\mathcal{N}(G))))$ is called the *Lovász complex* of G and is denoted by $\mathcal{L}o(G)$.

Since the map N is order-reversing, we see that the map N^2 is order-preserving. Furthermore, it follows directly from Definition 9.10 that for any vertex set $A \subseteq V(G)$, we have $N^2(A) \supseteq A$. Indeed, if $v \in A$, then for all $w \in N(A)$ we have $(v,w) \in E(G)$, and therefore $v \in N^2(A)$.

Since N is order-reversing, the inclusion $N^2(A) \supseteq A$ implies the inclusion $N^3(A) \subseteq N(A)$. On the other hand, substituting $N(A)$ instead of A into that first inclusion, we obtain $N^3(A) \supseteq N(A)$. Thus, we can conclude that $N^3(A) = N(A)$; in particular, $N(\mathcal{F}(\mathcal{N}(G))) = N^2(\mathcal{F}(\mathcal{N}(G)))$.

Proposition 13.16. *The simplicial complex* $\mathrm{Bd}\,(\mathcal{N}(G))$ *collapses onto its subcomplex* $\mathcal{L}o(G)$.

Proof. Define the map $\varphi : \mathcal{F}(\mathcal{N}(G)) \to \mathcal{F}(\mathcal{N}(G))$ by simply setting $\varphi := N^2$. From our previous comments, it is clear that φ is an order-preserving map, that $\varphi^2 = \varphi$ (i.e., that $N^4 = N^2$), and that $A \leq \varphi(A)$, for any $A \subseteq V(G)$. We conclude that φ is an ascending closure operator.

By Theorem 13.12 we conclude that the complex $\mathrm{Bd}\,(\mathcal{N}(G))$ collapses onto
$$\Delta(\varphi(\mathcal{F}(\mathcal{N}(G)))) = \Delta(N^2(\mathcal{F}(\mathcal{N}(G)))) = \Delta(N(\mathcal{F}(\mathcal{N}(G)))) = \mathcal{L}o(G). \quad \square$$

Application 3.

The next definition describes yet another object, which is standard in Combinatorial Algebraic Topology. Recall, that in order theory, elements that cover the minimal element $\hat{0}$ are called *atoms*.

Definition 13.17. *Let* $\mathcal{L}$ *be a lattice. The* **atom crosscut complex** $\Gamma(\mathcal{L})$ *associated to* $\mathcal{L}$ *is defined as follows:*

- *the set of vertices of* $\Gamma(\mathcal{L})$ *is equal to the set of atoms of* $\mathcal{L}$, *denoted by* $\mathcal{A}(\mathcal{L})$;
- *a subset* $\sigma \subseteq \mathcal{A}(\mathcal{L})$ *is a simplex in* $\Gamma(\mathcal{L})$ *if and only if the join of elements in* σ *is different from* $\hat{1}$.

Recall that a lattice $\mathcal{L}$ is called *atomic* if all elements of $\mathcal{L}$ can be represented as joins of atoms.

Theorem 13.18. *For any atomic lattice* $\mathcal{L}$, *the simplicial complex* $\mathrm{Bd}\,(\Gamma(\mathcal{L}))$ *collapses onto* $\Delta(\bar{\mathcal{L}})$.

Proof. Define a map $\varphi : \mathcal{F}(\Gamma(\mathcal{L})) \to \mathcal{F}(\Gamma(\mathcal{L}))$ as follows: a simplex σ maps to $\mathcal{A}_{\leq J(\sigma)}$, where $J(\sigma)$ is the join of all elements in σ. Clearly, φ is order-preserving, $\varphi^2 = \varphi$, and $\varphi(\sigma) \supseteq \sigma$.

By Theorem 13.12 we conclude that the complex $\Delta(\mathcal{F}(\Gamma(\mathcal{L}))) = \mathrm{Bd}\,(\Gamma(\mathcal{L}))$ collapses onto $\Delta(\varphi(\mathcal{F}(\Gamma(\mathcal{L}))))$. On the other hand, since the lattice is atomic, we have $\varphi(\mathcal{F}(\Gamma(\mathcal{L}))) = \bar{\mathcal{L}}$, and so $\mathrm{Bd}\,(\Gamma(\mathcal{L}))$ collapses onto $\Delta(\bar{\mathcal{L}})$. $\square$

Remark 13.19. Application 1 above is a special case of Theorem 13.18, since $\mathrm{DG}_n = \Gamma(\Pi_n)$.

For an arbitrary lattice $\mathcal{L}$, let $\mathcal{L}_a$ denote the sublattice consisting of all the elements that are joins of atoms. By the argument above, we see that $\mathrm{Bd}\,(\Gamma(\mathcal{L}))$ collapses onto $\Delta(\bar{\mathcal{L}}_a)$. On the other hand, the map $\psi : \mathcal{L} \to \mathcal{L}$ mapping x to the join of the elements of $\mathcal{A}(\mathcal{L})_{\leq x}$ is a descending closure map, and its image is precisely equal to $\mathcal{L}_a$.

We summarize: for an arbitrary lattice $\mathcal{L}$, both $\mathrm{Bd}\,(\Gamma(\mathcal{L}))$ and $\Delta(\bar{\mathcal{L}})$ collapse to $\Delta(\bar{\mathcal{L}}_a)$.

13.2.3 Monotone Poset Maps

Next, we would like to relax conditions on closure operators somewhat.

Definition 13.20. *Let P be a poset. An order-preserving map $\varphi : P \to P$ is called a* **monotone map** *if for every $x \in P$ either $x \geq \varphi(x)$ or $x \leq \varphi(x)$. If $x \geq \varphi(x)$ for all $x \in P$, then we call φ a* **decreasing map**; *analogously, if $x \leq \varphi(x)$ for all $x \in P$, then we call φ an* **increasing map**.

We remark here on the fact that while a composition of two decreasing (resp. increasing) maps is again a decreasing (resp. an increasing) map, the composition of two monotone maps is not necessarily a monotone map. To see a simple example, let P be the lattice of all subsets of $\{1,2\}$, and define $\varphi(S) = S \cup \{2\}$ and $\gamma(T) = T \setminus \{1\}$, for all $S, T \subseteq \{1,2\}$. The composition $T \circ S$ maps all the subsets to $\{2\}$; in particular, it is not a monotone map.

However, any power of a monotone map is again monotone. Indeed, let $\varphi : P \to P$ be monotone, let $x \in P$, and say $x \leq \varphi(x)$. Since φ is order-preserving, we conclude that $\varphi(x) \leq \varphi^2(x)$, $\varphi^2(x) \leq \varphi^3(x)$, etc. Hence $x \leq \varphi^N(x)$ for arbitrary N.

The following proposition shows that monotone maps have a canonical decomposition in terms of increasing and decreasing maps.

Proposition 13.21. *Let P be a poset, and let $\varphi : P \to P$ be a monotone map. There exist unique maps $\alpha, \beta : P \to P$ such that*

- *the map φ can be represented as a composition $\varphi = \alpha \circ \beta$;*
- *α is an increasing map, whereas β is a decreasing map;*
- *we have $\mathrm{Fix}\,\alpha \cup \mathrm{Fix}\,\beta = P$.*

Proof. Set

$$\alpha(x) = \begin{cases} \varphi(x), & \text{if } \varphi(x) > x; \\ x, & \text{otherwise,} \end{cases}$$

and

$$\beta(x) = \begin{cases} \varphi(x), & \text{if } \varphi(x) < x; \\ x, & \text{otherwise.} \end{cases}$$

Clearly, $\varphi = \alpha \circ \beta$, and $\operatorname{Fix}\alpha \cup \operatorname{Fix}\beta = P$. To see that α is an increasing map, we just need to see that it is order-preserving. Since α either fixes an element or maps it to a larger one, the only situation that needs to be considered is $x, y \in P$, $x < y$, and $\alpha(x) = \varphi(x)$, $\alpha(y) = y$. However, under these conditions we must have $\varphi(y) \leq y$; thus we get $\alpha(y) = y \geq \varphi(y) \geq \varphi(x) = \alpha(x)$, and so α is order-preserving. That β is a decreasing map can be seen analogously. Finally, the uniqueness follows from the fact that each $x \in P$ must be fixed by either α or β, and the value $\varphi(x)$ determines which one will fix x. $\quad\square$

13.2.4 The Reduction Theorem and Implications

The next theorem strengthens and generalizes Theorem 13.12.

Theorem 13.22. *Let P be a poset, and let $\varphi : P \to P$ be a monotone map.*
(a) Assume $x \in P$ such that $\varphi(x) \neq x$, and $P_{<x} \cup P_{>x}$ is finite. Then the abstract simplicial complex $\Delta(P_{<x}) * \Delta(P_{>x})$ is nonevasive.
(b) Assume that $P \supseteq Q \supseteq \operatorname{Fix}\varphi$, that $P \setminus Q$ is finite, and that for every $x \in P \setminus Q$, the set $P_{<x} \cup P_{>x}$ is finite. Then $\Delta(P) \searrow_{NE} \Delta(Q)$; in particular, the simplicial complex $\Delta(P)$ collapses onto the subcomplex $\Delta(Q)$.

Remark 13.23.
(1) Note that when P is finite, the conditions of Theorem 13.22(b) simply reduce to $P \supseteq Q \supseteq \operatorname{Fix}\varphi$.

(2) Under the conditions of Theorem 13.22(b), the simplicial complex $\Delta(P)$ collapses onto the simplicial complex $\Delta(Q)$. This implies Theorem 13.12 as a special case. In particular, the complexes $\Delta(P)$ and $\Delta(Q)$ have the same simple homotopy type.

(3) Under the conditions of Theorem 13.22(b), the topological space $\Delta(Q)$ is a strong deformation retract of the topological space $\Delta(P)$.

(4) Any poset Q satisfying $P \supseteq Q \supseteq \varphi(P)$ will also satisfy $P \supseteq Q \supseteq \operatorname{Fix}\varphi$; hence Theorem 13.22 will apply. In particular, for finite P, we have the following corollary: $\Delta(P) \searrow_{NE} \Delta(\varphi(P))$.

The proof of Theorem 13.22 follows the general lines of the proof of Theorem 13.12. However, there are some further technicalities to be dealt with.

Proof of Theorem 13.22.

We can assume that $P \neq Q$. The proof is by induction, and to start with, some explanation is in order. For finite P the proof is by induction on $|P|$, and both statements are proved in parallel, with statement (a) being proved first. For infinite P, we can first prove statement (a) using statement (b) for finite posets, and then prove statement (b) by induction on $|P| - |Q|$, which is assumed to be finite.

Let us now proceed with the induction. We start with statement (a). Since the expression $\Delta(P_{<x}) * \Delta(P_{>x})$ is symmetric with respect to inverting the partial order of P, without loss of generality, it is enough to consider only the case $\varphi(x) < x$. Let us show that in this case, $\Delta(P_{<x})$ is nonevasive.

Let $\psi : P_{<x} \to P_{<x}$ denote the restriction of φ. It is easy to see that ψ is a monotone map of $P_{<x}$. By the induction hypothesis we see that $\Delta(P_{<x}) \searrow_{NE} \Delta(P_{\leq \varphi(x)})$, since $\mathrm{Fix}\, \psi \subseteq \psi(P_{<x}) \subseteq P_{\leq \varphi(x)}$, and $P_{<x}$ is finite. On the other hand, $\Delta(P_{\leq \varphi(x)})$ is a cone; hence by Proposition 13.7(1), it is nonevasive, and therefore $\Delta(P_{<x})$ is nonevasive as well. Again by Proposition 13.7(1), it follows that $\Delta(P_{<x}) * \Delta(P_{>x})$ is nonevasive. Thus (a) for P follows from (b) for $P_{<x} \supseteq P_{\leq \varphi(x)}$.

Let us now prove statement (b). To start with, we replace the monotone map φ with a monotone map γ satisfying $\gamma(P) \subseteq Q$ and $\mathrm{Fix}\, \gamma = \mathrm{Fix}\, \varphi$. To achieve that objective we can set $\gamma := \varphi^N$, where $N = |P \backslash Q|$. With this choice of γ, the inclusion $\gamma(P) \subseteq Q$ follows from the assumption that $\mathrm{Fix}\, \varphi \subseteq Q$.

Take an arbitrary $x \in P \setminus Q$. Since $x \notin \gamma(P)$, we have $x \neq \gamma(x)$; hence by (a) we know that $\mathrm{lk}_{\Delta(P)} x = \Delta(P_{<x}) * \Delta(P_{>x})$ is nonevasive. This means that $\Delta(P) \searrow_{NE} \Delta(P \setminus \{x\})$.

Let the map $\psi : P \backslash \{x\} \to P \backslash \{x\}$ be the restriction of the map γ. Clearly, ψ is a monotone map, and $\mathrm{Fix}\, \psi = \mathrm{Fix}\, \gamma$. This implies $\mathrm{Fix}\, \psi \subseteq Q$; hence by the induction hypothesis, $\Delta(P \setminus \{x\}) \searrow_{NE} \Delta(Q)$. Summarizing, we conclude that $\Delta(P) \searrow_{NE} \Delta(Q)$. This shows that (b) for $P \supseteq Q$ follows from (a) for P together with (b) for $P \setminus \{x\} \supseteq Q$. $\square$

13.3 Further Facts About Nonevasiveness

13.3.1 NE-Reduction and Collapses

NE-reduction can be used to define an interesting equivalence relation on the set of all abstract simplicial complexes.

Definition 13.24. *Let X and Y be abstract simplicial complexes. Recursively, we say that $X \simeq_{NE} Y$ if $X \searrow_{NE} Y$ or $Y \nearrow_{NE} X$, or if there exists an abstract simplicial complex Z such that $X \simeq_{NE} Z$ and $Y \simeq_{NE} Z$.*

Clearly, if X is nonevasive, then $X \simeq_{NE}$ pt; but is the opposite true? The answer to that is no. As one example, consider the standard instance of a space

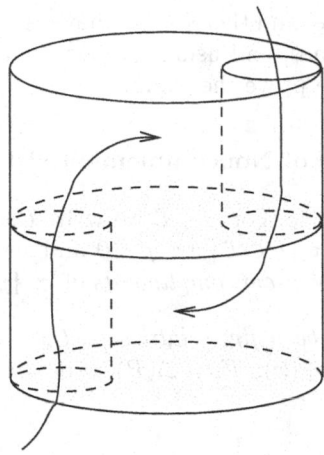

Fig. 13.5. A house with two rooms.

that is contractible but not collapsible: let H be the so-called *house with two rooms*; see Figure 13.5.

Independently of a particular triangulation, the space H is not collapsible; hence it is evasive. On the other hand, we leave it to the reader to see that it is possible to triangulate the filled cylinder C given by the equations $|z| \leq 1$, $x^2 + y^2 \leq 1$, so that $C \searrow_{\text{NE}} H$.

Recall from Section 6.4 that the analogous equivalence relation, where $\searrow_{\text{NE}}$ and $\nearrow_{\text{NE}}$ are replaced by $\searrow$ and $\nearrow$, is called the simple homotopy type, and that the celebrated Whitehead theorem implies that the simplicial complexes with the simple homotopy type of a point are precisely those that are contractible; see Theorem 6.16. Therefore, the class of simplicial complexes that are NE-equivalent to a point relates to nonevasiveness in the same way as contractibility relates to collapsibility. Clearly, this means that this class should constitute an interesting object of study.

We conjecture that NE-equivalence is much coarser than Whitehead's simple homotopy type.

Conjecture 13.25. There exists an infinite family of finite simplicial complexes $\{X_i\}_{i=1}^{\infty}$ all of which have the same simple homotopy type such that $X_i \not\simeq_{\text{NE}} X_j$, for all $i \neq j$.

Finally, let us remark that whenever we have abstract simplicial complexes $X \simeq_{\text{NE}} Y$, there exists an abstract simplicial complex Z such that $X \nearrow_{\text{NE}} Z \searrow_{\text{NE}} Y$. Indeed, assume that $A \searrow_{\text{NE}} B \nearrow_{\text{NE}} C$, for some abstract simplicial complexes A, B, and C. Set $S := V(A) \setminus V(B)$, and set $T := V(C) \setminus V(B)$. Let D be the abstract simplicial complex obtained by attaching to A the vertices from T in the same way as they would be attached to $B \subseteq A$. Clearly, since the links of the vertices from S did not change, they can still be removed in the same fashion as before, and therefore we have $A \nearrow_{NE} D \searrow_{NE} C$. Repeating this operation several times, and using the fact that the reductions $\nearrow_{NE}$ (as well as $\searrow_{NE}$) compose, we prove the claim.

13.3.2 Nonevasiveness of Noncomplemented Lattices

Let $\mathcal{L}$ denote a finite lattice. For $x \in \bar{\mathcal{L}}$ we write $\mathcal{C}o_{\mathcal{L}}(x)$ for the set of *complements* of x, i.e., the set $\{y \in \bar{\mathcal{L}} \,|\, x \wedge y = \hat{0} \text{ and } x \vee y = \hat{1}\}$, while $\mathcal{C}o^{\vee}{}_{\mathcal{L}}(x)$ will denote the set of *upper semicompliments* of x: $\{y \in \bar{\mathcal{L}} \,|\, x \vee y = \hat{1}\}$.

Theorem 13.26. *Let $\mathcal{L}$ be a finite lattice. Let $x \in \bar{\mathcal{L}}$ and let $P = \mathcal{L} \setminus B$, where $\mathcal{C}o^{\vee}{}_{\mathcal{L}}(x) \supseteq B \supseteq \mathcal{C}o_{\mathcal{L}}(x)$. Then $\Delta(\bar{P})$ is nonevasive; in particular, it is collapsible.*

Remark 13.27.
(1) A particularly interesting special case is $B = \mathcal{C}o_{\mathcal{L}}(x) = \emptyset$, that is, x has no complements. Then $\Delta(\bar{\mathcal{L}})$ is nonevasive.

(2) For symmetry reasons, the theorem remains true if upper semicomplements are replaced by lower semicomplements.

(3) The result of Theorem 13.26 can be translated to algorithmic language in the following way. Let P be as in the formulation of Theorem 13.26. Assume that $\mathcal{A}$ is a subset of $\bar{P}$ that is not known in advance. One is allowed to ask questions of the type, "*Is y in $\mathcal{A}$?*" where $y \in \bar{P}$. Then there exists a strategy that determines whether the set $\mathcal{A}$ is a chain in $\bar{P}$, using at most $|\bar{P}| - 1$ questions.

Proof of Theorem 13.26.
We use induction on the number of elements in the poset $\bar{\mathcal{L}}$. Since $x \in \bar{P}$, we know that $\Delta(\bar{P})$ is not empty. Furthermore, if $\bar{P}$ consists of only one element, then $\Delta(\bar{P})$ is a simplicial complex consisting of only one point, and hence is nonevasive by definition.

Assume now that $|\bar{\mathcal{L}}| \geq |\bar{P}| > 1$. In order to show that $\Delta(\bar{P})$ is nonevasive, we shall find a suitable element $y \in \bar{P}$ for which we shall prove the following two claims:

Claim 1. The order complex $\Delta\left(\overline{P \setminus \{y\}}\right)$ is nonevasive.

Claim 2. The order complex $\Delta\left(\overline{P_{\leq y}} \oplus \overline{P_{\geq y}}\right)$ is nonevasive.

One way to show that $\Delta(Q)$ is nonevasive (for Q taken from the claims above) will be to write Q in the form $\mathcal{L}' \setminus B'$, where $\mathcal{L}'$ is some lattice such that $|\mathcal{L}'| < |\mathcal{L}|$ and $\mathcal{C}o^{\vee}{}_{\mathcal{L}'}(z) \supseteq B' \supseteq \mathcal{C}o_{\mathcal{L}'}(z)$, for some $z \in \mathcal{L}' \setminus B'$, and to use induction.

We divide the main part of the proof into two cases, depending on the choice of y.

Case 1: the primary case.
There exists an element y such that y is an atom of P and $x \not\geq y$.

Proof of Claim 1.
Since y is an atom, $\mathcal{L}' = \mathcal{L} \setminus \{y\}$ is a lattice. Also, it is clear that $x \in \mathcal{L}'$. Set $B' = B \cap \mathcal{L}'$. Then $\mathcal{L}' \setminus B' = (\mathcal{L} \setminus B) \setminus \{y\} = P \setminus \{y\}$. If $t \in \mathcal{C}o_{\mathcal{L}'}(x)$, then $x \vee t = \hat{1}$ and $x \wedge t = \hat{0}$ in $\mathcal{L}'$. However, the identity $x \wedge t = y$ is impossible in $\mathcal{L}$, since then it would imply $y \leq x$, which we have assumed to be false. Hence $x \wedge t = \hat{0}$ in $\mathcal{L}$, so $t \in \mathcal{C}o_{\mathcal{L}}(x)$ and therefore $B' \supseteq \mathcal{C}o_{\mathcal{L}'}(x)$. Furthermore, if $t \in \mathcal{C}o^{\vee}{}_{\mathcal{L}}(x)$, $t \neq y$, then $t \in \mathcal{C}o^{\vee}{}_{\mathcal{L}'}(x)$, i.e., $\mathcal{C}o^{\vee}{}_{\mathcal{L}'}(x) \supseteq \mathcal{C}o^{\vee}{}_{\mathcal{L}}(x) \cap \mathcal{L}'$.

So we have shown that $\mathcal{C}o^{\vee}{}_{\mathcal{L}'}(x) \supseteq B' \supseteq \mathcal{C}o_{\mathcal{L}'}(x)$, and hence $\Delta\left(\overline{P \setminus \{y\}}\right)$ is nonevasive by the induction hypothesis.

Proof of Claim 2.
We refer to Figure 13.6 for an illustration of our argument. We see that $\mathcal{L}' = [y, \hat{1}]$ is a lattice, since it is an interval in the lattice $\mathcal{L}$. Let $z = x \vee y$. Since y is an atom and $y \not\leq x$, we know that $x \wedge y = \hat{0}$. On the other hand, we have $y \notin \mathcal{C}o_{\mathcal{L}}(x)$, so $z \neq \hat{1}$.

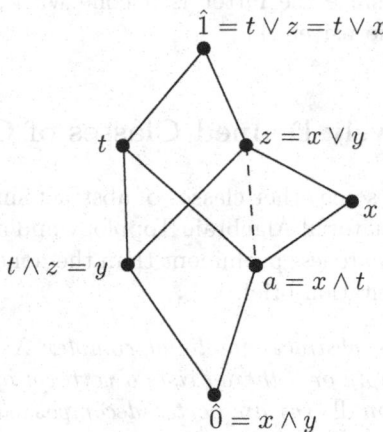

Fig. 13.6. The comparison diagram for Case 1, Claim 2.

Set $B' = \mathcal{L}' \cap B$. First we show that $\mathcal{C}o_{\mathcal{L}'}(z) \subseteq \mathcal{C}o_{\mathcal{L}}(x) \cap \mathcal{L}'$. Take $t \in \mathcal{C}o_{\mathcal{L}'}(z)$. Then $t \wedge z = y$ and $t \vee z = \hat{1}$. But

$$\hat{1} = t \vee z = t \vee (x \vee y) = (t \vee y) \vee x = t \vee x.$$

Let $a = t \wedge x$. Clearly

$$z = x \vee y \geq x \geq x \wedge t = a$$

and

$$t \geq x \wedge t = a;$$

hence

$$y = t \wedge z \geq a.$$

Since $x \geq a$ we get $\hat{0} = x \wedge y \geq a$, hence $a = \hat{0}$, and so $t \in Co_{\mathcal{L}}(x) \cap \mathcal{L}'$. Therefore we get

$$Co_{\mathcal{L}'}(z) \subseteq Co_{\mathcal{L}}(x) \cap \mathcal{L}' \subseteq B \cap \mathcal{L}' = B'.$$

Furthermore,

$$B' = B \cap \mathcal{L}' \subseteq Co^{\vee}{}_{\mathcal{L}}(x) \cap \mathcal{L}' \subseteq Co^{\vee}{}_{\mathcal{L}'}(z),$$

where to prove the last inclusion we just note that if $t \vee x = \hat{1}$, then $t \vee z = t \vee x \vee y = \hat{1} \vee y = \hat{1}$. Claim 2 now follows, since $\mathcal{L}' \setminus B' = P_{\geq y}$.

Case 2: the secondary case.
For every atom $y \in P$ we have $x \geq y$.

Consider the map $\varphi : \overline{P} \to \overline{\mathcal{L}}_{\leq x}$, mapping $y \mapsto y \wedge x$. Clearly, φ is a descending closure operator, whose image is $\overline{\mathcal{L}}_{\leq x}$. By a previous theorem it follows that $\Delta(\overline{P}) \searrow_{NE} \Delta(\overline{\mathcal{L}}_{\leq x})$, and since the latter is a cone with x as an apex, we conclude that $\Delta(\overline{P})$ is nonevasive. $\square$

13.4 Other Recursively Defined Classes of Complexes

In this section we mention some other classes of abstract simplicial complexes that are studied in Combinatorial Algebraic Topology and are defined in a recursive way. These families are less prominent than the nonevasive complexes, so we shall keep our presentation brief.

Definition 13.28. *A pure abstract simplicial complex X is called* **vertex-decomposable** *if it is empty or if there exists a vertex v of X such that both the link* $\mathrm{lk}_X(v)$ *and deletion* $\mathrm{dl}_X(v)$ *are vertex-decomposable.*

For example, the simplicial complex consisting of a single vertex is vertex-decomposable, since both the link and deletion of this vertex are empty. Furthermore, any simplex is vertex-decomposable.

Definition 13.29. *A pure abstract simplicial complex X is called* **constructible** *if it is a simplex, including the empty simplex, or if it has constructible subcomplexes Y and Z such that*

- *we have $X = Y \cup Z$;*
- *the complex $Y \cap Z$ is constructible;*
- *we have $\dim X = \dim Y = \dim Z = \dim(Y \cap Z) + 1$.*

We finish by stating without proof the following important proposition.

Proposition 13.30. *For an arbitrary pure abstract simplicial complex we have the following implications:*

$$nonevasive \Rightarrow vertex\text{-}decomposable \Rightarrow shellable \Rightarrow constructible,$$

with all implications being strict.

13.5 Bibliographic Notes

The Evasiveness Conjecture for the case when n is a prime power and the case $n = 6$ is due to Kahn, Saks, and Sturtevant; see [KSS84]. Theorem 13.5 is due to Oliver; for its proof we refer the interested reader to the original paper of Oliver, [Ol75].

The proof of the Generalized Aanderaa–Rosenberg Conjecture for the case when n is a prime power is due to Rivest & Vuillemin, [RV75]. The first counterexample to the Generalized Aanderaa–Rosenberg Conjecture is due to Illies, [Il78]. Furthermore, an infinite family of counterexamples can be found in [Lu01].

That ascending and descending closure operators induce strong deformation retractions is well known in topological combinatorics; see, e.g., [Bj96, Corollary 10.12], where it is proved using Quillen's theorem A, [Qu73, p. 85]. It was also proved in [Bj96] that if the map φ satisfies the additional condition $\varphi^2 = \varphi$, then φ induces a strong deformation retraction from $\Delta(P)$ to $\Delta(\varphi(P))$. The latter result was strengthened in [Ko06b, Theorem 2.1], where it was shown that whenever φ is an ascending (or descending) closure operator, $\Delta(P)$ collapses onto $\Delta(\varphi(P))$. Our presentation follows the modern approach from [Ko06b].

As was mentioned already, the complex DG_n appeared in the work of Vassiliev on knot theory, [Va93], whereas $\Delta(\bar{\Pi}_n)$ encodes the geometry of the braid arrangement by means of the Goresky–MacPherson theorem; see [GoM88]. Vassiliev was the first to prove that the complex of disconnected graphs is homotopy equivalent to a wedge of spheres, though his beautiful ad hoc argument is quite different from the one presented here.

Proposition 13.16 was proved by the author in [Ko06c], though the fact that the neighborhood complex is homotopy equivalent to the Lovász complex was well known before that. Theorem 13.18 was also proved by the author in [Ko06c].

Finally, Theorem 13.26 was proved in [Ko98, Theorem 2.4], and we essentially follow that argument here.

14

Colimits and Quotients

14.1 Quotients of Nerves of Acyclic Categories

14.1.1 Desirable Properties of the Quotient Construction

Assume that we have a finite group G acting on a finite poset P in an order-preserving way. The purpose of this chapter is to study a construction of the quotient associated with this action. One structural approach is to view P as an acyclic category and to view the group action of G as a functor from the one-element category associated to the group G to **AC**. Then, as we have pointed out earlier in Subsection 4.4.3, it is natural to define P/G to be the colimit of this functor. As a result, P/G can in general turn out to be an acyclic category, which is not necessarily a poset.

Our plan for this chapter is as follows. After describing the formal setting in Section 14.2 we proceed in Section 14.3 with imposing different conditions on the group action. We shall give conditions for each of the following properties to be satisfied:

(1) the morphisms of P/G are exactly the orbits of the morphisms of P; we call it *regularity*;
(2) the quotient construction commutes with the nerve functor;
(3) the category P/G is again a poset.

Since the quotient of the group action on a poset is in general an acyclic category, whereas the quotient of the group action on an acyclic category, as we shall see, is always acyclic itself, the class of acyclic categories appears here as a natural category that contains the category of posets, and is at the same time closed under taking quotients.

14.1.2 Quotients of Simplicial Actions

Before we proceed with the general setup, let us see what makes things complicated in the simplicial situation. To do that, let us consider a simplicial

G-action on X, where G is a finite group and X is a finite abstract simplicial complex. What can in general be said about the quotient X/G? It is easy to take the topological quotient, but the simplicial (or cell) structure on it may not be derived from the simplicial structure of X in a very nice way.

Consider, for instance, the reflection action of the additive group $\mathbb{Z}_2$ on an interval. The topological quotient is again an interval. One of its vertices is an orbit of the original $\mathbb{Z}_2$-action on the vertices of the interval, whereas the other one is not. This is not a very pleasing situation; therefore we would like to require that the action satisfy further properties.

Condition (S1). *For any group element $g \in G$ and any simplex $\sigma \in X$ such that $g(\sigma) = \sigma$, we have $g|_\sigma = \mathrm{id}_\sigma$.*

In words, Condition (S1) says that if a simplex is preserved by a group element, then it is fixed by that group element pointwise. Note that this condition is not satisfied in the example of the $\mathbb{Z}_2$-action above.

Condition (S2). *For any group element $g \in G$, and any simplex $\sigma \in X$, we know that g fixes $g(\sigma) \cap \sigma$ pointwise.*

Clearly Condition (S1) is a special case of Condition (S2), when $g(\sigma) = \sigma = g(\sigma) \cap \sigma$. Here is an example of a situation in which Condition (S1) holds, whereas Condition (S2) does not.

Example 14.1. Let X be the hollow triangle, i.e., the abstract simplicial complex consisting of three vertices 1, 2, and 3 and three connecting edges $(1,2)$, $(1,3)$, and $(2,3)$. Let G be the cyclic group with three elements, and let the generator of this group act on X by cyclic shifting: $1 \mapsto 2$, $2 \mapsto 3$, $3 \mapsto 1$. Clearly, Condition (S1) is true, since no element other than the identity ever preserves a simplex, whereas Condition (S2) fails, for example, if one takes g to be the generator of the group, and takes $\sigma = (1,2)$.

When Condition (S1) or both Conditions (S1) and (S2) are valid, we can say something useful about the quotient.

Proposition 14.2. *Assume that we have a simplicial action of a finite group G on a finite abstract simplicial complex X. Then the following hold:*

(1) If this action satisfies Condition (S1), then X/G can be viewed as a CW complex whose cells are indexed by G-orbits on the set of the simplices of X.

(2) If, in addition, this action satisfies Condition (S2), then the quotient complex is a generalized simplicial complex.

Proof. If our action satisfies Condition (S1), then there will be no self-identifications of open cells. Hence the quotient X/G can be obtained by gluing in the orbits of open cells along the quotients of the original attaching maps. This proves part (1) of the proposition.

If the action additionally satisfies Condition (S2), then there will be no self-identifications of the closed cells either. In other words, the new attaching maps are the same as the original ones. Hence the quotient X/G is a generalized simplicial complex; see Definition 2.41. $\square$

Let us next consider a similar, yet somewhat different, situation, where X is a finite trisp and the group G acts by trisp maps.

Proposition 14.3. *Assume that we have a trisp action of a finite group G on a finite trisp X. Then the following hold:*

(1) The quotient X/G is again a trisp, whose simplices are indexed by the G-orbits of the set of simplices of X.

(2) If, furthermore, X is a finite regular trisp and the G-action satisfies Condition (S2), then the quotient trisp X/G is again regular.

Proof. To see (1), notice that a trisp map always preserves the order of the vertices in each simplex; hence Condition (S1) is always satisfied. This is basically the reason why (1) holds for arbitrary trisp actions. The formal argument is as follows. For every nonnegative integer n, set $S_n(X/G)$ to be the set of G-orbits of $S_n(X)$. Furthermore, for an arbitrary simplex $\sigma \in S_n(X)$ and arbitrary order-preserving injection $f : [m+1] \hookrightarrow [n+1]$, we set

$$B_f(X/G)([\sigma]) := [B_f(X)(\sigma)], \tag{14.1}$$

where we use square brackets to denote the G-orbits. The value of $B_f(X/G)$ is well-defined, since the commutation relation (2.14) tells us that $[\sigma_1] = [\sigma_2]$ implies $[B_f(X)(\sigma_1)] = [B_f(X)(\sigma_2)]$. We also obviously have $B_{f \circ g}(X/G) = B_g(X/G) \circ B_f(X/G)$ and $B_{\mathrm{id}_n}(X/G) = \mathrm{id}_{S_n(X/G)}$, and hence the quotient X/G has a well-defined trisp structure.

To see (2), we just need to verify that no two vertices belonging to the same simplex are identified by the group action. This follows from Condition (S2), since if σ is a simplex and both v and $g(v)$ are vertices of σ, for some $g \in G$, then $g(v) \in \sigma \cap g(\sigma)$, and hence $g(g(v)) = g(v)$, implying $g(v) = v$. $\square$

Proposition 14.3 can now be applied in the case of a group action on a finite acyclic category.

Proposition 14.4. *Let C be a finite acyclic category, and let G be a finite group acting on C. Then $\Delta(C)/G$ is a regular trisp, whose simplices are indexed by the G-orbits on the set of simplices of $\Delta(C)$.*

Proof. First, by our discussion in Chapter 10, $\Delta(C)$ is a trisp, and the induced G-action on $\Delta(C)$ is a trisp action. Hence it follows from Proposition 14.3(1) that $\Delta(C)/G$ is also a trisp, whose simplices are indexed by the G-orbits on the set of simplices of $\Delta(C)$.

Furthermore, since the acyclic category C is assumed to be finite, we know that if there is a morphism between x and $g(x)$, then $g(x) = x$, for any $g \in G$, $x \in \mathcal{O}(C)$. It follows that for any simplex σ of $\Delta(C)$ and any group element $g \in G$, the intersection $g(\sigma) \cap \sigma$ consists of objects and morphisms that are all fixed by g. Thus, our action satisfies Condition (S2), and the regularity of $\Delta(C)/G$ follows from Proposition 14.3(2). □

Let us look at a few classical examples to see what may happen.

Example 14.5. The symmetric group $\mathcal{S}_n$ acts on the Boolean algebra $\mathcal{B}_n$ by permuting the ground set. Every $\mathcal{S}_n$-orbit can be uniquely encoded by the sequence of the cardinalities of the sets in the chain. Analyzing how these are glued together, we see that $\Delta(\mathcal{B}_n)/\mathcal{S}_n$ is an n-simplex.

Example 14.6. The group $\mathbb{Z}_2$ acts on $P = \underbrace{\mathbf{2} \oplus \cdots \oplus \mathbf{2}}_{n+1}$, by simultaneously swapping elements in each copy of the antichain $\mathbf{2}$. We have seen that $\Delta(P)$ is homeomorphic to $\mathbb{S}^n$, and we can see that the induced $\mathbb{Z}_2$-action is antipodal; hence the quotient $\Delta(P)/\mathbb{Z}_2$ is homeomorphic to the projective space $\mathbb{RP}^n$.

Example 14.7. Recall that Π_n denotes the poset of all set partitions of $[n]$, ordered by refinement. Again, the symmetric group $\mathcal{S}_n$ acts on Π_n by permuting the ground set, and we can consider the quotient complex $\Delta(\bar{\Pi}_n)/\mathcal{S}_n$. This space has been considered in Sections 9.3 and 11.2.

14.2 Formalization of Group Actions and the Main Question

14.2.1 Definition of the Quotient and Formulation of the Main Problem

Our main object of study is described in the following definition.

Definition 14.8. *We say that a finite group G **acts on** a finite acyclic category C if there is a functor $\mathcal{A}_C : G \to \mathbf{AC}$ that takes the unique object of G to C. The colimit of $\mathcal{A}_C$ is called the **quotient** of C by the action of G and is denoted by C/G.*

To simplify notation, we identify $\mathcal{A}_C(g)$ with g itself. Furthermore, in Definition 14.8 the category $\mathbf{AC}$ could be replaced with a different one. For example, when considering the group actions on posets, one could ask to take the colimit with the category of posets instead. The resulting notion of quotient would then be different; see Example 14.9.

Main Problem.   *Understand the relation between the topological and the categorical quotients, that is, between $\Delta(C/G)$ and $\Delta(C)/G$.*

To start with, by the universal property of colimits, see Definition 4.33, there exists a canonical map $\lambda : \Delta(C)/G \to \Delta(C/G)$. In Section 14.3 we shall give combinatorial conditions under which this map is an isomorphism.

14.2.2 An Explicit Description of the Category C/G

It can be shown in general that if a group G acts on the category C, then the colimit C/G exists. In combinatorial situations it is useful to have an explicit description, which we now proceed to give.

When x is an object or a morphism of C, we denote by $G(x)$ the orbit of x under the action of G. For objects we have $\mathcal{O}(C/G) = \{G(a) \,|\, a \in \mathcal{O}(C)\}$. The situation with morphisms is more complicated. Define a relation $\leftrightarrow$ on the set $\mathcal{M}(C)$ by setting $x \leftrightarrow y$ if there are decompositions $x = x_1 \circ \cdots \circ x_t$ and $y = y_1 \circ \cdots \circ y_t$ with $G(y_i) = G(x_i)$ for all $i \in [t]$. The relation $\leftrightarrow$ is reflexive and symmetric, since G has an identity and inverses; however, it is not in general transitive. Let $\sim$ be the transitive closure of $\leftrightarrow$; it is clearly an equivalence relation. Denote the $\sim$ equivalence class of x by $[x]$.

Perhaps a better way to understand the relation $\sim$ is to note that it is the minimal equivalence relation on $\mathcal{M}(C)$ closed under the G-action and under the composition. That is, we require that $a \sim g(a)$ for any $g \in G$, and that if $x \sim x'$ and $y \sim y'$, and $x \circ x'$ and $y \circ y'$ are defined, then $x \circ x' \sim y \circ y'$. It is not difficult to check that the set $\{[x] \,|\, x \in \mathcal{M}(C)\}$ with the relations $\partial_\bullet[x] = [\partial_\bullet x]$, $\partial^\bullet[x] = [\partial^\bullet x]$ and $[x] \circ [y] = [x \circ y]$ (whenever the composition $x \circ y$ is defined) are the morphisms of the category C/G.

Let us now look at a few more examples illustrating various points that we have considered so far.

Example 14.9. Let P be the poset in the middle of Figure 14.1. Let $\mathbb{Z}_2$ act on P by simultaneously permuting a with b and c with d. Arrow (I) shows $P/\mathbb{Z}_2$ in $\mathbf{P}$, whereas arrow (II) shows $P/\mathbb{Z}_2$ in $\mathbf{AC}$. Note that in this case the quotient in $\mathbf{AC}$ commutes with the functor Δ (the canonical surjection λ is an isomorphism), whereas the quotient in $\mathbf{P}$ does not.

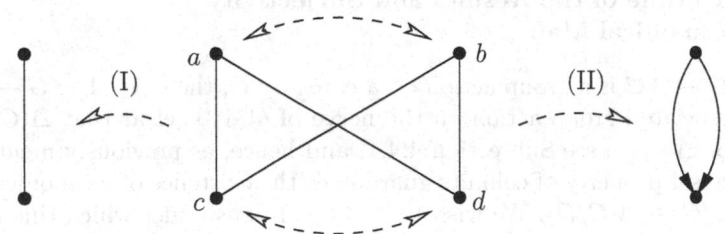

Fig. 14.1. Quotient taken in the category of acyclic categories versus the one taken in the category of posets.

Example 14.10. Let P be the poset in Figure 14.2. Let $\mathbb{Z}_2$ act on P by simultaneously permuting a with b and c with d, while fixing x. The quotient $P/\mathbb{Z}_2$

is a fully ordered set with three elements; hence $\Delta(P/\mathbb{Z}_2)$ is 2-simplex. On the other hand, the nerve $\Delta(P)$ is a union of four simplices that is homeomorphic to a 2-dimensional disk, with the induced action given by the antipodal map. The quotient $\Delta(P)/\mathbb{Z}_2$ is a simplicial complex consisting of two 2-simplices, also shown in Figure 14.2. In this case, the canonical map λ maps both 2-simplices of $\Delta(P)/\mathbb{Z}_2$ onto the 2-simplex of $\Delta(P/\mathbb{Z}_2)$, identifying two edges.

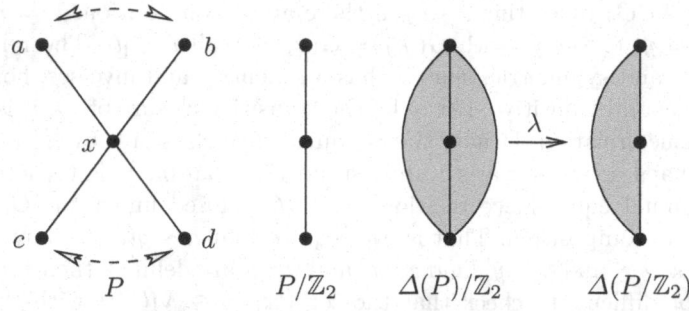

Fig. 14.2. An example in which the map λ is not an isomorphism.

14.3 Conditions on Group Actions

In this section we consider combinatorial conditions for a finite group G acting on a finite acyclic category C that ensure that taking the quotient of this group action commutes with the nerve functor.

14.3.1 Outline of the Results and Surjectivity of the Canonical Map

If $\mathcal{A}_C : G \to \mathbf{AC}$ is a group action on a category C, then $\Delta \circ \mathcal{A}_C : G \to \mathbf{RTS}$ is the associated group action on the nerve of C. It is clear that $\Delta(C/G)$ is a sink for $\Delta \circ \mathcal{A}_C$, see Subsection 4.4.1, and hence, as previously mentioned, the universal property of colimits guarantees the existence of a canonical map $\lambda : \Delta(C)/G \to \Delta(C/G)$. We wish to find conditions under which this map is an isomorphism.

First, we prove in Proposition 14.11 that λ is always surjective. Furthermore, $G(a) = [a]$ for $a \in \mathcal{O}(C)$, which means that, restricted to 0-skeletons, λ is an isomorphism. If the two regular trisps were abstract simplicial complexes (only one face for any fixed vertex set), this would suffice to show isomorphism. Neither one is an abstract simplicial complex in general, but while the quotient of a complex $\Delta(C)/G$ can have simplices with fairly arbitrary face sets in common, $\Delta(C/G)$ has only one face for any fixed edge set, since it is a nerve of an acyclic category. Thus for λ to be an isomorphism it is necessary and sufficient to find conditions under which

(1) the map λ is an isomorphism restricted to 1-skeletons;
(2) the simplicial complex $\Delta(C)/G$ has only one face with any given set of edges.

We will give conditions equivalent to λ being an isomorphism, and then give some stronger conditions that are often easier to check. The strongest of these conditions is also inherited by the action of any subgroup H of G acting on C, a fact that is useful when we want to take quotients for the induced subgroup actions.

To start with, recall that a simplex of $\Delta(C/G)$ is a composable morphism chain, i.e., $([m_1], \ldots, [m_t])$, with $m_i \in \mathcal{M}(C)$ and $\partial_{\bullet}[m_{i-1}] = \partial^{\bullet}[m_i]$. On the other hand, a simplex of $\Delta(C)/G$ is an orbit of a composable morphism chain $(n_1, \ldots, n_t)$, $n_i \in \mathcal{M}(C)$, which we denote by $G(n_1, \ldots, n_t)$. The canonical map λ is given by $\lambda(G(n_1, \ldots, n_t)) = ([n_1], \ldots, [n_t])$.

Proposition 14.11. *Let C be a category and G a finite acyclic group acting on C. Then the canonical map $\lambda : \Delta(C)/G \to \Delta(C/G)$ is surjective.*

Proof. By the above description of λ it suffices to fix a composable morphism chain $([m_1], \ldots, [m_t])$ and to find a composable morphism chain $(n_1, \ldots, n_t)$, with $[n_i] = [m_i]$. The proof proceeds by induction on t. The case $t = 1$ is obvious; just take $n_1 = m_1$.

Assume now that we have found $n_1, \ldots, n_{t-1}$ such that $[n_i] = [m_i]$, for $i = 1, \ldots, t-1$, and $n_1, \ldots, n_{t-1}$ compose, i.e., $\partial^{\bullet} n_i = \partial_{\bullet} n_{i+1}$, for $i = 1, \ldots, t-2$. Since $[\partial_{\bullet} n_{t-1}] = [\partial_{\bullet} m_{t-1}] = [\partial^{\bullet} m_t]$, we can find $g \in G$ such that $g(\partial^{\bullet} m_t) = \partial_{\bullet} n_{t-1}$. If we now take $n_t = g(m_t)$, we see that n_{t-1} and n_t compose, and $[n_t] = [m_t]$, which provides a proof for the induction step. $\square$

14.3.2 Condition for Injectivity of the Canonical Projection

Let C be a finite acyclic category and assume that a finite group G acts on C. We impose the following condition on our action.

Condition (R). *If $x, y_a, y_b \in \mathcal{M}(C)$ such that $\partial_{\bullet} x = \partial^{\bullet} y_a = \partial^{\bullet} y_b$ and $G(y_a) = G(y_b)$, then $G(x \circ y_a) = G(x \circ y_b)$.*

Condition (R) is illustrated by Figure 14.3.

Definition 14.12. *Let C be a finite acyclic category and let G be a finite group acting on C. We say that the action is* **regular** *if it satisfies Condition (R).*

The next proposition shows why Condition (R) is the right one for our purposes.

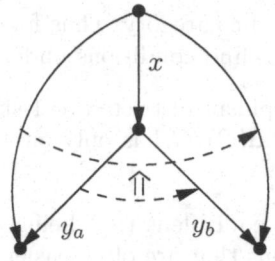

Fig. 14.3. Condition (R).

Proposition 14.13. *Let C be a finite acyclic category and G a finite group acting on C. This action satisfies Condition (R) if and only if the canonical surjection $\lambda : \Delta(C)/G \to \Delta(C/G)$ is injective on 1-skeletons.*

Proof. The injectivity of λ on 1-skeletons is equivalent to requiring that $G(m) = [m]$, for all $m \in \mathcal{M}(C)$, while Condition (R) is equivalent to requiring that $G(m \circ G(n)) = G(m \circ n)$, for all $m, n \in \mathcal{M}(C)$ with $\partial_\bullet m = \partial^\bullet n$; here $m \circ G(n)$ means the set of all $m \circ g(n)$ for which the composition is defined.

Assume that λ is injective on 1-skeletons. Then we have the following computation:

$$G(m \circ G(n)) = G(m) \circ G(n) = [m] \circ [n] = [m \circ n] = G(m \circ n),$$

which shows that Condition (R) is satisfied.

Conversely, assume that the Condition (R) is satisfied, that is, we have $G(m \circ G(n)) = G(m \circ n)$. Since the equivalence class $[m]$ is generated by G and composition, it suffices to show that orbits are preserved by composition, which is precisely $G(m \circ G(n)) = G(m \circ n)$. $\square$

14.3.3 Conditions for the Canonical Projection to be an Isomorphism

The following theorem is the main result of this section. It provides us with combinatorial conditions that are equivalent to λ being an isomorphism.

Theorem 14.14. *Let C be a finite acyclic category and let G be a finite group acting on it. The following two conditions are equivalent:*

Condition (C1). *Let $t \geq 2$, and assume that both $(m_1, \ldots, m_{t-1}, m_a)$ and $(m_1, \ldots, m_{t-1}, m_b)$ are composable morphism chains. Furthermore, assume that $G(m_a) = G(m_b)$. Then there exists some $g \in G$ such that $g(m_a) = m_b$ and $g(m_i) = m_i$, for all $1 \leq i \leq t - 1$; see Figure 14.4.*

Condition (C2). *The canonical surjection $\lambda : \Delta(C)/G \to \Delta(C/G)$ is injective.*

For language convenience, if one of Conditions (C1) and (C2) is satisfied, we shall also say that Condition (C) is satisfied. In this case, the map λ is an isomorphism.

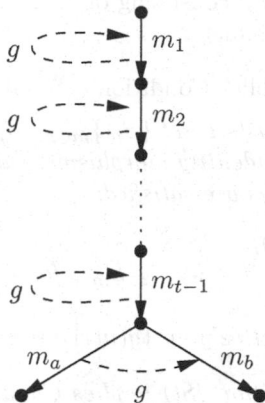

Fig. 14.4. Condition (C1).

Proof of Theorem 14.14. Condition (C1) is equivalent to requiring that $G(m_1, \ldots, m_t) = G(m_1, \ldots, m_{t-1}, G(m_t))$, where this notation is used, as before, to describe all sequences $(m_1, \ldots, m_{t-1}, g(m_t))$ that are composable morphism chains. Condition (C2) implies Condition (R) above, and so it can be restated as $G(m_1, \ldots, m_t) = (G(m_1), \ldots, G(m_t))$.

Condition (C2) implies Condition (C1) by the following computation:

$$G(m_1, \ldots, m_t) = (G(m_1), \ldots, G(m_t)) = G(G(m_1), \ldots, G(m_t))$$
$$\supseteq G(m_1, \ldots, m_{t-1}, G(m_t)) \supseteq G(m_1, \ldots, m_t).$$

Let us now also see that Condition (C1) implies Condition (C2). To show that $(G(m_1), \ldots, G(m_t)) = G(m_1, \ldots, m_t)$, we need to see that if $(m_1, \ldots, m_t)$ is a composable morphism chain, and $g_1, \ldots, g_t \in G$ such that $(g_1(m_1), \ldots, g_t(m_t))$ is again a composable morphism chain, then there exists $g \in G$ such that $g_i(m_i) = g(m_i)$, for all $i = 1, \ldots, t$.

Set $h_1 := g_1$. Condition (C1) applies for $t = 2$ to the composable morphism chains $(h_1(m_1), h_1(m_2))$ and $(h_1(m_1), g_2(m_2))$, implying that there exists an element $h_2 \in G$ such that $h_2 h_1(m_1) = h_1(m_1) = g_1(m_1)$ and $h_2 h_1(m_2) = g_2(m_2)$. Next, Condition (C1) applies for $t = 3$ to the composable morphism chains $(h_2 h_1(m_1), h_2 h_1(m_2), h_2 h_1(m_3))$ and $(h_2 h_1(m_1), h_2 h_1(m_2), g_3(m_3))$, yielding an element $h_3 \in G$ such that $h_3 h_2 h_1(m_1) = g_1(m_1)$, $h_3 h_2 h_1(m_2) = g_2(m_2)$ and $h_3 h_2 h_1(m_3) = g_3(m_3)$. Continuing this procedure for higher values of t, we will eventually find the element $g := h_t \cdots h_2 h_1 \in G$ that satisfies $g_i(m_i) = g(m_i)$ for all $i = 1, \ldots, t$. $\square$

Example 14.15. We give an example of a group action that satisfies Condition (C) for $t = 1, \ldots, k$ but does not satisfy Condition (C) for $t = k + 1$. Let P_{k+1} be the order sum of $k + 1$ copies of the 2-element antichain. The automorphism group of P_{k+1} is the direct product of $k + 1$ copies of $\mathbb{Z}_2$. Take G to be the subgroup of index 2 consisting of elements with an even number of nonidentity terms in the product.

The following condition implies Condition (C), and is often easier to check.

Condition (St). *There exists a set $\{S_m\}_{m \in \mathcal{M}(C)}$, where we include objects of C in the indexing set as identity morphisms, such that for any $m \in \mathcal{M}(C)$ the following three properties are satisfied:*

(1) we have $S_m \subseteq \text{Stab}(m)$;
(2) we have inclusions $S_m \subseteq S_{\partial \bullet m} \subseteq S_{m'}$, for any $m' \in \mathcal{M}(C)$, such that $\partial_\bullet m' = \partial^\bullet m$;
(3) the set $S_{\partial \bullet m}$ acts transitively on $\{g(m) \,|\, g \in \text{Stab}(\partial^\bullet m)\}$.

Proposition 14.16. *Condition (St) implies Condition (C).*

Proof. Let C be a finite acyclic category, and assume that G is a finite group acting on C. Let morphisms $m_1, \ldots, m_{t-1}, m_a, m_b \in \mathcal{M}(C)$ and a group element $g \in G$ be as in Condition (C1). Then, since $g \in \text{Stab}(\partial^\bullet m_a)$, by part (3) of Condition (St), there must exist $\tilde{g} \in S_{\partial \bullet m_a}$ such that $\tilde{g}(m_a) = m_b$. Now, by part (2) of Condition (St), one can conclude that $\tilde{g}(m_i) = m_i$, for all $i \in [t-1]$. $\square$

We say that the **strong** Condition (St) is satisfied if Condition (St) is satisfied with the specific choice $S_a := \text{Stab}(a)$. Clearly, in such a case parts (1) and (3) of Condition (St) are obsolete, and part (2) is reduced to saying that $\text{Stab}(\partial_\bullet m) \subseteq \text{Stab}\, m$, for all $m \in \mathcal{M}(C)$.

Example 14.17. Here is an example of a group action satisfying Condition (St) but not the strong Condition (St). Let $C = \mathcal{B}_n$, the lattice of all subsets of $[n]$ ordered by inclusion, and let $G = \mathcal{S}_n$ act on $\mathcal{B}_n$ by permuting the ground set $[n]$. Clearly, for $A \subseteq [n]$, we have $\text{Stab}(A) = \mathcal{S}_A \times \mathcal{S}_{[n] \setminus A}$, where for $X \subseteq [n]$, $\mathcal{S}_X$ denotes the subgroup of $\mathcal{S}_n$ that fixes elements of $[n] \setminus X$ and acts as a permutation group on the set X. Since $A > B$ means $A \supset B$, part (2) of Condition (St) is not satisfied for $S_A = \text{Stab}(A)$, since $\mathcal{S}_A \times \mathcal{S}_{[n] \setminus A} \not\supseteq \mathcal{S}_B \times \mathcal{S}_{[n] \setminus B}$. However, we can set $S_A := \mathcal{S}_A$ instead. It is easy to check that for this choice of $\{S_A\}_{A \in \mathcal{B}_n}$, Condition (St) is satisfied.

We close the discussion of the conditions stated above with the following proposition.

Proposition 14.18.

(1) The sets of group actions that satisfy Condition (C) or Condition (St) are closed under taking the restriction of the group action to a subcategory.

(2) *Assume that a finite group G acts on a finite acyclic category C so that Condition (St) is satisfied. Let $x \in \mathcal{O}(C)$, and take a subgroup H of G such that $S_x \subseteq H \subseteq \operatorname{Stab}(x)$. Then Condition (St) is satisfied for the action of H on $C_{\leq x}$.*

(3) *Assume that a finite group G acts on a finite acyclic category C so that the strong Condition (St) is satisfied, and assume that H is a subgroup of G. Then the strong version of Condition (St) is again satisfied for the action of H on C.*

Proof. Parts (1) and (3) of the proposition are obvious.

Let us now show part (2). Recall that objects of $C_{\leq x}$ are all pairs (a, m) such that $\partial^{\bullet} m = x$ and $\partial_{\bullet} m = a$. Set $\widetilde{S}(a, m) := S_a$. Furthermore, a morphism between (a, m_a) and (b, m_b) in $C_{\leq x}$ is a morphism of C such that $m_b = m \circ m_a$. For the sake of precision, we denote such a morphism by the 5-tuple (m, a, m_a, b, m_b). Set $\widetilde{S}(m, a, m_a, b, m_b) := S_m$.

To check part (1) of Condition (St) for the action of H on $C_{\leq x}$, we note that $\widetilde{S}(a, m) = S_a \subseteq S_m \subseteq \operatorname{Stab}_G(m) \cap H = \operatorname{Stab}_H(a, m)$, where the second inclusion follows from $S_m \subseteq S_x \subseteq H$. Furthermore, we note that $\widetilde{S}(m, a, m_a, b, m_b) = S_m \subseteq \operatorname{Stab}_G(m) \cap S_{m_a} \cap S_x \subseteq \operatorname{Stab}_G(m) \cap \operatorname{Stab}_G(m_a) \cap H = \operatorname{Stab}_H(m, a, m_a, b, m_b)$, where the first inequality follows from $S_m \subseteq S_a \subseteq S_{m_a} \subseteq S_x$, and the last equality follows from the fact that $m_b = m \circ m_a$; hence $\operatorname{Stab}_G(m) \cap \operatorname{Stab}_G(m_a) \subseteq \operatorname{Stab}_G(m_b)$.

Part (2) of Condition (St) remains true, since we have $\widetilde{S}(b, m_b) = S_b \subseteq \widetilde{S}(m, a, m_a, b, m_b) = S_m \subseteq S_a = \widetilde{S}(a, m_a)$.

We finish by checking part (3) of Condition (St). Let (a, m_a) be an object of $C_{\leq x}$, and let (m, a, m_a, b, m_b) be a morphism in that category. Let furthermore $g \in \operatorname{Stab}_H(a, m_a) \subseteq \operatorname{Stab}_G(a)$. By our assumption, there exists $h \in S_a$ such that $h(m) = g(m)$. On the other hand, we have $g(m_a) = m_a$, and since $S_a \subseteq S_{m_a} \subseteq \operatorname{Stab}_G(m_a)$, we also have $h(m_a) = m_a$. Finally, we have $h(m_b) = h(m \circ m_a) = h(m) \circ h(m_a) = g(m) \circ g(m_a) = g(m_b)$, and part (3) is completely checked. $\square$

14.3.4 Conditions for the Categories to be Closed Under Taking Quotients

Recall that if C is a finite acyclic category and T is a functor from C to itself, it is easy to show that whenever $x \neq T(x)$, for some $x \in \mathcal{O}(C)$, we must have $\mathcal{M}_C(x, T(x)) = \mathcal{M}_C(T(x), x) = \emptyset$.

Proposition 14.19. *Let C be a finite acyclic category and let T be a functor from C to itself. Then $\Delta(C_T) = \Delta(C)_{\Delta(T)}$, where C_T denotes the subcategory of C fixed by T, while $\Delta(C)_{\Delta(T)}$ denotes the subcomplex of $\Delta(C)$ fixed by $\Delta(T)$.*

Proof. Obviously, we have $\Delta(C_T) \subseteq \Delta(C)_{\Delta(T)}$. On the other hand, if for some $x \in \Delta(C)$ we have $\Delta(T)(x) = x$, then the minimal simplex σ that contains x is fixed as a set. Since the order of simplices is preserved by T, the simplex σ must be fixed by T pointwise; thus $x \in \Delta(C_T)$. $\square$

The category of finite acyclic categories can be seen as the closure of the category of finite posets under the operation of taking the quotient of group actions. More precisely, we have the following statement.

Proposition 14.20. *The quotient of a finite acyclic category by a group action is again a finite acyclic category. In particular, the quotient of a finite poset by a group action is a finite acyclic category.*

Proof. This follows directly from our direct description of the corresponding colimit in Subsection 14.2.2. $\square$

Next, we shall state a condition under which the quotient of an acyclic category is a poset.

Condition (SR). *If $x, y \in \mathcal{M}(C)$ such that $\partial^\bullet x = \partial^\bullet y$, and $G(\partial_\bullet x) = G(\partial_\bullet y)$, then $G(x) = G(y)$.*

Condition (SR) is illustrated by Figure 14.5.

Proposition 14.21. *Let C be a finite acyclic category and let G be a finite group acting on C. The following are equivalent:*

(1) The group action satisfies Condition (SR);
(2) The group G acts regularly on the category C and C/G is a poset.

Proof. The implication $(2) \Rightarrow (1)$ follows immediately from the regularity of the action of G and the fact that there must be only one morphism between $[\partial^\bullet x](= [\partial^\bullet y])$ and $[\partial_\bullet x](= [\partial_\bullet y])$.

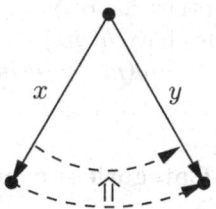

Fig. 14.5. Condition (SR).

Let us prove the implication $(1) \Rightarrow (2)$. Obviously, Condition (SR) implies Condition (R); hence the action of G is regular. Furthermore, if $x, y \in \mathcal{M}(C)$ and there exist $g_1, g_2 \in G$ such that $g_1(\partial^\bullet x) = \partial^\bullet y$ and $g_2(\partial_\bullet x) = \partial_\bullet y$, then we can replace x by $g_1(x)$ and reduce the situation to the one described in Condition (SR), namely that $\partial^\bullet x = \partial^\bullet y$. Applying Condition (SR) and acting with g_1^{-1} yields the result. $\square$

When P is a poset, Condition (SR) can be stated in simpler terms.

Condition (SRP). *If $a, b, c \in P$ such that $a \geq b$, $a \geq c$, and there exists $g \in G$ such that $g(b) = c$, then there exists $\tilde{g} \in G$ such that $\tilde{g}(a) = a$ and $\tilde{g}(b) = c$.*

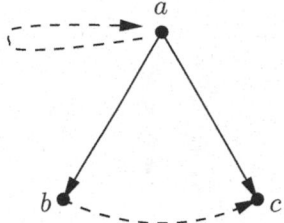

Fig. 14.6. Condition (SRP).

That is, for any $a, b \in P$ such that $a \geq b$, we require that the stabilizer of a act transitively on Gb; see Figure 14.6.

Proposition 14.22. *Let P be a finite poset and assume that G is a finite group acting on P. The action of G on P induces an action of G on the barycentric subdivision $\mathrm{Bd}\, P$. Then $(\mathrm{Bd}\, P)/G$ is a poset and the simplicial complexes $\Delta(\mathrm{Bd}\, P)/G$ and $\Delta((\mathrm{Bd}\, P)/G)$ are isomorphic.*

Proof. Let $a = (a_1 > \cdots > a_t)$, $a \in \mathrm{Bd}\, P$, and assume $b \in \mathrm{Bd}\, P$ and $a > b$. Clearly, if $g \in G$ fixes a, then it fixes each a_i, for $i = 1, \ldots, t$; hence it fixes b as well. In other words, $\mathrm{Stab}\,(a) \subseteq \mathrm{Stab}\,(b)$. It follows that the G-action on $(\mathrm{Bd}\, P)^{\mathrm{op}}$ satisfies the strong Condition (St); hence by Proposition 14.11, Theorem 14.14, and Proposition 14.16, the simplicial complexes $\Delta(\mathrm{Bd}\, P)/G$ and $\Delta((\mathrm{Bd}\, P)/G)$ are isomorphic.

Furthermore, if also $c \in \mathrm{Bd}\, P$ such that $a > c$, then it is clear that if $g(b) = c$, then $b = c$, since the action preserves the order of the a_i's. Hence Condition (SRP) is satisfied as well, and so by Proposition 14.21, $(\mathrm{Bd}\, P)/G$ is a poset. $\square$

14.4 Bibliographic Notes

The material of this chapter is partially based on the paper of Babson and the author, [BK05]. The reader is encouraged to consult the original text for a more general discussion, including the generalization to the infinite categories.

15

Homotopy Colimits

15.1 Diagrams over Trisps

15.1.1 Diagrams and Colimits

We start directly by defining the main character of this chapter.

Definition 15.1. *The following data constitute what is called a* **diagram of topological spaces** $\mathcal{D}$ *over a trisp* Δ:

- *for each vertex v of Δ we have a topological space $\mathcal{D}(v)$;*
- *for each edge $(v \to w)$ we have a continuous map $\mathcal{D}(v \to w) : \mathcal{D}(v) \to \mathcal{D}(w)$.*

Additionally, we require that these maps commute over each triangle in Δ.

We remark that if the maps commute over each triangle, then they commute over each simplex in the trisp Δ. Definition 15.1 describes the notion of a diagram of topological spaces, though clearly any other category can be substituted instead of the category of topological spaces and continuous maps. This construction is also functorial, meaning that any functor $\mathcal{F} : \mathcal{A} \to \mathcal{B}$ will turn diagrams of objects from $\mathcal{A}$ into diagrams of objects from $\mathcal{B}$.

Example 15.2.
(1) To any continuous map f from a topological space X to itself there is a standard way to associate a diagram of the kind described in Definition 15.1. Namely, let Δ be a trisp with one vertex v and one edge $(v \to v)$. Then one sets $\mathcal{D}(v) := X$ and $\mathcal{D}(v \to v) := f$; see the first diagram in Figure 15.1.

(2) Furthermore, to any continuous map $f : X \to Y$ between topological spaces one can associate such a diagram as well. Namely, let Δ be a trisp with two vertices v and w, and an edge $(v \to w)$. Then one sets $\mathcal{D}(v) := X$, $\mathcal{D}(w) := Y$, and $\mathcal{D}(v \to w) := f$; see the second diagram in Figure 15.1.

(3) There is also another diagram associated to a continuous map $f : X \to Y$ between topological spaces. This one is over a trisp with three vertices a, b,

and c, and two edges $(c \to a)$ and $(c \to b)$. We set $\mathcal{D}(c) := X$, $\mathcal{D}(b) := Y$, and $\mathcal{D}(c \to b) := f$. Furthermore, we let $\mathcal{D}(a)$ be a point. This defines the map $\mathcal{D}(c \to a)$ uniquely; see the third diagram in Figure 15.1.

(4) Similarly, when X, Y, and Z are topological spaces such that $Z = X \cup Y$, there is also a standard diagram $\mathcal{D}$ of this type. Namely, again let Δ be the trisp described in (2). We set $\mathcal{D}(a) := X$, $\mathcal{D}(b) := Y$ and $\mathcal{D}(c) := X \cap Y$, and the maps $\mathcal{D}(c \to a) := (X \cap Y \hookrightarrow X)$ and $\mathcal{D}(c \to b) := (X \cap Y \hookrightarrow Y)$, where the latter two are the corresponding inclusion maps; see the rightmost diagram in Figure 15.1.

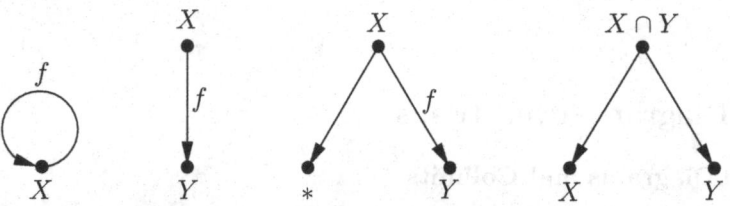

Fig. 15.1. Examples of diagrams over trisps.

The first construction associated to such a diagram is the operation of "gluing along all maps."

Definition 15.3. *Let $\mathcal{D}$ be a diagram of topological spaces over a trisp Δ. A* **colimit** *of $\mathcal{D}$ is the quotient space* $colim\, \mathcal{D} = \coprod_{v \in \Delta^{(0)}} \mathcal{D}(v)/ \sim$, *where the equivalence relation $\sim$ is generated by $x \sim \mathcal{D}(v \to w)(x)$, for all $v, w \in \Delta^{(0)}$ and all $x \in \mathcal{D}(v)$.*

When objects and morphisms of a category $\mathcal{C}$ are considered as a 1-skeleton of a trisp and the diagram is given by a functor $\mathcal{F} : \mathcal{C} \to \mathbf{Top}$, Definition 15.3 gives the same construction as the previously defined colimit of functors.

It can be instructive to go through Example 15.2. In (1) we get the quotient space $X/ \sim$ with the induced quotient topology, where the equivalence relation is generated by $x \sim f(x)$, for all $x \in X$. In (2) the colimit is equal to Y. In (3) the colimit is the quotient space $Y/f(X)$, meaning that all points in the image $f(X)$ get identified and we have the induced quotient topology. Finally, in (4) we have $\mathbf{colim}\, \mathcal{D} = X \cup Y$.

15.1.2 Arrow Pictures and Their Nerves

Often, the base trisp is generated by some smaller combinatorial data.

Definition 15.4. *An* **arrow picture** *$\mathcal{C}$ is a pair of sets (O, A), where O is the set of objects and $A \subseteq O \times O$ is the set of arrows equipped with the*

*partial rule of composition, meaning that to some pairs of arrows $\alpha : a \to b$
and $\beta : b \to c$ one associates a third arrow $\gamma : a \to c$, which is called their
composition, and is denoted by $\gamma = \beta \circ \alpha$.*

For example, any category is an arrow picture. Another example is obtained if we take a poset P, take its elements as objects, and take all pairs
(x, y) such that $x > y$, $x \neq y$ as arrows, so that if $x > y > z$, then the arrows
$x \to y$ and $y \to z$ compose to give $x \to z$ (since we asked for only one arrow
between comparable objects, there is no other choice).

A sequence of arrows $a_0 \xrightarrow{m_1} a_1 \xrightarrow{m_2} \cdots \xrightarrow{m_n} a_n$ is called *composable* if
any subsequence can be composed and the end result will not depend on the
order of composition. For example, when $n = 3$ this condition translates to the
following statements: m_1 and m_2 are composable, m_2 and m_3 are composable,
$m_2 \circ m_1$ and m_3 are composable, m_1 and $m_3 \circ m_2$ are composable, and finally
$(m_3 \circ m_2) \circ m_1 = m_3 \circ (m_2 \circ m_1)$.

Definition 15.5. *Let $\mathcal{C}$ be an arrow picture. The* **nerve** *of $\mathcal{C}$, denoted by $\mathcal{B}\mathcal{C}$,
is a trisp whose vertices are objects of $\mathcal{C}$ and whose n-simplices are all com-
posable sequences of arrows $a_0 \to a_1 \to \cdots \to a_n$. The simplex attachments
are described by skipping a vertex a_i, and, in case $i \neq 0, n$, composing the
arrows $a_{i-1} \to a_i$ and $a_i \to a_{i+1}$ after that.*

One example of an arrow picture and its nerve is shown in Figure 15.2.
In the special case when $\mathcal{C}$ is a category, the space $\mathcal{B}\mathcal{C}$ coincides with the
classifying space of $\mathcal{C}$.

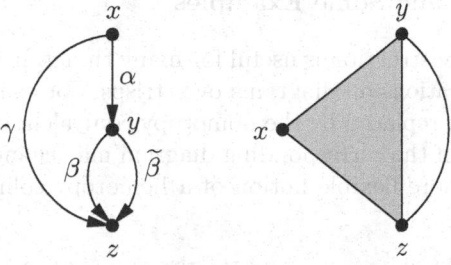

Fig. 15.2. An arrow picture on the left and its nerve on the right. In this arrow
picture we assume that $\gamma = \beta \circ \alpha$, and that arrows α and $\widetilde{\beta}$ do not compose.

Definition 15.6. *Let $\mathcal{C} = (A, O)$ be an arrow picture. A* **diagram of topological spaces over** $\mathcal{C}$ *is a rule associating a topological space $\mathcal{D}(v)$ to every
object $v \in O$, and a continuous map $\mathcal{D}(v \to w)$ for every arrow $(v \to w) \in A$
such that whenever $\alpha, \beta \subset A$ and the composition $\alpha \circ \beta$ is defined, we have
$\mathcal{D}(\alpha \circ \beta) = \mathcal{D}(\alpha) \circ \mathcal{D}(\beta)$.*

Again, the topological spaces and continuous maps can be replaced with any other category. In fact, when $\mathcal{C}$ is a category, a diagram of topological spaces over $\mathcal{C}$ is nothing but a functor $\mathcal{D} : \mathcal{C} \to \mathbf{Top}$.

We note that when considering a poset P, in order to specify a diagram of spaces over $\Delta(P)$, by the commutativity condition we need only give a topological space $\mathcal{D}(x)$ for every $x \in P$ and to describe a continuous map $\mathcal{D}(x \to y) : \mathcal{D}(x) \to \mathcal{D}(y)$ for each pair $x, y \in P$ such that x *covers* y.

It is easy to see that a diagram over an arrow picture $\mathcal{C}$ will give rise to a diagram over its nerve $\mathcal{BC}$. This is often used as a compact way to describe a diagram over trisps, and some literature considers only diagrams of this type. It is straightforward to define the colimit of a diagram over an arrow picture. As a matter of fact, one gets the same space as the colimit of the associated diagram over its nerve.

Example 15.7. We would like to specifically mention a very important example. Let X be a topological space, and let a group G act on X by continuous maps. Recall that in Chapter 4 we described how to view G as a category with one object o, and with one arrow corresponding to each group element so that the arrow composition corresponds to multiplication in G. Now we can let $\mathcal{D}$ be the diagram over G (and hence over $\mathcal{B}G$) defined as follows: $\mathcal{D}(o) = X$, and the continuous maps for arrows encode the G-action on X. In this case we have $\mathbf{colim}\,\mathcal{D} = X/G$.

15.2 Homotopy Colimits

15.2.1 Definition and Some Examples

While the colimit construction is useful for many things, it is too rigid to allow for topological operations on diagrams over trisps. For example, when spaces over the vertices are replaced by the homotopy equivalent ones, the homotopy type of the colimit of the corresponding diagram may change. For this reason, one considers the more flexible notion of a homotopy colimit, which we now proceed to define.

Definition 15.8. *The **homotopy colimit**, denoted by **hocolim** $\mathcal{D}$, of a diagram $\mathcal{D}$ of topological spaces over a trisp Δ, is the quotient space*

$$\mathbf{hocolim}\,\mathcal{D} = \coprod_{\sigma = v_0 \to \cdots \to v_n} (\sigma \times \mathcal{D}(v_0)) / \sim,$$

where the disjoint union is taken over all simplices in Δ. The equivalence relation $\sim$ is generated by the following condition: for $\tau_i \in \partial\sigma$, $\tau_i = v_0 \to \cdots \to \hat{v}_i \to \cdots \to v_n$, let $f_i : \tau_i \hookrightarrow \sigma$ be the inclusion map; then

- *for $i > 0$, $\tau_i \times \mathcal{D}(v_0)$ is identified with the subset of $\sigma \times \mathcal{D}(v_0)$ by the map induced by f_i;*

- *for $\tau_0 = v_1 \rightarrow \cdots \rightarrow v_n$, we have $f_0(\alpha) \times x \sim \alpha \times \mathcal{D}(v_0 \rightarrow v_1)(x)$, for any $\alpha \in \tau_0$, and $x \in \mathcal{D}(v_0)$.*

In other words, to construct homotopy colimit, consider the disjoint union of spaces $\mathcal{D}(v)$, for $v \in \Delta^{(0)}$; then for any directed edge $v \rightarrow w$ glue in the mapping cylinder of the map $\mathcal{D}(v \rightarrow w)$, taking $\mathcal{D}(v)$ as the source and $\mathcal{D}(w)$ as the base of it; furthermore, for every simplex $v \rightarrow w \rightarrow u$ glue in the "mapping triangle" of maps $\mathcal{D}(v \rightarrow w)$, $\mathcal{D}(w \rightarrow u)$, and $\mathcal{D}(v \rightarrow u)$, and so on for the whole Δ.

Example 15.9. Let us see a few examples of homotopy colimits involving some of the diagrams that have previously appeared in the text.

(0) If $\mathcal{D}(v)$ is a point, for any $v \in \Delta^{(0)}$, then **hocolim** $\mathcal{D} = \Delta$. We call such a diagram a *point diagram* over Δ.

(1) The homotopy colimit of the diagram from Example 15.2(1) is obtained by attaching a cylinder $X \times [0,1]$ to X by two ends, using id_X as the attaching map on one end, and using f as the attaching map on the other end; see the first space in Figure 15.3.

(2) The homotopy colimit of the diagram from Example 15.2(2) is precisely the mapping cylinder of the map f, M_f.

(3) The homotopy colimit of the diagram from Example 15.2(3) is the mapping cone of the map f, Cone_f; see the space in the middle of Figure 15.3.

(4) The homotopy colimit of the diagram in Example 15.2(4) consists of two spaces X, Y, and a cylinder connecting the two copies of $X \cap Y$ inside X and Y with each other; see the last space in Figure 15.3.

(5) Let X and Y be two given topological spaces and let us take another diagram over the poset $a < c > b$: $\mathcal{D}(a) = X$, $\mathcal{D}(b) = Y$, $\mathcal{D}(c) = X \times Y$, and the morphisms are the standard projections. Then one can see that **hocolim** $(\mathcal{D}) = X * Y$.

(6) Consider the diagram in Example 15.7. In this case, **hocolim** $\mathcal{D}$ is the so-called *Borel construction*: **hocolim** $\mathcal{D} = (X \times E\Gamma)/G = X \times_G E\Gamma$. The cohomology of this space is by definition equal to the equivariant cohomology of X. In the special case that X is just a point, we get **hocolim** $\mathcal{D} = B\Gamma$, which is the classifying space of the group G and whose cohomology is by definition equal to the cohomology of the group G.

15.2.2 Structural Maps Associated to Homotopy Colimits

Let us now enrich the set of diagrams by introducing appropriate structure-preserving maps.

Definition 15.10. *Let $\mathcal{D}_1$ and $\mathcal{D}_2$ be two diagrams over the same triop Δ. A **diagram map** $\mathcal{F}: \mathcal{D}_1 \rightarrow \mathcal{D}_2$ is a collection of maps $\mathcal{F}(v): \mathcal{D}_1(v) \rightarrow \mathcal{D}_2(v)$,*

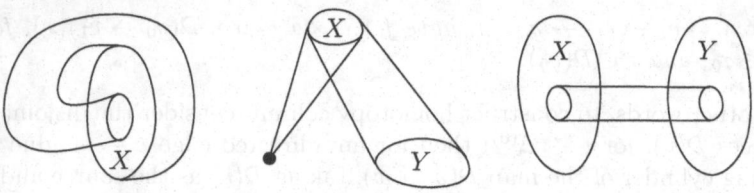

Fig. 15.3. Examples of homotopy colimits.

for each $v \in \Delta^{(0)}$ that commute with the arrow maps of diagrams, i.e., for any $v, w \in \Delta^{(0)}$, $(v \to w) \in \Delta_1$, the following diagram commutes:

$$
\begin{array}{ccc}
\mathcal{D}_1(v) & \xrightarrow{\ \mathcal{F}(v)\ } & \mathcal{D}_2(v) \\
{\scriptstyle \mathcal{D}_1(v \to w)}\Big\downarrow & & \Big\downarrow{\scriptstyle \mathcal{D}_2(v \to w)} \\
\mathcal{D}_1(w) & \xrightarrow{\ \mathcal{F}(w)\ } & \mathcal{D}_2(w)
\end{array}
\tag{15.1}
$$

We remark that for a fixed trisp Δ, the diagrams of objects of any fixed category over Δ, together with diagram maps form a category.

For a diagram $\mathcal{D}$ over Δ we have a *base projection map*

$$
\mathrm{proj} = p_b : \mathbf{hocolim}\,\mathcal{D} \longrightarrow \mathbf{hocolim}\,\widetilde{\mathcal{D}} = \Delta
$$

induced by the diagram map $\mathcal{D} \to \widetilde{\mathcal{D}}$, where $\widetilde{\mathcal{D}}$ is the point diagram. In general, a diagram map $\mathcal{F} : \mathcal{D}_1 \to \mathcal{D}_2$ induces a continuous map between the corresponding homotopy colimits $\mathbf{hocolim}\,\mathcal{F} : \mathbf{hocolim}\,\mathcal{D}_1 \to \mathbf{hocolim}\,\mathcal{D}_2$ induced by $\mathrm{id}_\sigma \times \mathcal{F}(v) : \sigma \times \mathcal{D}_1(v) \to \sigma \times \mathcal{D}_2(v)$.

Another structural map is the so-called *fiber projection*

$$
p_f : \mathbf{hocolim}\,\mathcal{D} \longrightarrow \mathbf{colim}\,\mathcal{D},
$$

which is induced by the map that simply forgets the first coordinate.

Let us see what these maps are for the diagrams of spaces considered in Examples 15.9.

Example 15.11.

(0) For the diagram from Example 15.9(0), we see that the base projection is just an identity map $p_b = \mathrm{id}_\Delta$, whereas $p_f : \mathbf{hocolim}\,\mathcal{D} \to *$ is the trivial map taking everything to a point.

(1) For the diagram from Example 15.9(1), we see that p_b is the canonical projection of the homotopy colimit to S^1, whereas p_f is the quotient map that collapses each "string" along the attached cylinder to a point.

(2) For the diagram from Example 15.9(2), we see that p_b is the canonical projection of the mapping cylinder to an interval, whereas p_f collapses the mapping cylinder to its base space.

(3) For the diagram from Example 15.9(3), we see that p_b is the canonical projection of the mapping cone onto an interval, whereas p_f collapses the actual cone inside the mapping cone to a point, which is the same as to collapse the image of X in Y to a point.

(4) For the diagram from Example 15.9(4), we see that p_b is the projection onto an interval, which can be thought of as schematically depicting the gluing process, whereas p_f collapses the cylinder so as to obtain $X \cup Y$.

(5) For the diagram from Example 15.9(5), we see that p_b is the canonical projection of the join onto an interval, whereas p_f collapses everything to a point.

(6) For the diagram from Example 15.9(6), both projections are quotients of the canonical product projections; namely, the maps $X \times \mathbf{E}\Gamma \to X$ and $X \times \mathbf{E}\Gamma \to \mathbf{E}\Gamma$ give rise to corresponding maps $X \times_G \mathbf{E}\Gamma \to X/G$ and $X \times_G \mathbf{E}\Gamma \to \mathbf{E}\Gamma/G = \mathbf{B}\Gamma$.

15.3 Deforming Homotopy Colimits

It is now time to formalize an important property of homotopy colimits: their flexibility with respect to homotopy type. It turns out that if the topological spaces in a diagram are replaced by homotopy equivalent ones in a coherent way (as a diagram map), then the homotopy type of the homotopy colimit does not change.

Theorem 15.12. (Homotopy lemma)
Let $\mathcal{F} : \mathcal{D}_1 \to \mathcal{D}_2$ be a diagram map between diagrams of spaces over Δ such that for each $v \in \Delta^{(0)}$, the map $\mathcal{F}(v) : \mathcal{D}_1(v) \to \mathcal{D}_2(v)$ is a homotopy equivalence. Then the induced map $\mathrm{hocolim}\,\mathcal{F} : \mathrm{hocolim}\,\mathcal{D}_1 \to \mathrm{hocolim}\,\mathcal{D}_2$ is a homotopy equivalence as well.

Proof. Let $\mathcal{M}$ be the diagram consisting of mapping cylinders derived from $\mathcal{F}$, i.e., for $v \in \Delta^{(0)}$ we let $\mathcal{M}(v) = \mathrm{M}_{\mathcal{F}(v)}$, and for $v \to w$ we let $\mathcal{M}(v \to w) : \mathrm{M}_{\mathcal{F}(v)} \to \mathrm{M}_{\mathcal{F}(w)}$ be the induced map given by the *naturality*[1] of mapping cylinders. It is not difficult to see that $\mathrm{M}_{\mathrm{hocolim}\,\mathcal{F}} \cong \mathrm{hocolim}\,\mathcal{M}$. Indeed, to start with, they both consist of the same two types of pieces, namely, $\tau \times \mathcal{D}_2(v)$ and $\sigma \times \mathcal{D}_1(v) \times I$, where $\sigma, \tau \in \Delta$ and $v \in \Delta^{(0)}$. Furthermore, one can also verify directly that the identification maps are the same. We leave this verification as an exercise.

It is obvious that $\mathrm{M}_{\mathrm{hocolim}\,\mathcal{F}}$ deformation retracts onto $\mathrm{hocolim}\,\mathcal{D}_2$, and it remains to be checked that it also deformation retracts onto $\mathrm{hocolim}\,\mathcal{D}_1$. We shall construct this retraction by induction on the dimension of the underlying trisp Δ.

---

[1] Naturality means that the construction behaves well with respect to maps

We start with $\dim \Delta = 0$. In this case, the space $\mathrm{M}_{\mathrm{hocolim}\,\mathcal{F}}$ is simply a disjoint union of several mapping cylinders, one for each vertex of Δ. Since for each $v \in \Delta^{(0)}$ the map $\mathcal{F}(v)$ is a homotopy equivalence, we can apply the mapping cylinder retraction, as in corollary 7.16, to each one of these mapping cylinders, and derive the necessary conclusion.

Assume now that $\dim \Delta = n > 0$. We already know by the induction hypothesis that the space $\mathrm{M}^{n-1}_{\mathrm{hocolim}\,\mathcal{F}} \cup \mathbf{hocolim}\,\mathcal{D}_1$ deformation retracts onto $\mathbf{hocolim}\,\mathcal{D}_1$, where $\mathrm{M}^{n-1}_{\mathrm{hocolim}\,\mathcal{F}}$ denotes the part of the mapping cylinder lying over the $(n-1)$th skeleton of Δ. Let X denote $\mathrm{M}_{\mathrm{hocolim}\,\mathcal{F}}$, and let A denote $\mathrm{M}^{n-1}_{\mathrm{hocolim}\,\mathcal{F}} \cup \mathbf{hocolim}\,\mathcal{D}_1$. We need to show that the space X deformation retracts onto A.

Clearly, since retractions happen in separate cells, it is enough to consider the case when Δ is a single n-simplex. Let us denote this simplex by $0 \to 1 \to \cdots \to n$ (with the indicated orientations), and let $A_i = \mathcal{D}_1(i)$, $B_i = \mathcal{D}_2(i)$, for $i = 0, \ldots, n$.

It is not difficult to see that the pair (X, A) is NDR. Therefore it is enough to show that the inclusion map $A \hookrightarrow X$ is a homotopy equivalence. As we have said, A deformation retracts onto $\mathbf{hocolim}\,\mathcal{D}_1$, which in turn, by the construction of homotopy colimit, deformation retracts onto A_n. On the other hand, again by the construction of homotopy colimit, the whole space X deformation retracts onto the mapping cylinder of the map $\mathcal{F}(n) : A_n \to B_n$. Since $\mathcal{F}(n)$ is a homotopy equivalence, we conclude that $A_n \hookrightarrow \mathrm{M}_{\mathcal{F}(n)}$ is a homotopy equivalence, which in turn implies that $A \hookrightarrow X$ is a homotopy equivalence as well. $\square$

Remark 15.13. It may be worthwhile to explicitly mention the following special case: when all the spaces $\mathcal{D}(v)$ in the diagram are contractible, then we have a diagram map from $\mathcal{D}$ to the point diagram defined in Example 15.9(0). It follows by Theorem 15.12 that the homotopy colimit of our diagram is homotopy equivalent to the base trisp, with the homotopy equivalence given by p_b.

15.4 Nerves of Coverings

15.4.1 Nerve Diagram

When X is a topological space, we shall say that a family of sets $\{X_i\}_{i \in I}$ is *a covering* of X if

- $X_i \subseteq X$, for all $i \in I$;
- $X_i \neq \emptyset$, for all $i \in I$;
- $\bigcup_{i \in I} X_i = X$.

We say that a covering has some specific property (such as open or locally finite) if every set in the covering has this property. Another covering $\{Y_j\}_{j \in J}$

is called a *refinement* of $\{X_i\}_{i \in I}$ if for every $j \in J$, there exists $i \in I$ such that $Y_j \subseteq X_i$.

Definition 15.14. *Let X be a topological space, and let $\mathcal{U} = \{X_i\}_{i \in I}$ be a covering. The **nerve** of $\mathcal{U}$ is the abstract simplicial complex, denoted by $\mathcal{N}(\mathcal{U})$, whose set of vertices is given by I and whose set of simplices is described as follows: the finite subset $S \in I$ gives a simplex of $\mathcal{N}(\mathcal{U})$ if and only if the intersection $\bigcap_{i \in S} X_I$ is nonempty.*

We shall see later that under further conditions on the covering spaces X_i and their intersections, there exists a close relationship between the topology of the space X and the topology of the nerve complex $\mathcal{N}(\mathcal{U})$.

Definition 15.15. *Let X be a topological space, and let $\mathcal{U} = \{X_i\}_{i \in I}$ be a covering of X. We define a diagram of spaces $\mathcal{D}$ over the space $\mathrm{Bd}\,\mathcal{N}(\mathcal{U})$ as follows:*

- $\mathcal{D}(S) = X_{i_1} \cap \cdots \cap X_{i_k}$, *for $S \in \mathcal{N}(\mathcal{U})$, $S = \{i_1, \ldots, i_k\}$;*
- $\mathcal{D}(S_1 \to S_2) : \mathcal{D}(S_1) \hookrightarrow \mathcal{D}(S_2)$ *is the inclusion map, for $S_1, S_2 \in \mathcal{N}(\mathcal{U})$, $S_2 \subset S_1$.*

*We call $\mathcal{D}$ the **nerve diagram** of $\mathcal{U}$.*

The importance of the nerve diagram will come to the fore in the rest of this section.

15.4.2 Projection Lemma

It is easy to see that when X is a topological space, $\mathcal{U} = \{X_i\}_{i \in I}$ is a covering of X, and $\mathcal{D}$ is the nerve diagram of $\mathcal{U}$, then $\mathbf{colim}\,\mathcal{D} = X$.

Definition 15.16. *Let $\mathcal{U} = \{X_i\}_{i \in I}$ be a locally finite open covering of X. A **partition of unity** for $\mathcal{U}$ is a collection of continuous maps $\{\varphi_i\}_{i \in I}$, $\varphi_i : X \to [0,1]$ such that*

(1) $\mathrm{supp}\,\varphi_i \subseteq X_i$, *for all $i \in I$;*
(2) $\sum_{i \in I} \varphi_i(x) = 1$, *for all $x \in X$.*

*If $\mathcal{U}$ is not assumed to be locally finite, then we define a **partition of unity subordinate to** $\mathcal{U}$ to be a partition of unity for some locally finite refinement of $\mathcal{U}$.*

Note that the condition (2) in Definition 15.16 can always be weakened in the following way: if we found functions $\{\varphi_i\}_{i \in I}$, $\varphi_i : X \to [0,1]$, such that $\mathrm{supp}\,\varphi_i \subseteq X_i$ and the sum function $\sum_{i \in I} \varphi_i(x)$ is nowhere zero, then dividing all these functions by $\sum_{i \in I} \varphi_i(x)$, we achieve the normalization.

Definition 15.17. *A topological space X is called **paracompact** if it is Hausdorff and every open covering $\{X_i\}_{i \in I}$ of X has a locally finite refinement.*

We cite the following theorem without proof.

Theorem 15.18. (Existence of a partition of unity)
If X is paracompact, and $\mathcal{U}$ is an open covering of X, then there exists a partition of unity that is subordinate to $\mathcal{U}$. Moreover, if $\mathcal{U}$ is locally finite, then there exists a partition of unity for $\mathcal{U}$.

Perhaps a different name for our next result could have been "lifting lemma," since de facto, the intersections get lifted over simplices of different dimensions. However, we choose to adopt the current terminology in the literature considering the projection map instead.

Theorem 15.19. (Projection lemma)
Let X be a paracompact topological space, let $\mathcal{U} = \{X_i\}_{i \in I}$ be a locally finite open covering of X, and let $\mathcal{D}$ be the nerve diagram of $\mathcal{U}$. Then the projection map $p_f : \mathbf{hocolim}\,\mathcal{D} \to \mathbf{colim}\,\mathcal{D} = X$ is a homotopy equivalence.

Remark 15.20. When X is a CW complex, the open covering in Theorem 15.19 may be replaced by a covering by subcomplexes. This is the formulation that turns out to be most useful for combinatorial applications. More generally, it works for coverings such that (X, X_i) is an NDR-pair, for all $i \in I$.

Proof of projection lemma. Choose a partition of unity $\{\varphi_i\}_{i \in I}$ for $\mathcal{U}$. We note that in our situation, a point in $\mathbf{hocolim}\,\mathcal{D}$ can be encoded as a pair (α, β), where α is a point in the interior of some $\sigma \in \mathcal{N}(\mathcal{U})$ (if $\dim \sigma = 0$, drop the word *interior*), $\sigma = (i_1, \ldots, i_k)$, and $\beta \in X_{i_1} \cap \cdots \cap X_{i_k}$. So, the preimage $p_f^{-1}(x)$ is the barycentric subdivision of a $(k-1)$-simplex when $x \in X_{i_1} \cap \cdots \cap X_{i_k}$ and $x \notin X_i$ if $i \notin \{i_1, \ldots, i_k\}$.

Now we define the map $g : X \to \mathbf{hocolim}\,\mathcal{D}$ as follows: for $x \in X$ we set

$$g(x) := \left(\sum_{i \in I} \varphi_i(x) v_i, x \right),$$

where v_i denotes the vertex of $\mathcal{N}(\mathcal{U})$ corresponding to X_i. Note that g is well-defined, since by definition of the partition of unity, if $x \notin X_i$, then $\varphi_i(x) = 0$.

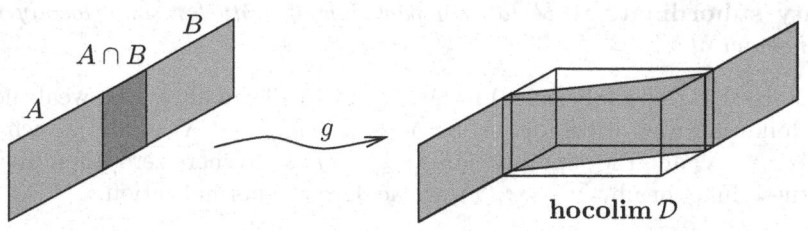

Fig. 15.4. The Segal map and the linear deformation.

By construction, we have $fg = \mathrm{id}_X$. Let us see that $\mathrm{id}_{\mathbf{hocolim}\,\mathcal{D}}$ is homotopic to gf. The homotopy is given by the linear deformation connecting $\left(\sum_{i \in I} \lambda_i(x)v_i, x\right)$ with $\left(\sum_{i \in I} \varphi_i(x)v_i, x\right)$; see Figure 15.4. We leave it as an exercise to verify that this gives a continuous deformation. $\square$

15.4.3 Nerve Lemmas

In combinatorial situations it often turns out that the space that we consider has a combinatorially motivated decomposition into simpler pieces. Often, these pieces and all of their intersections have similar combinatorial structure and can be analyzed by induction. This is one of the reasons why the next result is applied so often in Combinatorial Algebraic Topology.

Theorem 15.21. (Nerve lemma)
If $\mathcal{U}$ is a finite open cover of X such that every nonempty intersection of sets in $\mathcal{U}$ is contractible, then $X \simeq \mathcal{N}(\mathcal{U})$.

Remark 15.22. If X is a CW complex, the nerve lemma holds for covers by subcomplexes.

Proof of nerve lemma. Let $\mathcal{D}$ be the nerve diagram of $\mathcal{U}$, and let $\mathcal{D}^*$ be the trivial diagram over $\mathrm{Bd}\,\mathcal{N}(\mathcal{U})$, i.e., all associated spaces are points. Consider the unique diagram map $\mathcal{F} : \mathcal{D} \to \mathcal{D}^*$. Since all maps $\mathcal{F}(\sigma)$ for $\sigma \in \mathcal{N}(\mathcal{U})$ are homotopy equivalences, we can conclude by the homotopy lemma that also the map $\mathbf{hocolim}\,\mathcal{F} : \mathbf{hocolim}\,\mathcal{D} \to \mathbf{hocolim}\,\mathcal{D}^*$ is a homotopy equivalence.

Furthermore, by the projection lemma, the map $p_f : \mathbf{hocolim}\,\mathcal{D} \to \mathbf{colim}\,\mathcal{D}$ is a homotopy equivalence. Since $\mathbf{colim}\,\mathcal{D} = X$ and $\mathbf{hocolim}\,\mathcal{D}^* = \mathcal{N}(\mathcal{U})$, the nerve lemma follows. $\square$

As an application of the nerve lemma, let us prove the topological equivalence of the two product notions that we have so far defined for abstract simplicial complexes.

Proposition 15.23. *Let Δ_1 and Δ_2 be two arbitrary finite abstract simplicial complexes. Then the direct product $\Delta_1 \times \Delta_2$ is homotopy equivalent to the categorical product $\Delta_1 \prod \Delta_2$.*

Proof. For any two simplices $\sigma_1 \in \Delta_1$ and $\sigma_2 \in \Delta_2$, let $\sigma_1 \prod \sigma_2$ denote the simplex of $\Delta_1 \prod \Delta_2$ spanned by the set of vertices $V(\sigma_1) \times V(\sigma_2)$; cf. Definition 4.25. We see that when σ_1, resp. σ_2, ranges over all maximal simplices of Δ_1, resp. Δ_2, the family $\{\sigma_1 \prod \sigma_2\}_{\sigma_1, \sigma_2}$ is a simplicial subcomplex cover of $\Delta_1 \prod \Delta_2$. Since

$$\left(\sigma_1 \prod \sigma_2\right) \cap \left(\sigma_1' \prod \sigma_2'\right) = (\sigma_1 \cap \sigma_1') \prod (\sigma_2 \cap \sigma_2'), \qquad (15.2)$$

and each $\sigma_1 \prod \sigma_2$ is contractible, we can use Theorem 15.21 to conclude that $\Delta_1 \prod \Delta_2$ is homotopy equivalent to the nerve of this covering.

On the other hand, when σ_1, resp. σ_2, ranges over all maximal simplices of Δ_1, resp. Δ_2, the family $\{\sigma_1 \times \sigma_2\}_{\sigma_1, \sigma_2}$ is a CW subcomplex cover of $\Delta_1 \times \Delta_2$. Again, each $\sigma_1 \times \sigma_2$ is contractible, and

$$(\sigma_1 \times \sigma_2) \cap (\sigma_1' \times \sigma_2') = (\sigma_1 \cap \sigma_1') \times (\sigma_2 \cap \sigma_2'). \tag{15.3}$$

Therefore we can use Theorem 15.21 to conclude that $\Delta_1 \times \Delta_2$ is homotopy equivalent to the nerve of this covering.

The proposition follows now from the fact that the equations (15.2) and (15.3) imply that the nerves of the two coverings that we have considered are isomorphic to each other as abstract simplicial complexes. $\square$

Sometimes, we do not know that the covering complexes and their intersections are contractible. Instead, only partial information about their connectivity can be extracted. The more general nerve lemma then allows us to draw a weaker assumption about the connectivity of the total space.

Theorem 15.24. (Generalized nerve lemma).
Let X be a simplicial complex, and let $\mathcal{A} = \{A_i\}_{i=1}^n$ be a covering of X by its subcomplexes. Assume that for some $k \geq -1$, for every $\{i_1, \dots, i_t\} \subseteq [n]$, the complex $A_{i_1} \cap \cdots \cap A_{i_t}$ is either empty or $(k - t + 1)$-connected. Then X is k-connected if and only if $\mathcal{N}(\mathcal{A})$ is k-connected.

Proof. Passing to the barycentric subdivision, we can always achieve that the subcomplexes A_i are induced, i.e., for all $i \in [n]$, we have $A_i = X[V_i] = X \cap \Delta^{V_i}$, where V_i is the set of vertices of A_i. The proof now proceeds by deforming the complex X and the subcomplexes $\{A_i\}_{i=1}^n$.

Deformation Step. Choose an arbitrary $m \in [n]$. Replace X with $\widetilde{X} = X \cup \Delta^{V_m}$, and replace A_i with $\widetilde{A}_i = A_i \cup \Delta^{V_m \cap V_i}$, for all $i \in [n]$.

We now prove several properties of this deformation.

(1) The family $\{\widetilde{A}_i\}_{i=1}^n$ is a covering of $\widetilde{X}$.
Proof. We have $\widetilde{X} = X \cup \Delta^{V_m} = X \cup \widetilde{A}_m = (\bigcup_{i=1}^n A_i) \cup \widetilde{A}_m \subseteq \bigcup_{i=1}^n \widetilde{A}_i$.

(2) The subcomplexes $\widetilde{A}_i$ are induced.
Proof. This follows from the following calculation: $\widetilde{X} \cap \Delta^{V_i} = (X \cup \Delta^{V_m}) \cap \Delta^{V_i} = (X \cap \Delta^{V_i}) \cup (\Delta^{V_m} \cap \Delta^{V_i}) = A_i \cup \Delta^{V_m \cap V_i} = \widetilde{A}_i$.

(3) The nerve of the covering does not change: $\mathcal{N}(\mathcal{A}) = \mathcal{N}(\widetilde{\mathcal{A}})$.
Proof. The subcomplexes intersect if and only if their sets of vertices intersect. We see that their sets of vertices do not change.

(4) The complex $\widetilde{X}$ is k-connected if and only if X is k-connected.
Proof. We have $X \cap \Delta^{V_m} = A_m$, which is k-connected. So if X is k-connected, then by Corollary 6.29, $\widetilde{X} = X \cup \Delta^{V_m}$ is k-connected as well.

Conversely, assume that $\widetilde{X} = X \cup \Delta^{V_m}$ is k-connected. Since $X \cap \Delta^{V_m} = A_m$ is k-connected, we see that X is k-connected by Corollary 6.30.

(5) The conditions of the generalized nerve lemma are still satisfied, i.e., if the intersection of t of the $\widetilde{A}_i$'s is nonempty, then it has to be $(k-t+1)$-connected. **Proof.** We have the following verification:

$$
\begin{aligned}
\widetilde{A}_{i_1} \cap \cdots \cap \widetilde{A}_{i_t} &= \widetilde{X} \cap \Delta^{V_{i_1} \cap \cdots \cap V_{i_t}} = \left(X \cup \Delta^{V_m}\right) \cap \Delta^{V_{i_1} \cap \cdots \cap V_{i_t}} \\
&= \left(X \cap \Delta^{V_{i_1} \cap \cdots \cap V_{i_t}}\right) \cup \Delta^{V_m \cap V_{i_1} \cap \cdots \cap V_{i_t}} \\
&= \left(A_{i_1} \cap \cdots \cap A_{i_t}\right) \cup \Delta^{V_m \cap V_{i_1} \cap \cdots \cap V_{i_t}}.
\end{aligned}
$$

By our assumptions, $A_{i_1} \cap \cdots \cap A_{i_t}$ is $(k-t+1)$-connected, whereas $A_{i_1} \cap \cdots \cap A_{i_t} \cap \Delta^{V_m \cap V_{i_1} \cap \cdots \cap V_{i_t}} = A_m \cap A_{i_1} \cap \cdots \cap A_{i_t}$ is $(k-t)$-connected. By Corollary 6.29 we may conclude that $\widetilde{A}_{i_1} \cap \cdots \cap \widetilde{A}_{i_t}$ is $(k-t+1)$-connected.

Note that in particular, during such a deformation step, A_m gets replaced with the simplex $\widetilde{A}_m = \Delta^{V_m}$, and that if A_i was a simplex already, then $\widetilde{A}_i = A_i \cup \Delta^{V_m \cap V_i} = \Delta^{V_i}$ is a simplex again. It follows that applying this deformation step n times if necessary, we can produce the covering of the final deformed space X_f by simplices. The statement now follows by property (4) coupled with the regular nerve lemma. $\square$

15.5 Gluing Spaces

15.5.1 Gluing Lemma

The next result shows that if they are coherent, homotopy equivalences can be glued together. We shall give an explicit proof to illustrate the combined use of the projection and homotopy lemmas.

Theorem 15.25. (Gluing lemma)
Let A, B, C, and D be topological spaces such that A and B are open in $A \cup B$, and C and D are open in $C \cup D$. Let furthermore $f : A \cup B \to C \cup D$ be a map such that the restriction maps $f|_A : A \to C$, $f|_B : B \to D$, and $f|_{A \cap B} : A \cap B \to C \cap D$ are homotopy equivalences. Then f itself is a homotopy equivalence.

Remark 15.26. The gluing lemma holds when $A \cup B$ and $C \cup D$ are CW complexes and A, B, C, and D are subcomplexes. More generally, it holds for cofibrations.

Remark 15.27. A straightforward argument allows one to generalize Theorem 15.25 to the case of covering by finitely many spaces. Simply proceed by induction on the number of covering spaces, using the formula $A_n \cap (A_1 \cup \cdots \cup A_{n-1}) = (A_n \cap A_1) \cup \cdots \cup (A_n \cap A_{n-1})$.

Proof of the gluing lemma. If $A \cap B = C \cap D = \emptyset$, then the statement is trivial, so assume that these are nonempty. Let $\mathcal{D}_1$ be the nerve diagram of the covering of $A \cup B$ by A and B, and analogously, let $\mathcal{D}_2$ be the nerve diagram of the covering of $C \cup D$ by C and D. The continuous map f induces a diagram map $\mathcal{F} : \mathcal{D}_1 \to \mathcal{D}_2$, and hence a map between homotopy colimits **hocolim** $\mathcal{F}$: **hocolim** $\mathcal{D}_1 \to$ **hocolim** $\mathcal{D}_2$. By the homotopy lemma, the map **hocolim** $\mathcal{F}$ is a homotopy equivalence.

Let s_1 : **colim** $\mathcal{D}_1 \to$ **hocolim** $\mathcal{D}_1$ be the Segal map defined in the proof of the projection lemma, where it was also shown that it is a homotopy equivalence. Finally, the projection map p_f : **hocolim** $\mathcal{D}_2 \to$ **colim** $\mathcal{D}_2$ is a homotopy equivalence by the projection lemma, and therefore we conclude that the composition

$$p_f \circ \textbf{hocolim}\,\mathcal{F} \circ s_1 : A \cup B = \textbf{colim}\,\mathcal{D}_1 \to \textbf{colim}\,\mathcal{D}_2 = C \cup D$$

is a homotopy equivalence. Tracing the definition of the Segal map, we see that the following diagram commutes:

$$
\begin{array}{ccc}
\textbf{hocolim}\,\mathcal{D}_1 & \xrightarrow{\ \textbf{hocolim}\,\mathcal{F}\ } & \textbf{hocolim}\,\mathcal{D}_2 \\
{\scriptstyle s_1}\Big\uparrow & & \Big\downarrow{\scriptstyle p_f} \\
A \cup B & \xrightarrow{\quad f \quad} & C \cup D
\end{array}
$$

i.e., that this composition map is actually equal to f.   □

15.5.2 Quillen Lemma

As an application of the gluing lemma, let us prove the so-called Quillen lemma, which turned out to be a useful tool in Combinatorial Algebraic Topology.

Theorem 15.28. (Quillen lemma)
Let $\varphi : P \to Q$ be an order-preserving map such that for any $x \in Q$ the complex $\Delta(\varphi^{-1}(Q_{\geq x}))$ is contractible. Then the induced map between the simplicial complexes $\Delta(\varphi) : \Delta(P) \to \Delta(Q)$ is a homotopy equivalence.

Proof. We use induction on $|Q|$. If $|\min Q| = 1$, then $\varphi^{-1}(Q_{\geq x}) = P$; hence $\Delta(P)$ is contractible, and therefore $\Delta(\varphi) : \Delta(P) \to \Delta(Q)$ is a homotopy equivalence. Before we proceed with the induction, let us verify a simple fact.

Fact. *If A is a proper upper ideal in Q, then $\varphi : \varphi^{-1}(A) \to A$ induces a homotopy equivalence $\Delta(\varphi) : \Delta(\varphi^{-1}(A)) \to \Delta(A)$.*

Proof of the fact. For $x \in A$ we have $A_{\leq x} = Q_{\leq x}$. Therefore $\Delta(A_{\leq x}) = \Delta(Q_{\leq x})$ is contractible. The conditions of the Quillen lemma are thereby satisfied, and the fact follows by induction, since $|A| < |Q|$.   □

Assume now that $|Q| \geq 2$. Let x be a minimal element of Q. Since the posets $Q_{>x}$, $Q_{\geq x}$ and $Q \setminus \{x\}$ all are upper ideals in Q, we may conclude that the maps $\Delta(\varphi) : \Delta(\varphi^{-1}(Q_{>x})) \to \Delta(Q_{>x})$, $\Delta(\varphi) : \Delta(\varphi^{-1}(Q_{\geq x})) \to \Delta(Q_{\geq x})$, and $\Delta(\varphi) : \Delta(\varphi^{-1}(Q \setminus \{x\})) \to \Delta(Q \setminus \{x\})$ all are homotopy equivalences. Since additionally we have

$$\Delta(\varphi^{-1}(Q_{\geq x})) \cap \Delta(\varphi^{-1}(Q \setminus \{x\})) = \Delta(\varphi^{-1}(Q_{>x})),$$

$$\Delta(\varphi^{-1}(Q_{\geq x})) \cup \Delta(\varphi^{-1}(Q \setminus \{x\})) = \Delta(\varphi^{-1}(Q)) = \Delta(P),$$

$$\Delta(Q_{\geq x}) \cap \Delta(Q \setminus \{x\}) = \Delta(Q_{>x}), \text{ and } \Delta(Q_{\geq x}) \cup \Delta(Q \setminus \{x\}) = \Delta(Q),$$

we can apply the gluing lemma to conclude that $\Delta(\varphi) : \Delta(P) \to \Delta(Q)$ is a homotopy equivalence. $\square$

15.6 Bibliographic Notes

Homotopy colimits is a subject of several books and surveys. A classical source is provided by [BoK72]; see also [Vo73]. Probably the first use of homotopy colimits in combinatorics can be found in [ZZ93]. Our treatment of the general theory in this chapter has an introductory character. We recommend [WZZ99] for a more in-depth approach.

A good general reference on gluing spaces is the textbook [Hat02]. Several arguments here are essentially borrowed form excellent research articles. For example, the proof of the projection lemma, Theorem 15.19, is adopted from [Se68], while our approach to the generalized nerve lemma follows [BKL85]. The reader is invited to consult these sources for further details and the general context.

Our treatment is self-contained with the exception of the standard Theorem 15.18, whose proof can be found in many textbooks, see, for example, [Bre93, Chapter I, Section 12].

The Quillen lemma originally appeared in [Qu73] and [Qu78], with both sources being well worth a further look.

16
Spectral Sequences

Spectral sequences constitute an important tool for concrete combinatorial calculations of homology groups. In this chapter we shall give a short introduction, which is aimed at setting up the notation and at helping the reader to develop intuition. Our presentation will be purely algebraic, using the topological picture only as a source for the algebraic gadgets.

The spectral sequence setup comes in two flavors: for homology and for cohomology. For convenience and to suit our applications we shall describe here only the cohomology version. The homology is readily obtained by the usual process of "inverting all arrows."

16.1 Filtrations

In concrete situations it can be difficult to compute the cohomology groups of a cochain complex without using auxiliary constructions. The idea behind spectral sequences is to break up this large task into smaller subtasks, with the formal machinery to support the bookkeeping. This "break-up" is usually phrased in terms of a filtration.

Definition 16.1. *A (finite)* **filtration** *on a cochain complex* $\mathcal{C} = (C^*, \partial^*)$ *is a nested sequence of cochain complexes*

$$\mathcal{C}_j = \cdots \xrightarrow{\partial^{i-2}} C_j^{i-1} \xrightarrow{\partial^{i-1}} C_j^i \xrightarrow{\partial^i} C_j^{i+1} \xrightarrow{\partial^{i+1}} \cdots ,$$

for $j = 0, 1, 2, \ldots, t$, *such that* $\mathcal{C} = \mathcal{C}_0 \supseteq \mathcal{C}_1 \supseteq \cdots \supseteq \mathcal{C}_{t-1} \supseteq \mathcal{C}_t$ *(that is why we suppressed the lower index in the differential).*

In general, infinite filtrations can be considered, but in this book we limit ourselves to the finite ones. Given a filtration $\mathcal{C} = \mathcal{C}_0 \supseteq \mathcal{C}_1 \supseteq \cdots \supseteq \mathcal{C}_t$, for the convenience of notation we set $\mathcal{C}_{-1} = \mathcal{C}_{t+1} = 0$.

There are many standard filtrations of cochain complexes. For example, if a cochain complex is bounded, say $C^i = 0$, for $i < 0$, or $i > t$, then the standard skeleton filtration is defined as follows:

$$C_j^i = \begin{cases} C^i, & \text{if } i \geq j; \\ 0, & \text{otherwise.} \end{cases}$$

This filtration turns out to be not very interesting, since computing the cohomology groups with its help is canonically equivalent to computing the cohomology groups from the cochain complex directly, thus nothing new is gained from this approach.

For a CW complex X, a classical way to define a filtration on its cellular cochain complex is to choose a cell filtration on X, i.e., a sequence of cell subcomplexes $X = X_t \supseteq X_{t-1} \supseteq \cdots \supseteq X_0$ (again for the convenience of notation, we set $X_{-1} = \emptyset$). The cellular cochain complexes $\mathcal{C}_j = C^*(X, X_{j-1})$, for $j = 0, \ldots, t+1$, form a sequence of nested cochain subcomplexes. Recall that $C^*(X, A)$ denotes the set of all functions that take the value 0 on all cells of A.

If the CW complex X is finite-dimensional, then, taking X_i to be the ith skeleton of X, we recover the standard skeleton filtration on the cochain groups $C^*(X)$. A much more interesting situation is described in the following definition.

Definition 16.2. *Assume that we have a cell map $\varphi : X \to Y$ and a filtration $Y = Y_t \supseteq Y_{t-1} \supseteq \cdots \supseteq Y_0$. Define a filtration on X as follows: set $X_i := \varphi^{-1}(Y_i)$, for $i = 0, \ldots, t$. This filtration on X is called the **pullback** of the filtration on Y along φ.*

Of special interest is the pullback filtration of the topological skeleton filtration on Y. A similar filtration arising from fiber bundles is called the *Leray–Serre filtration*. We extend the use of this name to our more general case, and we also use it for the corresponding filtration on the cellular cochain complex of X.

16.2 Contriving Spectral Sequences

16.2.1 The Objects to be Constructed

Once we have fixed a filtration $\mathcal{C} = \mathcal{C}_0 \supseteq \mathcal{C}_1 \supseteq \cdots \supseteq \mathcal{C}_t$, $\mathcal{C}_i = (C_i^*, \partial^*)$, on a cochain complex, we can proceed to compute its cohomology groups by studying auxiliary algebraic gadgets derived from that filtration following the blueprint that we now describe.

Namely, rather than working with the 1-dimensional cochain complex directly, we study a sequence of 2-dimensional tableaux $E_n^{*,*}$, $n = 0, 1, 2, \ldots$. Recall that initially, our cochain complex had the usual differential, going up one in degree $\partial^* : C^* \to C^{*+1}$. Here, each tableau $E_n^{*,*}$ is equipped with an almost diagonal differential:

$$d_n^{*,*} : E_n^{*,*} \to E_n^{*+n,*+1-n}.$$

One expresses this fact by saying that d_n is *a differential of bidegree* $(n, -n + 1)$. The reader may want to think of it as a "generalized knight move"; see Figure 16.1.

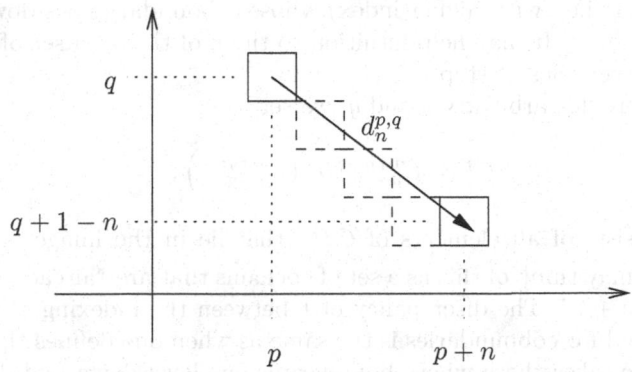

Fig. 16.1. The differential of bidegree $(n, -n + 1)$.

Each subsequent differential d_n is in a way derived from the original differential ∂^*, and furthermore, $E_{n+1}^{*,*}$ is the cohomology tableau of $E_n^{*,*}$ in the appropriate sense. The idea is then to compute the tableaux $E_n^{*,*}$ one by one, until they stabilize. The stabilized tableau is usually called $E_\infty^{*,*}$.

We would like to alert the reader at this point that even after the tableau $E_\infty^{*,*}$ is computed, it can still require additional work to determine the cohomology groups of the original cochain complex. Surely, if $\mathcal{R}$ is a field, the situation is easy. Namely, one has

$$H^d(\mathcal{C}) = \bigoplus_{p+q=d} E_\infty^{p,q}. \tag{16.1}$$

However, if $\mathcal{R}$ is an arbitrary ring (for example $\mathcal{R} = \mathbb{Z}$), then one may need to solve a number of extension problems before obtaining the final answer. This has to do with the fact that, as mentioned above, in a short exact sequence of $\mathcal{R}$-modules

$$0 \longrightarrow A \xrightarrow{\alpha} B \xrightarrow{\beta} C \longrightarrow 0,$$

the $\mathcal{R}$-module B does not necessarily split as a direct sum of the submodule A and the quotient module C. We take a closer look at this question in Subsection 16.2.3.

16.2.2 The Actual Construction

Let us now describe more precisely how the tableaux $E_n^{*,*}$ and the differentials d_n are constructed. As auxiliary modules, for arbitrary p and q we set

$$Z_n^{p,q} := C_p^{p+q} \cap (\partial^*)^{-1} \left(C_{p+n}^{p+q+1} \right), \tag{16.2}$$

where whenever S is a set, we use $(\partial^*)^{-1}(S)$ to denote the set of all elements whose coboundary belongs to S. Thus, $Z_n^{p,q}$ is the set of all $(p+q)$-cochains on level p or lower (i.e., with higher index) whose coboundary goes down to the level $p + n$ or lower. It may help intuition to think of this as a set of cochains that are "still cocycles at step n."

Furthermore, for arbitrary p and q, we set

$$B_n^{p,q} := C_p^{p+q} \cap \partial^* \left(C_{p-n}^{p+q-1} \right), \tag{16.3}$$

i.e., $B_n^{p,q}$ consists of all elements of C_p^{p+q} that lie in the image of ∂^* from C_{p-n}^{p+q-1}. One may think of this as a set of cochains that are "already coboundaries at step $n + 1$." The discrepancy of 1 between the indexing of the steps for cocycles and for coboundaries is the same as when one defines the relative homology: one takes those whose boundary is one level down and divides by boundaries of those at this level.

These are the settings for $n \geq 0$. Finally, for $n = -1$, we use the following convention:

$$Z_{-1}^{p,q} := C_p^{p+q} \qquad \text{and} \qquad B_{-1}^{p,q} := \partial^* \left(C_{p+1}^{p+q-1} \right).$$

Finally, we set

$$Z_\infty^{p,q} := \operatorname{Ker}\partial^* \cap C_p^{p+q} \qquad \text{and} \qquad B_\infty^{p,q} := \operatorname{Im}\partial^* \cap C_p^{p+q}.$$

Clearly, we have

$$B_{-1}^{p,q} \subseteq B_0^{p,q} \subseteq B_1^{p,q} \subseteq \cdots \subseteq B_\infty^{p,q} \subseteq Z_\infty^{p,q} \subseteq \cdots \subseteq Z_1^{p,q} \subseteq Z_0^{p,q} \subseteq Z_{-1}^{p,q}.$$

Let us check various identities. To start with, as a direct consequence of the identity $\partial^* \circ \partial^* = 0$, we have

$$\partial^*(Z_n^{p,q}) \subseteq Z_n^{p+n,q-n+1}. \tag{16.4}$$

Furthermore, since $\partial^*(Z_{n-1}^{p+1,q-1}) \subseteq B_{n-1}^{p+n,q-n+1}$ and $\partial^*(B_{n-1}^{p,q}) = 0$, we get

$$\partial^*(Z_{n-1}^{p+1,q-1} + B_{n-1}^{p,q}) \subseteq Z_{n-1}^{p+n+1,q-n} + B_{n-1}^{p+n,q-n+1}. \tag{16.5}$$

It is now time to define the tableau entries. These are given by certain quotients as follows:

$$E_n^{p,q} := Z_n^{p,q} \Big/ \left(Z_{n-1}^{p+1,q-1} + B_{n-1}^{p,q} \right),$$ (16.6)

for all $0 \le n \le \infty$. Again the analogy with the relative homology is in order: we take those whose boundary is one level lower, the "relative cycles," and divide by those that are boundaries plus those that are one level lower themselves.

Inclusions (16.4) and (16.5) imply that the differential ∂^* induces a map from $E_n^{p,q}$ to $E_n^{p+n,q-n+1}$, via the quotient maps, which we choose to call $d_n^{p,q}$, or just d_n if it is clear what the indices p and q are. It seems reasonable to call the map d_n a differential as well.

One can view the tableau $(E_n^{*,*}, d_n)$ as a collection of nearly diagonal cochain complexes. This allows one to compute the cohomology groups, just as for the usual cochain complexes, by setting

$$H^{p,q}(E_n^{*,*}, d_n) := \mathrm{Ker}\left(E_n^{p,q} \xrightarrow{d_n} E_n^{p+n,q-n+1} \right) \Big/ \mathrm{Im}\left(E_n^{p-n,q+n-1} \xrightarrow{d_n} E_n^{p,q} \right).$$

In this context we can make the meaning of the words "$E_{n+1}^{*,*}$ is the cohomology tableau of $E_n^{*,*}$" precise; namely, we have

$$E_{n+1}^{p,q} = H^{p,q}\left(E_n^{*,*}, d_n \right).$$ (16.7)

One can verify equation (16.7) formally, using equation (16.6). For brevity, we omit this verification, since it is not needed for our combinatorial computations. Instead we observe that the intuitive meaning of the identity (16.7) is rather clear.

Indeed, passing from $E_n^{p,q}$ to $E_{n+1}^{p,q}$, means expanding our scope of vision by one level in each direction. That is we now look at those cochains whose coboundary lies (modulo lower coboundaries) $n+1$ levels down, which is one deeper than before, and divide out those that are discovered to be boundaries when we look n levels up; again this is one more than previously. This, on the other hand, is precisely what one does when computing the cohomology groups of this "knight move" skew cochain complex.

It could be instructive to unwind these definitions for some special cases. Let us start with $n = 0$. It follows from our conventions for $n = -1$ that

$$E_0^{p,q} = \left(C_p^{p+q} \cap (\partial^*)^{-1} \left(C_p^{p+q+1} \right) \right) \Big/ \left(C_{p+1}^{p+q} + \partial^* \left(C_{p+1}^{p+q-1} \right) \right)$$
$$= C_p^{p+q} / C_{p+1}^{p+q}.$$ (16.8)

Furthermore, the differential $\partial^* : C_p^{p+q} \to C_p^{p+q+1}$ induces the differential $d_0 : E_0^{p,q} \to E_0^{p,q+1}$, which is nothing else but the differential of the relative cochain complex (C_p, C_{p+1}). By equation (16.7), this yields

$$E_1^{p,q} = H^{p+q}(C_p, C_{p+1}).$$ (16.9)

Moreover, in the case $t - 1$, i.e., when the filtration consists in choosing a single cochain subcomplex, one can show that the only nontrivial differential

$d_1^{0,q} : E_1^{0,q} \longrightarrow E_1^{1,q}$ is, in fact, nothing but the connecting homomorphism $\partial^* : H^q(\mathcal{C}_0, \mathcal{C}_1) \longrightarrow H^{q+1}(\mathcal{C}_1)$ in the long exact sequence associated to the pair of cochain complexes $(\mathcal{C}_0, \mathcal{C}_1)$.

Usually, unless some additional specific information is available, it is hard to say what happens in the tableaux for $n \geq 2$. The important thing is that with the setup above, the spectral sequence runs its course and eventually converges (modulo the extension difficulties outlined above) to the cohomology groups of the original cochain complex.

16.2.3 Questions of Convergence and Interpretation of the Answer

There are two important questions associated with general spectral sequences: the question of convergence – *when does the process terminate?* and the question of interpretation of the final tableau – *what shall one do to extract the cohomology groups from the final tableau?*

The first question is not of much concern if one assumes, as we do, that the filtration is finite. In our case the nonzero entries are all concentrated to a finite rectangle, so, once n grows large, the differentials will no longer fit into the rectangle, and so will not cause any further changes. We will end up with the stable tableau that is denoted by $E_\infty^{*,*}$.

The entries of this tableau are $E_\infty^{p,q} = Z_\infty^{p,q}/(Z_\infty^{p+1,q-1} + B_\infty^{p,q})$. These filter the cohomology of the original cochain complex as follows:

$$E_\infty^{p,q} = F^p H^{p+q}(\mathcal{C}, \partial^*)/F^{p+1} H^{p+q}(\mathcal{C}, \partial^*), \qquad (16.10)$$

where we set $F^p H^*(\mathcal{C}, \partial^*) := \mathrm{Im}\, \iota^*$, for the inclusion map $\iota : \mathcal{C}_p \hookrightarrow \mathcal{C}$. This means that if we are working over a field, then the cohomology $H^*(\mathcal{C}, \partial^*)$ can be recovered as a direct sum (16.1). Otherwise, one can proceed as follows. Fix the dimension $d = p + q$, and consider all $E_\infty^{p,q}$ with $p + q = d$ and $p = 0, \ldots, t$. By (16.10) we have $E_\infty^{t,q} = F^t H^d(\mathcal{C}, \partial^*)$, and $E_\infty^{t-1,q} = F^{t-1} H^d(\mathcal{C}, \partial^*)/F^t H^d(\mathcal{C}, \partial^*)$. This means that we can find $F^{t-1} H^d(\mathcal{C}, \partial^*)$ by solving the extension problem

$$0 \longrightarrow E_\infty^{t-1,q} \longrightarrow F^{t-1} H^d(\mathcal{C}, \partial^*) \longrightarrow E_\infty^{t,q} \longrightarrow 0.$$

Inductively, solving further extension problems for lower values of p, we shall eventually compute $H^d(\mathcal{C}, \partial^*)$.

16.2.4 An Example

Let us illustrate how a typical computation may run with a simple example. Let X be the CW complex shown in Figure 16.2, which is homeomorphic to $\mathbb{RP}^2$. The arrows indicate the cell orientations.

Consider the subcomplex filtration $X = X_2 \subset X_1 \subset X_0$, where $X_0 = \{x, y\}$ and $X_1 = \{x, y, a\}$. Then we have

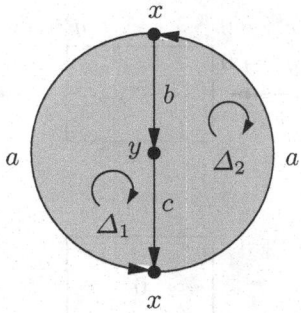

Fig. 16.2. A cell subdivision of the real projective plane.

$$C^*(X) = \mathbb{Z}^2 \to \mathbb{Z}^3 \to \mathbb{Z}^2,$$
$$C^*(X, X_0) = 0 \to \mathbb{Z}^3 \to \mathbb{Z}^2,$$
$$C^*(X, X_1) = 0 \to \mathbb{Z}^2 \to \mathbb{Z}^2.$$

The four consecutive tableaux are shown in Figures 16.3, 16.4, 16.5, and 16.6, where as usual for a cell σ we use σ^* to denote the cochain that evaluates to 1 on σ, and to 0 on all other cells. We conclude that, as expected, $H^0(X) = \mathbb{Z}$, $H^1(X) = 0$, and $H^2(X) = \mathbb{Z}_2$.

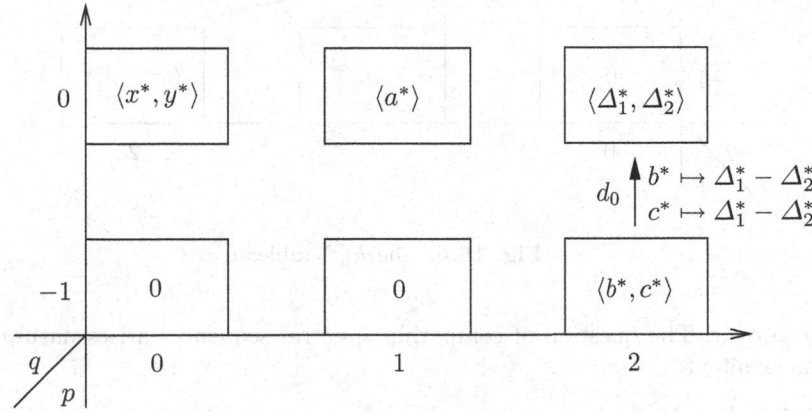

Fig. 16.3. The $E_0^{*,*}$-tableau.

16.3 Maps Between Spectral Sequences

Sometimes, instead of understanding the spectral sequence of the cochain complex of interest, one succeeds in analyzing the one that is in some sense

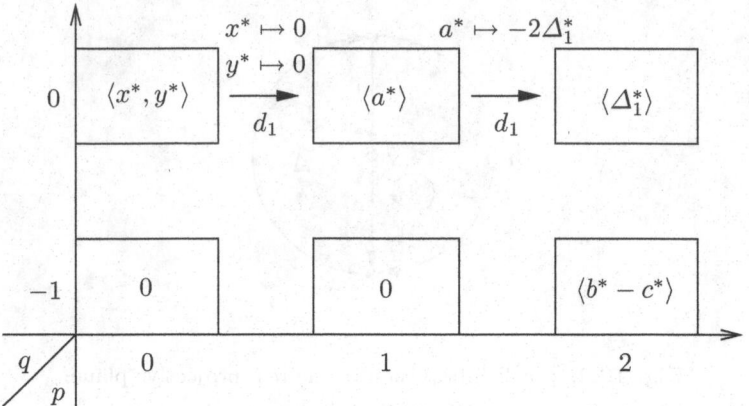

Fig. 16.4. The $E_1^{*,*}$-tableau.

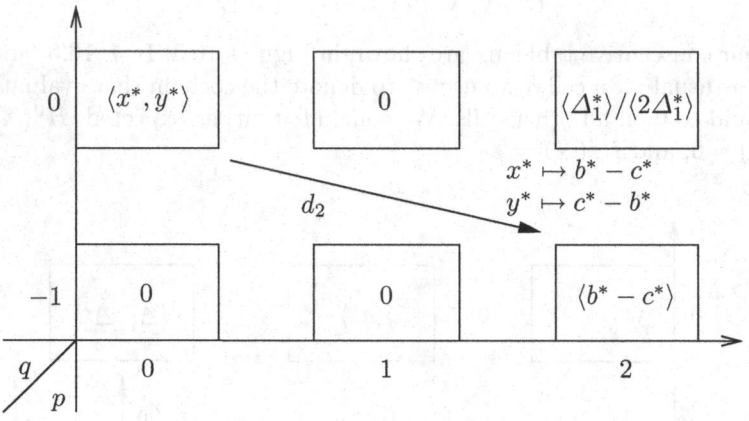

Fig. 16.5. The $E_2^{*,*}$-tableau.

close to it. The question of comparing spectral sequences arises naturally in that context.

Definition 16.3. *A* **morphism between spectral sequences** $(E_*^{*,*}, d_*)$ *and* $(\widetilde{E}_*^{*,*}, \tilde{d}_*)$ *is a collection of homomorphisms* $\{f_n^{p,q}\}_{n,p,q}$, $f_n^{p,q} : E_n^{p,q} \to \widetilde{E}_n^{p,q}$ *that commute with the tableau differentials, i.e.,* $f_n^{*,*} \circ d_n = \tilde{d}_n \circ f_n^{*,*}$, *and each* $f_{n+1}^{*,*}$ *is as a map on cohomology of "knight-move" cochain complexes, induced by* $f_n^{*,*}$.

In fact, one can show that spectral sequences together with morphisms between them form a category.

The standard situation in which a morphism between spectral sequences arises is that in which we have a map between the underlying CW complexes

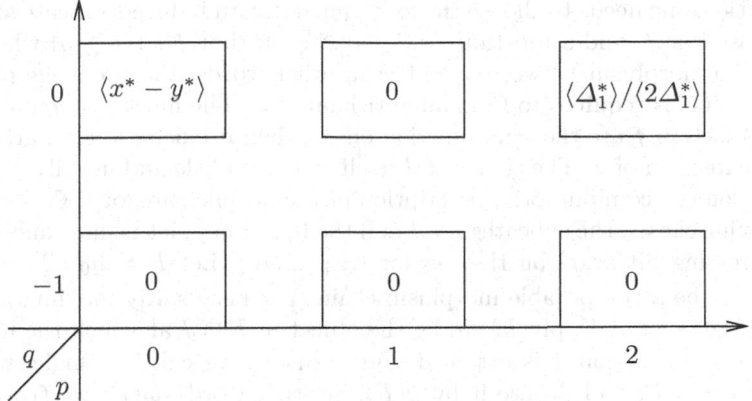

Fig. 16.6. The $E_3^{*,*}$-tableau.

that respects filtration, i.e., for two filtered CW complexes

$$X = X_t \supseteq X_{t-1} \supseteq \cdots \supseteq X_0 \text{ and } \widetilde{X} = \widetilde{X}_t \supseteq \widetilde{X}_{t-1} \supseteq \cdots \supseteq \widetilde{X}_0,$$

we have a cellular map $\varphi : X \to \widetilde{X}$ such that $\varphi(X_i) \subseteq \widetilde{X}_i$, for all i. This is in full analogy with the fact that a map of pairs induces a homomorphism of the corresponding long exact sequences.

Theorem 16.4.
(1) A cellular map between filtered CW complexes that respects the filtration induces a morphism between the spectral sequences that are defined by these filtrations.
(2) If for some n, all the homomorphisms $f_n^{*,*}$ are isomorphisms, then the homomorphisms $f_s^{*,*}$ are isomorphisms for all $s \geq n$, including $s = \infty$. Furthermore, the map induces an isomorphism of the cohomology groups $H^*(X) \cong H^*(\widetilde{X})$.

Again, for the sake of brevity, we omit the proof of Theorem 16.4. The required argument is fairly straightforward, and the only part that requires a little bit of work is the very last statement. This can be done by the repeated use of the five lemma.

16.4 Spectral Sequences and Nerves of Acyclic Categories

16.4.1 A Class of Filtrations

Let now K be a finite acyclic category. We would like to describe a special class of filtrations on the cochain complex $C^*(\Delta(K))$. In order to construct such a filtration, one needs to choose the following data: an induced subcategory K, which we call J, and a function $f : J \longrightarrow \mathbb{N}$ such that $f(x) \neq f(y)$ whenever there is a morphism between x and y; in other words, the preimage of each element in $\mathbb{N}$ is required to form an antichain in J. The most frequent choices of the function f are the rank function on J (when it exists) and an arbitrary linear extension of J. The choice of J itself is more subtle and usually depends heavily on the combinatorial description of the acyclic category K.

Having chosen the subcategory J and the function f, let us now first define an increasing filtration on the regular trisp $\Delta(K)$. Let $\Gamma = a_0 \xrightarrow{m_1} a_1 \xrightarrow{m_2} \cdots \xrightarrow{m_k} a_k$ be a composable morphism chain (not necessarily maximal) in K. Define the *pivot* of Γ, $\mathrm{piv}(\Gamma)$, to be the object in $\Gamma \cap J$ at which the highest value of the function f is attained. Furthermore, we call this highest value the *weight* of Γ, and denote it by $\omega(\Gamma)$; in other words, $\omega(\Gamma) := f(\mathrm{piv}(\Gamma))$. By construction, we know that f takes different values on different elements in $\Gamma \cap J$, and hence the notion of pivot and its weight is well defined. If $\Gamma \cap J = \emptyset$, we say that the chain Γ has weight 0. This assignment of weights gives us a filtration of the regular trisp $\Delta(K)$ as follows: for all $k \geq 0$ and $i \geq 0$ we set $\Delta_i := \Delta(K_i)$, where K_i is the full subcategory of K consisting of all elements of weight $\leq i$. For convenience, we also set $\Delta_{-1} := \emptyset$. Clearly, we have $\Delta_{-1} \subseteq \Delta_0 \subseteq \Delta_1 \subseteq \cdots \subseteq \Delta_t$, where t is the maximal value attained by the function f.

By definition of the nerve of an acyclic category, the differential skips one of the objects of the composable morphism chain. Omitting an object other than the pivot does not alter the weight of this chain, while omitting the pivot turns another element into the new pivot, on which f takes a lower value than on $\mathrm{piv}(\Gamma)$. Thus the resulting chain has a strictly lower weight. Hence $\partial_*(\Delta_i) \subseteq \Delta_i$, i.e., the differential operator ∂_* respects the filtration.

Now we set $\mathcal{C}_j := C^*(X, X_{j-1})$ to get a filtration on the cochain complex $C^*(\Delta(K)) = \mathcal{C}_0 \supseteq \mathcal{C}_1 \supseteq \cdots \supseteq \mathcal{C}_t = 0$. The following general fact helps us understand the first tableau of the corresponding spectral sequence.

Fact. *Assume that X is a regular trisp, and that Y is a subtrisp of X such that the vertices in $X \setminus Y$ have no edges in between them. Then we have*

$$C^*(X, Y) = \bigoplus_{i=1}^{l} C^{*-1}(\mathrm{lk}_X v_i), \tag{16.11}$$

where $\{v_1, \ldots, v_l\}$ is the set of the vertices in $X \setminus Y$, and the notation $* - 1$ is used to indicate a shift (in this case by one) in indexing.

For $a \in J$, we let S_a be the induced subcategory of C with the set of objects $(K \setminus J) \cup \{b \in J \mid f(b) < f(a)\}$. Since the objects of J that are added to the complex in the same filtration step form an antichain, we can use the general fact above and derive

$$E_1^{p,q} = H^{p+q}(\mathcal{C}_p, \mathcal{C}_{p+1}) = \bigoplus_{a \in f^{-1}(p)} H^{p+q-1}(\mathrm{lk}_{\Delta(S_a)} a), \tag{16.12}$$

for $p \geq 1$. For $p = 0$ we simply have

$$E_1^{0,q} = H^q(\Delta(K \setminus J)). \tag{16.13}$$

In the next subsection we will see that one can in general relate the Möbius function in an acyclic category to the values of the tableau entries in a spectral sequence computing the cohomology groups of the associated nerve. Here, Proposition 16.5 specializes to

$$\mu_K(s) = \mu_{K \setminus J}(s) + \sum_{a \in J} \mu_{S_a}(m_1) \cdot \mu_{S_a}(m_2), \tag{16.14}$$

where m_1 is the unique morphism from a to s, and m_2 is the unique morphism from t to a.

Let us now consider several special cases of the construction above. First, assume that P is a lattice. Let x be any element of $\bar{P}$, and set $J := \bar{P} \setminus \bar{P}_{\geq x}$. Finally, let f be an arbitrary order-preserving function on $J \cup \{\hat{0}\}$. Then equation (16.14) gives Weisner's theorem, which says that

$$\mu_P(\hat{0}, \hat{1}) = - \sum_{a \vee x = \hat{1}} \mu_P(\hat{0}, a). \tag{16.15}$$

Another standard special case arises when P is a poset, J is a lower ideal in P, and f is an order-preserving function. In this case, the formula (16.12) specializes to

$$E_1^{p,q} = \bigoplus_{a \in f^{-1}(p)} H^{p+q-1}(\Delta(P_{<a}) * \Delta(P_{>a} \cap (P \setminus J))). \tag{16.16}$$

Finally, let us take P to be a ranked poset, and let us choose $J = \bar{P}$ and $f(x) = \mathrm{rk}\,(x)$. In this case, equations (16.16) and (16.13) specialize to

$$E_1^{p,q} = \bigoplus_{\mathrm{rk}\,(a)=p} H^{p+q-1}(\Delta(P_{<a})).$$

We can read off the so-called *Whitney cohomology groups* of P from the $E_1^{*,*}$-tableau:

$$W^d(P) := \bigoplus_{p+q=d} E_1^{p,q} = \bigoplus_{a \in P} \tilde{H}^{d-1}(\Delta(P_{<a})), \quad \text{for } d \in \mathbb{Z}.$$

16.4.2 Möbius Function and Inequalities for Betti Numbers

In this subsection we restrict ourselves to the field coefficients. The homology and cohomology spectral sequences are then just the duals of each other. We formulate the results using cohomology.

When specializing to a spectral sequence for the cohomology of the nerve of an acyclic category, we immediately observe that its Möbius function can be read off from the $E_n^{*,*}$-tableau, for any nonnegative integer n.

Proposition 16.5. *Let C be a finite acyclic category with a terminal object t and initial object s, and let $(E_n^{*,*})_{n=0}^{\infty}$ be a spectral sequence converging to the cohomology of $\Delta(\bar{C})$. Then, for any n we have*

$$\mu_C(s) = \sum_{p,q \in \mathbb{Z}} (-1)^{p+q} rk\, E_n^{p,q} - 1. \tag{16.17}$$

Proof. For arbitrary n, we define the Euler characteristic $\chi(E_n^{*,*})$ to be the sum $\sum_{p,q \in \mathbb{Z}} (-1)^{p+q} \mathrm{rk}\, E_n^{p,q}$. By (16.7), we know that $E_{n+1}^{p,q} = H^{p,q}(E_n^{*,*}, d_n)$, and hence, fixing $p+q$ and using the Euler–Poincaré formula, we get

$$\sum_{p+q=d} (-1)^d \mathrm{rk}\, E_n^{p,q} = \sum_{p+q=d} (-1)^d \mathrm{rk}\, E_{n+1}^{p,q},$$

for all d. Summing over all d, we obtain the equality $\chi(E_n^{*,*}) = \chi(E_{n+1}^{*,*})$, for all n.

Furthermore, since $\chi(E_\infty^{*,*}) = \chi(\Delta(\bar{C}))$, we get $\chi(E_n^{*,*}) = \chi(\Delta(\bar{C}))$, for all n. Finally, Theorem 10.26 says that $\mu_C(s) = \tilde{\chi}(\Delta(C))$, and so formula (16.17) follows. $\square$

As we have seen earlier, formula (16.17) specializes to several well-known formulas for Möbius function computations, once the spectral sequence is specified.

Before the proof of the next proposition let us recall the following useful fact from group theory.

Fact. *If G is an abelian group and H is a subgroup of G, then $rk(G) = rk(H) + rk(G/H)$.*

This fact is fairly standard and we shall not give an argument here.

Proposition 16.6. *Let C be a finite acyclic category, and let $(E_n^{*,*})_{n=0}^{\infty}$ be a spectral sequence converging to $H^*(\Delta(C))$. If for some n, we happen to have $E_{p,q}^n = 0$, for all $p,q \in \mathbb{Z}$ such that $p+q=d$, then we can conclude that $H_d(\Delta(C)) = 0$.*

Furthermore, for any d and for any n we have

$$\beta_d(\Delta(C)) \le \sum_{p+q=d} rk\, E_n^{p,q} \tag{16.18}$$

and

$$\beta_d(\Delta(C)) - \beta_{d-1}(\Delta(C)) - \beta_{d+1}(\Delta(C))$$
$$\ge \sum_{p+q=d} rk\, E_n^{p,q} - \sum_{p+q=d-1} rk\, E_n^{p,q} - \sum_{p+q=d+1} rk\, E_n^{p,q}. \tag{16.19}$$

Proof. From the identity (16.7) we can conclude that $\mathrm{rk}\, E_{n+1}^{p,q} \le \mathrm{rk}\, E_n^{p,q}$, for all p, q, and n. Hence $E_n^{p,q} = 0$ would imply $E_{n+1}^{p,q} = 0$, and so the first statement of the proposition follows. Furthermore, we have

$$\beta_d(\Delta(C)) = \operatorname{rk} H_d(\Delta(C)) = \sum_{p+q=d} \operatorname{rk} E_\infty^{p,q} \leq \sum_{p+q=d} \operatorname{rk} E_n^{p,q},$$

for all d and all n.

We shall now prove the inequality (16.19). Let us fix d and n, and let us set $d^0 := d_n|_{E_n^{p-2n,q+2n-1}}$, $d^1 := d_n|_{E_n^{p-n,q+n-1}}$, $d^2 := d_n|_{E_n^{p,q}}$ and $d^3 := d_n|_{E_n^{p+n,q-n+1}}$. Then we have the following cochain complex:

$$\cdots \longrightarrow E_n^{p-2n,q+2n-1} \xrightarrow{d_0} E_n^{p-n,q+n-1} \xrightarrow{d_1} E_n^{p,q} \xrightarrow{d_2}$$

$$\xrightarrow{d_2} E_n^{p+n,q-n+1} \xrightarrow{d_3} E_n^{p+2n,q-2n+2} \longrightarrow \cdots.$$

From (16.7) we know that $E_{n+1}^{p,q} = \operatorname{Ker} d_2 / \operatorname{Im} d_1$, $E_{n+1}^{p-n,q+n-1} = \operatorname{Ker} d_1 / \operatorname{Im} d_0$, and $E_{n+1}^{p+n,q-n+1} = \operatorname{Ker} d_3 / \operatorname{Im} d_2$; hence we have

$$\operatorname{rk} E_{n+1}^{p,q} - \operatorname{rk} E_{n+1}^{p-n,q+n-1} - \operatorname{rk} E_{n+1}^{p+n,q-n+1}$$
$$= (\operatorname{rk} \operatorname{Ker} d_2 - \operatorname{rk} \operatorname{Im} d_1) - (\operatorname{rk} \operatorname{Ker} d_1 - \operatorname{rk} \operatorname{Im} d_0) - (\operatorname{rk} \operatorname{Ker} d_3 - \operatorname{rk} \operatorname{Im} d_2)$$
$$= (\operatorname{rk} \operatorname{Ker} d_2 + \operatorname{rk} \operatorname{Im} d_2) - (\operatorname{rk} \operatorname{Ker} d_1 + \operatorname{rk} \operatorname{Im} d_1) - (\operatorname{rk} \operatorname{Ker} d_3 + \operatorname{rk} \operatorname{Im} d_3)$$
$$+ \operatorname{rk} \operatorname{Im} d_0 + \operatorname{rk} \operatorname{Im} d_3 \geq \operatorname{rk} E_n^{p,q} - \operatorname{rk} E_n^{p-n,q+n-1} - \operatorname{rk} E_n^{p+n,q-n+1}. \quad (16.20)$$

Summing inequalities (16.20) over all pairs (p,q) such that $p+q = d$, we obtain

$$\sum_{p+q=d} \operatorname{rk} E_n^{p,q} - \sum_{p+q=d-1} \operatorname{rk} E_n^{p,q} - \sum_{p+q=d+1} \operatorname{rk} E_n^{p,q}$$
$$\leq \sum_{p+q=d} \operatorname{rk} E_{n+1}^{p,q} - \sum_{p+q=d-1} \operatorname{rk} E_{n+1}^{p,q} - \sum_{p+q=d+1} \operatorname{rk} E_{n+1}^{p,q}; \quad (16.21)$$

hence using formula (16.7), we get

$$\beta_d(\Delta(C)) - \beta_{d-1}(\Delta(C)) - \beta_{d+1}(\Delta(C))$$
$$= \sum_{p+q=d} \operatorname{rk} E_\infty^{p,q} - \sum_{p+q=d-1} \operatorname{rk} E_\infty^{p,q} - \sum_{p+q=d+1} \operatorname{rk} E_\infty^{p,q}$$
$$\geq \sum_{p+q=d} \operatorname{rk} E_n^{p,q} - \sum_{p+q=d-1} \operatorname{rk} E_n^{p,q} - \sum_{p+q=d+1} \operatorname{rk} E_n^{p,q}.$$

This finishes the proof. $\square$

When we are dealing with the nerve of a lattice, this additional structure is sufficient to make a stronger statement about the Betti numbers of its nerve.

Theorem 16.7. *Let P be a finite lattice, x an atom in P. Then the following inequalities hold:*

$$\beta_k(\hat{0}, \hat{1}) \le \sum_{y \vee x = \hat{1}} \beta_{k-1}(\hat{0}, y), \tag{16.22}$$

$$\sum_{y \vee x = \hat{1}} (\beta_{k-1}(\hat{0}, y) - \beta_{k-2}(\hat{0}, y) - \beta_k(\hat{0}, y))$$

$$\le \beta_k(\hat{0}, \hat{1}) - \beta_{k-1}(\hat{0}, \hat{1}) - \beta_{k+1}(\hat{0}, \hat{1}). \tag{16.23}$$

In particular, if $\beta_{k-2}(\hat{0}, y) = \beta_k(\hat{0}, y) = 0$, for all $y \in P$, such that $y \vee x = \hat{1}$, then

$$\beta_k(\hat{0}, \hat{1}) = \sum_{y \vee x = \hat{1}} \beta_{k-1}(\hat{0}, y).$$

Proof. Let $J = \bar{P} \setminus \bar{P}_{\ge x}$ and let $x_1, \ldots, x_k$ be any linear extension of J. Consider the spectral sequence E that is associated to the ideal J, where we filtrate using the given linear extension of J. Observe first that $\bar{P} \setminus J = \bar{P}_{\ge x}$ is contractible. Also, for any $a \in J$, we have $(a, \hat{1}) \cap (\bar{P} \setminus J) = (a, \hat{1}) \cap \bar{P}_{\ge x} = \bar{P}_{\ge x \vee a}$.

This means that $(a, \hat{1}) \cap (\bar{P} \setminus J)$ is contractible (actually a cone with apex $x \vee a$) unless $x \vee a = \hat{1}$. When $x \vee a = \hat{1}$ we get $(a, \hat{1}) \cap (\bar{P} \setminus J) = \emptyset$. So, using formulas (16.13) and (16.16), we obtain

$$E_1^{p,q} = \begin{cases} H^{p+q-1}(\Delta(P_{<x_p})), & \text{if } x_p \vee x = \hat{1}; \\ 0, & \text{otherwise.} \end{cases}$$

The inequalities (16.22) and (16.23) follow from inequalities (16.18) and (16.19). □

16.5 Bibliographic Notes

The existing literature on spectral sequences is rich, and our introduction here is just the tip of the iceberg. We encourage the interested reader to consult some of these excellent sources, e.g., [FFG86, McC01], for further reading. In particular, the verification of equation (16.7) can be found in [McC01, Section 2.2]. Furthermore, we refer to [McC01, Section 3.1] for the complete proof of Theorem 16.4.

The material of Section 16.4 is inspired by [Ko01]. The new contribution of the current presentation is doing everything in the generality of acyclic categories. We also phrase everything in terms of cohomology instead of homology.

There are several papers that use the approach of Section 16.4. For example, in [FK00] the case $P = \mathcal{D}_{n,k}$, where $\mathcal{D}_{n,k}$ is the intersection lattice of the k-equal arrangement of type $\mathcal{D}$, has been considered. In this situation it turned out to be appropriate to take J to be the set of all the elements without unbalanced component. Phil Hanlon, in [Han91], considers the case in which J is a lower-order ideal and f is the rank function (he considers pure posets only).

The uniform treatment of general formulas in Section 16.4 has many specializations within combinatorics. For example, special cases of formula (16.14) can be found [Sta97]. For more information on Weisner's theorem itself the reader may want to consult [Sta97, Corollary 3.9.3]. Baclawski has introduced in [Bac75] *Whitney homology groups* of a poset P; these are dual to the Whitney cohomology groups from Subsection 16.4.1.

Finally, we mention for the sake of completeness that the reader who is interested in a proof of the fact that whenever G is an abelian group and H a subgroup, one has $\mathrm{rk}\,(G) = \mathrm{rk}\,(H) + \mathrm{rk}\,(G/H)$, may, for example, consult [KM79, Exercise 7.2.2.].

Complexes of Graph Homomorphisms

Chromatic Numbers and the Kneser Conjecture

As mentioned in the introduction, one of the main themes of Combinatorial Algebraic Topology is to study problems on the borderline between discrete mathematics and algebraic topology, whose solutions benefit from the interaction of the two fields. Usually, this implies constructing a topological space starting with a discrete object as an input, or, conversely, providing a discrete model for an already existing geometric or topological setting.

In this part of the book we describe a topological approach to the classical problem of vertex-colorings in graphs, which roughly can be summarized by the following scheme:

$$\text{Graph } G \quad \longrightarrow \quad \text{Topological space } X(G)$$

$$\downarrow$$

$$\begin{array}{ccc} \text{Combinatorial} & \longleftarrow & \text{Topological} \\ \text{properties of } G & & \text{properties of } X(G) \end{array}$$

17.1 The Chromatic Number of a Graph

17.1.1 The Definition and Applications

To keep this part as self-contained as possible, we repeat some of our conventions. In our considerations all the graphs are undirected, unless explicitly stated otherwise. We do not allow multiple edges, but we do allow loops.

For a graph G, $V(G)$ denotes the set of its vertices, and $E(G)$ denotes the set of its edges. If convenient, we think of $E(G)$ as a $\mathbb{Z}_2$-equivariant subset of $V(G) \times V(G)$, where $\mathbb{Z}_2$ acts on $V(G) \times V(G)$ by switching the coordinates: $(x, y) \mapsto (y, x)$.

Definition 17.1. *Let G be a graph. A* **vertex-coloring** *of G is a set map $c : V(G) \to S$ such that $(x, y) \in E(G)$ implies $c(x) \neq c(y)$.*

Clearly, a vertex coloring exists if and only if G has no loops.

Definition 17.2. *The* **chromatic number** *of G, $\chi(G)$, is the minimal cardinality of a set S such that there exists a vertex-coloring $c : V(G) \to S$.*

If G has loops, we use the convention $\chi(G) = \infty$.

The literature devoted to the applications of computing the chromatic number of a graph is very extensive. Two of the basic applications are the *frequency assignment problem* and the *task scheduling problem.*

The first one concerns a collection of transmitters, with certain pairs of transmitters required to have different frequencies (e.g., because they are too close). Clearly, the minimal number of frequencies required for such an assignment is precisely the chromatic number of the graph, whose vertices correspond to the transmitters, with two connected by an edge if and only if they are required to have different frequencies.

The second problem concerns a collection of tasks that need to be performed. Each task has to be performed exactly once, and the tasks are to be performed in regularly allocated slots (e.g., hours). The only constraint is that certain tasks cannot be performed simultaneously. Again, the minimal number of slots required for the task scheduling is equal to the chromatic number of the graph whose vertices correspond to tasks, with two vertices connected by an edge if and only if the corresponding tasks cannot be performed simultaneously.

17.1.2 The Complexity of Computing the Chromatic Number

The problem of computing the chromatic number of a graph is NP-complete, implying that the worst-case performance of any algorithm is, most likely, exponential in the number of vertices. In fact, several seemingly much more special problems are NP-complete as well. Two examples are the problem of deciding whether a given planar graph is 3-colorable, and the problem of finding a coloring with 4 colors for a 3-colorable graph.

On the positive side, deciding whether $\chi(G) = 2$ (i.e., whether the graph is bipartite) is computationally much easier: plain depth-first search yields $O(|V(G)| + |E(G)|)$ performance time. Also, one can 4-color a planar graph in polynomial (in fact quadratic) time.

In the general case, however, the picture remains bleak even if we switch to considering approximations. For example, it is known that if a polynomial-time approximate algorithm for graph coloring exists (in precise formulation, meaning that the output of the algorithm does not differ by more than a constant factor, which is smaller than 2, from the actual value of the chromatic number), then there exists a polynomial-time algorithm for graph coloring, which, of course, is not very likely.

Much of the same can be said about *running tests* to yield lower bounds for the chromatic number. Even the most immediate one, computing the clique number, is not good, since this is an NP-complete problem as well. The existence of cliques of a fixed size can be decided in polynomial time, but does not provide a very interesting test.

The original Lovász test, which was based on computing the connectivity of a certain abstract simplicial complex associated to the graph in question, also has high computational complexity, since determining the triviality of the homotopy groups is an extremely hard problem, even in low dimensions.

It is then a positive surprise that the tests based on the Stiefel–Whitney characteristic classes are polynomially computable if we fix the test graph and the tested dimension and consider the computational complexity with respect to the number of vertices.

17.1.3 The Hadwiger Conjecture

A special place in the theory of vertex-colorings of a graph is occupied by the so-called four-color problem – the question whether there is a four-coloring of a planar map such that every pair of countries that share a (nonpoint) boundary segment receive different colors. Let us show the weaker five-color theorem. Before we can prove it, we need a standard fact, which is a special case of the Euler–Poincaré formula.

Theorem 17.3. *Let G be a nonempty finite connected graph, drawn in a plane without self-intersections. Let $R(G)$ denote the set of regions into which the plane is divided. Then we have the following formula:*

$$|V(G)| + |R(G)| - |E(G)| = 2. \tag{17.1}$$

Proof. We shall prove (17.1) by induction on the number of edges in G. If G has no edges, then, since G is connected, $|V(G)| = 1$ and $|R(G)| = 1$; hence the formula (17.1) follows.

Assume now that we have at least one edge. If G has no cycles (in particular no loops), then G is a tree. It is then well known that in this case, $|V(G)| = |E(G)| + 1$. Since also $|R(G)| = 1$, we get (17.1).

If G has a cycle (for example, if some vertex has a loop), let e be an arbitrary edge of this cycle. Consider the graph G' obtained from G by deleting e. Since e is a part of a cycle, the graph G' is again connected. Furthermore, we have $V(G) = V(G')$ and $|G'| = |G| - 1$. Finally by the Jordan curve theorem, the edge e borders two different regions, which are merged when e is removed; thus $|R(G')| = |R(G)| - 1$, and (17.1) follows by the induction hypothesis.   □

Lemma 17.4. *An arbitrary planar simple graph G has a vertex with valency at most 5.*

Proof. Assume the contrary. Then we have $2|E(G)| = \sum_{x \in V(G)} \mathrm{val}(x) \geq \sum_{x \in V(G)} 6 = 6|V(G)|$. Also, since the graph is simple, every region is bound by at least 3 edges, so the double counting of incidences (edge, region) gives $2|E(G)| = \sum_{r \in R(G)} s(r) \geq \sum_{r \in R(G)} 3 = 3|R(G)|$, where $s(r)$ denotes the number of bounding edges of the region r. Summarizing, we get

$$|E(G)| = |E(G)|/3 + 2|E(G)|/3 \geq |V(G)| + |R(G)|,$$

which contradicts the formula (17.1). □

Theorem 17.5. (Five-color theorem).
Every loopless planar graph is five-colorable.

Proof. Let G be the graph under consideration. Since having multiple edges does not change the chromatic number, we can assume that G is a simple graph.

According to Lemma 17.4, there exists a vertex x of G with valency at most 5. If the valency of x is at most 4, then we can color $G - x$ by the induction hypothesis with at most 5 colors, and then color x with whichever color was not used for one of its neighbors.

So we can assume that the valency of x is exactly 5. If there exists a coloring of $G - x$ with at most 5 colors such that not all neighbors of x have different colors, then we can again color x with whichever color was not used for one of its neighbors.

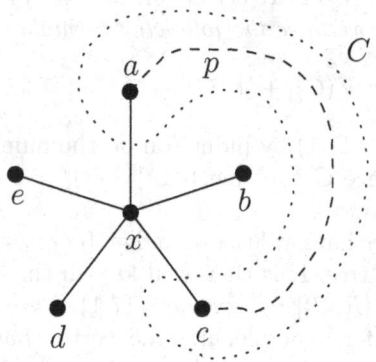

Fig. 17.1. Finding a color-alternating path.

Consider now a 5-coloring of $G - x$, with all neighbors of x having different colors. Let a, b, c, d, e index the neighbors of x in clockwise order, starting from an arbitrary neighbor, and assume that these are colored with colors 1, 2, 3, 4, and 5 respectively. Let C be the maximal induced subgraph of G satisfying the following conditions:

- C is connected;
- C contains the vertex a;
- all vertices C are colored either with color 1 or with color 3 (in particular, C is bipartite).

If we recolor all the vertices of C simultaneously, by swapping colors 1 and 3, then we obtain a new vertex-coloring of G. If c does not belong to C, then such a recoloring will produce a 5-coloring of G such that vertices a and c have color 3; hence we can color the vertex x with the color 1.

Assume now that c belongs to C. Let D be the maximal induced subgraph of G defined in a way identical to how we defined C, with the difference being that we start with the vertex b instead of a, and follow the colors 2 and 4 instead of the colors 1 and 3. Again, if the vertex d does not belong to D, then we re-color D by swapping colors 2 and 4 and then color x with the color 2.

Finally, assume that d belongs to D. Take an arbitrary non-self-intersecting path p connecting a and c without going through x, together with the edges (a, x) and (c, x). This is a non-self-intersecting loop l. Vertices b and d lie on different sides of this loop; thus, by the Jordan curve theorem, any path connecting them should cross l. But, we can find a path $\tilde{p}$ connecting b and d inside D, so we obtain a contradiction, since the paths p and $\tilde{p}$ are colored with different color sets, and therefore they cannot intersect; see Figure 17.1 for an illustration. $\square$

As a result of various contributions, the four-color theorem was reduced to the analysis of the large but finite set of "unavoidable" configurations. Following that, the original conjecture was proved 1976 by Appel & Haken, using extensive computer computations. A new, shorter, and more structural proof (though still relying on computers) was obtained only rather recently, in 1997, by Robertson, Sanders, Seymour & Thomas. The usual way to formulate this theorem is to dualize the map to obtain a planar graph, coloring vertices instead of the countries.

Theorem 17.6. (The four-color theorem).
Every loopless planar graph is four-colorable.

In 1943, Hadwiger stated a conjecture closely related to the four-color theorem. Recall that a graph H is called a *minor* of another graph G if H can be obtained from a subgraph of G by a sequence of edge-contractions.

Conjecture 17.7. **(Hadwiger conjecture).**
For every positive integer t, if a loopless graph has no K_{t+1} minor, then it has a t-coloring, in other words, every graph G has $K_{\chi(G)}$ as its minor.

The Hadwiger conjecture is trivial for $\mathcal{H}(G) = 1$, since K_1 is a minor of any graph. For $\chi(G) = 2$ it just says that K_2 is a minor of an arbitrary graph containing an edge. If $\chi(G) = 3$, then G contains an odd cycle, and in particular it has K_3 as a minor. Currently, the Hadwiger conjecture has been proved for $\chi(G) \leq 5$.

17.2 State Graphs and the Variations of the Chromatic Number

17.2.1 Complete Graphs as State Graphs

Clearly, for any two positive integers m and n, a graph homomorphism $\varphi : K_m \to K_n$ exists if and only if $m \leq n$. More generally, we can now restate Definition 17.2 in the language of graph homomorphisms.

Definition 17.8. *The* **chromatic number** *of G, $\chi(G)$, is the minimal positive integer n such that there exists a graph homomorphism $\varphi : G \to K_n$.*

In fact, we can see the following.

Corollary 17.9. *If there exists a graph homomorphism $\varphi : T \to G$, then $\chi(T) \leq \chi(G)$.*

Proof. If G can be colored with n colors, then there exists a graph homomorphism $\psi : G \to K_n$. Considering the composition $T \overset{\varphi}{\to} G \overset{\psi}{\to} K_n$, we conclude that T can be colored with n colors as well. $\square$

In this sense, the problem of vertex-colorings and computing chromatic numbers corresponds to choosing a particular family of graphs, namely unlooped complete graphs; fixing a valuation on this family, here we are mapping K_n to n, and then searching for a graph homomorphism from a given graph to the chosen family that would minimize the fixed valuation. Using intuition from statistical mechanics, we call such a family of graphs *state graphs*.

A natural question arises: are there any other choices of families of state graphs and valuations that correspond to other natural and well-studied classes of graph problems? The answer is yes, and we shall describe two examples in the following subsections.

17.2.2 Kneser Graphs as State Graphs and Fractional Chromatic Number

Let us now consider the family of Kneser graphs $\{KG_{n,k}\}_{n \geq 2k}$, and choose the evaluation $KG_{n,k} \mapsto n/k$.

Definition 17.10. *Let n, k be positive integers, $n \geq 2k$. The* **Kneser graph** *$KG_{n,k}$ is defined to be the graph whose set of vertices is the set of all k-subsets of $[n]$, and the set of edges is the set of all pairs of disjoint k-subsets.*

Example 17.11.

- $KG_{2k,k}$ is a matching on $\binom{2k}{k}$ vertices;
- $KG_{n,1}$ is the unlooped complete graph K_n;
- $KG_{5,2}$ is the Petersen graph.

Definition 17.12. *Let G be a graph. The* **fractional chromatic number** *of G, $\chi_f(G)$, is defined by*

$$\chi_f(G) = \inf_{(n,k)} \frac{n}{k} \ ,$$

where the infimum is taken over all pairs (n, k) such that there exists a graph homomorphism from G to $KG_{n,k}$.

An equivalent definition reads as follows.

Definition 17.13. *Let G be a graph. The number $\widetilde{\chi}_f(G)$ is defined by*

$$\widetilde{\chi}_f(G) = \inf_{(n,k)} \frac{n}{k} \ ,$$

where the infimum is taken over all covers of $V(G)$ by n independent sets $I_1, \ldots, I_n$ such that each vertex is covered at least k times, i.e., $|\{i \mid v \in I_i\}| \geq k$, for all $v \in V(G)$.

Of course, by making the independent sets smaller if necessary, we can always achieve the precise covering of each vertex by k sets.

Next we give yet another equivalent definition, which might help to explain the usage of the word *fractional*.

Definition 17.14. *The* **fractional coloring** *is a function $f : I(G) \to \mathbb{R}_{\geq 0}$ such that for every $v \in V(G)$ we have $\sum_{I \in I(G), v \in I} f(I) \geq 1$; here $I(G)$ denotes the collection of all the independent sets of G.*

The **weight** *of a fractional coloring is defined to be the sum of the values of f over all independent sets: $\mathrm{weight}(f) = \sum_{I \in I(G)} f(I)$.*

The number $\widehat{\chi}_f(G)$ is defined to be the infimum of the weight, taken over the set of all fractional colorings.

Note that the usual vertex coloring is a fractional coloring defined by $f : I(G) \to \{0, 1\}$, where the maximal monochromatic independent sets map to 1, and all other sets map to 0. The weight of this fractional coloring is equal to the number of colors; hence we have $\widehat{\chi}_f(G) \leq \chi(G)$.

Proposition 17.15. *We have $\widehat{\chi}_f(G) = \widetilde{\chi}_f(G)$.*

Proof. First, we can interpret the choice of n independent sets $I_1, \ldots, I_n$ as a function $f : I(G) \to \mathbb{Z}_{\geq 0}$ such that $\sum_{I \in I(G)} f(I) = n$. Therefore, we obtain a fractional coloring $g : I(G) \to \mathbb{R}_{\geq 0}$, defined by $g(I) := f(I)/k$. Then g is well-defined, since $\sum_{I \in I(G), v \in I} g(I) = 1/k \cdot \sum_{I \in I(G), v \in I} f(I) \geq k/k = 1$. Clearly, the weight of g is equal to n/k. Thus $\widehat{\chi}_f(G) \leq \widetilde{\chi}_f(G)$.

On the other hand, consider a fractional coloring with *rational* values $f : I(G) \to \mathbb{Q}_{\geq 0}$. Choose $k \in \mathbb{Z}_{\geq 0}$ such that for all independent sets I, we have $kf(I) \in \mathbb{Z}_{\geq 0}$. Then $g : I(G) \to \mathbb{Z}_{\geq 0}$, defined by $g(I) := kf(I)$, is a covering of $V(G)$ with $k \cdot \text{weight}(f)$ independent sets, so that each element of $V(G)$ belongs to at least k of these independent sets.

Summarizing, we can conclude that if we limit our consideration to the rational-valued fractional colorings, then we get $\widehat{\chi}_f^{\mathbb{Q}}(G) = \widetilde{\chi}_f(G)$. Since we are considering the infimum, we get $\widehat{\chi}_f(G) = \widetilde{\chi}_f(G)$. □

We note that by the theory of linear programming, the infimum of the weight of a fractional coloring can always be achieved by some rational-valued fractional coloring.

Proposition 17.16. *We have* $\widetilde{\chi}_f(G) = \chi_f(G)$.

Proof. We establish a bijection between coverings by n independent sets $I_1, \ldots, I_n$ such that every point is covered k times and graph homomorphisms $G \to KG_{n,k}$.

Let A be a $|V(G)| \times n$ matrix with entries from the set $\{0, 1\}$ defined as follows:

- if we start with a covering by n independent sets $I_1, \ldots, I_n$, then the column C_i is the characteristic vector of the set I_i, and we read the rows to find a graph homomorphism $\varphi : G \to KG_{n,k}$;
- if we start with a graph homomorphism $\varphi : G \to KG_{n,k}$, then the row R_v is the characteristic vector of the set $\varphi(v)$, and we read the columns to find a covering by n independent sets.

In both cases, the condition on the matrix A is the same: if $(v, w) \in E(G)$, then for all $i \in [n]$, $A_{w,i}$ and $A_{v,i}$ cannot both be equal to 1. It is easy to see that this establishes the desired bijection. □

17.2.3 The Circular Chromatic Number

Again, we start by defining the appropriate family of graphs.

Definition 17.17. *Let* r *be a real number,* $r \geq 2$. *Then* R_r *is defined to be the graph whose set of vertices is the set of unit vectors in the plane pointing from the origin, and two vertices* x *and* y *are connected by an edge if and only if* $2\pi/r \leq \alpha$, *where* α *is the sharper of the two angles between* x *and* y *(or* π *if these two angles are equal).*

Note that both the number of vertices and valencies of the vertices (if $r > 2$) are infinite.

Definition 17.18. *Let* G *be a graph. The* **circular chromatic number** *of* G *is* $\chi_c(G) = \inf r$, *where the infimum is taken over all positive reals* r *such that there exists a graph homomorphism from* G *to* R_r.

In other words, the family of the state graphs is $\{R_r\}_{r \geq 2}$, and the chosen valuation is $R_r \mapsto r$.

It is possible to define $\chi_c(G)$ using only finite state graphs.

Definition 17.19. *Let n, k be positive integers, $n \geq 2k$. Then $R_{n,k}$ is defined to be the graph whose set of vertices is $[n]$, and two vertices $x, y \in [n]$ are connected by an edge if and only if*

$$k \leq |x - y| \leq n - k.$$

Examples:

- $R_{2k,k}$ is a complete matching on $2k$ vertices;
- $R_{2k+1,k}$ is a cycle with $2k + 1$ vertices;
- $R_{n,1} = K_{n,1} = K_n$;
- $R_{n,2}$ is the unlooped complement of a cycle with n vertices.

The equivalent definition of $\chi_c(G)$ in terms of finite-state graphs is also functorial.

Proposition 17.20. *Let G be a graph. We have the equality*

$$\chi_c(G) = \inf_{(n,k)} \frac{n}{k} \,,$$

where the infimum is taken over all pairs (n, k) such that there exists a graph homomorphism from G to $R_{n,k}$.

Here, the state graphs are $\{R_{n,k}\}_{n \geq 2k}$, and the chosen valuation on this family is again $R_{n,k} \mapsto n/k$.

We remark that for any graph G, we have

$$\chi(G) - 1 < \chi_c(G) \leq \chi(G).$$

17.3 Kneser Conjecture and Lovász Test

17.3.1 Formulation of the Kneser Conjecture

Considering the elementary definition of the Kneser graphs, it turned out to be surprisingly difficult to determine their chromatic numbers.

To start with, it is easy to see that $\chi(KG_{n,k}) \leq n - 2k + 2$. Indeed, the following is an example of a vertex coloring: for fixed $i = 1, \ldots, n - 2k + 1$, let the set $C_i \subseteq V(KG_{n,k})$ consist of all k-subsets of $[n]$ that contain the element i. Clearly, for fixed i, each C_i is an independent set, and therefore all vertices in C_i can be colored with the same color. On the other hand, the set $\{n - 2k + 2, \ldots, n\}$ has $n - (n - 2k + 1) = 2k - 1$ elements, so any of its k-subsets must intersect, which means that $V(KG_{n,k}) \backslash (C_1 \cup \cdots \cup C_{n-2k+1})$ is an independent set as well. Thus we have decomposed the vertex set $V(KG_{n,k})$ into $n - 2k + 2$ independent sets, which is just another way of saying that $\chi(KG_{n,k}) \leq n - 2k + 2$.

The Kneser conjecture states that in fact equality holds. This was proved in 1978 by L. Lovász, who used geometric obstructions of Borsuk–Ulam type to show the nonexistence of certain graph colorings.

Theorem 17.21. (Kneser–Lovász)
For arbitrary positive integers n, k such that $n \geq 2k$, we have $\chi(KG_{n,k}) = n - 2k + 2$.

Theorem 17.21 was rather influential for further developments in this field. We shall sketch the modern version of its proof in the next three subsections.

17.3.2 The Properties of the Neighborhood Complex

Recall that in Subsection 9.1.4, more precisely in Definition 9.9, to an arbitrary graph G we have associated an abstract simplicial complex, called the neighborhood complex, which we denoted by $\mathcal{N}(G)$. Note that when $A \subseteq V(G)$ is a simplex of $\mathcal{N}(G)$, then so is $N(A)$. However, mapping A to $N(A)$ would not give a simplicial map from $\mathcal{N}(G)$ to itself. Instead, we need to proceed to the face poset of $\mathcal{N}(G)$, which in our notation is called $\mathcal{F}(\mathcal{N}(G))$.

Clearly, if $A, B \subseteq V(G)$ and $A \subseteq B$, then $N(A) \supseteq N(B)$. This means that we have an order-preserving map $N : \mathcal{F}(\mathcal{N}(G)) \to \mathcal{F}^{\mathrm{op}}(\mathcal{N}(G))$; recall here that for an arbitrary poset P we denote by P^{op} the poset that has the same set of elements as P and whose partial order is defined by $x \leq y$ in P^{op} if and only if $x \geq y$ in P.

Proposition 10.6 implies that we get a simplicial map

$$\Delta(N) : \Delta(\mathcal{F}(\mathcal{N}(G))) \longrightarrow \Delta(\mathcal{F}^{\mathrm{op}}(\mathcal{N}(G))),$$

or, rephrasing, $\Delta(N) : \mathrm{Bd}\,(\mathcal{N}(G)) \longrightarrow \mathrm{Bd}\,(\mathcal{N}(G))$, where we have used that $\Delta(\mathcal{F}(K)) = \mathrm{Bd}\,(K)$ is true for any abstract simplicial complex K.

Example 17.22.
(1) Taking $G = K_3$, we see that $\Delta(N)$ is an antipodal map on a hexagon. In general, for $G = K_n$, $\Delta(N)$ is an antipodal map on the barycentric subdivision of the boundary of the $(n-1)$-dimensional simplex.
(2) Take $G = L_3$, which is a connected graph with three vertices and two edges. Then $\mathcal{N}(G)$ is a disjoint union of a point and an interval, and $\Delta(N)$ maps the isolated point to the barycenter of the interval, and it maps the entire interval to the isolated point.

We see from our examples that for $G = K_n$, the map $\Delta(N)$ gives a $\mathbb{Z}_2$-action on $\mathcal{N}(G)$, whereas for $G = L_3$ it does not.

Proposition 17.23. *The map $\Delta(N)$ is an involution when restricted to the image of $\Delta(N)$, in other words, we have $\Delta(N)^3 = \Delta(N)$.*

Proof. This has been proved in Subsection 13.2.2, Application 2. □

Example 17.24. Let G be the graph defined by the set of vertices $V(G) = \{1, 2\}$ and the set of edges $E(G) = \{(1,1), (1,2), (2,1)\}$. The complex $\mathcal{N}(G)$ is equal to the 1-simplex $\Delta^{[2]}$. The map $\Delta(N)$ maps the barycenter of the

1-simplex to the vertex 1, it maps the vertex 2 to the vertex 1, and it maps the vertex 1 to the barycenter of the 1-simplex. We see that the image of $\Delta(N)$ is the 1-simplex spanned by the vertex 1 and the barycenter of $\mathcal{N}(G)$, and that the $\mathbb{Z}_2$-action on it is the reflection map.

We see that the $\mathbb{Z}_2$-action induced by $\Delta(N)$ does not have to be free.

Proposition 17.25. *The $\mathbb{Z}_2$-action on $\mathrm{Im}(\Delta(N))$ induced by $\Delta(N)$ is free if and only if G has no loops.*

Proof. It is a general fact that a finite simplicial group action on a simplicial complex is free if and only if there is no simplex whose set of vertices is preserved by the action.

The simplices of $\mathrm{Bd}\,(\mathcal{N}(G)) = \Delta(\mathcal{F}(\mathcal{N}(G)))$ are chains of $\mathcal{F}(\mathcal{N}(G))$. Since the action reverses the order, we see that the $\mathbb{Z}_2$-action on $\mathrm{Im}(\Delta(N))$ induced by $\Delta(N)$ is not free if and only if there exists a chain C such that $\Delta(N)(C) = C^{\mathrm{op}}$.

This is equivalent to saying that there exist $A, B \in \mathcal{F}(\mathcal{N}(G))$ such that $A \subseteq B$, $N(A) = B$, and $N(B) = A$. Indeed, such a pair would produce the required chain (consisting of one or two elements); on the other hand, the middle edge, or element of the chain C with the properties above, would give such a pair.

If such a pair (A, B) exists, then all the vertices in A have loops, since $N(A) \supseteq A$. On the other hand, if there exists $v \in V(G)$ with a loop, $(v, v) \in E(G)$, then $A = N(N(v))$, $B = N(v)$ is an appropriate pair, since $N(v) \supseteq \{v\}$ implies $N(N(v)) \subseteq N(v)$. $\square$

17.3.3 Lovász Test for Graph Colorings

Lovász has introduced the neighborhood complex $\mathcal{N}(G)$ as a part of his topological approach to the resolution of the Kneser conjecture. The hard part of the proof is to show the inequality $\chi(K_{n,k}) \geq n - 2k + 2$, and Lovász's idea was to use the connectivity information of the topological space $\mathcal{N}(G)$ to find obstructions to the vertex-colorability of G. More precisely, he proved the following statement.

Theorem 17.26. *Let G be a graph such that $\mathcal{N}(G)$ is k-connected for some $k \in \mathbb{Z}$, $k \geq -1$. Then $\chi(G) \geq k + 3$.*

The main topological tool that Lovász employed was the Borsuk–Ulam theorem. Since then, topological equivariant methods have gained ground and become part of the standard repertoire in combinatorics.

It will be shown in Theorem 18.3 that the complexes $\mathcal{N}(G)$ and $\mathrm{Bip}\,(G)$ have the same simple homotopy type. This fact leads one to consider the family of Hom complexes as a natural context in which to look for further obstructions to the existence of graph homomorphisms.

Before proceeding with the proof of Lovász's theorem, we need the following useful lemma.

Lemma 17.27. *Let X be an arbitrary $\mathbb{Z}_2$-space and let k be an integer, $k \geq -1$. If X is k-connected, then there exists a $\mathbb{Z}_2$-map $f : \mathbb{S}_a^d \to X$, for all $0 \leq d \leq k + 1$.*

Proof. This follows immediately by repeated application of Proposition 8.25. □

We are now ready to give a sketch of the proof of Theorem 17.26, which will become complete once the details of the structure theory of Hom complexes are presented in Chapter 18.

Proof of Theorem 17.26. First of all, since Bip (G) is homotopy equivalent to $\mathcal{N}(G)$, we can replace the simplicial complex $\mathcal{N}(G)$ by the prodsimplicial complex Bip (G). Now assume $\chi(G) \leq k + 2$. Then there exists a graph homomorphism $\varphi : G \to K_{k+2}$. In Subsection 18.3.5 it will be shown that this induces a $\mathbb{Z}_2$-map

$$\varphi^{K_2} : \mathrm{Bip}\,(G) \to \mathrm{Bip}\,(K_{k+2}),$$

where the $\mathbb{Z}_2$-action on Bip (G) is induced by the nontrivial $\mathbb{Z}_2$-action on K_2.

It will be shown in Proposition 19.8 that Bip (K_n) is $\mathbb{Z}_2$-homeomorphic to $\mathbb{S}_a^{n-2}$, for all integers $n \geq 2$, where as we recall from Chapter 8, $\mathbb{S}_a^d$ denotes the d-dimensional sphere with the antipodal $\mathbb{Z}_2$-action. Hence, we have actually obtained a $\mathbb{Z}_2$-map $\varphi^{K_2} : \mathrm{Bip}\,(G) \to \mathbb{S}_a^k$.

On the other hand, by Lemma 17.27, there exists a $\mathbb{Z}_2$-map $f : \mathbb{S}_a^{k+1} \to$ Bip (G). Combining these, we obtain a sequence of $\mathbb{Z}_2$-maps

$$\mathbb{S}_a^{k+1} \xrightarrow{\;f\;} \mathrm{Bip}\,(G) \xrightarrow{\;\varphi^{K_2}\;} \mathbb{S}_a^k.$$

This, however, contradicts the Borsuk–Ulam theorem (Theorem 8.22).   □

17.3.4 Simplicial and Cubical Complexes Associated to Kneser Graphs

To finish the proof of the Kneser conjecture (Theorem 17.21), we still need to see that the neighborhood complex of Kneser graphs is sufficiently connected. In fact, it turns out that the homotopy type of these complexes can be determined precisely.

Proposition 17.28. *For arbitrary positive integers n and k such that $n \geq 2k$, the abstract simplicial complex $\mathcal{N}(KG_{n,k})$ is homotopy equivalent to a wedge of spheres of dimension $n - 2k$. In particular, the complex $\mathcal{N}(KG_{n,k})$ is $(n - 2k - 1)$-connected.*

Proof. Consider the order-preserving map

$$\varphi : \mathcal{F}(\mathcal{N}(KG_{n,k})) \to \mathcal{F}(\mathcal{N}(KG_{n,k})),$$

defined by

$$\varphi : \Sigma \mapsto \bigcup \Sigma,$$

where Σ is an arbitrary nonempty collection of k-subsets of the set $[n]$, which forms a simplex in the neighborhood complex $\mathcal{N}(KG_{n,k})$, and $\bigcup \Sigma$ denotes the union of all the subsets from the collection Σ. Clearly, we have $\varphi^2 = \varphi$, and $\varphi(\Sigma) \geq \Sigma$, and hence φ is an ascending closure operator. By Theorem 13.12 we see that the simplicial complex $\Delta(\mathcal{F}(\mathcal{N}(KG_{n,k}))) = \mathrm{Bd}\,\mathcal{N}(KG_{n,k})$ collapses to the simplicial complex $\Delta(\mathrm{Im}(\varphi))$.

On the other hand, it is easy to see that the image of the operator φ consists of all subsets $S \subseteq [n]$ such that $k \leq |S| \leq n - k$. In other words, the poset $\mathrm{Im}(\varphi)$ is a certain rank selection of the Boolean algebra $\mathcal{B}_n$. By Proposition 12.6, see also specifically Example 12.10(2), we know that the order complex of that poset is homotopy equivalent to a wedge of spheres of dimension $(n - k) - k = n - 2k$. $\square$

Theorem 17.21 now follows. Though this is a short and clarifying proof of Proposition 17.28, we shall investigate the complexes associated to Kneser graphs at some further length to uncover some interesting cubical constructions and to illustrate some other techniques that we developed in Part II.

By Theorem 18.3 we can replace the neighborhood complexes $\mathcal{N}(KG_{n,k})$ by the Hom complexes $\mathrm{Bip}\,(KG_{n,k})$ as far as the homotopy type is concerned. Again, we define an order-preserving map

$$\psi : \mathcal{F}(\mathrm{Bip}\,(KG_{n,k})) \to \mathcal{F}(\mathrm{Bip}\,(KG_{n,k})),$$

defined by

$$\psi : (\Sigma_1, \Sigma_2) \mapsto \left(\bigcup \Sigma_1, \bigcup \Sigma_2 \right),$$

where Σ_1 and Σ_2 are both arbitrary nonempty collections of k-subsets of the set $[n]$ such that $A \cap B$ whenever $A \in \Sigma_1$ and $B \in \Sigma_2$, and as before, the symbol $\bigcup$ denotes taking the union of all the subsets in the corresponding collection. As is easily seen, the map ψ is an ascending closure operator, and hence by Theorem 13.12, the simplicial complex $\mathrm{Bd}\,\mathrm{Bip}\,(KG_{n,k})$ collapses onto its subcomplex $\Delta(\mathrm{Im}(\psi))$, to which we now restrict our attention.

For the rest of this subsection we let $P_{n,k}$ denote the poset $\mathrm{Im}(\psi)$ with the order reversed. It can be described as the poset consisting of all ordered pairs (S_1, S_2) of subsets of the set $[n]$ such that $|S_1| \geq k$, $|S_2| \geq k$, and $S_1 \cap S_2 = \emptyset$. The partial order is given by the rule $(T_1, T_2) \geq (S_1, S_2)$ if and only if $T_1 \subseteq S_1$ and $T_2 \subseteq S_2$. The crucial observation now is that this poset $P_{n,k}$ is actually a face poset of a cubical complex.

Definition 17.29. *We call the cubical complex whose face poset is given by $P_{n,k}$ as above the Kneser cubical complex and denote it by $KC_{n,k}$.*

The dimension of $KC_{n,k}$ is equal to $n - k$. Its maximal cells are indexed by pairs (S_1, S_2) of disjoint subsets of $[n]$ such that $|S_1| = |S_2| = k$, and its vertices are indexed by subsets $S \subseteq [n]$ such that $k \leq |S| \leq n - k$, the actual index being $(S, [n] \setminus S)$. Though we already know at this point that the cubical complex $KC_{n,k}$ is homotopy equivalent to a wedge of $(n - k)$-dimensional spheres, we would now like to describe an acyclic matching on the set of its cells that is maximal in the following sense: it has one 0-dimensional critical cell, and all other critical cells have dimension $n - k$.

Let Q denote the chain $n > n - 1 > \cdots > n - k$, and consider the order-preserving map

$$\varphi : P_{n,k} \longrightarrow Q,$$

defined by

$$(S_1, S_2) \mapsto \max([n] \setminus S_2).$$

According to Theorem 11.10, it is enough if we construct appropriate acyclic matchings on the preimages $\varphi^{-1}(m)$, for $n-k \leq m \leq n$. Let us now fix such an m. For any element (S_1, S_2) in $\varphi^{-1}(m)$ we know that $m+1, \ldots, n \in S_2$, and of course that $m \notin S_2$.

Whenever possible, we now match the elements (S_1, S_2) and $(S_1 \oplus \{m\}, S_2)$, where $\oplus$ denotes the symmetric sum. This is clearly an acyclic matching, and the critical cells are indexed by all pairs (S_1, S_2) such that $|S_1| = k$ and $m \in S_1$. These elements form an upper ideal in $\varphi^{-1}(m)$; hence we are free to extend our acyclic matching in this set in any way we want.

Set now $t := \max([m] \setminus S_1)$, and match, whenever possible, the old critical elements (S_1, S_2) and $(S_1, S_2 \oplus \{t\})$. Again, clearly the total matching is an acyclic one. The final critical cells are indexed by all pairs (S_1, S_2) such that $|S_1| = |S_2| = k$, $m \in S_1$, and $t \in S_2$. These cells are all of dimension $n-k$. Additionally, if $m = k$, there is one more critical cell of dimension 0 indexed by $(\{1, \ldots, k\}, \{k+1, \ldots, n\})$. This concludes the description of our matching and provides an alternative proof of the fact that the cubical complex $KC_{n,k}$ is homotopy equivalent to a wedge of $(n - k)$-dimensional spheres.

17.3.5 The Vertex-Critical Subgraphs of Kneser Graphs

Recall that for a graph G and a vertex $v \in V(G)$, the notation $G - v$ denotes the graph that is obtained from G by deleting the vertex v and all edges adjacent to v, i.e., the new set of vertices is given by $V(G - v) := V(G) \setminus \{v\}$, and the new set of edges is given by $E(G-v) := E(G) \cap (V(G-v) \times V(G-v))$.

Shortly after it was published, Lovász's resolution of the Kneser conjecture was complemented by finding a maximal subgraph having the same chromatic number as the original Kneser graph. To formulate this result, we recall that a graph G is called *vertex-critical* if for any vertex $v \in V(G)$, we have $\chi(G) = \chi(G - v) + 1$.

Definition 17.30. *Let n, k be positive integers, $n \geq 2k$. The* **stable Kneser graph** $KG_{n,k}^{Stab}$ *is defined to be the graph whose set of vertices is the set of all k-subsets S of $[n]$ such that if $i \in S$, then $i + 1 \notin S$, and if $n \in S$, then $1 \notin S$. Two subsets are joined by an edge if and only if they are disjoint.*

Clearly, $KG_{n,k}^{Stab}$ is an induced subgraph of $KG_{n,k}$.

Theorem 17.31. *The graph $KG_{n,k}^{Stab}$ is a vertex-critical subgraph of $KG_{n,k}$, i.e., $KG_{n,k}^{Stab}$ is a vertex-critical graph, and $\chi(KG_{n,k}^{Stab}) = n - 2k + 2$.*

17.3.6 Chromatic Numbers of Kneser Hypergraphs

Theorem 17.21 was generalized in 1986 to the case of hypergraphs. To start with, recall the standard way to extend the notion of the chromatic number to hypergraphs.

Definition 17.32. *For a hypergraph $\mathcal{H}$, the chromatic number $\chi(\mathcal{H})$ is, by definition, the minimal number of colors needed to color the vertices of $\mathcal{H}$ so that no hyperedge is monochromatic.*

Next, there is a standard way to generalize Definition 17.10 to the case of hypergraphs.

Definition 17.33. *Let n, k, r be positive integers such that $r \geq 2$ and $n \geq rk$. The **Kneser r-hypergraph** $KG_{n,k}^r$ is the r-uniform hypergraph whose ground set consists of all k-subsets of $[n]$, and the set of hyperedges consists of all r-tuples of disjoint k-subsets.*

Using the introduced notations we can now formulate a generalization of Theorem 17.21.

Theorem 17.34. *For arbitrary positive integers n, k, r such that $r \geq 2$ and $n \geq rk$, we have*

$$\chi(KG_{n,k}^r) = \left\lceil \frac{n - rk + r}{r - 1} \right\rceil.$$

17.4 Bibliographic Notes

For the proof of the fact that deciding whether a given planar graph is 3-colorable is NP-complete, we refer to [GJ79], whereas NP-completeness of coloring a 3-colorable graph with 4 colors was first shown in [KLS93]. The quartic time for the 4-coloring of a planar graph was obtained in [AH89]. This was later improved to quadratic time in [RSST]. We refer to Garey & Johnson, [GJ76], for an in-detail elaboration on the subject of approximate algorithms for graph colorings.

The question of computing $\chi(G)$ for the planar graph G has a long history. The question was formulated in 1852 by F. Guthrie; see [Gut80]. The first time this question appeared in print was in a paper by Cayley, [Cay78], after which it became known as the four-color problem, one of the most famous questions in graph theory, as well as a popular brain-teaser. There is a very extensive literature on the subject; see, e.g., [Har69, KS77, MSTY, Ore67, Th98].

The apparently first proof, offered by A. Kempe in 1880, [Kem79], turned out to be false, as did many later ones. The flaw was noticed in 1890 by P. Heawood, [Hea90], who also proved the weaker five-color theorem. Important contributions to the four-color problem, were made (among others) by G. Birkhoff and H. Heesch, [Hee69]. The original announcement of the resolution of the four-color problem by Appel & Haken can be found in [AH76], and [AH89] is the last reprint. The reference to the work of Robertson, Sanders, Seymour & Thomas on this subject is [RSST].

Hadwiger has stated his conjecture in [Had43]. The case $\chi(G) = 4$ is reasonably easy, and was shown by Hadwiger, [Had43], and Dirac, [Dir52]. Furthermore, it was shown in 1937 by Wagner, [Wag37], that the case $\chi(G) = 5$ of the Hadwiger conjecture is equivalent to the four-color theorem.

We recommend an excellent and comprehensive textbook by Godsil and Royle, [GR01], where more about fractional chromatic number can be found. As for the circular chromatic number of a graph, we refer to nice articles [Vi88, Zhu01] for rather extensive information.

The Kneser conjecture was posed in 1955, see [Kn55], and solved in 1978 by L. Lovász; see [Lov78]. Shorter proofs of the Lovász theorem were obtained by Bárány; see [Bar78], and Greene; see [Gr02], both using some versions of the Borsuk–Ulam theorem; see also [GR01]. A nice brief survey of these has been written by De Longueville; see [dL03].

The vertex-critical subgraphs of Kneser graphs, complementing the proof of Lovász's theorem (Theorem 17.21), were found by Schrijver, in [Sr78], who proved Theorem 17.31. Alon, Frankl, and Lovász, [AFL86], have generalized Theorem 17.21 and proved Theorem 17.34. Their argument makes use of the generalization of the Borsuk–Ulam theorem that has previously appeared in the work of Bárány, Shlosman, and A. Szűcs; see [BSS81].

There has been a substantial body of further important work; some of the references are [Dol88, Kr92, Kr00, Ma04, MZ04, Sar90, Zie02]. There have also been multiple constructions, such as box complexes, designed to generalize the original Lovász neighborhood complexes. However, as later research showed, the tests obtained in that way were completely convertible, since the $\mathbb{Z}_2$-homotopy types of these complexes were very closely related, either by simply being the same, or by means of one being the suspension of another, or something close to that. This means that all these constructions are avatars of the same object, as explained in [Ziv05a]. In contrast to that, the Hom complexes have been shown to have an intricate and interesting behavior, going substantially beyond the original Lovász complexes.

Structural Theory of Morphism Complexes

18.1 The Scope of Morphism Complexes

18.1.1 The Morphism Complexes and the Prodsimplicial Flag Construction

In this section we would like to prove a property that holds for general morphism complexes, which were described in Definition 9.24. A crucial fact about this family of prodsimplicial complexes is that $\text{Hom}_-(-,-)$ complexes are fully determined by the low-dimensional data; in fact, it turns out that already knowing the 1-skeleton suffices; cf. Definition 9.16.

Proposition 18.1. *All complexes* $\text{Hom}_-(-,-)$ *with isomorphic 1-skeletons are isomorphic to each other as polyhedral complexes. More precisely, the complexes* $\text{Hom}_-(-,-)$ *are prodsimplicial flag complexes.*

Proof. Let us consider a complex X of the type $\text{Hom}_M(A, B)$, where A, B and M are a priori unknown. Trivially, the 0-skeleton of X is the set M itself. Furthermore, the 1-skeleton tells us which pairs of set maps $\varphi, \psi : A \to B$ differ precisely in one element of A.

Clearly, for $\sigma = \prod_{x \in A} \sigma_x \in C(A, B)$ to belong to $\text{Hom}_M(A, B)$, it is required that the 1-skeleton of σ be a subgraph of the 1-skeleton of $\text{Hom}_M(A, B)$. Let us show that the converse of this statement is true as well.

Let Γ be the 1-skeleton of X. For every edge e in Γ let $\lambda(e) \in A$ denote the element in which the value of the function is changed along e. Since we do not know the set A, we can only make statements of the type, *the labels of these two edges are the same/different*.

Assume that we have $S \subseteq M$ such that S can be written as a direct product $S = S_1 \times S_2 \times \cdots \times S_t$. Assume furthermore that the subgraph of Γ induced by S is precisely the 1-skeleton of the corresponding cell.

First, consider three elements $a, b, c \in S_1 \times \cdots \times S_t$ that have the same indices in all S_i's except for exactly one. Then, by our assumption on S, the subgraph of Γ induced on the vertices a, b, and c is a triangle. Clearly, if three changes of a value of a function result in the same function, then the changes were done in the same element of A, i.e., $\lambda(a, b) = \lambda(a, c) = \lambda(b, c)$.

Next, consider four elements $a, b, c, d \in S_1 \times \cdots \times S_t$ such that pairs of vertices $(a, b), (b, c), (c, d)$, and (a, d) have the same indices in all S_i's except for exactly one. Assume further that this index is not the same for (a, b) and (b, c): say a and b differ in S_1, and b and c differ in S_2.

According to our assumption on S, $(a, b), (b, c), (c, d)$, and (a, d) are edges of Γ. If $\lambda(a, b) = \lambda(b, c)$, then Γ contains the edge (a, c), and $\lambda(a, b) = \lambda(a, c)$, which contradicts our choice of S.

If, on the other hand, $\lambda(a, b) \neq \lambda(b, c)$, then, since changes of functions along the paths $a \to b \to c$ and $a \to d \to c$ should give the same answer, we are left with only one possibility, namely, that $\lambda(a, b) = \lambda(c, d)$ and $\lambda(b, c) = \lambda(a, d)$.

Let $a \in S_1 \times \cdots \times S_t$, $a = (a_1, \ldots, a_t)$. By our first argument, if $b \in S_1 \times \cdots \times S_t$, $b = (a_1, \ldots, \tilde{a}_i, \ldots, a_t)$, then $\lambda(a, b)$ does not depend on a_i and $\tilde{a}_i$. Furthermore, let $c, d \in S_1 \times \cdots \times S_t$, $d = (a_1, \ldots, \tilde{a}_j, \ldots, a_t)$, $c = (a_1, \ldots, \tilde{a}_i, \ldots, \tilde{a}_j, \ldots, a_t)$, for $i \neq j$. By our second argument, applied to a, b, c, d, we get that $\lambda(a, b) = \lambda(c, d)$. If iterated for various j, this implies that $\lambda(a, b)$ does not depend on $a_1, \ldots, a_{i-1}, a_{i+1}, \ldots, a_t$ either; thus it may depend only on the index i.

Finally, this label should be different for different i's, since otherwise, by the same argument as above, we would get more edges in the subgraph of Γ induced by S than what we allowed by our assumptions.

Summarizing, we have shown that the cell $\sigma = \prod_{x \in A} \sigma_x \in C(A, B)$ belongs to $\text{Hom}_M(A, B)$ if and only if the 1-skeleton of σ is a subgraph of Γ. This implies that $\text{Hom}_M(A, B)$ is uniquely determined by its 1-skeleton. $\square$

Proposition 18.1 comes in handy when we need to show that certain prodsimplicial complexes cannot be represented as Hom complexes, since it imposes the rather rigid restriction of being a prodsimplicial flag complex.

We finish this section by noting that not every prodsimplicial flag complex is representable as a Hom complex. For example, let us consider a (hollow) pentagon. Clearly, it is a prodsimplicial flag complex. On the other hand, if it can be represented as a Hom complex then we can trace the changes in the set maps following around the pentagon once. Since we have five changes, and no value can be changed only once, we see that at most two different values will be changing as we go around the pentagon; hence one of the values will be changed at least three times. This implies that there exists a value that will be changed along two adjacent edges, which implies that our complex must contain a triangle. This is a contradiction, and thus a pentagon cannot be represented as a Hom complex.

18.1.2 Universality

It happens very often that a family of combinatorially defined complexes is universal with respect to the invariants that one is interested in computing. This is also the case not only for general Hom complexes, but even for the Bip$(-)$ complexes.

Theorem 18.2. *For each finite abstract simplicial complex X with a free $\mathbb{Z}_2$-action, there exists a graph G such that $\mathrm{Bip}(G)$ is $\mathbb{Z}_2$-homotopy equivalent to X.*

Proof. We shall satisfy ourselves here with proving a weaker statement, finding G such that $\mathrm{Bip}(G)$ is homotopy equivalent to X. To start with, since $\mathrm{Bip}(G)$ is homotopy equivalent to $\mathcal{N}(G)$, we can deal with the neighborhood complex instead.

For an arbitrary finite abstract simplicial complex X with a free $\mathbb{Z}_2$-action γ we associate a graph G_X as follows:

- the set of vertices of G_X is the same as the set of vertices of X;
- two vertices v and w are connected by an edge if either $w = \gamma v$ or $(w, \gamma v)$ is an edge in the 1-skeleton of X.

Let us now consider the graph associated to the barycentric subdivision of X. In our notation, this graph is called $G_{\mathrm{Bd}\,X}$. We claim that the neighborhood complex of this graph $\mathcal{N}(G_{\mathrm{Bd}\,X})$ is homotopy equivalent to $\mathrm{Bd}\,X$, and hence to X.

Consider the so-called closed star covering of $\mathrm{Bd}\,X$, that is, its covering by all closed stars of its vertices. A subset of these stars has a nonempty intersection if and only if it contains a vertex, which means that there is a vertex that is connected with the centers of all these stars, and possibly coincides with one of them. Therefore, we see that the nerve of this covering coincides with the neighborhood complex $\mathcal{N}(G_{\mathrm{Bd}\,X})$. To prove the homotopy equivalence we will show that whenever a set of these stars has a nonempty intersection, then it must be contractible, and then invoke Theorem 15.21.

Recall that the vertices of $\mathrm{Bd}\,X$ are indexed with subsets of the set of vertices of X. Let $A_1, \ldots, A_t$ be a collection of such subsets such that the corresponding closed stars intersect, and let S be any vertex in this intersection. After possible reindexing we may assume that for some $0 \leq k \leq t$, we have $A_i \subseteq S$, for all $1 \leq i \leq k$, and $A_i \supseteq S$, for all $k + 1 \leq i \leq t$. We shall consider two cases.

Assume first that $k \geq 1$. In this case, set $B := \cup_{i=1}^k A_i$. We have $B \neq \emptyset$. We claim that the intersection of the closed stars of the A_i's is a cone with apex in B. To see that, take an arbitrary simplex $\sigma = (B_1 \subseteq \cdots \subseteq B_r)$ in this intersection. The sets A_i can all be inserted into this inclusion chain. Let τ be the simplex obtained from σ by adding B. This simplex is well-defined, since we can insert B in the same spot where we can insert A_{i_0}. The set B is possibly larger than A_k, but since the subsequent B_j's contain all the sets

$A_1, \ldots, A_k$, they will also contain B. On the other hand, the simplex τ is still in the intersection of the considered closed stars, because B contains all the sets $A_1, \ldots, A_k$ and is contained in $A_{k+1}, \ldots, A_t$. This proves that the intersection of the closed stars in this case is a cone, and hence contractible.

The remaining case is $k = 0$. In this case, we can set $B := \cap_{i=1}^{t} A_i$, and using a completely analogous argument, we see that the intersection of the closed stars will be a cone with apex in B, and hence contractible as well.   □

18.2 Special Families of Hom Complexes

18.2.1 Coloring Complexes of a Graph

It is useful to think of $\mathrm{Hom}\,(G, K_n)$ as a way to put a topology on the space of all n-colorings of G. We call these complexes the *coloring complexes of G*. The fact that $\mathrm{Hom}\,(G, K_n)$ is nonempty encodes the fact that that G is n-colorable. The connectivity of $\mathrm{Hom}\,(G, K_n)$ expresses a certain flexibility property of the set of all n-colorings of the graph G. Intuitively, one should think that the higher the connectivity (that is, the greater the number of the initial homotopy groups that are trivial), the more flexible the set of all n-colorings of the graph G is. The significance of the other properties of $\mathrm{Hom}\,(G, K_n)$ is at present the subject of an active field of research.

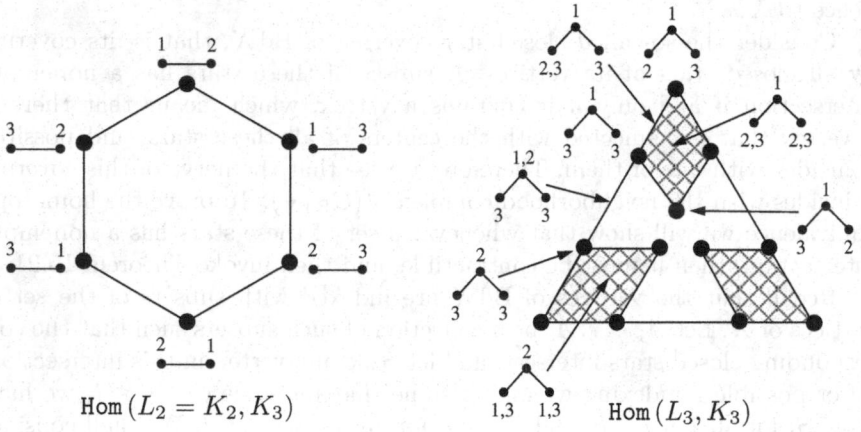

$$\mathrm{Hom}\,(L_2 = K_2, K_3) \qquad\qquad \mathrm{Hom}\,(L_3, K_3)$$

Fig. 18.1. Coloring complexes of strings I.

Some examples of such complexes are shown in Figures 18.1, 18.2, and 18.3. In these figures we used the following notation: L_n denotes an n-string, i.e., a tree with n vertices and no branching points, C_m denotes a cycle with m vertices, i.e., $V(C_m) = \mathbb{Z}_m$, $E(C_m) = \{(x, x+1), (x+1, x) \mid x \in \mathbb{Z}_m\}$.

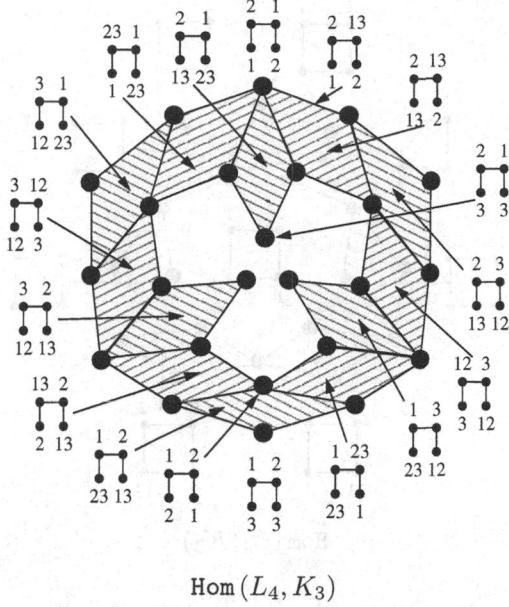

$$\mathrm{Hom}\,(L_4, K_3)$$

Fig. 18.2. Coloring complexes of strings II.

18.2.2 Complexes of Bipartite Subgraphs and Neighborhood Complexes

As mentioned above, Hom complexes were introduced as a one parameter expansion of the family of neighborhood complexes. The next theorem makes this statement precise.

Theorem 18.3. *For an arbitrary graph G, the neighborhood complex $\mathcal{N}(G)$ and the polyhedral complex $\mathrm{Bip}\,(G)$ have the same simple homotopy type.*

Proof. Set $P := \mathcal{F}^{\mathrm{op}}(\mathrm{Bip}\,(G)) \cup \{\hat{0}, \hat{1}\}$. As was mentioned before, P is a lattice, and $\Delta(\bar{P}) = \mathrm{Bd}\,\mathrm{Bip}\,(G)$. By Theorem 13.18(b), we see that both simplicial complexes $\mathrm{Bd}\,\mathrm{Bip}\,(G)$ and $\mathrm{Bd}\,\Gamma(P)$ collapse onto the simplicial complex $\Delta(\bar{P}_a)$. See Figure 18.4 for an example of the poset $\bar{P}_a$.

Description of $\Gamma(P)$. The vertices of $\Gamma(P)$ are all the pairs (A, B), $A, B \subseteq V(G)$ such that $N(A) = B$ and $N(B) = A$. These can be indexed with the simplices $A \in \mathcal{N}(G)$, $A \in \mathrm{Im}(N)$, which is the same as to take the elements of $N(\mathcal{F}(\mathcal{N}(G)))$ or the vertices of $\Delta(N(\mathcal{F}(\mathcal{N}(G)))) = \mathcal{L}o(G)$.

The simplices of $\Gamma(P)$ are all sets of pairs $\{(A_1, B_1), \ldots, (A_t, B_t)\}$ such that $\bigcap_{i=1}^{t} A_i \neq \emptyset$, and $\bigcap_{i=1}^{t} B_i \neq \emptyset$. Since $N(A) \cap N(B) = N(A \cup B)$, for arbitrary subsets $A, B \subseteq V(G)$, and since $B_i = N(A_i)$, for $1 \leq i \leq t$, the second condition amounts to saying that $N(\bigcup_{i=1}^{t} A_i) \neq \emptyset$.

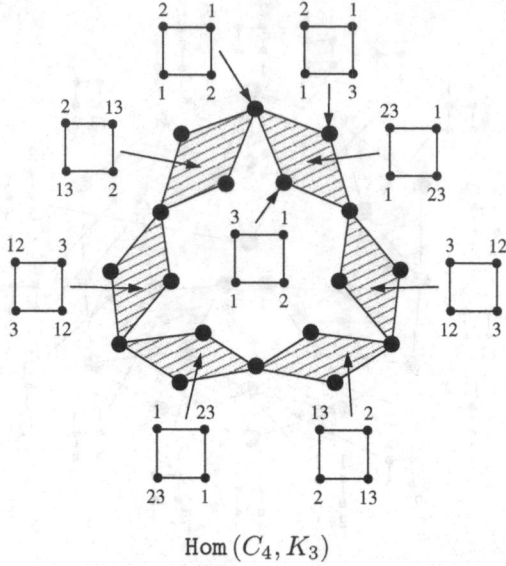

Hom (C_4, K_3)

Fig. 18.3. Coloring complexes of cycles.

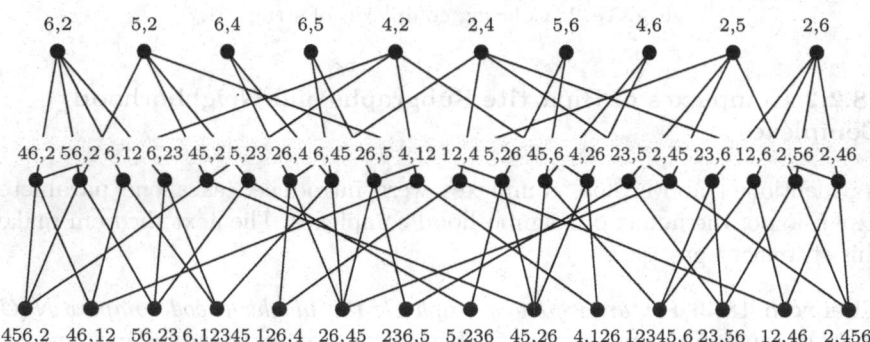

Fig. 18.4. The poset $\bar{P}_a$ for $P = \mathcal{F}^{\mathrm{op}}(\mathrm{Bip}(G)) \cup \{\hat{0}, \hat{1}\}$.

Let $\mathcal{L}$ denote the poset of all $A \in \mathcal{N}(G)$, $A \in \mathrm{Im}(N)$, ordered by inclusion, with a minimal and a maximal element attached. Clearly, $\Delta(\bar{\mathcal{L}}) = \mathcal{L}o(G)$. From the description of $\Gamma(P)$ above, we see that $\Delta(\bar{\mathcal{L}})$ is a subcomplex of $\Gamma(P)$. On the other hand, by Theorem 11.21, the simplicial complex $\mathcal{J}(\mathcal{L})$ collapses onto $\Delta(\bar{\mathcal{L}})$.

Let μ be the acyclic matching from the proof of Theorem 11.21 that gives the collapsing sequence. We claim that the restriction of μ to $\mathcal{F}(\Gamma(P))$ is again an acyclic matching. Since $\mathcal{F}(\Gamma(P))$ is a lower ideal in $\mathcal{F}(\mathcal{J}(\mathcal{L}))$, the only thing that has to be checked is that if $S \in \mathcal{F}(\Gamma(P)) \cap \Sigma$, then $\mu(S) \in \mathcal{F}(\Gamma(P))$; here Σ is as in the proof of the Theorem 11.21.

Assume that $S = \{A_1, \ldots, A_t\}$, where the sets are listed in the linear extension order, i.e., if $1 \leq i < j \leq t$, then $A_i \not\supseteq A_j$. Let $a(S)$ be the subset of $V(G)$ defined as in the proof of Theorem 11.21. Clearly, $a(S) \subseteq A_t$. This implies that $a(S) \cup \bigcup_{i=1}^{t} A_i = \bigcup_{i=1}^{t} A_i$, and therefore, the set of pairs

$$\mu(S) = \{(A_1, B_1), \ldots, (A_t, B_t), (a(S), N(a(S)))\}$$

is a simplex of $\Gamma(P)$. We conclude that the restriction of μ to $\mathcal{F}(\Gamma(P))$ gives a collapsing sequence from $\Gamma(P)$ to $\mathcal{L}o(G)$.

Let us summarize our findings in the following concatenation of sequences of collapses and expansions:

$$\mathrm{Bd}\,\mathrm{Bip}\,(G) \searrow \Delta(\bar{P}_a) \nearrow \mathrm{Bd}\,\Gamma(P), \quad \Gamma(P) \searrow \mathcal{L}o(G) \nearrow \mathrm{Bd}\,\mathcal{N}(G), \quad (18.1)$$

where the first two sequences are given by Theorem 13.18(b), the third sequence is given by the restriction of the acyclic matching μ as above, and the fourth sequence is given by Proposition 13.16.

The discussion in Section 6.5 implies now that the polyhedral complex of all bipartite subgraphs of G, $\mathrm{Bip}\,(G)$, and the neighborhood complex $\mathcal{N}(G)$, have the same simple homotopy type, and yields an explicit formal deformation between these two complexes. $\square$

18.3 Functoriality of Hom $(-,-)$

Recall from Subsection 4.1.2 that **Graphs** denote the category of all finite graphs.

18.3.1 Functoriality on the Right

Let T, G, and K be three arbitrary graphs, and let φ be a graph homomorphism from G to K. For any graph homomorphism $v : T \to G$, i.e., a vertex of $\mathrm{Hom}\,(T, G)$, we have the composition $\varphi \circ v : T \to G \to K$, i.e., a vertex of $\mathrm{Hom}\,(T, K)$. In general, the composition induces a poset map

$$f : \mathcal{F}(\mathrm{Hom}\,(T, G)) \to \mathcal{F}(\mathrm{Hom}\,(T, K)),$$

namely, for $\eta : V(T) \to 2^{V(G)} \setminus \{\emptyset\}$, we have $f(\eta) = 2^{\varphi} \circ \eta$, where $2^{\varphi} : 2^{V(G)} \setminus \{\emptyset\} \to 2^{V(K)} \setminus \{\emptyset\}$ is the map induced on the subsets.

One can check that the poset map $f : \mathcal{F}(\mathrm{Hom}\,(T, G)) \to \mathcal{F}(\mathrm{Hom}\,(T, K))$ comes from a cellular map from $\mathrm{Hom}\,(T, G)$ to $\mathrm{Hom}\,(T, K)$, which we denote by φ^T.

Moreover, a detailed analysis of the polyhedral structure of $\mathrm{Hom}\,(T, G)$, using the explicit point description of $\mathrm{Hom}\,(T, G)$ from Subsection 9.2.3, shows that cells (direct products of simplices) map surjectively to other cells, and that this map is a product map induced by the corresponding maps on the simplices. Therefore, φ^T is a polyhedral map.

Example 18.4. Recall that Bip (K_3) is a hexagon, and Bip (K_4) is the polyhedron in Figure 9.6. Consider $\varphi : K_3 \hookrightarrow K_4$ mapping i to i, for any $i \in \{1, 2, 3\}$. Then the cellular map φ^{K_2} includes the hexagon as one of the meridians of Bip (K_4).

Now we can draw the following conclusion.

Proposition 18.5. *For any fixed graph T, the* Hom *construction yields a covariant functor* Hom $(T, -)$.

Proof. The only property that has to be checked is that for any graph homomorphisms $\varphi : G \to K$ and $\psi : K \to L$, we have

$$(\psi \circ \varphi)^T = \psi^T \circ \varphi^T.$$

This equality is true on the level of poset maps, since it just amounts to taking a composition of two relabelings. Hence it is also true for the corresponding cellular maps. □

18.3.2 Aut (G) Action on Hom (T, G)

The functoriality on the right has the following important consequence.

Proposition 18.6. *The map given by $\varphi \mapsto \varphi^T$ is a group homomorphism from* Aut (G) *to* Aut $(\text{Hom}(T, G))$; *in particular,* Aut (G) *acts cellularly on* Hom (T, G).

Proof. Assume $\varphi \in$ Aut (G). We have induced maps $\varphi^T :$ Hom $(T, G) \to$ Hom (T, G) and $(\varphi^{-1})^T :$ Hom $(T, G) \to$ Hom (T, G). Since

$$(\varphi^{-1})^T \circ \varphi^T = (\varphi^{-1} \circ \varphi)^T = (\text{id}_G)^T = \text{id}_{\text{Hom}(T,G)},$$

we have $(\varphi^{-1})^T = (\varphi^T)^{-1}$, and therefore $\varphi^T \in$ Aut $(\text{Hom}(T, G))$. □

Example 18.7.
(1) The automorphism group of K_3 is the symmetric group $\mathcal{S}_3$. The transpositions act on the hexagon Bip (K_3) as reflections that fix the barycenters of two opposite sides.

(2) For an arbitrary graph G, the symmetric group $\mathcal{S}_n$ acts cellularly on Hom (G, K_n), which is the topological space of all n-colorings of G.

18.3.3 Functoriality on the Left

The situation is slightly more complicated if one considers the functoriality in the first argument. Let us choose some graph homomorphism ψ from T to G, and let K be some graph.

Again, using composition we can define an order-preserving map

$$g : \mathcal{F}(\text{Hom}\,(G,K)) \longrightarrow \mathcal{F}(\text{Hom}\,(T,K)).$$

Namely, for $\eta : V(G) \to 2^{V(K)} \setminus \{\emptyset\}$ and $v \in V(T)$, we set

$$g(\eta)(v) := \eta(\psi(v)).$$

This map is well-defined, since if $v, w \in V(T)$ and $(v, w) \in E(T)$, then $(\psi(v), \psi(w)) \in E(G)$, and therefore, for any $x \in \eta(\psi(v))$ and $y \in \eta(\psi(w))$, we have $(x, y) \in E(K)$. Furthermore, this map is order-preserving: if $\tau \geq \eta$, i.e., if $\tau(w) \supseteq \eta(w)$, for any $w \in V(T)$, then $g(\tau)(w) = \tau(\psi(w)) \supseteq \eta(\psi(w)) = g(\eta)(w)$.

Intuitively, one can think of the map g as the pullback map. It is important to remark that if ψ is not injective, it may happen that $\dim g(\eta) > \dim \eta$. Since g is an order-preserving map, the induced map between abstract simplicial complexes $\Delta(g) : \text{Bd}\,(\text{Hom}\,(G,K)) \to \text{Bd}\,(\text{Hom}\,(T,K))$ is simplicial and gives the corresponding map of topological spaces, which we denote by ψ_K. It is important to notice that the map g does not always come from a cellular map.

Example 18.8. Consider a graph homomorphism $\psi : L_3 \to K_2$. The induced map $\psi_{K_3} : \text{Bip}\,(K_3) \to \text{Hom}\,(L_3, K_3)$ is not cellular, but it becomes simplicial once we pass to the barycentric subdivisions. This is illustrated in Figure 18.5.

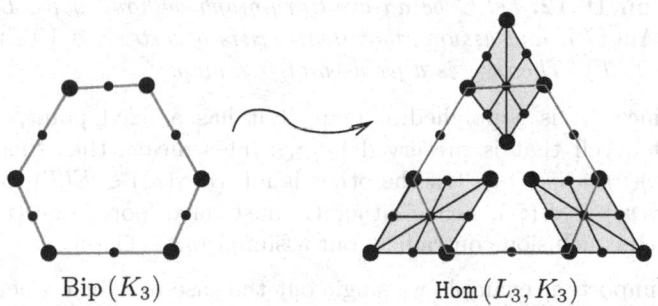

$$\text{Bip}\,(K_3) \hspace{5cm} \text{Hom}\,(L_3, K_3)$$

Fig. 18.5. Mapping $\text{Bip}\,(K_3)$ to $\text{Hom}\,(L_3, K_3)$.

However, on the positive side, we have the following proposition.

Proposition 18.9. *If the graph homomorphism $\psi : T \to G$ is injective on the vertices of T, then there exists a polyhedral map $h : \text{Hom}\,(G,K) \to \text{Hom}\,(T,K)$ such that topologically, $h = \psi_K$.*

Proof. This follows from an ad hoc argument by a direct analysis of point descriptions; see Subsection 9.2.3. The crucial fact is that the cells are mapped to cells of equal or smaller dimension. $\square$

We are now ready to draw a conclusion that is dual to Proposition 18.5.

Proposition 18.10. . *For any fixed graph K, the* Hom *construction yields a contravariant functor* $\mathrm{Hom}\,(-, K)$.

Proof. We only need to check the following fact: for any sequence of graph homomorphisms $\psi : T \to G$ and $\varphi : G \to L$, the induced maps $\psi_K : \mathrm{Hom}\,(G, K) \to \mathrm{Hom}\,(T, K)$ and $\varphi_K : \mathrm{Hom}\,(L, K) \to \mathrm{Hom}\,(G, K)$ satisfy the identity

$$\psi_K \circ \varphi_K = (\varphi \circ \psi)_K.$$

However, this identity is immediate on the level of poset maps. □

18.3.4 Aut (T) Action on $\mathrm{Hom}\,(T, G)$

Again, the functoriality on the left has concrete and useful implications.

Proposition 18.11. *The group* $\mathrm{Aut}\,(T)$ *acts on* $\mathrm{Hom}\,(T, G)$ *by polyhedral transformations.*

Proof. Let $\varphi \in \mathrm{Aut}\,(T)$. Then $\varphi_G : \mathrm{Hom}\,(T, G) \to \mathrm{Hom}\,(T, G)$. As before, we have

$$(\varphi^{-1})_G \circ \varphi_G = (\varphi \circ \varphi^{-1})_G = (\mathrm{id}_T)_G = \mathrm{id}_{\mathrm{Hom}\,(T,G)}.$$

Therefore, we get an action of $\mathrm{Aut}\,(T)$ on $\mathrm{Hom}\,(T, G)$. By Proposition 18.9 this action is polyhedral. □

Proposition 18.12. *Let G be an arbitrary graph without loops. Let furthermore $\gamma \in \mathrm{Aut}\,(T)$, and assume that there exists a vertex $v \in V(T)$ such that $(v, \gamma(v)) \in E(T)$. Then γ_G is a fixed-point-free map.*

Proof. Since γ_G is a polyhedral map, if it has a fixed point, then there must exist a cell that is preserved by γ_G. This means that there exists η such that $\eta(v) = \eta(\gamma(v))$. On the other hand, $(v, \gamma(v)) \in E(T)$ implies that $\eta(v) \times \eta(\gamma(v)) \in E(G)$; hence either G must have loops, or $\eta(v)$ must be empty. Each conclusion contradicts our assumptions. □

As an important example we single out the case in which γ actually *flips* some edge of the graph T. In this situation, $\mathrm{Hom}\,(T, G)$ is actually a $\mathbb{Z}_2$-space, meaning a topological space with a free $\mathbb{Z}_2$-action. The reflection actions on K_2, C_{2r+1}, and C_{2r} (flipping two edges) are special cases of that.

18.3.5 Commuting Relations

Proposition 18.13. *Let $\varphi : T \to T'$ and $\psi : G \to G'$ be graph homomorphisms. Then the following diagram commutes:*

$$\begin{CD}
\mathrm{Hom}\,(T', G) @>{\psi^{T'}}>> \mathrm{Hom}\,(T', G') \\
@V{\varphi_G}VV @VV{\varphi_{G'}}V \\
\mathrm{Hom}\,(T, G) @>{\psi^T}>> \mathrm{Hom}\,(T, G')
\end{CD} \tag{18.2}$$

Proof. The easiest way to see that the diagram (18.2) commutes is on the level of posets. The elements of $\mathcal{F}(\mathrm{Hom}\,(T',G))$ may be encoded as t-tuples of sets $(A_1,\dots,A_t)$, where $[t] = V(T')$ and $A_i \subseteq V(G)$ for each $i \in [t]$. Following the diagram either way, we obtain the $|V(T)|$-tuple $(\dots,\psi(A_{\varphi(v)}),\dots)$, where the set $\psi(A_{\varphi(v)})$ is in the position indexed by $v \in V(T)$. $\square$

Obviously, the fact that the above diagram (18.2) commutes implies that the map $\psi^T : \mathrm{Hom}\,(T,G) \to \mathrm{Hom}\,(T,G')$ preserves the $\mathrm{Aut}\,(T)$ action, as well as that the map $\varphi_G : \mathrm{Hom}\,(T',G) \to \mathrm{Hom}\,(T,G)$ preserves $\mathrm{Aut}\,(G)$ action.

Combined with the previous remarks, we see that, if γ is an involution of T that flips an edge, G and G' have no loops, and $\varphi : G \to G'$ is a graph homomorphism, then $\varphi^T : \mathrm{Hom}\,(T,G) \to \mathrm{Hom}\,(T,G')$ is a $\mathbb{Z}_2$-map (meaning a $\mathbb{Z}_2$-equivariant map of $\mathbb{Z}_2$-spaces).

Example 18.14. An arbitrary graph homomorphism of loop-free graphs $\varphi : G \to G'$ will induce $\mathbb{Z}_2$-maps $\varphi^{K_2} : \mathrm{Bip}\,(G) \to \mathrm{Bip}\,(G')$ and $\varphi^{C_{2r+1}} : \mathrm{Hom}\,(C_{2r+1},G) \to \mathrm{Hom}\,(C_{2r+1},G')$.

Furthermore, one can see that if γ acts both on the graph T and on the graph T', and we have a Γ-equivariant map $\varphi : T \to T'$, then the map $\varphi_G : \mathrm{Hom}\,(T',G) \to \mathrm{Hom}\,(T,G)$ is Γ-equivariant as well; and the same is true for the map $\varphi^G : \mathrm{Hom}\,(G,T) \to \mathrm{Hom}\,(G,T')$.

More generally, if the diagram

$$
\begin{array}{ccc}
T & \xrightarrow{\ \alpha\ } & \widetilde{T} \\
{\scriptstyle \varphi}\downarrow & & \downarrow{\scriptstyle \beta} \\
T' & \xrightarrow{\ \psi\ } & \widetilde{T}'
\end{array}
\qquad (18.3)
$$

commutes, then the diagram

$$
\begin{array}{ccc}
\mathrm{Hom}\,(T,G) & \xleftarrow{\ \alpha_G\ } & \mathrm{Hom}\,(\widetilde{T},G) \\
{\scriptstyle \varphi_G}\uparrow & & \uparrow{\scriptstyle \beta_G} \\
\mathrm{Hom}\,(T',G) & \xleftarrow{\ \psi_G\ } & \mathrm{Hom}\,(\widetilde{T}',G)
\end{array}
\qquad (18.4)
$$

commutes as well, and the same is true for the right argument.

Example 18.15. A $\mathbb{Z}_2$-equivariant inclusion $i : K_2 \hookrightarrow C_{2r+1}$ induces a $\mathbb{Z}_2$-map $i_G : \mathrm{Hom}\,(C_{2r+1},G) \to \mathrm{Bip}\,(G)$.

Remark 18.16. All of the results in this section hold for general morphism complexes as long as the set maps compose accordingly.

18.4 Products, Compositions, and Hom Complexes

18.4.1 Coproducts

For any three graphs G, H, and K, we have

$$\text{Hom}\,(G \textstyle\coprod H, K) = \text{Hom}\,(G, K) \times \text{Hom}\,(H, K), \qquad (18.5)$$

and if G is connected and $G \neq K_1$, then also

$$\text{Hom}\,(G, H \textstyle\coprod K) = \text{Hom}\,(G, H) \textstyle\coprod \text{Hom}\,(G, K),$$

where the equality denotes isomorphism of polyhedral complexes.

The first formula is obvious. To verify the second one, note that for any graph homomorphism $\eta : V(G) \to 2^{V(H) \cup V(K)} \setminus \{\emptyset\}$ and any $x, y \in V(G)$ such that $(x, y) \in E(G)$, if $\eta(x) \cap V(H) \neq \emptyset$, then $\eta(y) \subseteq V(H)$, which under the assumptions on G implies that $\bigcup_{x \in V(G)} \eta(x) \subseteq V(H)$.

18.4.2 Products

For any three graphs G, H, and K, we have the following homotopy equivalence:

$$\text{Hom}\,(G, H \times K) \simeq \text{Hom}\,(G, H) \times \text{Hom}\,(G, K). \qquad (18.6)$$

This can be strengthened as follows.

Proposition 18.17. *The left-hand side of the formula* (18.6) *is simple homotopy equivalent to the right-hand side.*

Proof. Consider the following three maps:

$$2^{p_H} : 2^{V(H) \times V(K)} \to 2^{V(H)},$$

$$2^{p_K} : 2^{V(H) \times V(K)} \to 2^{V(K)},$$

and

$$c : 2^{V(H)} \times 2^{V(K)} \to 2^{V(H) \times V(K)},$$

where 2^{p_H} and 2^{p_K} are induced by the standard projection maps $p_H : V(H) \times V(K) \to V(H)$ and $p_K : V(H) \times V(K) \to V(K)$, and c is given by $c(A, B) = A \times B$. We let $\psi : 2^{V(H) \times V(K)} \to 2^{V(H) \times V(K)}$ denote the composition map $\psi(S) = c(2^{p_H}(S), 2^{p_K}(S)) = 2^{p_H}(S) \times 2^{p_K}(S)$.

Given a cell of $\text{Hom}\,(G, H \times K)$ indexed by $\eta : V(G) \to 2^{V(H) \times V(K)} \setminus \{\emptyset\}$, one can see that the composition function $\psi \circ \eta : V(G) \to 2^{V(H) \times V(K)} \setminus \{\emptyset\}$ will also index a cell. Indeed, for any $(x, y) \in E(G)$ we know that $(\eta(x), \eta(y))$ is a complete bipartite subgraph of $H \times K$, which is the same as to say that for any $\alpha \in \eta(x)$ and $\beta \in \eta(y)$, we have $(p_H(\alpha), p_H(\beta)) \in E(H)$ and $(p_K(\alpha), p_K(\beta)) \in E(K)$. If we now choose $\tilde{\alpha} \in \psi(\eta(x))$ and $\tilde{\beta} \in \psi(\eta(y))$, we have $p_H(\tilde{\alpha}) = p_H(\alpha)$ for some $\alpha \in \eta(x)$, and $p_H(\tilde{\beta}) = p_H(\beta)$ for some $\beta \in \eta(y)$, hence verifying that $(p_H(\tilde{\alpha}), p_H(\tilde{\beta})) \in E(H)$. The fact that $(p_K(\tilde{\alpha}), p_K(\tilde{\beta})) \in E(K)$ can be proved analogously.

This means that we have a map

$$\varphi : \mathcal{F}(\mathrm{Hom}\,(G, H \times K)) \to \mathcal{F}(\mathrm{Hom}\,(G, H \times K)).$$

It is easy to see that φ is order-preserving and ascending. It follows from Theorem 13.12 that the complex $\Delta(\mathcal{F}(\mathrm{Hom}\,(G, H \times K))) = \mathrm{Bd}\,(\mathrm{Hom}\,(G, H \times K))$ collapses onto the complex $\Delta(\mathrm{Im}(\varphi))$.

On the other hand, we see that $\mathcal{F}(\mathrm{Hom}\,(G, H)) \times \mathcal{F}(\mathrm{Hom}\,(G, K)) \cong \mathrm{Im}(\varphi)$ with the isomorphism given by the map $(\eta_1, \eta_2) \mapsto \eta$, where $\eta(x) = \eta_1(x) \times \eta_2(x)$, for any $x \in V(G)$. Thus we conclude that the complex $\mathrm{Bd}\,(\mathrm{Hom}\,(G, H \times K))$ collapses onto the complex $\Delta(\mathcal{F}(\mathrm{Hom}\,(G, H)) \times \mathcal{F}(\mathrm{Hom}\,(G, K)))$. Our argument is now complete, since we know that $\Delta(\mathcal{F}(\mathrm{Hom}\,(G, H)) \times \mathcal{F}(\mathrm{Hom}\,(G, K))) \cong \Delta(\mathcal{F}(\mathrm{Hom}\,(G, H))) \times \Delta(\mathcal{F}(\mathrm{Hom}\,(G, K))) = \mathrm{Bd}\,(\mathrm{Hom}\,(G, H)) \times \mathrm{Bd}\,(\mathrm{Hom}\,(G, K)) \cong \mathrm{Hom}\,(G, H) \times \mathrm{Hom}\,(G, K)$. $\square$

For the analogue of the formula (18.6), where the direct product is taken on the left, we need the following additional standard notion.

Definition 18.18. *For two graphs H and K, the* **power graph** K^H *is defined by*

- $V(K^H)$ *is the set of all set maps* $f : V(H) \to V(K)$;
- $(f, g) \in E(K^H)$, *for* $f, g : V(H) \to V(K)$ *if and only if whenever* $(v, w) \in E(H)$, *we also have* $(f(v), g(w)) \in E(K)$.

It is easy to see that the power graph notion is introduced precisely so that for any triple of graphs the following adjunction relation holds:

$$\mathrm{Hom}^{(0)}(G \times H, K) = \mathrm{Hom}^{(0)}(G, K^H). \tag{18.7}$$

In our topological situation the formula (18.7) generalizes up to homotopy. More precisely, we have the following homotopy equivalence:

$$\mathrm{Hom}\,(G \times H, K) \simeq \mathrm{Hom}\,(G, K^H). \tag{18.8}$$

Proposition 18.19. *The left-hand side of formula (18.8) is simple homotopy equivalent to the right-hand side.*

Proof. Define a map $\psi : 2^{V(K^H)} \to 2^{V(K^H)}$, $\psi : \Omega \mapsto \psi(\Omega)$, as follows: $g \in \psi(\Omega)$ if and only if $g(x) \in \{f(x) \mid f \in \Omega\}$, for all $x \in V(H)$. In other words, we use the collection of functions Ω to specify the sets of values that functions from $\psi(\Omega)$ are allowed to take. Clearly, we have $\psi(\Omega) \supseteq \Omega$. Take a cell of $\mathrm{Hom}\,(G, K^H)$, $\eta : V(G) \to 2^{V(K^H)} \setminus \{\emptyset\}$, and consider the composition map $\psi \circ \eta : V(G) \to 2^{V(K^H)} \setminus \{\emptyset\}$. Since η is a cell, we know that if $(x, y) \in E(G)$

and $\alpha \in \eta(x)$, $\beta \in \eta(y)$, then $(\alpha, \beta) \in E(K^H)$, i.e., whenever $(v, w) \in E(H)$, we have $(\alpha(v), \beta(w)) \in E(K)$.

Choose $\tilde{\alpha} \in \psi(\eta(x))$ and $\tilde{\beta} \in \psi(\eta(y))$. To check that $(\tilde{\alpha}, \tilde{\beta}) \in E(K^H)$, we need to check that for any $(v, w) \in E(H)$, we have $(\tilde{\alpha}(v), \tilde{\beta}(w)) \in E(K)$. However, by the definition of ψ, we know that $\tilde{\alpha}(v) = \alpha(v)$ for some $\alpha \in \eta(x)$, and $\tilde{\beta}(w) = \beta(w)$ for some $\beta \in \eta(y)$. It follows that $(\tilde{\alpha}(v), \tilde{\beta}(w)) = (\alpha(v), \beta(w)) \in E(K)$, and hence $\psi \circ \eta$ is again a cell.

As a consequence, the composition gives us an order-preserving ascending map $\varphi : \mathcal{F}(\mathrm{Hom}\,(G, K^H)) \to \mathcal{F}(\mathrm{Hom}\,(G, K^H))$. The image of this map is isomorphic to $\mathcal{F}(\mathrm{Hom}\,(G \times H, K))$. The isomorphism map takes the poset element $\eta : V(G) \times V(H) \to 2^{V(K)} \setminus \{\emptyset\}$ to the poset element $\tilde{\eta} : V(G) \to 2^{V(K^H)} \setminus \{\emptyset\}$ defined by

$$\tilde{\eta}(x) = \{f : V(H) \to V(K) \mid f(v) \in \eta(x, v), \text{ for all } v \in V(H)\},$$

for all $x \in V(G)$. By Theorem 13.12, we conclude that the simplicial complex $\Delta(\mathcal{F}(\mathrm{Hom}\,(G, K^H))) = \mathrm{Bd}\,(\mathrm{Hom}\,(G, K^H))$ collapses onto its subcomplex $\Delta(\mathrm{Im}(\varphi)) = \mathrm{Bd}\,(\mathrm{Hom}\,(G \times H, K))$. $\square$

We obtain an interesting special case of the formula (18.8) when substituting $G = K_1^o$ (which means a graph with one looped vertex). Since $K_1^o \times H = H$, for any graph H, we conclude that $\mathrm{Hom}\,(H, K) \simeq \mathrm{Hom}\,(K_1^o, K^H)$ for any two graphs H and K. As seen directly for an arbitrary graph G, $\mathrm{Hom}\,(K_1^o, G)$ is the clique complex of the looped part of G, i.e., of the subgraph induced by the set of vertices that have loops. In particular, the complex $\mathrm{Hom}\,(K_1^o, G)$ is simplicial. On the other hand, a vertex $f \in V(K^H)$ has a loop if and only if f is a graph homomorphism. We can therefore conclude that for arbitrary graphs H and K, the complex $\mathrm{Hom}\,(H, K)$ is homotopy equivalent to the clique complex of the subgraph of K^H induced by the set of all graph homomorphisms from H to K.

18.4.3 Composition of Hom Complexes

For three arbitrary graphs T, G, and K, there is a composition map

$$\xi : \mathcal{F}(\mathrm{Hom}\,(T, G)) \times \mathcal{F}(\mathrm{Hom}\,(G, K)) \longrightarrow \mathcal{F}(\mathrm{Hom}\,(T, K)),$$

whose detailed description is as follows: for graph homomorphisms $\alpha : V(T) \to 2^{V(G)} \setminus \{\emptyset\}$ and $\beta : V(G) \to 2^{V(K)} \setminus \{\emptyset\}$, define the map $\tilde{\beta} : 2^{V(G)} \setminus \{\emptyset\} \to 2^{V(K)} \setminus \{\emptyset\}$ by

$$\text{for } S \in 2^{V(G)} \setminus \{\emptyset\}, \quad \tilde{\beta}(S) := \cup_{x \in S} \beta(x),$$

and then set $\xi(\alpha, \beta) := (\tilde{\beta} \circ \alpha : V(T) \to 2^{V(K)} \setminus \{\emptyset\})$. It is easy to check that this map is well-defined. Indeed, let $x, y \in V(T)$ be such that $(x, y) \in E(T)$, choose arbitrary $a \in \alpha(x)$ and $b \in \alpha(y)$, and then choose arbitrary $\tilde{a} \in \beta(a)$ and $\tilde{b} \in \beta(b)$. Clearly, $(x, y) \in E(T)$ implies $(a, b) \in E(G)$, since α is a graph homomorphism, which then implies $(\tilde{a}, \tilde{b}) \in E(K)$, since β is a graph homomorphism.

Applying the nerve functor Δ to the poset map ξ, we get a simplicial map

$$\Delta(\xi) : \Delta(\mathcal{F}(\mathrm{Hom}\,(T, G)) \times \mathcal{F}(\mathrm{Hom}\,(G, K))) \longrightarrow \mathrm{Bd}\,(\mathrm{Hom}\,(T, K)),$$

and hence, since for any posets P_1 and P_2, the simplicial complex $\Delta(P_1 \times P_2)$ is homeomorphic to the polyhedral complex $\Delta(P_1) \times \Delta(P_2)$ (in fact it is its subdivision), we have a corresponding topological map

$$\mathrm{Hom}\,(T, G) \times \mathrm{Hom}\,(G, K) \longrightarrow \mathrm{Hom}\,(T, K).$$

18.5 Folds

18.5.1 Definition and First Properties

It became clear early on that Hom complexes behave well with respect to the following standard operation from graph theory.

Definition 18.20. *For a graph G and a vertex $v \in V(G)$, the graph $G - v$ is called a* **fold** *of G if there exists a vertex $u \in V(G)$ such that $u \neq v$ and $\mathcal{N}(u) \supseteq \mathcal{N}(v)$.*

Let $G - v$ be a fold of G. We let $i : G - v \hookrightarrow G$ denote the inclusion homomorphism, and let $f : G \to G - v$ denote the folding homomorphism defined by

$$f(x) = \begin{cases} u, & \text{for } x = v; \\ x, & \text{for } x \neq v. \end{cases}$$

Note that i is a graph homomorphism for an arbitrary choice of $v \in V(G)$, whereas f is a graph homomorphism if and only if $G - v$ is a fold, so in particular, this could be taken as an alternative definition of the fold.

Example 18.21.

- A tree folds to any one of its edges.
- A 4-cycle folds to any one of its edges.
- Let F be a forest. Its complement $\bar{F}$ folds to a complete graph K_n, where n is the maximal cardinality of an independent set in F.

Theorem 18.22. *Let $G - v$ be a fold of G and let H be some graph. Then*

(1) the simplicial complex $\mathrm{Bd}\,\mathrm{Hom}\,(G, H)$ collapses onto the simplicial complex $\mathrm{Bd}\,\mathrm{Hom}\,(G - v, H)$;

(2) the prodsimplicial complex $\mathrm{Hom}\,(H, G)$ collapses onto the prodsimplicial complex $\mathrm{Hom}\,(H, G - v)$.

The maps i_H and f^H are strong deformation retractions.

Corollary 18.23.
(1) *If T is a finite tree with at least one edge, then the map i_{K_n} : $\mathrm{Hom}\,(T, K_n) \to \mathrm{Bip}\,(K_n)$ induced by any inclusion $i : K_2 \hookrightarrow T$ is a homotopy equivalence; in particular, $\mathrm{Hom}\,(T, K_n) \simeq \mathbb{S}^{n-2}$.*
(2) *If F is a finite forest, and $T_1, \ldots, T_k$ are all its connected components consisting of at least two vertices, then $\mathrm{Hom}\,(F, K_n) \simeq \prod_{i=1}^{k} \mathbb{S}^{n-2}$.*

Curiously, another computable special case is that of an unlooped complement of a forest.

Proposition 18.24. *Let F be a finite forest, and let G be an arbitrary graph. Then $\mathrm{Hom}\,(\overline{F}, G) \simeq \mathrm{Hom}\,(K_m, G)$, where m is the maximal cardinality of an independent set in F.*

18.5.2 Proof of the Folding Theorem

Proof. First we show that $\mathrm{Bd}\,\mathrm{Hom}\,(G, H)$ collapses onto $\mathrm{Bd}\,\mathrm{Hom}\,(G - v, H)$. Identify $\mathcal{F}(\mathrm{Hom}\,(G - v, H))$ with the subposet of $\mathcal{F}(\mathrm{Hom}\,(G, H))$ consisting of all η such that $\eta(v) = \eta(u)$. Let X be the subposet consisting of all $\eta \in \mathcal{F}(\mathrm{Hom}\,(G, H))$ satisfying $\eta(v) \supseteq \eta(u)$. Then $\mathcal{F}(\mathrm{Hom}\,(G - v, H)) \subseteq X \subseteq \mathcal{F}(\mathrm{Hom}\,(G, H))$. Consider order-preserving maps

$$\mathcal{F}(\mathrm{Hom}\,(G, H)) \xrightarrow{\alpha} X \xrightarrow{\beta} \mathcal{F}(\mathrm{Hom}\,(G - v, H))$$

defined by

$$\alpha\eta(x) = \begin{cases} \eta(u) \cup \eta(v), & \text{for } x = v; \\ \eta(x), & \text{otherwise}; \end{cases} \qquad \beta\eta(x) = \begin{cases} \eta(u), & \text{for } x = v; \\ \eta(x), & \text{otherwise}; \end{cases}$$

for all $x \in V(G)$; see Figure 18.6. Maps α and β are well-defined because $G - v$ is a fold of G. Clearly $\beta \circ \alpha = \mathcal{F}(i_H)$, α is an ascending closure operator, and β is a descending closure operator. Since $\mathrm{Im}\,\mathcal{F}(i_H) = \mathcal{F}(\mathrm{Hom}\,(G - v, H))$, the statement follows from Theorem 13.12.

We show that $\mathrm{Hom}\,(H, G)$ collapses onto $\mathrm{Hom}\,(H, G - v)$ by presenting a sequence of elementary collapses. Define $V(H) = \{x_1, \ldots, x_t\}$. For $\eta \in \mathcal{F}(\mathrm{Hom}\,(H, G))$, let $1 \leq i(\eta) \leq t$ be the minimal index such that $v \in \eta(x_{i(\eta)})$. Write $\mathcal{F}(\mathrm{Hom}\,(H, G))$ as a disjoint union $A \cup B \cup \mathcal{F}(\mathrm{Hom}\,(H, G - v))$, defined as follows: for $\eta \in A \cup B$ we have $\eta \in A$ if $u \notin \eta(x_{i(\eta)})$, and we have $\eta \in B$ otherwise.

There is a bijection $\varphi : A \to B$ that adds u to $\eta(x_{i(\eta)})$ without changing other values of η. Adding u to $\eta(x_{i(\eta)})$ yields an element in $\mathcal{F}(\mathrm{Hom}\,(H, G))$, since $G - v$ is a fold of G. Clearly, $\varphi(\alpha)$ covers α, for all $\alpha \in A$. We take the set $\{(\alpha, \varphi(\alpha)) \,|\, \alpha \in A\}$ to be our collection of elementary collapses. These are ordered lexicographically after the pairs of integers $(i(\alpha), -\dim \alpha)$.

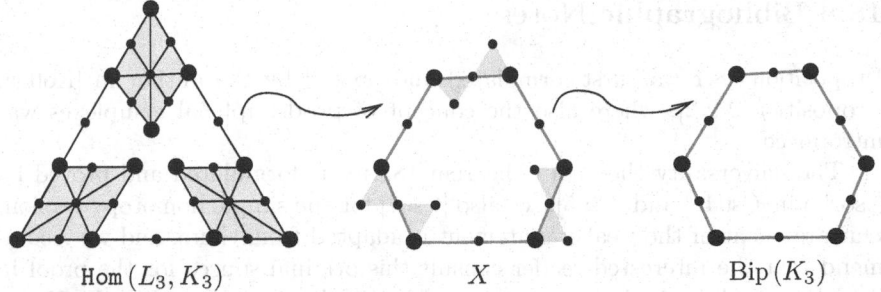

$$\text{Hom}\,(L_3, K_3) \qquad\qquad X \qquad\qquad \text{Bip}\,(K_3)$$

Fig. 18.6. A two-step folding of the first argument in $\text{Hom}\,(L_3, K_3)$.

Let us see that these collapses can be performed in this lexicographic order. Take $\eta > \alpha$, $\eta \neq \varphi(\alpha)$. Assume $i(\eta) = i(\alpha)$. If $\eta \in B$, then $\eta = \varphi(\tilde{\alpha})$, $i(\tilde{\alpha}) = i(\alpha)$, and $\dim \tilde{\alpha} > \dim \alpha$. Otherwise, $\eta \in A$ and $\dim \eta > \dim \alpha$. The third possibility is that $i(\eta) < i(\alpha)$. In either case, η was removed before α was. $\square$

Instead of verifying that the sequence of collapses is correct in the last paragraph of the proof, we could simply notice that the defined matching is acyclic and derive the result by discrete Morse theory; see Theorem 11.13.

Remark 18.25. In analogy with the first part of the proof, we can show that $\text{Bd}\,\text{Hom}\,(H, G)$ collapses onto $\text{Bd}\,\text{Hom}\,(H, G - v)$ by rewriting $\mathcal{F}(f^H)$ as a composition of two closure operators.

Indeed, let Y be the subposet consisting of all $\eta \in \mathcal{F}(\text{Hom}\,(H, G))$ such that for all $x \in V(H)$, $\eta(x) \cap \{u, v\} \neq \{v\}$, i.e., for any $x \in V(H)$ we have that if $v \in \eta(x)$, then $u \in \eta(x)$. Then $\mathcal{F}(\text{Hom}\,(H, G - v)) \subseteq Y \subseteq \mathcal{F}(\text{Hom}\,(H, G))$. Consider order-preserving maps

$$\mathcal{F}(\text{Hom}\,(H, G)) \xrightarrow{\varphi} Y \xrightarrow{\psi} \mathcal{F}(\text{Hom}\,(H, G - v))$$

defined by

$$\varphi\eta(x) = \begin{cases} \eta(x) \cup \{u\}, & \text{if } v \in \eta(x); \\ \eta(x), & \text{otherwise;} \end{cases} \qquad \psi\eta(x) = \begin{cases} \eta(x) \setminus \{v\}, & \text{if } v \in \eta(x); \\ \eta(x), & \text{otherwise;} \end{cases}$$

for all $x \in V(H)$.

The map ϕ is well-defined because $G - v$ is a fold of G, and the map ψ is well-defined by the construction of Y. We see that $\psi \circ \varphi = \mathcal{F}(f^H)$, φ is an ascending closure operator, and ψ is a descending closure operator. Since $\text{Im}\,\mathcal{F}(f_H) = \mathcal{F}(\text{Hom}\,(H, G - v))$, the statement follows from Theorem 13.12.

18.6 Bibliographic Notes

Proposition 18.1 was first formulated and proved by the author in [Ko05a, Proposition 2.2.2], where also the concept of prodsimplicial complexes was introduced.

The universality theorem (Theorem 18.2) was formulated and proved by Csorba in [Cs04a] and [Cs05]; see also [Cs07] for the simple homotopy version. Our argument of the weaker statement is adapted from there, and we recommend that the interested reader consult this original source for the proof in its full generality. An independent proof by Živaljević appeared in [Ziv05a].

Theorem 18.3 was proved in [Ko06c]. That the neighborhood complex $\mathcal{N}(G)$ is homotopy equivalent to $\mathrm{Bip}\,(G)$ was known earlier; see, e.g., [BK06] for an argument.

Most of the contents of Sections 18.3 and 18.4 was used explicitly or implicitly in the series of papers [BK03, BK06, BK04]. Our presentation here follows in large part the author's survey article [Ko05a], where a comprehensive structure theory for the Hom complexes was fixed on paper.

Our treatment of the topological implications of folds largely follows [Ko06a], where in particular the general Theorem 18.22 was proved. Various special cases of this theorem were known before; see [BK06, Cs05, Doc04].

Using Characteristic Classes to Design Tests for Chromatic Numbers of Graphs

In principle, all sorts of characteristic classes carry obstructions to graph colorings. Here we shall look at the applications of Stiefel–Whitney classes associated to free involutions.

19.1 Stiefel–Whitney Characteristic Classes and Test Graphs

19.1.1 Powers of Stiefel–Whitney Classes and Chromatic Numbers of Graphs

The following theorem describes the standard way to use the nonnullity of the powers of Stiefel–Whitney characteristic classes associated to $\mathbb{Z}_2$-spaces as tests for graph colorings.

Theorem 19.1. *Let T and G be two arbitrary graphs such that T has a $\mathbb{Z}_2$-action that flips some edge in T, whereas G has no loops. Assume that $\varpi_1^k(\mathrm{Hom}\,(T,G)) \neq 0$, and that $\varpi_1^k(\mathrm{Hom}\,(T,K_m)) = 0$, for some integers $k \geq 0$, $m \geq 1$. Then we can conclude that $\chi(G) \geq m+1$.*

Proof. By Proposition 18.12, we know that under the assumptions of the theorem, the prodsimplicial complex $\mathrm{Hom}\,(T,H)$ is a $\mathbb{Z}_2$-space for any loop-free graph H.

Assume now that the graph G is m-colorable, i.e., that there exists a graph homomorphism $\varphi : G \to K_m$. By functoriality of the Hom construction, it will induce a $\mathbb{Z}_2$-map $\varphi^T : \mathrm{Hom}\,(T,G) \to \mathrm{Hom}\,(T,K_m)$.

Since the Stiefel–Whitney characteristic classes are functorial and we assumed that $\varpi_1^k(\mathrm{Hom}\,(T,K_m)) = 0$, the existence of the $\mathbb{Z}_2$-map φ^T implies that also $\varpi_1^k(\mathrm{Hom}\,(T,G)) = 0$, thus yielding a contradiction to the assumption of the theorem. $\square$

Examples of graphs that appear as the graph T in Theorem 19.1 include the complete graphs and the cycles.

Remark 19.2. It is easy to see that the proof of Theorem 19.1 works just the same if the family of complete graphs is replaced with some other family, and we are interested in obtaining obstructions to the maps from some graph G into this new family.

19.1.2 Stiefel–Whitney Test Graphs

The next definition describes the family of test graphs that are most useful when it comes to looking for characteristic class obstructions to graph colorings.

Definition 19.3. *Let T be a graph with a $\mathbb{Z}_2$-action that flips an edge. Then T is called a **Stiefel–Whitney n-test graph** if we have*

$$\mathrm{h}(\mathrm{Hom}\,(T, K_n)) = n - \chi(T).$$

*Furthermore, T is called a **Stiefel–Whitney test graph** if it is a Stiefel–Whitney n-test graph for every integer $n \geq \chi(T)$.*

A direct application of Theorem 19.1 yields the next corollary, which also serves as an explanation for our terminology.

Corollary 19.4. *Assume that T is a Stiefel–Whitney test graph. Then for an arbitrary graph G, we have*

$$\chi(G) \geq \chi(T) + \mathrm{h}(\mathrm{Hom}\,(T, G)). \tag{19.1}$$

This property can be taken as a blueprint for the homotopy version of Stiefel–Whitney test graphs.

Definition 19.5. *A graph T is called a **homotopy test graph** if for an arbitrary graph G, the following equation is satisfied:*

$$\chi(G) > \chi(T) + \mathrm{conn}\,\mathrm{Hom}\,(T, G). \tag{19.2}$$

Note that by Corollary 8.26, we have $\mathrm{h}(X) \geq \mathrm{conn}\,X + 1$ for an arbitrary $\mathbb{Z}_2$-space X. Therefore, comparing equations (19.2) and (19.1), we see that if a graph T is a Stiefel–Whitney test graph, then, it is also a homotopy test graph.

Let us stress again that in analogy to the fact that the *height* is defined for $\mathbb{Z}_2$-spaces, the term *Stiefel–Whitney test graph* actually refers to a pair (T, γ), where T is a graph and γ is an involution of T that flips an edge. The following question arises naturally in this context.

We describe an important extension property of the class of Stiefel–Whitney test graphs.

Proposition 19.6. *Let T be an arbitrary graph, and let A and B be Stiefel–Whitney test graphs such that $\chi(T) = \chi(A) = \chi(B)$. Assume further that there exist $\mathbb{Z}_2$-equivariant graph homomorphisms $\varphi : A \to T$ and $\psi : T \to B$. Then T is also a Stiefel–Whitney test graph.*

Proof. Let n be an arbitrary positive integer. By the functoriality of Stiefel–Whitney characteristic classes, we have

$$\mathrm{h}(\mathrm{Hom}\,(A, K_n)) \leq \mathrm{h}(\mathrm{Hom}\,(T, K_n)) \leq \mathrm{h}(\mathrm{Hom}\,(B, K_n)).$$

Hence $n - \chi(A) \leq \mathrm{h}(\mathrm{Hom}\,(T, K_n)) \leq n - \chi(B)$, which, by the assumptions of the proposition, implies $\mathrm{h}(\mathrm{Hom}\,(T, K_n)) = n - \chi(T)$. $\square$

The next corollary describes a simple but instructive example of the situation in Proposition 19.6.

Corollary 19.7. *Any connected bipartite graph T with a $\mathbb{Z}_2$-action that flips an edge is a Stiefel–Whitney test graph.*

Indeed, we have $\mathbb{Z}_2$-equivariant graph homomorphisms $K_2 \hookrightarrow T \to K_2$, where the first one is the inclusion of the flipped edge, and the second one is the arbitrary coloring map. Since by Proposition 19.8, K_2 is a Stiefel–Whitney test graph, we conclude that T is also a Stiefel–Whitney test graph. In particular, any even cycle with the $\mathbb{Z}_2$-action that flips an edge is a Stiefel–Whitney test graph.

19.2 Examples of Stiefel–Whitney Test Graphs

19.2.1 Complexes of Complete Multipartite Subgraphs

We have seen in Theorem 18.3 that the complex $\mathrm{Bip}\,(G)$ has the same simple homotopy type as the neighborhood complex of G. In particular, the complex $\mathrm{Bip}\,(K_n)$ is homotopy equivalent to the sphere $\mathbb{S}^{n-2}$. The following proposition summarizes more complete information.

Proposition 19.8.
(a) *The prodsimplicial complex $\mathrm{Bip}\,(K_{n+1})$ is isomorphic as a polyhedral complex to the boundary complex of the Minkowski sum $\Delta^n + (-\Delta^n)$.*
(b) *The $\mathbb{Z}_2$-action on $\mathrm{Bip}\,(K_{n+1})$ induced by the flip action of $\mathbb{Z}_2$ on K_2 corresponds under this isomorphism to the central symmetry.*

Proof. Let M_n stand for the Minkowski sum

$$[-1/2, 1/2]^n + [(-1/2, -1/2, \ldots, -1/2), (1/2, 1/2, \ldots, 1/2)],$$

where $[-1/2, 1/2]^n$ denotes the cube in $\mathbb{R}^n$ whose vertices are all the points with coordinates having absolute value $1/2$. The polytope M_n is a zonotope in $\mathbb{R}^n$. Its dual, M_n^*, is the polytope associated to the hyperplane arrangement $\mathcal{A} = \{\mathcal{A}_1, \ldots, \mathcal{A}_{n+1}\}$ defined by

$$\mathcal{A}_i = \begin{cases} (x_i = 0), & \text{for } 1 \leq i \leq n; \\ (\sum_{j=1}^n x_j = 0), & \text{for } i = n+1. \end{cases}$$

Let us identify each cell $\eta : V(K_2) \rightarrow 2^{V(K_n)} \setminus \{\emptyset\}$ of Bip (K_2) with the ordered pair (A, B) of nonempty subsets of $[n]$ by taking $A = \eta(1)$ and $B = \eta(2)$. Set $P := \mathcal{F}(\text{Bip}(K_{n+1}))^{op}$. We shall see that P is isomorphic to the face poset of M_n, where we set $Q := \mathcal{F}(M_n)$. The future isomorphism will be denoted by ρ.

First, note that faces of the cube $[-1/2, 1/2]^n$ are encoded by all n-tuples of $1/2$, $-1/2$, and the symbol $*$, where the latter symbol denotes the coordinate where the value can be chosen arbitrarily from the interval $[-1/2, 1/2]$. For an arbitrary n-tuple x, we let supp $(x) \subseteq [n]$ denote the set of the indices of coordinates that are either nonzero or are denoted by the symbol $*$. Additionally, for an arbitrary number k, we let supp $(x, k) \subseteq [n]$ denote the set of the indices of the coordinates that are equal to k (in particular, they cannot be denoted by the symbol $*$).

Vertices of M_n are labeled by all n-tuples of 1, -1, and 0 such that 1 and -1 are not present simultaneously, and not all the coordinates are equal to 0; that is, v is a vertex of M_n if and only if $v \in \{0, 1\}^n$ or $v \in \{0, -1\}^n$, and $v \neq (0, \ldots, 0)$. These vertices correspond to atoms in P as follows:

$$v \quad \overset{\rho}{\longleftrightarrow} \quad \begin{cases} (\text{supp}(v), [n+1] \setminus \text{supp}(v)), & \text{if } v \in \{0, 1\}^n; \\ ([n+1] \setminus \text{supp}(v), \text{supp}(v)), & \text{if } v \in \{0, -1\}^n. \end{cases}$$

Clearly, restricted to atoms, ρ is a bijection.

Those faces of M_n that are contained in the closed star of $(1, \ldots, 1)$ can be indexed by $f \in \{0, 1, *\}^n$, where $|\text{supp}(f, 1)| \geq 1$. Symmetrically, those faces of M_n that are contained in the closed star of $(-1, \ldots, -1)$ can be indexed by $f \in \{0, -1, *\}^n$, where $|\text{supp}(f, -1)| \geq 1$. For these faces ρ can be defined as follows:

$$f \quad \overset{\rho}{\longleftrightarrow} \quad \begin{cases} (\text{supp}(f, 1), \text{supp}(f, 0) \cup \{n+1\}), & \text{if } f \in \overline{\text{St}}(1, \ldots, 1); \\ (\text{supp}(f, 0) \cup \{n+1\}, \text{supp}(f, -1)), & \text{if } f \in \overline{\text{St}}(-1, \ldots, -1). \end{cases}$$

Finally, we consider the faces of M_n that are not in $\overline{\text{St}}(1, \ldots, 1) \cup \overline{\text{St}}(-1, \ldots, -1)$. Each such face is the convex hull of the union of two faces, $f \cup \tilde{f}$, such that $f \in \overline{\text{St}}(1, \ldots, 1)$, $\tilde{f} \in \overline{\text{St}}(-1, \ldots, -1)$, with the condition that $\text{supp}(f, 0) = \text{supp}(\tilde{f}, -1)$, $\text{supp}(f, 1) = \text{supp}(\tilde{f}, 0)$. The element of P associated to such a face under ρ is $(\text{supp}(f, 1), \text{supp}(f, 0)) = (\text{supp}(\tilde{f}, 0), \text{supp}(\tilde{f}, -1))$.

It is an easy exercise to check that ρ defines a poset isomorphism between P and Q, which in turn induces the required cell complex isomorphism.

Finally, a brief scanning through the definition of ρ in different cases reveals that ρ is equivariant with respect to the described $\mathbb{Z}_2$-actions on both sides. Hence the last part of the proposition follows. □

The following notion provides an alternative way for describing the complexes of all n-colorings of complete graphs.

Definition 19.9. *Let $X_1, \ldots, X_t$ be a family of abstract simplicial complexes with isomorphic sets of vertices. The* **deleted product** *of this family is the subcomplex of the direct product of $X_1, \ldots, X_t$ consisting of all cells $\tau_1 \times \cdots \times \tau_t$ satisfying $\tau_i \cap \tau_j = \emptyset$, for any $i \neq j$.*

Clearly, the complex $\mathrm{Hom}\,(K_m, K_n)$ can be viewed as a deleted product of m copies of $(n-1)$-dimensional simplices. In this context, the special case $m = 2$, which is dealt with in Proposition 19.8, is well known.

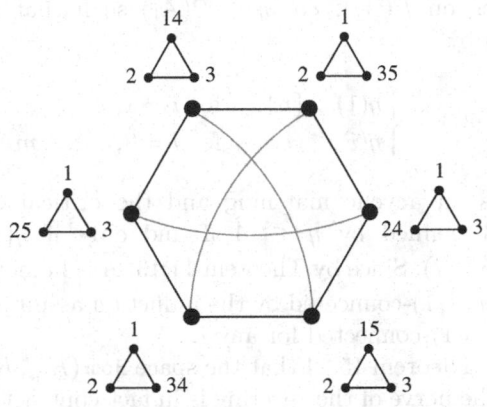

Link of a vertex in $\mathrm{Hom}\,(K_3, K_5)$

Fig. 19.1. Complexes of graph homomorphisms between complete graphs.

For $m \geq 3$, the prodsimplicial complexes $\mathrm{Hom}\,(K_m, G)$ can be thought of as consisting of all complete m-partite subgraphs of G. Even in the case $G = K_n$, it seems complicated to understand $\mathrm{Hom}\,(K_m, G)$ up to homeomorphism; see Figure 19.1. However, we still obtain a good description of the homotopy type.

Theorem 19.10. *Let us assume that m and n are positive integers and that $n \geq m$. The prodsimplicial complex of all n-colorings of a complete graph with m vertices, $\mathrm{Hom}\,(K_m, K_n)$, is homotopy equivalent to a wedge of $(n-m)$-dimensional spheres.*

Proof. We use induction on m and on $n - m$. The base is provided by the cases $\mathrm{Hom}\,(K_1, K_n)$, which is a simplex with n vertices, hence contractible, and

$\text{Hom}(K_n, K_n)$, which consists of $n!$ points, that is, a wedge of $n! - 1$ spheres of dimension 0. We assume now that $m \geq 2$ and $n \geq m + 1$.

For $i \in [m]$ let A_i be the subcomplex of $\text{Hom}(K_m, K_n)$ defined by

$$A_i = \{\eta : [m] \to 2^{[n]} \setminus \{\emptyset\} \mid n \notin \eta(j), \text{ for } j \in [m], j \neq i\}.$$

Since any two vertices of K_m are connected by an edge, n cannot be in $\eta(i_1) \cap \eta(i_2)$, for $i_1 \neq i_2$. This implies that $\bigcup_{i=1}^{m} A_i = \text{Hom}(K_m, K_n)$.

Clearly, for any $i \neq j$, $i, j \in [m]$, we have

$$A_i \cap A_j = \{\eta : [m] \to 2^{[n]} \setminus \{\emptyset\} \mid n \notin \eta(k), \text{ for all } k \in [m]\},$$

so $A_i \cap A_j$ is isomorphic to $\text{Hom}(K_m, K_{n-1})$; hence by induction, it is $(n - m - 2)$-connected.

We shall now see that each A_i is $(n - m - 1)$-connected. Since all A_i's are isomorphic to each other, it is enough to consider A_1. Let us describe a partial matching on $\mathcal{P}(A_1)$. For $\eta \in \mathcal{P}(A_1)$ such that $n \notin \eta(1)$, we set $\mu(\eta) := \tilde{\eta}$, defined by

$$\tilde{\eta}(i) = \begin{cases} \eta(1) \cup \{n\}, & \text{for } i = 1; \\ \eta(i), & \text{for } i = 2, 3, \ldots, m. \end{cases}$$

Obviously, this is an acyclic matching and the critical cells form a subcomplex $\widetilde{A} \subseteq A$ defined by $\eta \in \widetilde{A}$ if and only if $\eta(1) = \{n\}$. Thus $\widetilde{A} = \text{Hom}(K_{m-1}, K_{n-1})$. Since by Theorem 11.13, $\widetilde{A}$ is homotopy equivalent to A_1, and $\widetilde{A}$ is $(n - m - 1)$-connected by the induction assumption, we conclude that A_i is $(n - m - 1)$-connected for any i.

It follows from Theorem 15.24 that the space $\text{Hom}(K_m, K_n)$ is $(n - m - 1)$-connected, since the nerve of the covering is in fact contractible. On the other hand, the dimension of the complex $\text{Hom}(K_m, K_n)$ is $n - m$, and therefore it follows from Proposition 6.35 that $\text{Hom}(K_m, K_n)$ is homotopy equivalent to a wedge of spheres. $\square$

Introducing a new piece of notation, let us say that the complex $\text{Hom}(K_m, K_n)$ is homotopy equivalent to a wedge of $f(m, n)$ spheres. Let $S(-, -)$ denote the *Stirling numbers* of the second kind, and $SF_k(x) = \sum_{n \geq k} S(n, k) x^n$ denote the generating function (in the first variable) for these numbers. It is a well known fact from enumerative combinatorics that

$$SF_k(x) = x^k / ((1 - x)(1 - 2x) \cdots (1 - kx)).$$

For $m \geq 1$, let $F_m(x) = \sum_{n \geq 1} f(m, n) x^n$ be the generating function (in the second variable) for the number of spheres. Clearly, $F_1(x) = 0$ and $F_2(x) = x^2 / (1 - x)$.

Proposition 19.11. *The numbers $f(m, n)$ satisfy the following recurrence relation:*

$$f(m,n) = mf(m-1, n-1) + (m-1)f(m, n-1), \qquad (19.3)$$

for $n > m \geq 2$, with the boundary values $f(n,n) = n! - 1$, $f(1,n) = 0$ for $n \geq 1$, and $f(m,n) = 0$ for $m > n$.

Then the generating function $F_m(x)$ is given by the equation

$$F_m(x) = (m! \cdot x \cdot SF_{m-1}(x) - x^m)/(1 + x). \qquad (19.4)$$

As a consequence, the following nonrecursive formulas are valid:

$$f(m,n) = (-1)^{m+n+1} + m!(-1)^n \sum_{k=m}^{n} (-1)^k S(k-1, m-1) \qquad (19.5)$$

and

$$f(m,n) = \sum_{k=1}^{m-1} (-1)^{m+k+1} \binom{m}{k+1} k^n, \qquad (19.6)$$

for $n \geq m \geq 1$.

Proof. Let $\chi(m,n)$ be shorthand notation for the nonreduced Euler characteristics of the complexes $\mathrm{Hom}\,(K_m, K_n)$, and for $i = 1, \ldots, m$, let the subcomplex A_i be as in the proof of Theorem 19.10. Since we know that $\mathrm{Hom}\,(K_m, K_n) = \bigcup_{i=1}^{m} A_i$, $A_i \cap A_j = \mathrm{Hom}\,(K_m, K_{n-1})$, for all $i \neq j$, and $A_i \simeq \mathrm{Hom}\,(K_{m-1}, K_{n-1})$, for $i \in [m]$, we can conclude using simple inclusion exclusion counting that

$$\chi(m,n) = m\chi(m-1, n-1) - (m-1)\chi(m, n-1), \qquad (19.7)$$

for $n > m \geq 2$, and that additionally, $\chi(n,n) = n!$, $\chi(1,n) = 1$, for $n \geq 1$. Since $\chi(m,n) = 1 + (-1)^{m-n} f(m,n)$, a simple computation shows the validity of the relation (19.3).

For $m \geq 1$, let $G_m(x) = \sum_{n \geq 1} \chi(m,n) x^n$. Multiplying each side of equation (19.7) by x^n and summing over all n yields $G_m(x) = m \cdot x \cdot G_{m-1}(x) - (m-1) \cdot x \cdot G_m(x)$, implying

$$G_m(x) = \frac{mx}{1 + (m-1)x} G_{m-1}(x),$$

for $m \geq 1$, and hence, since $G_0(x) = 1/(1-x)$, we get

$$G_m(x) = \frac{m! \cdot x^m}{(1-x)(1+x)(1+2x)\ldots(1+(m-1)x)}$$
$$= m! \cdot x \cdot (-1)^{m-1} \cdot SF_{m-1}(-x)/(1-x),$$

for $m \geq 0$. By multiplying the identity $f(m,n) = (-1)^{m+n}(\chi(m,n) - 1)$ by x^n and summing over all $n \geq m$, we get

$$F_m(x) = (-1)^m G_m(-x) - x^m/(1+x)$$
$$= (-1)^m \cdot m! \cdot (-x) \cdot (-1)^{m-1} \cdot SF_{m-1}(x)/(1+x) - x^m/(1+x)$$
$$= (m! \cdot x \cdot SF_{m-1}(x) - x^m)/(1+x).$$

Equation (19.5) follows now from comparing the coefficients in equation (19.4). To prove equation (19.6) we see that it fits the boundary values and satisfies the recurrence relation (19.3). Verifying relation (19.3) is straightforward, and so is checking that equation (19.6) holds for $m = 1$ and for $m = 2$. Finally, the validity of equation (19.6) for $n = m \geq 2$ can be seen by expanding the expression $(e^x - 1)^n \cdot e^{-x}$ by the binomial theorem and comparing the coefficient of x^n on both sides of the expansion. $\square$

In particular, for small values of m, we obtain the following explicit formulas: $f(2, n) = 1$, for $n \geq 2$, $f(3, n) = 2^n - 3$, for $n \geq 3$, $f(4, n) = 3^n - 4 \cdot 2^n + 6$, for $n \geq 4$, $f(5, n) = 4^n - 5 \cdot 3^n + 10 \cdot 2^n - 10$, for $n \geq 5$.

We can now use Theorem 19.10 to give lower bounds for chromatic numbers of graphs in terms of Stiefel–Whitney classes of complexes of graph homomorphisms from complete graphs.

Theorem 19.12. *Let G be a graph, and let $n, k \in \mathbb{Z}$ such that $n \geq 2$, $k \geq -1$. If $\varpi_1^k(\mathrm{Hom}\,(K_n, G)) \neq 0$, then $\chi(G) \geq k + n$.*

Proof. After substituting $T = K_n$ and $m = k + n - 1$ in Theorem 19.1, all we need to do is to see that $\varpi_1^k(\mathrm{Hom}\,(K_n, K_{k+n-1})) = 0$. By Theorem 19.10, the complex $\mathrm{Hom}\,(K_n, K_{k+n-1})$ is homotopy equivalent to a wedge of $(k-1)$-dimensional spheres. Hence, by dimensional reasons we conclude that $\varpi_1^k(\mathrm{Hom}\,(K_n, K_{k+n-1})) = 0$. $\square$

19.2.2 Odd Cycles as Stiefel–Whitney Test Graphs

Recall that for $r \in \mathbb{N}$, we let C_{2r+1} denote both the cyclic graph with $2r + 1$ vertices and the additive cyclic group with $2r + 1$ elements. The adjacent vertices of $v \in C_{2r+1}$ get labels $v + 1$ and $v - 1$. Taking the negative in the cyclic group gives an involution γ of the graph with a fixed vertex 0 and a flipped edge $(r, r + 1)$.

The study of the complexes $X_{r,n} := \mathrm{Hom}\,(C_{2r+1}, K_n)$, where $r \geq 1$ and $n \geq 3$, has been of special interest. Under the involution above, $X_{r,n}$ is a $\mathbb{Z}_2$-space; hence the Stiefel–Whitney characteristic class $\varpi_1(X_{r,n}) \in H^1(X_{r,n}/\mathbb{Z}_2; \mathbb{Z}_2)$ of the associated line bundle can be considered.

Theorem 19.13. (Babson–Kozlov conjecture).
For all integers r and n such that $r \geq 1$, $n \geq 3$, we have

$$\varpi_1^{n-2}(X_{r,n}) = 0. \tag{19.8}$$

We shall now give a short self-contained combinatorial proof of Theorem 19.13 by taking a concrete cochain representative of $\varpi_1^{n-2}(X_{r,n})$ and certifying that it is a coboundary of another cochain for which we shall provide a combinatorial description.

First we fix notation. For an arbitrary cell complex X, we let X^d denote the set of d-dimensional cells of X (this is different from taking a d-skeleton). Since we are working with coefficients from $\mathbb{Z}_2$, we may identify d-cochains with their support subsets of X^d. Under this identification, the cochain addition is replaced by the symmetric difference of sets, which we denote by the symbol $\oplus$. The coboundary operator translates to

$$\partial S = \bigoplus_{\sigma \in S} \{\tau \in X_{r,n}^{d+1} \mid \tau \supset \sigma\},$$

for an arbitrary subset $S \subseteq X_{r,n}^d$. For any $v \in C_{2r+1}$, we set

$$A_v := \{\sigma \in X_{r,n}^{n-2} \mid \sigma(v) = [n-1]\}$$

and

$$B_v := \{\sigma \in X_{r,n}^{n-3} \mid \sigma(v-1) \cup \sigma(v+1) = [n-1]\}.$$

For $S \subseteq X_{r,n}^d$, set

$$q(S) := \bigoplus_{\sigma \in S} \{\mathbb{Z}_2 \sigma\} \in C^d(X_{r,n}/\mathbb{Z}_2),$$

where $\mathbb{Z}_2 \sigma = \{\sigma, \gamma\sigma\}$. We see that $q(A_0) = \emptyset$, because A_0 is symmetric with respect to the $\mathbb{Z}_2$-action. Also, we clearly have $q(S \oplus T) = q(S) \oplus q(T)$, for any $S, T \subseteq X_{r,n}^d$. Finally, since $\tau \cap \gamma\tau = \emptyset$ for any cell $\tau \in X_{r,n}^{d+1}$, we have the commutation relation $q(\partial S) = \partial q(S)$ for any $S \subseteq X_{r,n}^d$.

It is easy to describe a cochain representing the cohomology class $\varpi_1^{n-2}(X_{r,n})$. To do that, let $\iota : K_2 \hookrightarrow C_{2r+1}$ be the graph homomorphism given by $\iota(1) = r$, $\iota(2) = r+1$, where $V(K_2) = \{1,2\}$. This induces an algebra homomorphism

$$\varphi : H^*(\mathrm{Bip}\,(K_n)/\mathbb{Z}_2; \mathbb{Z}_2) \to H^*(X_{r,n}/\mathbb{Z}_2; \mathbb{Z}_2).$$

It follows from Proposition 19.8 that $\mathrm{Bip}\,(K_n)/\mathbb{Z}_2 \cong \mathbb{RP}^{n-2}$. Let us choose $\tau \in \mathrm{Bip}\,(K_n)^{n-2}$, which is given by $\tau(1) = [n-1]$, $\tau(2) = \{n\}$. Since the cohomology class that is represented by the dual of an arbitrary cell generates the group $H^{n-2}(\mathbb{RP}^{n-2}; \mathbb{Z}_2)$, we have $\varpi_1^{n-2}(\mathrm{Bip}\,(K_n)) = [\{\mathbb{Z}_2 \tau\}]$. By functoriality of Stiefel–Whitney characteristic classes, we get

$$\varpi_1^{n-2}(X_{r,n}) = [\varphi(\{\mathbb{Z}_2 \tau\})].$$

Comparing this to our notations, we derive

$$\varpi_1^{n-2}(X_{r,n}) = [q(A_r)]. \tag{19.9}$$

The next lemma provides the crucial combinatorial ingredient needed for the proof of Theorem 19.13.

Lemma 19.14. *We have $\partial B_v = A_{v-1} \oplus A_{v+1}$, for any $v \in C_{2r+1}$.*

Proof. According to our construction, the cells in ∂B_v are obtained by taking a cell $\sigma \in B_v$ and adding x to $\sigma(w)$, for some $x \in [n]$, $w \in C_{2r+1}$. We consider two cases.

Case 1. When $w \neq v \pm 1$, we get a cell τ that appears in ∂B_v twice: once in $\partial\sigma_1$ and once in $\partial\sigma_2$, where σ_1, σ_2 are obtained from τ by deleting one of the elements from $\tau(w)$.

Case 2. When $w = v \pm 1$, we also get a cell τ that appears in ∂B_v twice: once in $\partial\sigma_1$ and once in $\partial\sigma_2$, where σ_1, σ_2 are obtained from τ by deleting $\{x\} = \tau(v-1) \cap \tau(v+1)$ either from $\tau(v-1)$ or from $\tau(v+1)$, unless of course we have $|\tau(v-1)| = 1$ or $|\tau(v+1)| = 1$.

The latter cells in Case 2 appear once in ∂B_v and yield the desired term $A_{v-1} \oplus A_{v+1}$. $\square$

We are now ready to give the proof of the theorem.

Proof of Theorem 19.13.
To start with, we take the shortest way to get from the vertex labeled r to the vertex labeled 0 in the graph C_{2r+1}, where each step consists in jumping over a vertex in the graph. The direction in which we should start jumping depends on the parity of r. Accordingly, if r is even, set $t := r/2$ and $v_i := r - 2i + 1$, for $i \in [t]$; otherwise, set $t := (r+1)/2$ and $v_i := r + 2i - 1$, for $i \in [t]$.

We find the cochain, whose coboundary equals the representative of the appropriate power of the Stiefel–Whitney characteristic class by setting

$$K := \bigoplus_{i=1}^{t} q(B_{v_i}). \tag{19.10}$$

Indeed, a straightforward coboundary computation yields

$$\partial K = \bigoplus_{i=1}^{t} \partial q(B_{v_i}) = \bigoplus_{i=1}^{t} q(\partial B_{v_i})$$

$$= \bigoplus_{i=1}^{t} (q(A_{v_i-1}) \oplus q(A_{v_i+1})) = q(A_r) \oplus q(A_0) = q(A_r),$$

hence the sequence of equalities

$$\varpi_1^{n-2}(X_{r,n}) = [q(A_r)] = [\partial K] = 0$$

finishes the proof. $\square$

An example of the cochain K defined by equation (19.10) is shown in Figure 19.2.

The following theorem is an important corollary of Theorem 19.13.

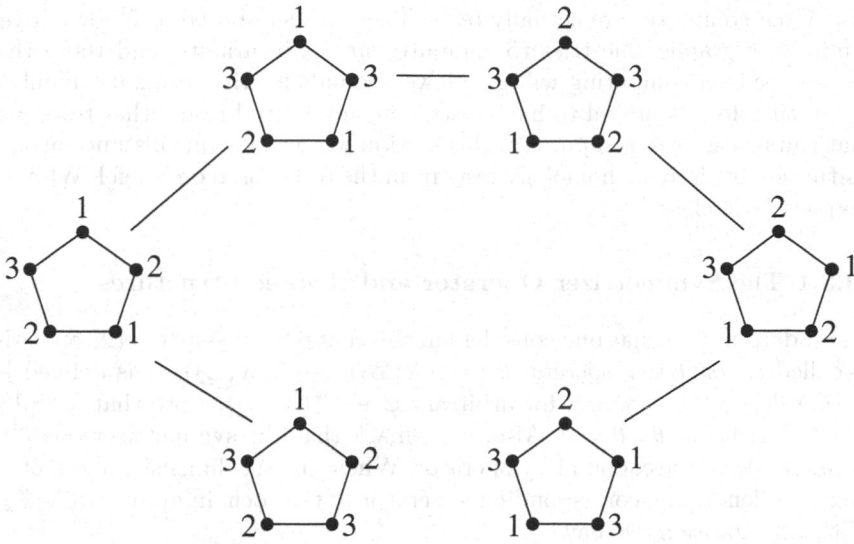

Fig. 19.2. A cochain whose coboundary equals the power of the characteristic class.

Theorem 19.15. (Lovász conjecture).
For an arbitrary graph G and any integers r and k such that $r \geq 1$, $k \geq -1$, we have the following implication:

if the complex $\mathrm{Hom}\,(C_{2r+1}, G)$ is k-connected, then $\chi(G) \geq k + 4$.

Proof. Assume that $\chi(G) \leq k + 3$, which means that there exists a graph homomorphism $\varphi : G \to K_{k+3}$. By functoriality of the Hom construction this yields a $\mathbb{Z}_2$-map

$$\psi : \mathrm{Hom}\,(C_{2r+1}, G) \to \mathrm{Hom}\,(C_{2r+1}, K_{k+3}).$$

On the other hand, if $\mathrm{Hom}\,(C_{2r+1}, G)$ is k-connected, then according to Corollary 8.26 there exists a $\mathbb{Z}_2$-map $\rho : \mathbb{S}_a^{k+1} \to \mathrm{Hom}\,(C_{2r+1}, G)$. For the characteristic classes this gives

$$0 = w^{k+1}(\mathrm{Hom}\,(C_{2r+1}, K_{k+3})) \longrightarrow w^{k+1}(\mathrm{Hom}\,(C_{2r+1}, G))$$
$$\longrightarrow w^{k+1}(\mathbb{S}_a^{k+1}) \neq 0,$$

thus yielding a contradiction. $\square$

19.3 Homology Tests for Graph Colorings

It is important to stress that most of the results that provide lower bounds for the chromatic numbers of graphs in terms of topological invariants for some associated complexes are actually tests. These tests are used as follows: given a family of graphs, one tries to compute various invariants, and those that one succeeds in computing will give lower bounds for the chromatic number. It can therefore be useful to have tests that are derived from other tests, but that can be easier to compute. In this section we describe one instance of such a situation by deriving homology tests from the tests based on Stiefel–Whitney characteristic classes.

19.3.1 The Symmetrizer Operator and Related Structures

A standard notion that one considers in the context of $\mathbb{Z}_2$-spaces (X, γ) is the so-called *symmetrizer operator* $\theta : C^*(X; \mathbb{Z}_2) \to C^*(X; \mathbb{Z}_2)$. It is defined by setting $\theta(\varphi) := \varphi + \gamma^*(\varphi)$, for arbitrary $\varphi \in C^*(X; \mathbb{Z}_2)$. Note that $\gamma^* \circ \theta = \theta \circ \gamma^* = \theta$; hence $\theta \circ \theta = 0$. Also, we remark that the symmetrizer operator commutes with the coboundary operator. When c is an i-dimensional cell of X, we let c^* denote the corresponding generator of the cochain group $C^i(X; \mathbb{Z}_2)$. With this notation, we have

$$q^*(q(c)^*) = \theta(c^*), \tag{19.11}$$

for an arbitrary i-cell c. Indeed, $\theta(c^*) = c^* + \gamma^*(c^*) = c^* + \gamma(c)^*$. The latter cochain evaluates to 1 on c and on $\gamma(c)$, and to 0 on other cells, which by definition is the same as the left-hand side of (19.11).

An easy but important observation for us is that for an arbitrary $\mathbb{Z}_2$-space X, we have

$$\mathrm{Im}(\theta) = \mathrm{Im}(q^*) = C^*_{\mathbb{Z}_2}(X; \mathbb{Z}_2). \tag{19.12}$$

The crucial fact needed to see (19.12) is that the involution γ is fixed-point-free. Indeed, for any nonnegative integer i, the vector space $C^i_{\mathbb{Z}_2}(X; \mathbb{Z}_2)$ has a basis consisting of $\theta(c^*)$, where c ranges over a set of i-dimensional cells of X, obtained by choosing exactly one cell from each orbit of the $\mathbb{Z}_2$-action on the set of all i-dimensional cells of X. This shows that $\mathrm{Im}(\theta) = C^*_{\mathbb{Z}_2}(X; \mathbb{Z}_2)$. On the other hand, equation (19.11) implies immediately that $\mathrm{Im}(q^*) \subseteq \mathrm{Im}(\theta)$. In fact, one sees further that $q^* : C^*(X/\mathbb{Z}_2; \mathbb{Z}_2) \to \mathrm{Im}(\theta)$ is an isomorphism. We let $p^* : \mathrm{Im}(\theta) \to C^*(X/\mathbb{Z}_2; \mathbb{Z}_2)$ denote its inverse.

19.3.2 The Topological Rationale for the Tests

Theorem 19.16. *Let X be a nonempty $\mathbb{Z}_2$-space with finite Stiefel–Whitney height. Then we have $\widetilde{H}^{h(X)}(X; \mathbb{Z}_2) \neq 0$.*

As a separate remark, we note that Theorem 19.16 does not get stronger if we, as would be natural to do in full parallel with the theorem of Walker, write the invariant cohomology $H^{h(X)}_{\mathbb{Z}_2}(X; \mathbb{Z}_2)$ instead of $H^{h(X)}(X; \mathbb{Z}_2)$. This is true because *if V is a nontrivial vector space over $\mathbb{Z}_2$, and γ is an involution of V, then γ fixes a nontrivial subspace of V.* To see the latter fact, simply take any nonzero element $x \in V$, and notice that either $x + \gamma(x)$ is a nonzero vector fixed by γ, or else $x + \gamma(x) = 0$, and hence $x = \gamma(x)$, and so x is a nonzero vector fixed by γ. As a matter of fact, the cohomology class that our proof of Theorem 19.16 produces will come out to be $\mathbb{Z}_2$-invariant anyway.

Proof of Theorem 19.16. For convenience of notation we set $d := \mathrm{h}(X)$. Let $\varphi : X \to \mathbb{S}_a^\infty$ be a cellular $\mathbb{Z}_2$-map, and consider the commuting diagram of topological spaces and continuous maps shown in Figure 19.3, where the vertical arrows correspond to quotient maps.

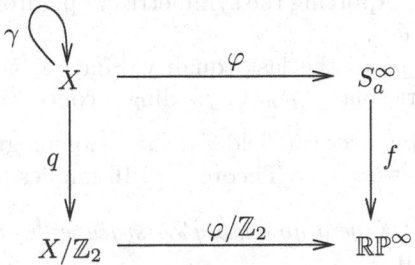

Fig. 19.3. The commuting diagram of topological spaces.

It induces a commuting diagram of cochain $\mathbb{Z}_2$-algebras, shown in Figure 19.4.

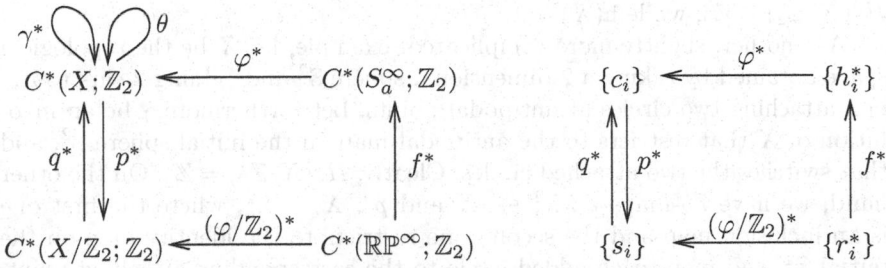

Fig. 19.4. The commuting diagram of cochain algebras, and the special cochain elements.

For all nonnegative integers i, we have $\theta(h_i^*) = f^*(r_i^*)$, and we set $c_i := \varphi^*(h_i^*)$ and $s_i := (\varphi/\mathbb{Z}_2)^*(r_i^*)$; see Figure 19.4. Then we have $\partial c_i = \theta c_{i+1}$, since φ^* is $\mathbb{Z}_2$-invariant, and furthermore, $q^*(s_i) = \theta c_i$ for all i, by commutativity of the diagram in Figure 19.4.

By our construction, $[s_{d+1}]$ is a trivial class in $H^{d+1}(X/\mathbb{Z}_2; \mathbb{Z}_2)$, i.e., $s_{d+1} = \partial\zeta$, for some $\zeta \in C^d(X/\mathbb{Z}_2; \mathbb{Z}_2)$. It follows that

$$\theta c_{d+1} = q^*(s_{d+1}) = q^*(\partial \zeta) = \partial(q^* \zeta) = \partial(\theta \tau),$$

for some $\tau \in C^d(X; \mathbb{Z}_2)$.

The calculation $\partial(c_d + \theta \tau) = \theta c_{d+1} + \partial(\theta \tau) = 0$ shows that $c_d + \theta \tau$ is a cocycle. On the other hand,

$$\gamma^*(c_d + \theta \tau) = \gamma^*(c_d) + \theta \tau = \theta c_d + c_d + \theta \tau = \partial c_{d-1} + (c_d + \theta \tau);$$

hence additionally, the cohomology class $[c_d + \theta \tau]$ is $\mathbb{Z}_2$-invariant.

If the class $[c_d + \theta \tau]$ is trivial, then there exists a cochain $\eta \in C^{d-1}(X; \mathbb{Z}_2)$ such that $\partial \eta = c_d + \theta \tau$. Applying the symmetrizer operator, we obtain $\partial(\theta \eta) = \theta(\partial \eta) = \theta(c_d + \theta \tau) = \theta c_d$.

Finally, we apply p^* to the last equality. Since $p^*(\theta c_d) = s_d$, we obtain $s_d = \partial(p^*(\theta \eta))$, in particular $[s_d] = 0$, yielding a contradiction. $\square$

Since we are working over the field $\mathbb{Z}_2$, the cohomology groups are isomorphic to the homology groups, so Theorem 19.16 implies the following result.

Corollary 19.17. *Let X be a nonempty $\mathbb{Z}_2$-space with finite height. Then we have $\widetilde{H}_{\mathrm{h}(X)}(X; \mathbb{Z}_2) \neq 0$.*

Furthermore, Theorem 19.16 is optimal in the sense that its converse does not hold. In other words, it is not true that the minimal dimension in which the reduced cohomology with $\mathbb{Z}_2$-coefficients of the space X is nontrivial is equal to the Stiefel–Whitney height of X.

As a first example, we may take the space X to consist of two points and a circle, and let γ swap the points and act antipodally on the circle. Clearly, $\widetilde{H}^0(X; \mathbb{Z}_2) = \mathbb{Z}_2^2$, while $\mathrm{h}(X) = 1$.

As another, slightly more complicated, example, let X be the topological space obtained by taking a 2-dimensional sphere $\mathbb{S}^2$ and "gluing 2 ears to it," i.e., attaching two circles at antipodal points. Let furthermore γ be an involution of X that restricts to the antipodal map on the initial sphere $\mathbb{S}^2$, and that switches the two attached circles. Clearly, $\widetilde{H}^1(X; \mathbb{Z}_2) = \mathbb{Z}_2^2$. On the other hand, we have $\mathbb{Z}_2$-maps $i : \mathbb{S}_a^2 \hookrightarrow X$ and $p : X \twoheadrightarrow \mathbb{S}_a^2$, where the first one is an inclusion map and the second one restricts to the identity map on the initial $\mathbb{S}^2$, and maps each added circle to the corresponding attaching point. It follows that $\mathrm{h}(X) = 2$.

19.3.3 Homology Tests

Theorem 19.16 implies the following homological test for graph colorings.

Theorem 19.18. *Assume that T is a graph with a $\mathbb{Z}_2$-action that flips an edge such that additionally,*

(1) T is a Stiefel–Whitney test graph,
(2) $\widetilde{H}_i(\mathrm{Hom}\,(T, G); \mathbb{Z}_2) = 0$, for $i \leq d$.

Then $\chi(G) \geq d + 1 + \chi(T)$.

Proof. If $\tilde{H}_i(\text{Hom}\,(T,G);\mathbb{Z}_2) = 0$, for all $i \leq d$, then by Corollary 19.17 we have $\text{h}(\text{Hom}\,(T,G)) \geq d + 1$. Substituting this into Corollary 19.4, we obtain $\chi(G) \geq \chi(T) + \text{h}(\text{Hom}\,(T,G)) \geq \chi(T) + d + 1$. $\square$

19.3.4 Examples of Homology Tests with Different Test Graphs

We shall now look at different examples of using homology tests.

Example 1. Let G be the disjoint union of an edge and a triangle.

The complex Bip (G) is a disjoint union of two isolated points and a circle. It follows that $\tilde{H}_0(\text{Bip}\,(G)) = \mathbb{Z}_2^2$; hence the best value of d for $T = K_2$ in Theorem 19.18 is $d = -1$. Thus the bound given by the homology test using K_2 is $d + 2 + 1 = 2$; see Figure 19.5.

On the other hand, the complex Hom (K_3, G) is a disjoint union of six isolated points. It follows that $\tilde{H}_0(\text{Hom}\,(K_3, G)) = \mathbb{Z}_2^5$; hence the best value of d for $T = K_3$ in Theorem 19.18 is $d = -1$. Thus the bound given by the homology test using K_3 is $d + 3 + 1 = 3$, which is in fact equal to the chromatic number of G. Again, we refer to Figure 19.5.

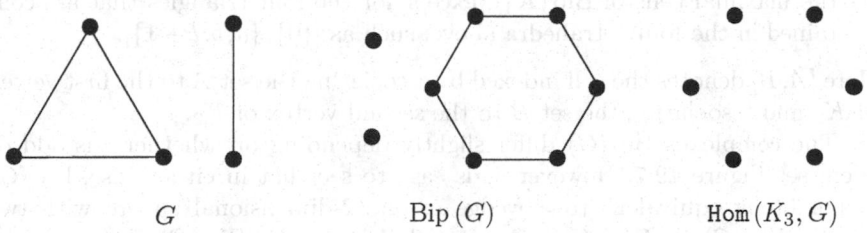

$$G \qquad\qquad \text{Bip}\,(G) \qquad\qquad \text{Hom}\,(K_3, G)$$

Fig. 19.5. The graph tested in Example 1 and the corresponding Hom complexes.

Furthermore, we remark that Hom (C_5, G) consists of two disjoint cycles; hence the homology test with C_5 as a test graph also yields the optimal bound $\chi(G) = 3$.

Example 2. Let G be obtained by taking a complete graph on four vertices K_4 and attaching a path of length $l + 1$ by its endpoints to two of the vertices of this K_4; see Figure 19.6.

Assume that $l \geq 4$, label the vertices of the added path sequentially 1 through l, label the vertices of the initial K_4 to which the path has been attached by 0 and $l + 1$ respectively, and finally label the two remaining vertices of K_4 by a and b.

Recall, that Bip (K_4) is the boundary of the 3-dimensional polytope shown in Figure 9.6. Let us understand the change in this complex that occurs when we pass from K_4 to G. The maximal cells of Bip (G) are:

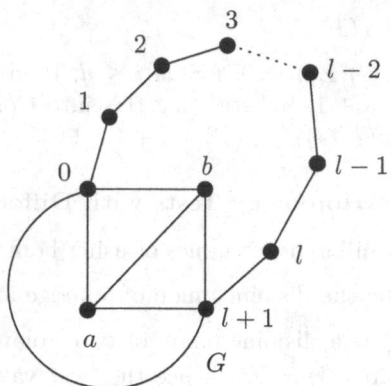

Fig. 19.6. The graph tested in Example 2.

- four tetrahedra $[\{0\}, \{a, b, l+1, 1\}]$, $[\{a, b, l+1, 1\}, \{0\}]$, $[\{l+1\}, \{0, a, b, l\}]$, $[\{0, a, b, l\}, \{l+1\}]$;
- $2l$ edges $[\{1\}, \{0, 2\}]$, $[\{0, 2\}, \{1\}]$, $[\{2\}, \{1, 3\}]$, $[\{1, 3\}, \{2\}]$, ..., $[\{l\}, \{l-1, l+1\}]$, $[\{l-1, l+1\}, \{l\}]$;
- the maximal cells of $\mathrm{Bip}\,(K_4)$, except for the four triangles that are contained in the four tetrahedra above, such as $[\{0\}, \{a, b, l+1\}]$.

Here $[A, B]$ denotes the cell indexed by associating the set A to the first vertex of K_2 and associating the set B to the second vertex of K_2.

The complexes $\mathrm{Bip}\,(G)$ differ slightly depending on whether l is odd or even; see Figure 19.7. However, it is easy to see that in either case, $\mathrm{Bip}\,(G)$ is homotopy equivalent to a wedge of one 2-dimensional sphere with two circles: $\mathrm{Bip}\,(G) \simeq \mathbb{S}^2 \vee \mathbb{S}^1 \vee \mathbb{S}^1$. It follows that $\widetilde{H}_0(\mathrm{Bip}\,(G); \mathbb{Z}_2) = 0$ and $\widetilde{H}_1(\mathrm{Bip}\,(G); \mathbb{Z}_2) = \mathbb{Z}_2^2$. Hence the best value of d for $T = K_2$ in Theorem 19.18 is $d = 0$. Thus the bound given by the homology test using K_2 is $d+2+1 = 3$.

Consider now $T = K_3$. Since the attached path is sufficiently long, we see that the complex $\mathrm{Hom}\,(K_3, G)$ is actually isomorphic to $\mathrm{Hom}\,(K_3, K_4)$. Also taking $T = C_5$, we see that the 5-cycle cannot wrap around the attached path, and that in fact, $\mathrm{Hom}\,(C_5, G)$ is isomorphic to $\mathrm{Hom}\,(C_5, G')$, where G' is obtained from K_4 by attaching two edges: one edge to 0 and one to $l+1$. Since the graph G' folds to the graph K_4, we may conclude by Theorem 18.22 that $\mathrm{Hom}\,(C_5, G) \simeq \mathrm{Hom}\,(C_5, K_4)$.

Clearly, both complexes $\mathrm{Hom}\,(K_3, K_4)$ and $\mathrm{Hom}\,(C_5, K_4)$ are connected (in fact, the prodsimplicial complex $\mathrm{Hom}\,(C_5, K_4)$ is homeomorphic to $\mathbb{RP}^3$). Therefore we have $\widetilde{H}_0(\mathrm{Hom}\,(K_3, G); \mathbb{Z}_2) = \widetilde{H}_0(\mathrm{Hom}\,(C_5, G); \mathbb{Z}_2) = 0$. Hence the best value of d for $T = K_3$ or $T = C_5$ in Theorem 19.18 is $d = 0$. Thus the bound given by the homology test using K_3 or C_5 is $d + 3 + 1 = 4$, which matches the chromatic number of G.

Example 3. Let G be the graph depicted in Figure 19.8.

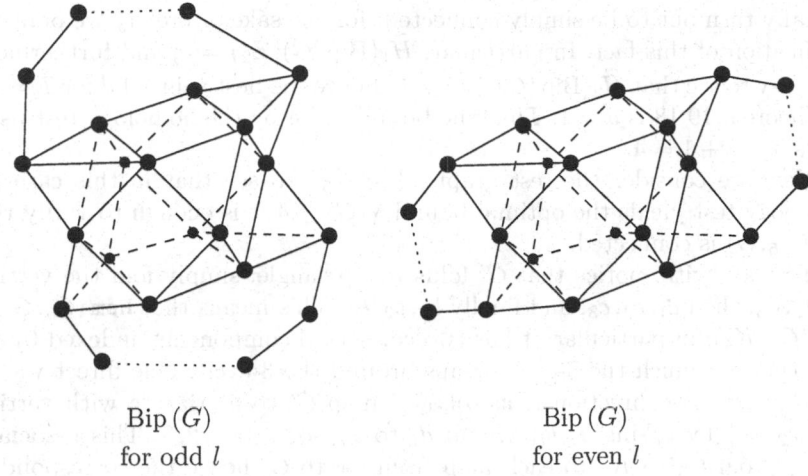

Bip (G)
for odd l

Bip (G)
for even l

Fig. 19.7. The Hom-complexes appearing in Example 2.

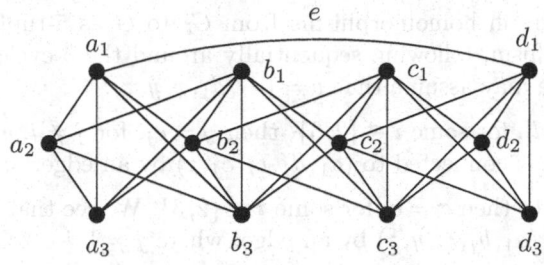

Fig. 19.8. The graph tested in Example 3.

It is easy to see that $\chi(G) = 4$. Let G' denote the graph obtained from G be deleting the edge e. We have $\chi(G') = 3$.

First we consider the homology test with the test graph $T = K_2$. Accordingly, let us analyze the structure of the cell complex Bip (G). Let v_1, v_2 denote the vertices of Bip (G) that are indexed by $[\{a_1\}, \{d_1\}]$ and $[\{d_1\}, \{a_1\}]$. Clearly, Bip (G) is obtained from Bip (G') by attaching two cones, with apexes in v_1 and v_2. The complex Bip (G') is connected, since G' is a connected graph with odd cycles.

Consider now the cone with apex v_1. It is attached to Bip (G') over the link of v_1 in Bip (G). It is easy to see that this link consists of two tetrahedra $[\{a_1\}, \{a_2, a_3, b_2, b_3\}]$, $[\{c_2, c_3, d_2, d_3\}, \{d_1\}]$, and two paths connecting these tetrahedra arising from the 2-cells $[\{a_1, c_3\}, \{b_2, d_1\}]$ and $[\{a_1, c_2\}, \{b_3, d_1\}]$. Similarly we can see the link of v_2 as well.

We see that to understand whether the first homology group of Bip (G) is trivial requires quite a bit of additional work. In this case the complex does actually turn out to be simply connected; for the sake of brevity we omit the verification of this fact. In particular, $\widetilde{H}_1(\text{Bip}\,(G); \mathbb{Z}_2) = 0$, and furthermore, it is easy to see that $\widetilde{H}_2(\text{Bip}\,(G); \mathbb{Z}_2) \neq 0$; hence the best value of d for $T = K_2$ in Theorem 19.18 is $d = 1$. Thus the bound given by the homology test using K_2 is $d + 2 + 1 = 4$.

Next we consider the test graph $T = C_5$. To see that in this case the homology test yields the optimal bound $\chi(G) \geq 4$ it is enough to verify that $\text{Hom}\,(C_5, G)$ is connected.

To start with, notice that G' folds to a triangle: simply fold the vertices d_1, d_2, d_3, then c_1, c_2, c_3, and finally b_1, b_2, b_3. This means that $\text{Hom}\,(C_5, G') \simeq \text{Hom}\,(C_5, K_3)$; in particular, it has two connected components, indexed by the directions in which the 5-cycle wraps around the 3-cycle. One direct way to see this winding direction is as follows: map G' to a triangle with vertices $\{x_1, x_2, x_3\}$ by taking a_i, b_i, c_i, and d_i to x_i, for $i = 1, 2, 3$. This associates a map from C_5 to K_3 to each map from C_5 to G; hence the corresponding winding direction is well-defined.

Consider now a vertex v of $\text{Hom}\,(C_5, G)$ (this is the same as a graph homomorphism from C_5 to G) that maps some edge of C_5 to the edge (a_1, d_1). We write the graph homomorphisms from C_5 to G as 5-tuples of values of this homomorphism, following sequentially around the 5-cycle. Without loss of generality we may assume that $v = (a_1, d_1, x, y, z)$.

Case 1. If $x = d_i$, for some $i \in \{2, 3\}$, then $y = c_j$, for $j \neq i$, and we see that (a_1, d_1, d_i, c_j, z) is connected to (a_1, d_1, c_i, c_j, z) by an edge.

Case 2. If $x \neq d_i$, then $x = c_i$ for some $i \in \{2, 3\}$. We see that (a_1, d_1, c_i, y, z) is connected to (a_1, b_j, c_i, y, z) by an edge, where $j \neq 1, i$.

In both cases we conclude that all the vertices of $\text{Hom}\,(C_5, G)$ are connected by a path to one of the vertices of $\text{Hom}\,(C_5, G')$. Since the latter has two connected components, we conclude that also the complex $\text{Hom}\,(C_5, G)$ has at most two connected components.

To see that the complex $\text{Hom}\,(C_5, G)$ is actually connected, we need to present a path that connects two vertices whose winding directions (which were defined above) are different. Such a path is given by the three vertices $(c_1, c_2, d_1, c_3, b_2)$, $(c_1, c_2, d_1, a_1, b_2)$, $(c_1, c_2, b_3, a_1, b_2)$; see Figure 19.9.

As mentioned above, it follows that the homology test with the test graph $T = C_5$ detects the chromatic number of G correctly.

As we have seen in the last example, even if the homology tests using an edge or an odd cycle yield the same bound for the chromatic number, it is often easier to verify this bound using the odd cycle, since the tests are done in one dimension lower, so we have to verify that something is connected instead of verifying that the all loops are boundaries, or we have to deal with loops instead of the 2-dimensional cycles, and so on. For these reasons, it appears in general to be preferable to test with graphs with high chromatic

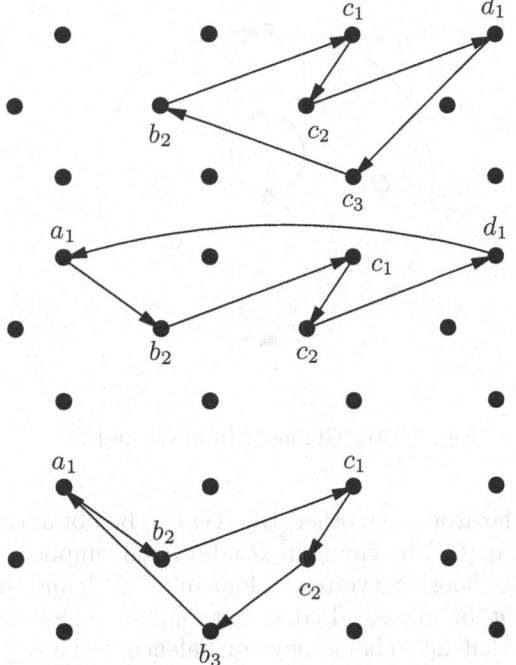

Fig. 19.9. The path connecting vertices with different winding directions.

number, although naturally there is no guarantee that these tests will give sharp bounds.

We finish by describing a large class of graphs for which the homology tests with the test graph K_3 produce a better answer than the ones with the test graph K_2. Assume that G_1 and G_2 are connected graphs such that $\chi(G_1) \geq \chi(G_2) \geq 3$ and G_2 is triangle-free. Let G be obtained by identifying the distinct vertices $v_1, \ldots, v_t \in V(G_1)$ with some t distinct vertices of G_2, where t is arbitrary; see Figure 19.10. Finally, assume that the shortest path in G connecting vertices v_i and v_j has at least three edges, for any $1 \leq i < j \leq t$.

Proposition 19.19. *Let G, G_1, and G_2 be graphs satisfying the conditions above. Then the homology test for G with the test graph K_3 gives the same bound as the homology test for G_1 with the test graph K_3. On the other hand, the homology test for G with the test graph K_2 gives the bound 3.*

Proof. First, we see that $\mathrm{Hom}\,(K_3, G) = \mathrm{Hom}\,(K_3, G_1)$, since G_2 is assumed to be triangle-free. This means that the homology test for G with K_3 as the test graph gives the same bound as the analogous test for G_1.

On the other hand, the prodsimplicial complex $\mathrm{Bip}\,(G)$ has nontrivial homology already in dimension 1. Indeed, since the vertices $v_1, \ldots, v_t$ are chosen

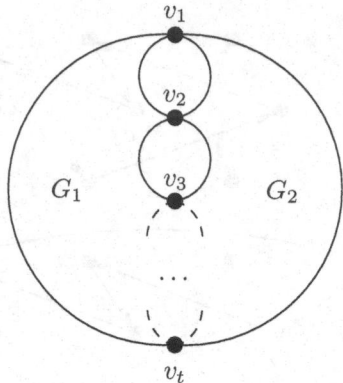

Fig. 19.10. Gluing G from G_1 and G_2.

to be sufficiently far from each other, Bip (G) can be obtained from the union of Bip (G_1) and Bip (G_2) by gluing in $2t$ additional simplices. Each such simplex is obtained by choosing a vertex v_i, for some $i \in [t]$, and then either taking all edges leaving v_i or taking all edges entering v_i. These simplices are disjoint, and we see that up to homotopy equivalence, the effect of adding these simplices is the same as that of attaching $2t$ cords connecting the complexes Bip (G_1) and Bip (G_2).

Since these initial complexes are connected, we can conclude that Bip (G) is homotopy equivalent to the wedge of the complexes Bip (G_1), Bip (G_2), and $2t-1$ copies of $\mathbb{S}^1$. Thus, no matter what the chromatic numbers of the graphs G, G_1, and G_2 are, the homology test using K_2 as the test graph will give only the bound 3 for $\chi(G)$. $\square$

19.4 Bibliographic Notes

Our terminology of various test graphs developed in Subsection 19.1 follows essentially the survey [Ko05a]. The reader is invited to consult this source for further details.

The proof Proposition 19.8 concerning the complex of bipartite subgraphs of a complete graph is taken from [BK06, Proposition 4.3], though the statement itself was known long before that. Theorem 19.10 was formulated and proved for the first time in [BK06, Proposition 4.5], where Theorem 19.12 was also derived as a corollary. In the same paper, Proposition 19.11 was proved as [BK06, Proposition 4.6].

The Lovász conjecture (here Theorem 19.15) was from the very beginning an important motivation for developing the theory of Hom complexes. It was originally settled by Babson and the author in a series of papers [BK03, BK06, BK04].

The Babson–Kozlov conjecture (here Theorem 19.13) stating that odd cycles are Stiefel–Whitney test graphs was first formulated in [BK04, Ko05a]. The case $r = 1$ of Theorem 19.13 was settled by Babson and the author already in [BK06]. For $r \geq 2$ and odd n, it was proved by the same authors in [BK04]; see also [Ko05a], where the remaining case, $r \geq 2$, $n \geq 4$, n is even, was conjectured. The latter was then proved by Schultz in [Su05a, Su06]. The proof presented here is adapted from [Ko06d], where an extremely short and purely combinatorial proof of the Babson–Kozlov conjecture was found by the author.

Our presentation in Section 19.3 is adapted from [Ko07], where the idea of running various topological tests for obtaining lower bounds for chromatic numbers of graphs was further developed. The topological result, Theorem 19.16, generalizes a result of Walker, [Wa83]. Even in the special case considered in [Wa83] the proof given here is simpler, bypassing the homotopy considerations and dealing directly with the cohomology groups.

Applications of Spectral Sequences to Hom Complexes

20.1 Hom$_+$ Construction

20.1.1 Various Definitions

We shall now define an abstract simplicial complex called $\text{Hom}_+(T,G)$, which is closely related to $\text{Hom}(T,G)$. It is easier to compute various algebro-topological invariants for this complex. We shall then connect the Hom and Hom$_+$ constructions by means of a spectral sequence.

A subcomplex of a total join

Let T and G be arbitrary graphs. We shall define $\text{Hom}_+(T,G)$ analogously to $\text{Hom}(T,G)$, replacing the direct product with the join. We note here that whenever we talk about $\text{Hom}_+(T,G)$, we always assume that the graph T is finite.

Let, as before, $\Delta^{V(G)}$ be a simplex whose set of vertices is $V(G)$. Let $J(T,G)$ denote the join $*_{x \in V(T)} \Delta^{V(G)}$, i.e., the copies of $\Delta^{V(G)}$ are indexed by vertices of T. A cell (simplex) in $J(T,G)$ is a join of (possibly empty) simplices $*_{x \in V(T)} \sigma_x$; the dimension of this simplex is $\sum_{x \in V(T)} (\dim \sigma_x + 1) - 1$. Observe that this number is finite, since we assumed that T is finite. Here we use the usual convention that $\dim \emptyset = -1$.

Definition 20.1. *For arbitrary graphs T and G, $\text{Hom}_+(T,G)$ is the simplicial subcomplex of $J(T,G)$ defined by the following condition: $\sigma = *_{x \in V(T)} \sigma_x \in \text{Hom}_+(T,G)$ if and only if for any $x,y \in V(T)$, if $(x,y) \in E(T)$, and both σ_x and σ_y are nonempty, then (σ_x, σ_y) is a complete bipartite subgraph of G.*

The intuition behind this definition is that we relax the conditions of the Hom case by allowing some of the "coloring lists" to be empty. One can think of $\text{Hom}_+(T,G)$ as a simplicial structure imposed on the set of all *partial graph homomorphisms* from T to G, i.e., graph homomorphisms from an induced subgraph of T to the graph G.

In analogy with the Hom case, we can describe the simplices of $\mathrm{Hom}_+(T, G)$ directly: they are indexed by all $\eta : V(T) \to 2^{V(G)}$ satisfying the same condition as in Definition 9.23. The closure of η is also defined identically to how it was defined for Hom. So the only difference is that $\eta(x)$ is allowed to be an empty set, for $x \in V(T)$.

A link of a vertex in an auxiliary Hom complex

The following construction is the graph analogue of the topological coning.

Definition 20.2. *For an arbitrary graph G, let G_+ be the graph obtained from G by adding an extra vertex a, called the apex vertex, and connecting it by edges to all the vertices of G_+ including a itself, i.e., $V(G_+) = V(G) \cup \{a\}$, and $E(G_+) = E(G) \cup \{(x, a), (a, x) \mid x \in V(G_+)\}$.*

We note that for an arbitrary polyhedral complex K such that all faces of K are direct products of simplices, and a vertex x of K, the *link* of x, $\mathrm{lk}_K(x)$, is a simplicial complex. This follows from the fact that a link of any vertex in a hypercube is a simplex, and the identity $\mathrm{lk}_{(A \times B)}(v, w) = \mathrm{lk}_A(v) * \mathrm{lk}_B(w)$, for arbitrary polyhedral complexes A and B (compare with identities (10.2) and (10.14)).

We are now ready to formulate another definition, which is equivalent to Definition 20.1.

Definition 20.3. *For arbitrary graphs T and G, the simplicial complex $\mathrm{Hom}_+(T, G)$ is defined to be the link in $\mathrm{Hom}(T, G_+)$ of the specific graph homomorphism α that maps all vertices of T to the apex vertex of G_+; in short, $\mathrm{Hom}_+(T, G) = \mathrm{lk}_{\mathrm{Hom}(T, G_+)}(\alpha)$.*

The equivalence of the definitions follows essentially from the following bijection: let $\eta \in \mathcal{F}(\mathrm{Hom}(T, G_+))_{>\alpha}$, and set $\tilde{\eta}(v) := \eta(v) \setminus \{a\}$, for any $v \in V(T)$. Clearly, $\tilde{\eta} \in \mathcal{F}(\mathrm{Hom}_+(T, G))$, and it is easily checked that this bijection produces an isomorphism of simplicial complexes. See Figure 20.1 for an illustration.

Functorial properties of the Hom$_+$ construction

Just as in the case of the $\mathrm{Hom}(-, -)$ construction, $\mathrm{Hom}_+(T, -)$ is a covariant functor from **Graphs** to **Top**. For two arbitrary graphs G and K and an arbitrary graph homomorphism φ from G to K, we have an induced simplicial map $\varphi^T : \mathrm{Hom}_+(T, G) \to \mathrm{Hom}_+(T, K)$.

Again, as in the case of $\mathrm{Hom}(-, -)$, the situation is somewhat more complicated with the functoriality in the first argument. Let T, G, and K be three arbitrary graphs. This time, for a graph homomorphism ψ from T to G to induce a topological map from $\mathrm{Hom}_+(G, K)$ to $\mathrm{Hom}_+(T, K)$, we must require that ψ be surjective on the vertices. We can define the topological map ψ_K

in the same way as for the Hom$(-,-)$ case, but if ψ is not surjective on the vertices, then we may end up mapping a nonempty cell to an empty one. If, in addition, we want a simplicial map $\psi_K : \text{Hom}_+(G,K) \to \text{Hom}_+(T,K)$, then, as before in Subsection 18.3.3, we must require that ψ be injective, hence bijective, on the vertices.

In particular, we still have that the group $\text{Aut}(T) \times \text{Aut}(G)$ acts on the complex $\text{Hom}_+(T,G)$ simplicially. The difference is that we do not have the freeness as easily as we had in the Hom$(-,-)$ case. For example, for an involution γ of T to induce a free action γ_G on $\text{Hom}_+(T,G)$ we need to require that all orbits of γ on $V(T)$ be of cardinality 2, and that the vertices in the same orbit be connected by an edge. For instance, the action of $\mathbb{Z}_2$ on $\text{Hom}_+(K_2,G)$ is free, whereas the reflection $\mathbb{Z}_2$-action on $\text{Hom}_+(C_{2r+1},G)$ is not.

20.1.2 Connection to Independence Complexes

Let us introduce some further graph terminology.

Definition 20.4. *For an arbitrary graph G, the* **strong complement** $\mathbb{C}G$ *is defined by $V(\mathbb{C}G) = V(G)$, and $E(\mathbb{C}G) = V(G) \times V(G) \setminus E(G)$.*

For example, $\mathbb{C}K_n$ is the disjoint union of n loops. This coincides with the notation $\mathbb{C}K_1$, which we have already used in Subsection 4.2.1 to denote the terminal object in the category **Graphs**.

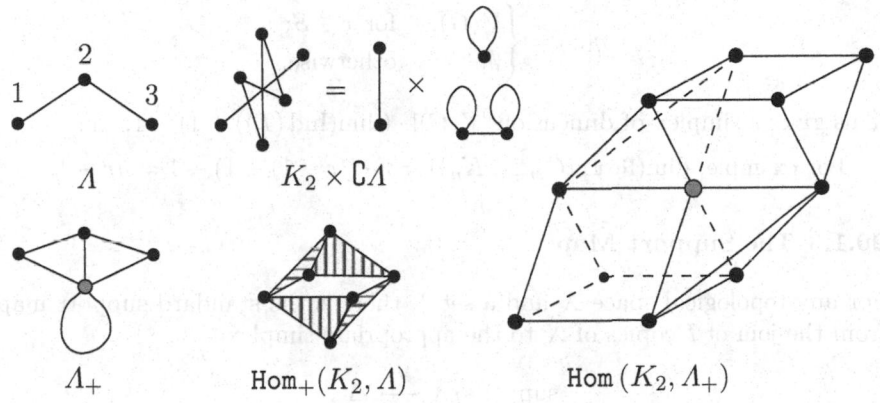

Λ $K_2 \times \mathbb{C}\Lambda$

Λ_+ $\text{Hom}_+(K_2, \Lambda)$ $\text{Hom}(K_2, \Lambda_+)$

Fig. 20.1. The +-construction.

We have defined the independence complexes of graphs in Subsection 9.1.1, Definition 9.2. Sometimes, it can be convenient to view the simplicial complex $\text{Hom}_+(G,H)$ as the independence complex of a certain graph.

Proposition 20.5. *For arbitrary graphs T and G, the simplicial complex $\text{Hom}_+(T,G)$ is isomorphic to the simplicial complex $\text{Ind}(T \times \mathbb{C}G)$.*

Specializing Proposition 20.5 to $G = K_n$, and taking into account $\mathsf{C}K_n = \coprod_{i=1}^{n} \mathsf{C}K_1$ (observed above) and the fact that for arbitrary graphs G_1 and G_2 we have

$$\text{Ind}\,(G_1 \coprod G_2) = \text{Ind}\,(G_1) * \text{Ind}\,(G_2),$$

we obtain the following corollary.

Corollary 20.6. *For an arbitrary graph T, the simplicial complex $\text{Hom}_+(T, K_n)$ is isomorphic to the n-fold join of the independence complex of T, which in our notation is called $\text{Ind}(T)^{*n}$.*

When G is loop-free, the dimension of the simplicial complex $\text{Hom}_+(T, G)$, unlike that of $\text{Hom}\,(T, G)$, is easy to find, once the size of the maximal independent set of G is computed.

Proposition 20.7. *For an arbitrary graph T and an arbitrary loop-free graph G, we have*

$$\dim(\text{Hom}_+(T, G)) = |V(G)| \cdot (\dim(\text{Ind}(T)) + 1) - 1.$$

Proof. Indeed, let $s = \dim(\text{Ind}\,(T)) + 1$ be the size of the maximal independent set in T. Since G is loop-free, every vertex of G occurs in at most s of the sets $\eta(x)$, for $x \in V(T)$. On the other hand, we can choose an independent set $S \subseteq V(T)$ such that $|S| = s$, and then assign

$$\eta(x) = \begin{cases} V(G), & \text{for } x \in S; \\ \emptyset, & \text{otherwise.} \end{cases}$$

This gives a simplex of dimension $|V(G)| \cdot (\dim(\text{Ind}\,(T)) + 1) - 1$. $\square$

For example, $\dim(\text{Hom}_+(C_{2r+1}, K_n)) = n \cdot ((r - 1) + 1) - 1 = nr - 1$.

20.1.3 The Support Map

For any topological space X and a set I, there is the standard support map from the join of I copies of X to the appropriate simplex

$$\text{supp} : *_I X \longrightarrow \Delta^I,$$

which "forgets" the coordinates in X.

Specializing to our situation, for arbitrary graphs T and G, we get the restriction map $\text{supp} : \text{Hom}_+(T, G) \to \Delta^{V(T)}$. Explicitly, for each simplex of $\text{Hom}_+(T, G)$, $\eta : V(T) \to 2^{V(G)}$, the support of η is given by $\text{supp}\,\eta = V(T) \setminus \eta^{-1}(\emptyset)$. See Figure 20.2 for an example.

An important property of the support map is that the preimage of the barycenter of $\Delta^{V(T)}$ is homeomorphic to $\text{Hom}\,(T, G)$. This is the crucial step in setting up a useful spectral sequence. The assumption that T is finite is

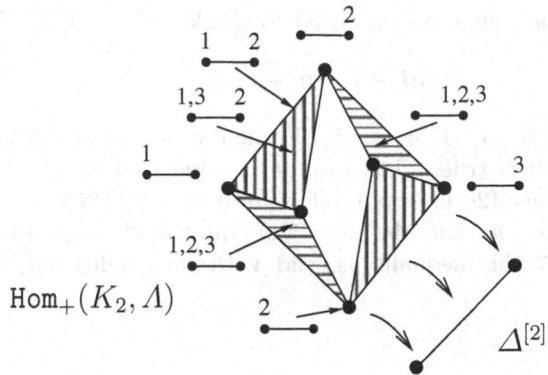

Fig. 20.2. The support map from $\mathrm{Hom}_+(K_2, \Lambda)$ to $\Delta^{[2]}$.

crucial at this point, since an infinite simplex does not have a well-defined barycenter.

An alternative concise way to phrase the definition of the map supp is to consider the map $t^T : \mathrm{Hom}_+(T, G) \to \mathrm{Hom}_+(T, \mathsf{C}K_1) \simeq \Delta^{V(T)}$ induced by the homomorphism $t : G \to \mathsf{C}K_1$. Then for each $\eta \in \mathrm{Hom}_+(T, G)$ we have supp $\eta = t^T(\eta)$, where the simplices in $\Delta^{V(T)}$ are identified with the finite subsets of $V(T)$.

20.1.4 An Example: Hom$_+(C_m, K_n)$

A special case that is useful for the computation that we will make in this chapter is the complex $\mathrm{Hom}_+(C_m, K_n)$. By Corollary 20.6, we have $\mathrm{Hom}_+(C_m, K_n) \simeq \mathrm{Ind}\,(C_m)^{*n}$. On the other hand, the homotopy type of the complexes $\mathrm{Ind}\,(C_m)$ was completely described by Proposition 11.17; hence we can derive the following explicit description.

Corollary 20.8. *For any $m \geq 2$, $n \geq 3$, we have*

$$\mathrm{Hom}_+(C_m, K_n) \simeq \begin{cases} \bigvee_{2^n\ copies} \mathbb{S}^{nk-1}, & \text{if } m = 3k; \\ \mathbb{S}^{nk-1}, & \text{if } m = 3k \pm 1. \end{cases}$$

The utility of this fact comes from the fact that the spectral sequence, that we will now proceed to set up, will be converging to the cohomology groups of $\mathrm{Hom}_+(C_m, K_n)$. For future reference, we also record the following specific observation.

Proposition 20.9. *For all integers m, n such that $m \geq 5$, $n \geq 4$, we have $\widetilde{H}^i(\mathrm{Hom}_+(C_m, K_n)) = 0$, for $i \leq m + n - 4$, with the one exception $(m, n, i) = (7, 4, 7)$.*

Proof. By Corollary 20.8, we just need to check that

$$nk - 1 > m + n - 4, \tag{20.1}$$

where $k = \lfloor (m+1)/3 \rfloor$. If $m \in \{5, 6\}$, then $k = 2$, and (20.1) reduces to $n + 3 > m$, which is true. Thus we can assume that $m \geq 7$. Notice that $k \geq (m-1)/3$; hence (20.1) would follow from $n(m-1)/3 + 3 > m + n$. This can be rewritten as $nm - 3m - 4n + 9 > 0$, or equivalently as $(n-3)(m-4) > 3$. Since $n \geq 4$, $m \geq 7$, this inequality is valid, with the equality only when $m = 7$, $n = 4$. $\square$

20.2 Setting up the Spectral Sequence

20.2.1 Filtration Induced by the Support Map

For simplicity, we shall for now consider $\mathbb{Z}_2$-coefficients. The additional issues arising in connection with switching to integer coefficients will be analyzed in Section 20.6.

For an arbitrary function $\eta : V(T) \to 2^{V(G)}$, we let η_+ denote the chain in $C_*(\mathrm{Hom}_+(T, G))$ consisting of a single simplex (with coefficient 1) indexed by η; when no confusion arises, we identify this chain with the simplex itself. Furthermore, we let η_+^* denote the corresponding dual cochain. We consider the Leray–Serre filtration of the cellular cochain complex $C^*(\mathrm{Hom}_+(T, G); \mathbb{Z}_2)$ associated with the support map. To describe the considered filtration explicitly, define the subcomplexes $F^p = F^p C^*(\mathrm{Hom}_+(T, G); \mathbb{Z}_2)$ of $C^*(\mathrm{Hom}_+(T, G); \mathbb{Z}_2)$ as follows:

$$F^p : \cdots \xrightarrow{\partial^{q-1}} F^{p,q} \xrightarrow{\partial^q} F^{p,q+1} \xrightarrow{\partial^{q+1}} \cdots ,$$

where

$$F^{p,q} = F^p C^q(\mathrm{Hom}_+(T, G); \mathbb{Z}_2)$$
$$= \mathbb{Z}_2 \left[\eta_+^* \mid \eta_+ \in \mathrm{Hom}_+^{(q)}(T, G), |\mathrm{supp}\,\eta| \geq p + 1 \right], \tag{20.2}$$

the map ∂^* is the restriction of the differential in $C^*(\mathrm{Hom}_+(T, G); \mathbb{Z}_2)$, and $\mathrm{Hom}_+^{(q)}(T, G)$ denotes the qth skeleton of $\mathrm{Hom}_+(T, G)$. In words: we filter by the preimages of the skeletons of $\Delta^{V(T)}$, and each $F^{p,q}$ is a vector space over $\mathbb{Z}_2$ generated by all elementary cochains corresponding to q-dimensional cells that are supported in at least $p + 1$ vertices of T. Note that this restriction defines a filtration, since the differential does not decrease the cardinality of the support set. We have

$$C^q(\mathrm{Hom}_+(T, G); \mathbb{Z}_2) = F^{0,q} \supseteq F^{1,q} \supseteq \cdots \supseteq F^{|V(T)|-1,q} \supseteq F^{|V(T)|,q} = 0.$$

20.2.2 The 0th and the 1st Tableaux

Next, we shall describe the 0th and the 1st tableaux of this spectral sequence, and then perform a partial analysis of the second tableau.

To start with, as an additional piece of notation, by writing the brackets $[-]$ after the name of a cochain complex we shall mean the index shifting (to the left); that is, for the cochain complex $\mathcal{C} = (C^*, d^*)$, the cochain complex $\mathcal{C}[s] = (C^*[s], d^*)$ is defined by $C^i[s] := C^{i+s}$. We sharpen the reader's attention on the fact that when considering integer coefficients, we choose not to change the sign of the differential.

Proposition 20.10. *For any p, we have*

$$F^p/F^{p+1} = \bigoplus_{\substack{S \subseteq V(T) \\ |S| = p+1}} C^*(\mathrm{Hom}\,(T[S], G); \mathbb{Z}_2)[-p]. \tag{20.3}$$

Hence, by (16.8), the zeroth tableau of the spectral sequence associated to the cochain complex filtration F^*, and converging to $H^{p+q}(\mathrm{Hom}_+(T, G); \mathbb{Z}_2)$, is given by

$$E_0^{p,q} = C^{p+q}(F^p, F^{p+1}) = \bigoplus_{\substack{S \subseteq V(T) \\ |S| = p+1}} C^q(\mathrm{Hom}\,(T[S], G); \mathbb{Z}_2). \tag{20.4}$$

Furthermore, using (16.9), we obtain the description of the first tableau as well:

$$E_1^{p,q} = H^{p+q}(F^p, F^{p+1}) = \bigoplus_{\substack{S \subseteq V(T) \\ |S| = p+1}} H^q(\mathrm{Hom}\,(T[S], G); \mathbb{Z}_2). \tag{20.5}$$

20.2.3 The First Differential

According to the formula (20.5), the first differential $d_1^{p,q} : E_1^{p,q} \longrightarrow E_1^{p+1,q}$ is actually a map

$$\bigoplus_{\substack{S \subseteq V(T) \\ |S| = p+1}} H^q(\mathrm{Hom}\,(T[S], G); \mathbb{Z}_2) \longrightarrow \bigoplus_{\substack{S \subseteq V(T) \\ |S| = p+2}} H^q(\mathrm{Hom}\,(T[S], G); \mathbb{Z}_2).$$

For $S_2 \subseteq S_1 \subseteq V(T)$, let $i[S_1, S_2] : T[S_2] \hookrightarrow T[S_1]$ be the inclusion graph homomorphism. Since $\mathrm{Hom}\,(-, G)$ is a contravariant functor, we have an induced map $i_G[S_1, S_2] : \mathrm{Hom}\,(T[S_1], G) \to \mathrm{Hom}\,(T[S_2], G)$, and hence an induced map on the cohomology groups

$$i_G^*[S_1, S_2] : H^*(\mathrm{Hom}\,(T[S_2], G); \mathbb{Z}_2) \to H^*(\mathrm{Hom}\,(T[S_1], G); \mathbb{Z}_2).$$

Let $\sigma \in H^q(\text{Hom}\,(T[S], G); \mathbb{Z}_2)$, for some q and some $S \subseteq V(G)$. The value of the first differential on σ is then given by

$$d_1^{p,q}(\sigma) = \sum_{x \in V(T) \setminus S} i_G^*[S \cup \{x\}, S](\sigma). \tag{20.6}$$

We shall next detail the formula (20.6) by introducing specific generators for the groups in the first tableau.

20.3 Encoding Cohomology Generators by Arc Pictures

20.3.1 The Language of Arcs

Let us now introduce some further notation. Let S be a proper subset of $V(C_m)$. We call the connected components of the graph $C_m[S]$ either *singletons* or *arcs* depending on whether they have 1 or at least 2 vertices. For $v \in S$ we let $a(S, v)$ denote the arc of S to which v belongs (assuming this arc exists). Furthermore, for an arbitrary arc a of S, we let $a = [a_\bullet, a^\bullet]_m$, and finally, we set $\widehat{a} := [a_\bullet - 1, a^\bullet + 1]_m$, so $|\widehat{a}| = |a| + 2$, if $|a| \le m - 2$. We stress that the arithmetic operations as well as intervals are taken modulo m; for example, $[m, 1]_m$ and $[1, 2]_m$ are arcs with two vertices each, while $\widehat{[m, 1]}_m = [m-1, 2]_m$ and $\widehat{[1, 2]}_m = [m, 3]_m$. To make the formulas easier, we also think of the cycle $a = C_m$ itself as an arc, in which case we use the convention $a_\bullet = a^\bullet = 1$. When A is a collection of arcs, we set $V(A) := \bigcup_{a \in A} a$, $A_\bullet := \bigcup_{a \in A} a_\bullet$, and we let $\widehat{A}$ denote the union $\bigcup_{a \in A} \widehat{a}$.

Definition 20.11. *Let $t \ge 1$. A t-arc picture is a pair (S, A), where S is some proper subset of $V(C_m)$, and A is a set consisting of t different arcs of $C_m[S]$. We refer to arcs in A as **marked arcs**, and sometimes simply call (S, A) an **arc picture**.*

20.3.2 The Corresponding Cohomology Generators

For $S \subset V(C_m)$ such that $|S| = p+1 < m$, the t-arc pictures (S, A) will index the generators of $E_1^{p,t(n-2)}$. To introduce these, let $V = \{v_a\}_{a \in A}$ be a set of vertex representatives of A, i.e., $v_a \in a$ for each $a \in A$. Now set

$$\sigma_V^S := \sum_\eta \eta_+^*, \tag{20.7}$$

where the sum is taken over all $\eta : S \to 2^{[n]} \setminus \{\emptyset\}$ such that $\eta(v_a) = [n-1]$, for all $a \in A$, and $|\eta(w)| = 1$, for all $w \in S \setminus V$.

It is easily checked that σ_V^S is a cocycle of $\text{Hom}_+(C_m[S], K_n)$. Indeed, taking the coboundary means adding an element from $[n]$ to one of the lists on S.

Since lists over vertices in V are already maximized, we can only add an element to one of the single-element lists over vertices in $S \setminus V$. Each such list assignment is obtained in exactly two ways: namely, from the list assignments that we get by removing one of the two elements from this 2-list. Since we are working over $\mathbb{Z}_2$, the total sum then equals 0.

When vertex representatives move along edges, the corresponding cohomology class $[\sigma_V^S]$ does not change. Therefore, we may limit our attention to $\sigma_{A_\bullet}^S$. The only reason we introduce the vertex representatives at all is for the later calculations with integer coefficients. In this case, the cohomology class $[\sigma_V^S]$ changes sign according to a certain pattern. Both the proof of the fact that the cohomology class does not depend on the choice of the vertex representative over $\mathbb{Z}_2$-coefficients and the description of the sign change pattern over $\mathbb{Z}$-coefficients can be found in Subsection 20.6.2.

It is now time to interpret the formula (20.6) in terms of our combinatorial generators. Since the differentials in the spectral sequence are induced by the differential in the original chain complex, we may conclude that

$$d_1([\sigma_{A_\bullet}^S]) = \sum_{w \notin S} \left[\sigma_{A_\bullet}^{S \cup \{w\}}\right], \tag{20.8}$$

where the sum is taken over all w not in S such that adding w to S does not unite two arcs from A. Note that (20.8) is also valid when $|S| = m - 1$, if we define σ_V^S in the same way for $S = [m]$. It might be worthwhile to observe at this point that the set of marked arcs in $S \cup \{w\}$ is equal to A if and only if $w \notin \widehat{A}$.

20.3.3 The First Reduction

By (16.7), in order to obtain the second tableau $\{E_2^{*,*}\}$ we need to calculate the cohomology groups of the complex $(E_1^{*,(n-2)t}, d_1)$. Let (A_t^*, d_1) be its subcomplex generated by all classes $[\sigma_{A_\bullet}^S]$ such that $\widehat{A} = V(C_m)$.

Proposition 20.12. *For all t, we have*

$$H^*(E_1^{*,(n-2)t}) = H^*(A_t^*). \tag{20.9}$$

Proof. We give a direct combinatorial matching argument using the chain complex version of the discrete Morse theory from Section 11.3.

The matchings can be done as follows: for each collection of arcs A such that $\widehat{A} \neq V(C_m)$, choose some element $x_A \in V(C_m) \setminus \widehat{A}$; then an element $[\sigma_{A_\bullet}^S]$ is matched with the element $\left[\sigma_{A_\bullet}^{S \oplus \{x_A\}}\right]$, where $\oplus$ denotes the symmetric difference (exclusive *or*) of sets. One can then check that this matching is acyclic, in the sense of the paragraph after Definition 11.22.

Indeed, assume that there exists a cycle $y_t \prec x_1 \succ y_1 \prec x_2 \succ \cdots \prec x_t \succ y_t$, where x_i and y_i are matched for all $i = 1, \ldots, t$. Consider the covering relation

$x_i \succ y_{i-1}$, for some $i = 2, \ldots, t$. Using (20.6) and (20.8), one can see that in this situation, x_i must be obtained from y_{i-1} by adding an element to its indexing set S that extends one of the arcs in A, since otherwise the set A does not change, hence the element x_A does not change, and so the element y_{i-1}, which contains x_A, cannot be matched upward. The same should hold for the covering relation $y_t \prec x_1$.

Clearly, this leads to a contradiction, since the total length of arcs in A does not change on the matching edges in the cycle (A itself does not change there), while it increases along other edges. The critical elements are precisely those $[\sigma_{A_\bullet}^S]$ for which $\widehat{A} = V(C_m)$, and therefore the identity (20.9) follows from Theorem 11.24. □

20.4 Topology of the Torus Front Complexes

20.4.1 Reinterpretation of $H^*(A_t^*, d_1)$ Using a Family of Cubical Complexes $\{\Phi_{m,n,g}\}$

Combining Proposition 20.12 with formula (20.8), we see that to compute the second tableau of our spectral sequence we need to calculate cohomology groups of a certain cochain complex, which we have combinatorially described by choosing an explicit basis of generators and then writing out the differentials in terms of these generators.

Our next idea is to reinterpret the cochain complex (A_t^*, d_1) as a chain complex that computes the $\mathbb{Z}_2$-homology of a certain cubical complex. We now proceed with defining these complexes. For a natural number $n \geq 2$, a circular n-set is the set of n points that are equidistantly arranged on the circle of length n.

Definition 20.13. *Let m, n, and g be natural numbers. The cubical complex $\Phi_{m,n,g}$ is defined as follows:*

- *vertices of $\Phi_{m,n,g}$ are indexed by all possible selections of m elements from the circular n-set, so that every two of the chosen vertices are at a distance at least g;*
- *to index higher-dimensional cubes we take all possible collections of m sets, where each set either contains a single element of the circular n-set or consists of a pair of two elements at distance 1; the sets are required to be at minimal distance g, where the distance between two sets is defined as the minimum of the distances between their elements.*

Clearly, the dimension of a cube indexed by a collection of sets is equal to the number of 2-element sets in this collection, and one cube contains another if and only if the corresponding collection of sets contains the collection of sets associated to the second cube. See Figure 20.3 for a graphic explanation.

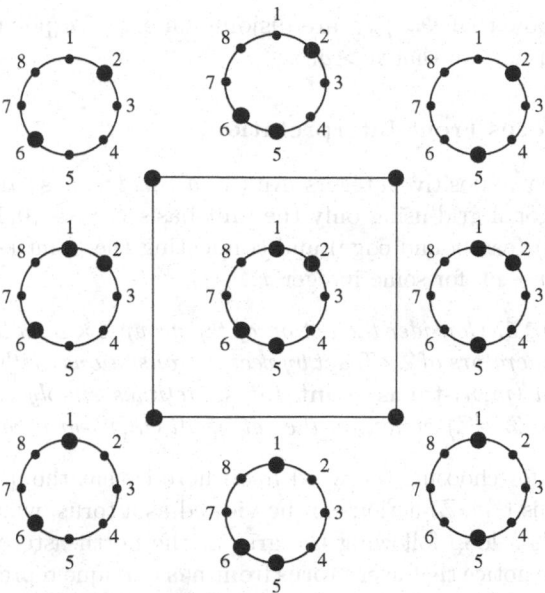

Fig. 20.3. Indexing of cubes in $\Phi_{2,8,2}$.

For convenience of formulation, we introduce the convention that the complex $\Phi_{0,n,g}$ is a point. One can quickly list some trivial cases of the complexes $\Phi_{m,n,g}$. To start with, a necessary and sufficient condition that these complexes be nonempty is that $n \geq gm$. The complex $\Phi_{n,n,1}$ is just a point, and more generally, $\Phi_{m,gm,g}$ is a disjoint union of g points. It is not difficult to check by hand that the complex $\Phi_{m,n,g}$ is connected whenever $n > gm$. The next case to consider is $\Phi_{m,gm+1,g}$. The interested reader is invited to check that $\Phi_{m,gm+1,g}$ is a cycle of length $gm + 1$. Finally, we remark that the dimension of $\Phi_{m,n,g}$ is equal to $\min(m, n - gm)$.

The usefulness of this family in our context becomes apparent from the following proposition.

Proposition 20.14. *The cochain complex $(C_*(\Phi_{t,m,3}; \mathbb{Z}_2), d)$ is isomorphic to the chain complex (A_t^*, d_1), with the module $C_i(\Phi_{t,m,3}; \mathbb{Z}_2)$ corresponding to the module $A_t^{m-t-i-1}$, for all i.*

Proof. We think of a collection of sets that indexes a cube in $\Phi_{t,m,3}$ as a collection of gaps between the arcs from A. This will clearly give a dimension-preserving bijection between the cubes of $\Phi_{t,m,3}$ and the collections of t arcs A such that $\widehat{A} = V(C_n)$. The condition that the distance between sets (read: between gaps) be at least 3 corresponds to the requirement that arcs must have at least two elements. It remains to compare the differential maps: they coincide by the formula (20.8) Note that we are using the fact that we are working over $\mathbb{Z}_2$-coefficients. □

Since we know that $\Phi_{m,gm,g}$ are disjoint unions of g points, it is enough to assume from now on that $n > gm$.

20.4.2 The Torus Front Interpretation

Let m and n be two positive integers. An (m, n)-*grid path* is a directed path on the unit orthogonal grid using only the unit basis vectors $(0, 1)$ (northbound edge) and $(1, 0)$ (eastbound edge) and connecting the point $(-x, x)$ with the point $(m - x, n + x)$, for some integer x.

Definition 20.15. *Consider the action of the group $\mathbb{Z} \times \mathbb{Z}$ on the plane, where the standard generators of $\mathbb{Z} \times \mathbb{Z}$ act by vector translations, with vectors $(-1, 1)$ and (m, n). An (m, n)-**torus front** (or sometimes simply **torus front**) is an orbit of this $(\mathbb{Z} \times \mathbb{Z})$-action on the set of all (m, n)-grid paths.*

The reason for choosing the word *torus* here is that the quotient space of the plane by this $(\mathbb{Z} \times \mathbb{Z})$-action can be viewed as a torus, where each (m, n)-grid path yields a loop following the grid in the northeasterly direction; see Figure 20.4. We notice that every torus front has a unique representative starting from the point $(0, 0)$. It will soon become clear why we choose to consider the orbits of the action rather than merely considering these representatives of the orbits.

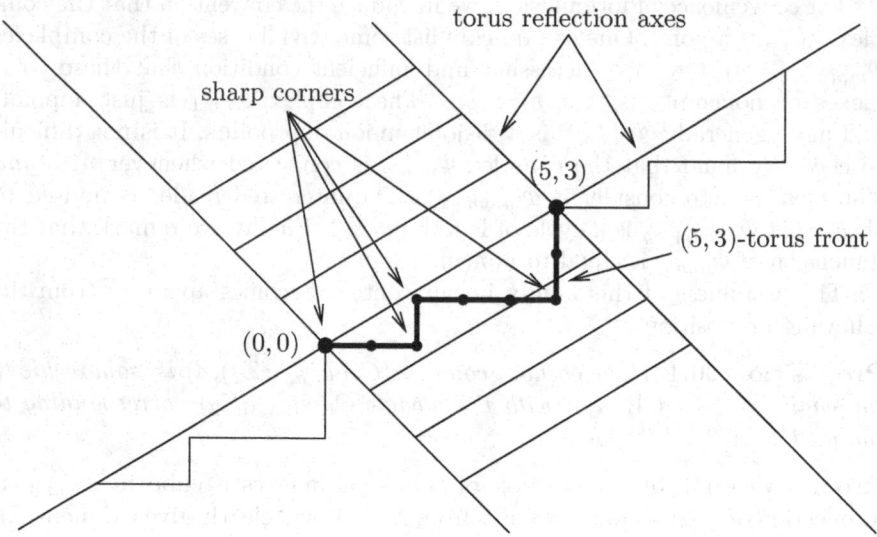

Fig. 20.4. $(5, 3)$-path and the associated torus front.

A *northwestern sharp corner* of a torus front is a vertex that is entered by a northbound edge and exited by an eastbound edge; in the same way, a *southeastern sharp corner* of a torus front is a vertex that is entered by an eastbound edge and exited by a northbound edge.

An *elementary flip* of a torus front is either a replacement of a northwestern sharp corner at (a, b) by a southeastern sharp corner at $(a + 1, b - 1)$ or vice versa. It may help to think of a torus front as a sort of flexible snake, where the sharp corners (meaning the vertex in the sharp corner, together with the two adjacent edges) can be flipped about (as if using ball-and-socket joints) the diagonal line connecting the neighbors of the sharp corner vertex.

Finally, A *flip* of a torus front is a collection of noninterfering elementary flips, i.e., the flipped vertices are at least at distance 2 from each other, where the distance is measured along the front; see Figure 20.5. The picture that we have in mind is that the torus front propagates through the torus by means of flips. Since we want flips to encode cells, we prefer to think of them as schemes of allowed elementary flips, rather than a concrete process of replacing one torus front with another. For a flip F, we call those torus fronts, which can be obtained by actually performing all the elementary flips from F (meaning here choosing one of the two positions for each flipped corner), the *members* of F.

Definition 20.16. *Let m, n, and g be natural numbers. The cubical complex $TF_{m,n,g}$ is defined as follows:*

- *vertices of $TF_{m,n,g}$ are indexed by all possible (m, n)-torus fronts whose horizontal legs have length at least g; here the length of the leg is taken to be the number of vertices it contains;*
- *the higher-dimensional cubes of $TF_{m,n,g}$ are indexed by all possible flips of (m, n)-torus fronts whose members are vertices of $TF_{m,n,g}$.*

Notice that for $g = 1$ there are no conditions, so $TF_{m,n,1}$ simply has all (m, n)-torus fronts as vertices and all flips as higher-dimensional cubes. It is also important to remark that the term "horizontal legs" includes the legs of length 1; in other words, for $g > 1$, it is prohibited that the torus front contain two consecutive northbound edges.

Proposition 20.17. *For any natural numbers m, n, and g, the cubical complexes $\Phi_{m,n,g}$ and $TF_{m,n-m,g}$ are isomorphic.*

Proof. Let $S \subseteq [n]$, $|S| = m$, be a vertex of $\Phi_{m,n,g}$. We can construct an $(m, n - m)$-grid path as follows. Start out from the point $(0,0)$; then, for all i from 1 to n perform the following step: either move along the northbound edge if $i \in S$, or move along the eastbound edge if $i \notin S$. Since $|S| = m$, we will eventually reach the point $(m, n - m)$. It is obvious that this gives a bijection between vertices of $\Phi_{m,n,g}$ and $(m, n - m)$-grid paths satisfying the extra restriction that the horizontal legs be of length at least g.

It is also clear that moving along edges in $\Phi_{m,n,g}$ corresponds to elementary flips in $TF_{m,n-m,g}$. This is completely obvious as long as the flip does not concern the vertex $(0,0)$ (or equivalently, the vertex $(m, n - m)$). If, on the

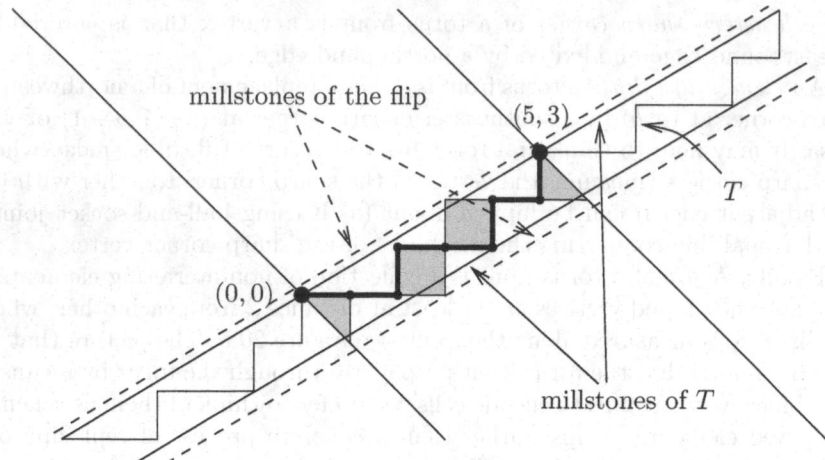

Fig. 20.5. A flip of a $(5, 3)$-torus front consisting of three noninterfering elementary flips.

other hand, the edge in $\Phi_{m,n,g}$ indicates that we should flip the sharp corner $(0, 0)$, then we can do that as well, but then we should also flip $(m, n-m)$, and we will get an $(m, n-m)$-path from $(1, -1)$ to $(m+1, n-m-1)$; translating back we will again get an $(m, n-m)$-path from $(0, 0)$ to $(m, n-m)$.

If the flips are done on torus fronts instead of grid paths, then we do not need to consider different special cases, which is the main reason why we chose to replace the paths by the orbits of the $(\mathbb{Z} \times \mathbb{Z})$-action. $\square$

Note that the complex $TF_{m,n-m,1}$ is isomorphic to the complex $TF_{n-m,m,1}$, and so the complex $\Phi_{m,n,1}$ is isomorphic to the complex $\Phi_{n-m,n,1}$.

To understand the complexes $\Phi_{m,n,g}$ satisfying $n > mg$, we shall study the complexes $TF_{m,n,g}$ with $n > m(g-1)$.

20.4.3 Grinding

We shall now drastically simplify complexes $TF_{m,n,g}$ by means of a collapsing procedure that we call *grinding*. To illustrate the idea, let us consider the case $m = n = 3$, $g = 1$. The complex $TF_{3,3,1}$ ($= \Phi_{3,6,1}$) consists of two cubes, indexed by the set collections $\{\{1, 2\}, \{3, 4\}, \{5, 6\}\}$ and $\{\{1, 6\}, \{2, 3\}, \{4, 5\}\}$, and six additional squares; see Figure 20.6.[1] Each of these squares contains a vertex that does not belong to any other maximal cell; for example, for the square indexed with $\{\{1, 2\}, \{3\}, \{4, 5\}\}$ this vertex is indexed with $\{\{2\}, \{3\}, \{4\}\}$. Therefore, we can collapse these six squares through these six vertices, and the only thing left will be the two cubes hanging together by their endpoints. This is the characteristic picture we shall now see in general.

---

[1] The figure is reproduced courtesy of Eva-Maria Feichtner.

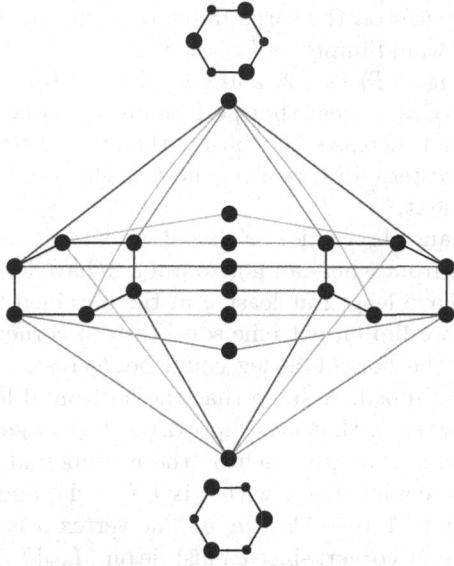

Fig. 20.6. The complex $TF_{3,3,1} = \Phi_{3,6,1}$.

Let T be an (m,n)-torus front, which we represent as an (m,n)-grid path starting from $(0,0)$. We associate to T two lines $M_{nw}(T)$ and $M_{se}(T)$ defined as follows: both lines are parallel to the line L passing through the points $(0,0)$ and (m,n), and they both are tangential to the grid path representing T, $M_{nw}(T)$ from the northwest, and $M_{se}(T)$ from the southeast; see Figure 20.5. We call these two lines the *millstones* of T.

The *width* of the torus front T, denoted by $w(T)$, is the distance between the millstones measured along the line $x = -y$. Clearly, the width cannot be less than $\sqrt{2}/2$, and the minimum is achieved only when $m = n$ by the torus front, where northbound and eastbound edges alternate at every step.

We let $E_{nw}(T)$ denote the set of points in $T \cap M_{nw}(T)$, which we call *northwestern extreme corners*; in the same way, $E_{se}(T) = T \cap M_{se}(T)$ denotes the set of southeastern extreme corners. Extreme corners are by construction always sharp corners.

As mentioned above, each sharp corner defines an elementary flip of T, during which it remains a sharp corner but changes orientation. If this sharp corner were extreme, then the new sharp corner would be extreme (on the opposite side) if and only if $w(T) \leq \sqrt{2}$. Clearly, for any torus front, the elementary flips of northwestern extreme corners are mutually noninterfering, and therefore form one flip; the same is true for the southeastern extreme corners. Furthermore, if $w(T) > \sqrt{2}/2$, then these two sets of elementary flips do not interfere with each other either. Indeed, if they did, it would mean that two neighboring vertices on the torus front were extreme corners of different orientations, which would imply $w(T) = \sqrt{2}/2$.

Assume now that $w(T) > \sqrt{2}$, and consider the total flip of all extreme corners. It is geometrically clear that performing any combination of these elementary flips will not increase the width of the torus front, and that performing all of the northwestern flips, or all of the southeastern flips, simultaneously will for sure decrease it.

Another important observation is that if an extreme northwestern corner is flipped on a torus front where all horizontal legs have length at least g, then all horizontal legs have length at least g in the obtained torus front as well; the same is true if we flip an extreme southeastern corner. To see this, note that the only time the horizontal leg could be decreased is when one of its endpoints would be flipped. Assume that the horizontal leg in question runs from vertex a to vertex b, that a is flipped (no loss of generality here), and denote by c the vertex that we reach by the northbound edge from b. Then the slope of the line connecting a with c is $1/(g-1)$, and since we assumed $m(g-1) < n$, we have $1/(g-1) > m/n$. The vertex a is flipped; hence it is a northwestern extreme corner, since c must lie on $M_{nw}(T)$ or to the southeast of it. This implies that $1/(g-1) \leq m/n$, yielding a contradiction.

20.4.4 Thin Fronts

Let us now describe the cells that will, in a sense to be specified later, form a core of the cubical complexes $TF_{m,n,g}$.

Definition 20.18. *We call a torus front T* **thin** *if $w(T) \leq \sqrt{2}$; we call it* **very thin** *if $w(T) < \sqrt{2}$.*

It is easy to construct a very thin torus front for fixed m and n: simply start from the point $(0,0)$ eastward and then approximate the line connecting $(0,0)$ with (m,n) as closely as possible, staying on one side of this line (touching the line is permitted). It is clear that if the approximation is not the closest possible, then $w(T) \geq \sqrt{2}$.

It is also important to notice that since we assumed $m(g-1) < n$, each horizontal leg in a thin torus front has length at least g; this can be proved with the same argument as the one we used to show that flipping all extreme corners preserves that property as well.

Let I denote the line segment connecting $(0,0)$ with (m,n). Counting from the origin in the northeasterly direction, the first point on I with integer coordinates, other than the origin itself, is $(m/\gcd(m,n), n/\gcd(m,n))$, which can be achieved in $(m+n)/\gcd(m,n)$ steps (it is customary to denote the greatest common divisor with simple round brackets, but since (m,n) is used to denote the point on the plane with these coordinates, we prefer here a somewhat longer notation).

The very thin torus front which we just constructed will have $(0,0)$ as a northwestern extreme corner, and after that the northwestern extreme corners will repeat with the period $(m+n)/\gcd(m,n)$. From this we see that in general, to determine a very thin torus front we just need to specify where on the part of the torus front connecting two subsequent northwestern extreme corners the origin will lie. Therefore, we conclude that there are precisely $(m+n)/\gcd(m,n)$ very thin torus fronts, and that the width of any very thin (m,n)-torus front depends only on m and n.

We now extend the definition of width. For any flip F, we define its width $w(F)$ to be the maximum of widths of all the torus fronts that are members of F. The concepts of millstones and extreme corners can be extended to flips as well; see Figure 20.5.

We call a flip F thin if $w(F) \leq \sqrt{2}$. The thin flips play an important role in our context, as the next theorem indicates.

Theorem 20.19. *For any natural numbers m, n, and g, there exists a sequence of collapses leading from the cubical complex $TF_{m,n,g}$ to its subcomplex consisting of all thin flips.*

Proof. To start with, we observe that if σ and τ are cells of $TF_{m,n,g}$, and σ is contained in τ, then $w(\sigma) \leq w(\tau)$. Therefore, it is enough to describe how to collapse away all the cells of a certain width $(> \sqrt{2})$, under the condition that the cells of larger width are already gone.

Let us fix this width and denote it by w. To every cell σ we can associate cells $\sigma^{\uparrow}$ and $\sigma_{\downarrow}$ as follows: $\sigma^{\uparrow}$ is obtained from σ by adding elementary flips on all extreme corners that do not yet have them, whereas $\sigma^{\downarrow}$ is obtained from σ by removing the elementary flips on all extreme corners, where in each elementary flip we keep the extreme corner (since $w > \sqrt{2}$, both corners of an elementary flip cannot be extreme; hence this is well-defined). Note that it is possible that $\sigma = \sigma^{\uparrow}$ or $\sigma = \sigma_{\downarrow}$, but not both.

By our discussion above, if all horizontal legs have length at least g in vertices of σ, then the same is true for all cells in the interval $[\sigma_{\downarrow}, \sigma^{\uparrow}]$. We also see that by our construction, all cells in the interval $[\sigma_{\downarrow}, \sigma^{\uparrow}]$ have the same width. Moreover, each such interval is isomorphic to a Boolean algebra B_t, where t is equal to the number of extreme corners. Clearly, the intervals $[\sigma_{\downarrow}, \sigma^{\uparrow}]$ are disjoint, and the removal of such an interval constitutes a collapse, as long as $\sigma^{\uparrow}$ is the only maximal cell containing $\sigma_{\downarrow}$ at the given time.

This observation allows us to do as follows: restrict the partial inclusion order to the cells $\sigma^{\uparrow}$, choose a linear extension of this partial order, and then remove intervals $[\sigma_{\downarrow}, \sigma^{\uparrow}]$ following the linear extension in decreasing order. The statement follows now from the fact that if τ contains σ, then $\tau^{\uparrow}$ contains $\sigma^{\uparrow}$, since at the moment the interval $[\sigma_{\downarrow}, \sigma^{\uparrow}]$ is being removed, this will translate to the condition that if τ contains $\sigma_{\downarrow}$, then $\tau^{\uparrow} = \sigma^{\uparrow}$. $\square$

Let us now finish our study of torus fronts by analyzing the subcomplex $\text{Thin}_{m,n,g}$ of $TF_{m,n,g}$ consisting of all thin flips.

Proposition 20.20. *For any natural numbers m, n, and g such that $n > m(g-1)$, the complex $\text{Thin}_{m,n,g}$ consists of $(m+n)/\gcd(m, n)$ cubes, each one having dimension $\gcd(m, n)$, which are connected together by their diagonally opposite end vertices, like a garland, to form one cycle.*

Proof. Let v be a thin torus front. Flipping a corner that is not extreme will give a flip of width strictly larger than $\sqrt{2}$; hence this is not allowed. If, on the other hand, an extreme northwestern corner is flipped, then the width of this flip is equal to $\sqrt{2}$, and flipping another extreme northwestern corner will not change the width, whereas flipping any other vertex will increase it. The same holds for flipping extreme southeastern corners. It follows that there are two ways to get a maximal cell adjacent to v: either by flipping all extreme northwestern corners or by flipping all extreme southeastern corners.

On the other hand, if F is a maximal cell, we have $w(F) = \sqrt{2}$. If we now choose a torus front T belonging to this cell, we will also have $w(T) = \sqrt{2}$, except for two cases: if in each elementary flip we either always choose the northwestern path, or if we always choose the southeastern path. We see that each maximal cell contains exactly two torus fronts of width strictly smaller than $\sqrt{2}$, and that these are opposite corners of the cube. Since $\text{Thin}_{m,n,g}$ is connected, the conclusion follows. □

We remark that since each cube can further individually be collapsed onto a path connecting two opposite vertices, Theorem 20.19 and Proposition 20.20 entail that $TF_{m,n,g}$ can be collapsed onto a cycle.

20.4.5 The Implications for the Cohomology Groups of Hom (C_m, K_n)

We now have enough information to complete the computation of the entire second tableau, except for the $(m-1)$th column, the latter containing the cohomology groups of $\text{Hom}(C_m, K_n)$. Since the cohomology groups of $\text{Hom}_+(C_m, K_n)$ are nontrivial in only one dimension, we can then use this to derive almost complete information about the cohomology groups of $\text{Hom}(C_m, K_n)$ with $\mathbb{Z}_2$-coefficients. In fact, with a little extra effort we can describe $H^*(\text{Hom}(C_m, K_n); \mathbb{Z}_2)$ completely already with what we have now. However, for the sake of brevity we postpone the complete determination to Section 20.6, where the general case of integer coefficients will be dealt with.

By Propositions 20.12 and 20.14, to compute the values $E_2^{p,q}$, for $p \neq m-1$, it is enough to calculate the $\mathbb{Z}_2$-cohomology groups of the complexes $\Phi_{t,n,3}$, for the required range of the parameter t. These, on the other hand, are isomorphic to $H^*(\text{Thin}_{t,n-t,3}; \mathbb{Z}_2)$, which follows from Proposition 20.17 and Theorem 20.19. Finally, by Proposition 20.20, we see that $H^0(\text{Thin}_{t,n-t,3}; \mathbb{Z}_2) = H^1(\text{Thin}_{t,n-t,3}; \mathbb{Z}_2) = \mathbb{Z}_2$, and the other cohomology groups are trivial. Tracing back all the indices, we obtain the following answer.

Proposition 20.21. *The nonzero entries $E_2^{p,q}$, for $p \neq m - 1$, are the following: $E_2^{m-3,n-2} = \mathbb{Z}_2$, $E_2^{m-t-2,t(n-2)} = E_2^{m-t-1,t(n-2)} = \mathbb{Z}_2$, for $2 \leq t \leq \lfloor (m-1)/3 \rfloor$, and $E_2^{0,0} = \mathbb{Z}_2$. If additionally $3|m$, then $E_2^{2m/3-1,m(n-2)/3} = \mathbb{Z}_2^3$.*

The nonzero entry $E_2^{2m/3-1,m(n-2)/3} = \mathbb{Z}_2^3$ stems from the fact that the complexes $\Phi_{m,mg,g}$ are disjoint unions of g points, and are not homotopy equivalent to circles.

It needs to be explained why $E_2^{m-2,n-2} = 0$. If the entries in the $(m-1)$th column were all 0, we would of course have $E_2^{m-2,n-2} = \mathbb{Z}_2$. However, by Proposition 20.9, we know that $H^{m+n-4}(\text{Hom}_+(C_m, K_n); \mathbb{Z}_2) = 0$, unless $(m,n) = (7,4)$. Since all the entries to the northwest from $E_2^{m-2,n-2}$ that could potentially eliminate this entry at a later stage of the spectral sequence computation are 0, we have no choice but to conclude that the map

$$d_1 : E_1^{m-2,n-2}/\text{Im}(d_1 : E_1^{m-3,n-2} \to E_1^{m-2,n-2}) \longrightarrow E_1^{m-1,n-2}$$

is injective, and so $E_2^{m-2,n-2} = 0$. The case $(m,n) = (7,4)$ can be computed directly.

Since the reduced cohomology of $\text{Hom}_+(C_m, K_n)$ is concentrated in one dimension, Proposition 20.21 allows us to compute almost all cohomology groups of $\text{Hom}(C_m, K_n)$. The complete computation is done in Section 20.6, see Theorem 20.27 for a full answer.

20.5 Euler Characteristic Formula

While the cohomology groups and even the Betti numbers of the Hom complexes are usually very hard to compute, the Euler characteristic may turn out to be a more accessible invariant.

To start with, let T and G be arbitrary graphs. A simple counting in the filtration that we imposed on the simplicial complex $\text{Hom}_+(T, G)$ yields the following formula:

$$\widetilde{\chi}(\text{Hom}_+(T, G)) = \sum_{\emptyset \neq S \subseteq V(T)} (-1)^{|S|+1} \widetilde{\chi}(\text{Hom}(T[S], G)); \tag{20.10}$$

compare also with Proposition 20.26.

Theorem 20.22. *Let T and G be arbitrary graphs. Then we have*

$$\widetilde{\chi}(\text{Hom}(T, G)) = \sum_{\emptyset \neq S \subseteq V(T)} (-1)^{|S|+1} \widetilde{\chi}(\text{Hom}_+(T[S], G)). \tag{20.11}$$

Proof. Take the formula (20.10), and apply Möbius inversion on the Boolean algebra without the minimal element of all subsets of $V(T)$. $\square$

Theorem 20.23. *For an arbitrary graph T, we have the following formula:*

$$\widetilde{\chi}(\text{Hom}\,(T, K_n)) = \sum_{\emptyset \neq S \subseteq V(T)} (-1)^{n+|S|} \widetilde{\chi}(\text{Ind}(T[S]))^n. \qquad (20.12)$$

Proof. By Corollary 20.6 we know that $\text{Hom}_+(T, K_n) = \text{Ind}\,(T)^{*n}$. On the other hand, for any two simplicial complexes X and Y, we have $\widetilde{\chi}(X * Y) = -\widetilde{\chi}(X)\widetilde{\chi}(Y)$. It follows that $\widetilde{\chi}(\text{Hom}_+(T, K_n)) = (-1)^{n-1}\widetilde{\chi}(\text{Ind}\,(T))^n$. Substituting this into identity (20.11) yields formula (20.12). $\square$

This gives a new proof of the following result.

Corollary 20.24. *For arbitrary positive integers m and n such that $n \geq m$, we have*

$$\widetilde{\chi}(\text{Hom}\,(K_m, K_n)) = \sum_{k=1}^{m-1} (-1)^{n+k+1} \binom{m}{k+1} k^n. \qquad (20.13)$$

Proof. Since $\text{Ind}\,(K_t)$ is just a set of t points, we have $\widetilde{\chi}(K_t) = t - 1$. Substituting this into equation (20.12) and noticing that the summands depend on the cardinality of S only, we obtain (20.13). $\square$

If we try to use formula (20.12) to compute the Euler characteristic of $\text{Hom}\,(C_m, K_n)$, we will need nontrivial reductions to get a nice answer. Using Proposition 20.21 instead yields a very simple formula directly. This reflects very well the reduction that often occurs in passing between tableaux in the spectral sequence computation, since using the filtration on the simplicial complex $\text{Hom}_+(T, G)$ as we did is the same as using the first tableau of our spectral sequence.

Corollary 20.25. *For arbitrary positive integers m and n such that $m \geq 5$, $n \geq 4$, we have*

$$\widetilde{\chi}(\text{Hom}\,(C_m, K_n)) = \begin{cases} (-1)^{nk-m}, & \text{if } m = 3k \pm 1, \\ (-1)^{nk-m}(2^n - 3), & \text{if } m = 3k. \end{cases} \qquad (20.14)$$

Proof. The entries $E_2^{m-t-2,t(n-2)}$ and $E_2^{m-t-1,t(n-2)}$ cancel out in pairs, except for $E_2^{m-3,n-2}$, which we can discount since $E_2^{m-2,n-2}$ already canceled out with $E_2^{m-1,n-2}$. Thus the only nontrivial contribution comes from the Betti numbers of the total complex $\text{Hom}_+(C_m, K_n)$ and in case $m = 3k$ from the entry $E_2^{2m/3-1,m(n-2)/3}$. $\square$

20.6 Cohomology with Integer Coefficients

20.6.1 Fixing Orientations on Hom and Hom$_+$ Complexes

In this section we work only with integer coefficients, so we shall suppress $\mathbb{Z}$ from the notation. To be able to work with integer coefficients we need to choose orientations on cells so that the differential maps can be determined.

Let T and G be two graphs, and let us fix an order of the vertices of T and of G. When convenient, we may identify vertices of T, resp. of G, with integers $1, \ldots, |V(T)|$, resp. $1, \ldots, |V(G)|$, according to the chosen orders. First we deal with the simplicial case of $\mathrm{Hom}_+(T, G)$, which is simpler. The vertices of $\mathrm{Hom}_+(T, G)$ are indexed with pairs (x, y), where $x \in V(T)$, $y \in V(G)$ such that if x is looped, then so is y. We order these pairs lexicographically: $(x_1, y_1) \prec (x_2, y_2)$ if either $x_1 < x_2$, or $x_1 = x_2$ and $y_1 < y_2$. Then we orient each simplex of $\mathrm{Hom}_+(T, G)$ according to this order on the vertices. We call this orientation *standard*, and call the oriented simplex η_+. We identify this simplex with the corresponding chain, and denote the dual cochain by η_+^*.

Next we deal with $C^*(\mathrm{Hom}(T, G))$. On each cell $\eta \in \mathrm{Hom}(T, G)$ we fix an orientation (also called standard) as follows: orient each simplex $\eta(i)$ according to the chosen order on the vertices of G; then order these simplices in the direct product according to the chosen order on the vertices of T. We call this oriented cell η; a choice of orders on the vertex sets of T and G is implicit.

Note that permuting the vertices of the simplex $\eta(i)$ by some $\sigma \in \mathcal{S}_{|\eta(i)|}$ changes the orientation of the cell η by $\mathrm{sgn}\,\sigma$, and that swapping the simplices with vertex sets $\eta(i)$ and $\eta(i+1)$ in the direct product changes the orientation by $(-1)^{\dim \eta(i) \cdot \dim \eta(i+1)}$. Furthermore, if $\tilde{\eta} \in \mathrm{Hom}^{(i+1)}(T, G)$ is obtained from $\eta \in \mathrm{Hom}^{(i)}(T, G)$ by adding a vertex v to the list $\eta(t)$, then we have

$$[\eta : \tilde{\eta}] = (-1)^{k+d-1}, \tag{20.15}$$

where k is the position of v in $\tilde{\eta}(t)$, and d is the dimension of the product of the simplices with the vertex sets $\eta(1), \ldots, \eta(t-1)$, i.e., $d = 1 - t + \sum_{j=1}^{t-1} |\eta(j)|$.

Let $\widetilde{C}^*$ be the subcomplex of $C^*(\mathrm{Hom}_+(T, G))$ generated by all η_+^*, for $\eta : V(T) \to 2^{V(G)}$ such that $\mathrm{supp}\,\eta = V(T)$. Set

$$X^*(T, G) := \widetilde{C}^*[|V(T)| - 1]. \tag{20.16}$$

Note that both $C^i(\mathrm{Hom}(T, G))$ and $X^i(T, G)$ are free $\mathbb{Z}$-modules with the bases $\{\eta^*\}_\eta$ and $\{\eta_+^*\}_\eta$ indexed by $\eta : V(T) \to 2^{V(G)} \setminus \{\emptyset\}$ such that $\sum_{j=1}^{|V(T)|} |\eta(j)| = |V(T)| + i$. Furthermore, when η is a cell of $\mathrm{Hom}(T, G)$, we set

$$c(\eta) := \sum_{\substack{i \text{ is even} \\ 1 \leq i \leq |V(T)|}} |\eta(i)|.$$

For any $\eta : V(T) \to 2^{V(G)} \setminus \{\emptyset\}$, set $\rho(\eta_+) := (-1)^{c(\eta)} \eta$. The induced map $\rho^* : X^i(T, G) \to C^i(\mathrm{Hom}(T, G))$ is of course an isomorphism of abelian groups for any i. It turns out that more is true.

Proposition 20.26. *For any two graphs T and G, the map $\rho^* : X^*(T, G) \to C^*(\mathrm{Hom}(T, G))$ is an isomorphism of the cochain complexes.*

Proof. We need to check that the map ρ^* commutes with the differentials, and it is enough to do this for a generator η_+^*, where $\operatorname{supp}\eta = V(T)$. Comparing the incidence number from (20.15) for $\operatorname{Hom}(T, G)$ with the one for the simplicial complex $\operatorname{Hom}_+(T, G)$, we see that the difference is the multiplicative functor $(-1)^{1-t}$, where t is the index of the simplex in the direct product where the new vertex is added. On the other hand, by the definition of $c(\eta)$ we know that $(-1)^{c(\eta)} = (-1)^{c(\tilde{\eta})\cdot(-1)^{1-t}}$, which proves that the incidence numbers coincide once $C^*(\operatorname{Hom}_+(T, G))$ is replaced with $X^*(T, G)$. $\square$

20.6.2 Signed Versions of Formulas for Generators $[\sigma_V^S]$

Let us now return to considering the cohomology classes $[\sigma_V^S]$ defined in Subsection 20.3.2, though this time we are working with integer coefficients. Clearly, these still index the generators of $E_1^{p,t(n-2)}$. The new feature appearing when we are working over integers is that the sign of the generator may change when vertex representatives move along edges.

Next, we describe the pattern of the sign change. Let S be a proper subset of $[m]$, so that $C_m[S]$ is a forest. Let (S, A) be an arc picture, and let V be the set of vertex representatives. Choose an arc $a \in A$ and denote its vertex representative by $v \in a$. Choose $w \in a$ such that $(v, w) \in E(C_m)$, and set $W := (V \cup \{w\}) \setminus \{v\}$. We would like to understand the sign change that occurs when v is replaced by w as the vertex representative of a. Let $\coprod_A K_2$ denote the disjoint union of copies of K_2 indexed by A. We define a graph homomorphism $\coprod_A K_2 \hookrightarrow C_m$ as follows: for the copy of K_2 indexed by a we choose one of the two graph homomorphisms $K_2 \to (v, w)$, for the other copies of K_2 we take any graph homomorphism mapping one of the vertices of K_2 to the vertex representative of the corresponding arc, and mapping the other vertex of K_2 to any of the neighboring vertices in S.

The graph homomorphism described above induces a $\mathbb{Z}_2$-equivariant map $H^*(\operatorname{Hom}(\coprod_A K_2, K_n)) \to H^*(\operatorname{Hom}(C_m[S], K_n))$, where the $\mathbb{Z}_2$-action on the left-hand side is induced by the antipodal action on the copy of K_2 indexed by a. On the other hand, we have $\operatorname{Hom}(\coprod_A K_2, K_n) \cong \prod_A \operatorname{Bip}(K_n) \cong \prod_A \mathbb{S}^{n-2}$, and the $\mathbb{Z}_2$-action on the last space is the antipodal action on the factor indexed by a. Recall that the antipodal action on $\mathbb{S}^n$ changes the sign of the n-dimensional cohomology generator if and only if n is even, and the same is true if we are acting on one of the factors in a direct product. Using already introduced terminology, we can take $\rho^*(\sigma_V^S)$ as a representative for the generator of $H^{t(n-2)}(\operatorname{Hom}(C_m[S], K_n))$ that we are considering. If $(v, w) \neq (1, m)$, then there is no further sign change, and we get

$$[\rho^*(\sigma_V^S)] = (-1)^{n+1}[\rho^*(\sigma_W^S)].$$

If, on the other hand, $(v, w) = (1, m)$, then the $\mathbb{Z}_2$-action does not interchange two neighboring simplices, but the first one and the last one instead. One of these simplices has dimension 0 and the other one has dimension $n - 2$.

Using the rules for the sign change described in Subsection 20.6.1, we see that we get the additional sign factor $(-1)^{n(|A|-1)}$, since we have to swap simplices of dimension $n-2$ in total $|A|-1$ times, and each swap yields the sign $(-1)^{(n-2)^2} = (-1)^n$; we ignore the swaps involving 0-dimensional simplices, since they do not influence the sign. So in this case we get

$$[\rho^*(\sigma_V^S)] = (-1)^{n|A|+1}[\rho^*(\sigma_W^S)].$$

Combining this with Proposition 20.26, we get

$$[\sigma_V^S] = -[\sigma_W^S] \tag{20.17}$$

if $(v,w) \neq (1,m)$, since the swapped vertices have different parity. In case $(v,w) = (1,m)$, the swapped vertices may or may not have different parity, so we have an additional sign factor $(-1)^{nb}$, where b is the number of vertices in S between w and v. So in this case we get

$$[\sigma_V^S] = (-1)^{n|A|+nb+n+1}[\sigma_W^S]. \tag{20.18}$$

Next, we look at the analogue of the formula (20.8). Combining the construction of generators σ_V^S with the choice of orientations on $\mathrm{Hom}_+(T, K_n)$, we get the following formula:

$$d_1([\sigma_V^S]) = \sum_{w \notin S}(-1)^{\mathrm{sgn}\,(w)}\left[\sigma_V^{S \cup \{w\}}\right], \tag{20.19}$$

where the sum is taken over all w such that adding w to S does not unite two arcs from A, and the sign $(-1)^{\mathrm{sgn}\,(w)}$ is given by the formula

$$\mathrm{sgn}\,(w) = |S \cap [w-1]| + n \cdot |V \cap [w-1]|. \tag{20.20}$$

The same way as for $\mathbb{Z}_2$-coefficients, equation (20.19) is also valid when $|S| = m - 1$.

20.6.3 Completing the Calculation of the Second Tableau

We shall now do the same computation as we did in Subsection 20.4.5, only this time with integer coefficients.

The first obstruction is that Proposition 20.14 may no longer be valid, since we have to take the signs into account. However, though the cochain complexes (A_t^*, d_1) and $(C_*(\Phi_{t,m,3}; \mathbb{Z}), d)$ may not be isomorphic, the only difference is that some of the incidence numbers differ by a sign.

The argument using torus fronts with which we computed $H_*(\Phi_{t,m,3}; \mathbb{Z}_2)$ consisted in presenting a sequence of collapses leading from $TF_{t,m-t,3}$ to $\mathrm{Thin}_{t,m-t,3}$. Due to isomorphism of cochain complexes over $\mathbb{Z}_2$, stated in Proposition 20.14, these collapses could have been performed directly on (A_t^*, d_1).

It is now crucial to realize that the same collapses can still be performed in the cochain complex (A_t^*, d_1), even though we are working over the integers. This follows from the version of discrete Morse theory for chain complexes from Section 11.3, since our matching is acyclic and the weights on the matching relations, which here are the incidence numbers, are ± 1, hence invertible over $\mathbb{Z}$.

Let $m - t > 2t$, i.e., $m > 3t$, and consider the complexes $\text{Thin}_{t,m-t,3}$. On each maximal cube, choose any of the shortest paths connecting the opposite vertices by which these cubes hang together. Such a path has $\gcd(t, m - t) = \gcd(t, m)$ edges. It is obvious that the collapsing process can be continued until the whole complex $\text{Thin}_{t,m-2,3}$ is collapsed onto this path P. Let us extend this collapsing to (A_t^*, d_1) as well, and let us denote the remaining chain complex by (Q_t, d_1).

The chain complex (Q_t, d_1) has only two nonzero entries, so let us write it as $0 \to Q_t^1 \to Q_t^0 \to 0$. If we worked over $\mathbb{Z}_2$, this chain complex would be simply computing the homology of a circle with coefficients in $\mathbb{Z}_2$. Over integers we have signs, and there are two possibilities: either $H_1(Q_t) = H_0(Q_t) = \mathbb{Z}$, or $H_1(Q_t) = 0$, and $H_0(Q_t) = \mathbb{Z}_2$. To distinguish between these two cases we shall now calculate the differential.

Both groups Q_t^1 and Q_t^0 have m generators. The boundary of each generator of Q_t^1 is a sum of exactly two generators of Q_t^0 with coefficients ± 1. Let α be the number of those generators of Q_t^1 for which these two coefficients are equal. It is easy to see that $H_1(Q_t) = H_0(Q_t) = \mathbb{Z}$ if α is even, and $H_1(Q_t) = 0$, $H_0(Q_t) = \mathbb{Z}_2$, if α is odd, so all that remains is to calculate the parity of α.

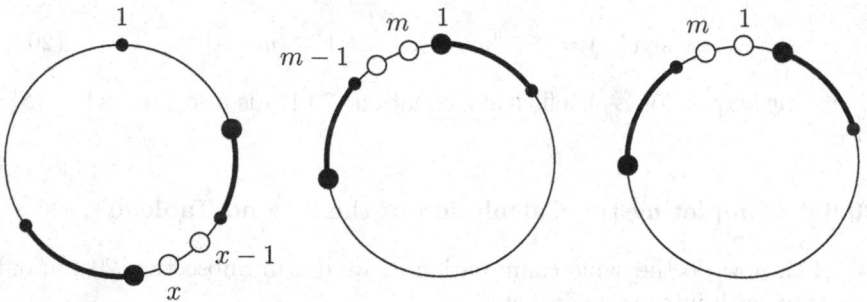

Fig. 20.7. Three cases of the sign calculation.

One may think of generators of Q_t^1 as being indexed by t-arc pictures (S, A), with all but one gap consisting of one element, and one 2-element gap, denoted by $[x - 1, x]_m$. Taking the boundary corresponds to narrowing this gap in two different ways: by extending one of the bordering arcs clockwise or counterclockwise. We break our calculation into three cases. See Figure 20.7.

Case 1. $2 \leq x \leq m - 1$. By formula (20.19), filling in $x - 1$ or x will have the same sign. Furthermore, when filling in x, we shall also need to move the vertex representative from $x + 1$ to x. These vertices have different parity in the order on $V(C_m[S \cup \{a\}])$; hence by (20.17) we get an additional factor -1 for this generator. We conclude that all these generators of Q_t^1 do not contribute to α.

Case 2. $x = m$. Again, by formula (20.19), filling in $x - 1$ or x will have the same sign. However, in this case we have to be careful about the sign caused by moving the vertex representative from 1 to m. By formula (20.18), the additional sign is $(-1)^{mn+n+1}$, since $b = m - t - 2$ and $|A| = t$. So the contribution to α is 1 if $mn + n$ is odd, and 0 otherwise.

Case 3. $x = 1$. In this case shifting the vertex representative gives the sign -1, by (20.17). The nontrivial contribution comes instead from the fact that filling x and $x - 1$ may give different signs. The difference in signs is $(-1)^{nt+m+t+1}$, since $m - t - 1 + t(n - 2)$ is the total number of vertices (if counted with multiplicities) in positions 2 through $m - 1$. Therefore, in this case the contribution to α is 1 if $nt + m + t + 1$ is even, and 0 otherwise.

By construction of the path P, the group Q_t^1 contains exactly one generator with $x = m$, andq exactly one generator with $x = 1$. Summarizing, we see that α is odd if and only if $mn + n + nt + m + t + 1$ is odd. Note that $mn + n + nt + m + t + 1 = (m + t + 1)(n + 1)$. Thus we obtain the extension of Proposition 20.21 to the case of integer coefficients. We present it in Table 20.1.

n, m, t	$E_2^{m-t-2, t(n-2)}$	$E_2^{m-t-1, t(n-2)}$
$t = 1$ $m(n + 1)$ is odd, $m(n + 1)$ is even,	0 $\mathbb{Z}$	0 0
n is odd, or $m + t$ is odd, $2 \leq t \leq \lfloor (m - 1)/3 \rfloor$	$\mathbb{Z}$	$\mathbb{Z}$
n is even, and $m + t$ is even, $2 \leq t \leq \lfloor (m - 1)/3 \rfloor$	0	$\mathbb{Z}_2$
n is any, and $3 \mid m$, $t = m/3$	0	$\mathbb{Z}^3$

Table 20.1. The entries of the second tableau in the case of integer coefficients.

20.6.4 Summary: the Full Description of the Groups $\widetilde{H}^*(\mathrm{Hom}\,(C_m, K_n); \mathbb{Z})$

For an abelian group Γ, we let $\Gamma(d)$ denote the copy of Γ in a graded $\mathbb{Z}$-algebra, placed in dimension d.

Theorem 20.27. *For any integers m, n such that $m \geq 5$, $n \geq 4$, we have*

$$\widetilde{H}^*(\mathrm{Hom}\,(C_m, K_n); \mathbb{Z}) = \left(\bigoplus_{t=1}^{\lfloor (m-2)/3 \rfloor} A_{t,m,n} \right) \oplus B_{m,n}, \qquad (20.21)$$

where

$$A_{t,m,n} = \begin{cases} \mathbb{Z}(tn - 3t) \oplus \mathbb{Z}(tn - 3t + 1), & \text{if } n \text{ is odd or } m + t \text{ is odd,} \\ \mathbb{Z}_2(tn - 3t + 1), & \text{if } n \text{ is even and } m + t \text{ is even,} \end{cases} \qquad (20.22)$$

and

$$B_{m,n} = \begin{cases} \mathbb{Z}^{2^n - 3}(nk - m), & \text{if } m = 3k, \\ \mathbb{Z}(nk - m + 2), & \text{if } m = 3k + 1, \\ \mathbb{Z}(nk - m), & \text{if } m = 3k - 1. \end{cases} \qquad (20.23)$$

For example, $\widetilde{H}^*(\mathrm{Hom}\,(C_6, K_4); \mathbb{Z}) = A_{1,6,4} \oplus B_{6,4} = \mathbb{Z}(1) \oplus \mathbb{Z}(2) \oplus \mathbb{Z}^{13}(2) = \mathbb{Z}(1) \oplus \mathbb{Z}^{14}(2)$, and $\widetilde{H}^*(\mathrm{Hom}\,(C_8, K_6); \mathbb{Z}) = A_{1,8,6} \oplus A_{2,8,6} \oplus B_{8,6} = \mathbb{Z}(3) \oplus \mathbb{Z}(4) \oplus \mathbb{Z}_2(7) \oplus \mathbb{Z}(10)$.

Proof of Theorem 20.27.
Recall that all cohomology groups of the simplicial complex $\mathrm{Hom}_+(C_m, K_n)$ are zero except for one; see Corollary 20.8. Since our spectral sequence converges to $H^*(\mathrm{Hom}_+(C_m, K_n); \mathbb{Z})$, we know that almost all entries in the second tableau should cancel out.

Since $n \geq 4$, the calculation summarized in Table 20.1 implies that there will be no nonzero differential between entries in columns $0, \ldots, m - 2$. If $n \geq 5$, then the differentials with entries in the $(m - 1)$st column as a target will never have the same target entry. If $n = 4$, the targets for two such differentials may be the same, but then the sources will simply sum up. This follows from the pattern of the entries and from the algebraic fact that if we have a group homomorphism $f : A \to G$ such that f is an injection, A is either $\mathbb{Z}$ or $\mathbb{Z}_2$, and $G/\mathrm{Im}(f)$ is isomorphic to $\mathbb{Z}$, then G is isomorphic to $A \oplus \mathbb{Z}$.

Since $\widetilde{H}^i(\mathrm{Hom}_+(C_m, K_n)) = 0$, for $i < nk - 1$, we obtain the "A part" of (20.21), with the exception of a single entry in the case $m = 3k + 1$, which is dealt with below. So we are done with the calculation of almost all cohomology groups of $\mathrm{Hom}\,(C_m, K_n)$ at this point. To get the remaining "B part," we need to see what happens on the diagonals $x + y = \mathrm{const} \geq nk - 2$. We shall consider three cases.

Assume first that $m = 3k - 1$. We see from Table 20.1 that the top (in terms of the sum of coordinates) nonzero element is $E_2^{m-k,(n-2)(k-1)}$. This element is on the diagonal $m - k + (n-2)(k-1) = m + nk - 3k - n + 2 = nk - n + 1 = (nk - 1) - (n - 2)$; hence it is $n - 2$ diagonals away from the diagonal $nk - 1$, and we conclude that $B_{m,n} = \mathbb{Z}(nk - m)$.

Assume now that $m = 3k$. The top nonzero element in columns $0, \ldots, m-2$ is $E_2^{m-k-1,k(n-2)}$, which lies precisely on the diagonal $x + y = nk - 1$. By the analysis in Subsection 20.3.2, the generators of $E_2^{m-k-1,k(n-2)}$ are given by those $[\sigma_{A_\bullet}^S]$ for which S is obtained from $[m]$ by deleting every third element, which can be done in three different ways, and A is the collection of all $m/3$ arcs of length 2. By construction, we see that in this case $d(\sigma_{A_\bullet}^S) = 0$, where the differential is taken in $C^*(\mathrm{Hom}_+(C_m, K_n))$. Additionally, there are no nonzero elements northwest from $E_2^{m-k-1,k(n-2)}$. It follows that the spectral sequence differential $d_k : E_2^{m-k-1,k(n-2)} \to E_2^{m-1,nk-m+1}$ is a zero-map, and therefore $E_2^{m-k-1,k(n-2)} = \mathbb{Z}^3$. Choosing a field $\mathcal{F}$ of arbitrary characteristic, we see that the Betti number of the entry $E_2^{m-1,nk-m}$ is equal to $2^n - 3$ if $n \geq 5$, and $2^n - 2$ if $n = 4$. Since this does not depend on the choice of $\mathcal{F}$, we conclude that $H^{nk-m}(\mathrm{Hom}(C_m, K_n)) = E_2^{m-1,nk-m}$ is equal to the direct sum of the corresponding number of copies of $\mathbb{Z}$, and hence $B_{m,n} = \mathbb{Z}^{2^n - 3}(nk - m)$ in this case.

Finally, assume that $m = 3k + 1$. In this case we have to deal with the differentials from $E_2^{m-k-2,k(n-2)} = E_2^{m-k-1,k(n-2)} = \mathbb{Z}$, and if $n = 4$, also from $E_2^{m-k,2k-2} = \mathbb{Z}_2$. Let X_+ denote $\mathrm{Hom}_+(C_m, K_n)$, let $X \subset X_+$ be the subcomplex consisting of all cells η such that $|\mathrm{supp}\,\eta| < m$, and consider the cohomology long exact sequence of the pair (X_+, X). Since the reduced cohomology groups $\widetilde{H}^*(X_+; \mathbb{Z})$ are trivial in all dimensions except for $nk - 1$, we find the following exact sequence inside the considered one:

$$0 \longleftarrow H^{nk}(X_+, X) \longleftarrow H^{nk-1}(X)$$
$$\xleftarrow{f} H^{nk-1}(X_+) \longleftarrow H^{nk-1}(X_+, X) \longleftarrow H^{nk-2}(X) \longleftarrow 0. \quad (20.24)$$

The crucial observation now is that the map f is an inclusion map to a direct summand. To see this, set $Y := \mathrm{Hom}_+(C_m[V(C_m) \setminus \{1\}], K_n) \subset X$. By the remark at the end of Example 1 in Subsection 11.2.3, and by the fact that $\mathrm{Hom}_+(C_m, K_n) \simeq \mathrm{Ind}\,(C_m)^{*n}$, we see that the inclusion map $i : Y \hookrightarrow X_+$ induces an isomorphism of the cohomology groups $i^* : H^*(X_+) \to H^*(Y)$. The isomorphism map i^{nk-1} factors as $H^{nk-1}(X_+) \xrightarrow{f} H^{nk-1}(X) \to H^{nk-1}(Y)$, since f itself is induced by the inclusion map $X \hookrightarrow X^+$. It follows from this factorization that f is an inclusion map to a direct summand, so the sequence (20.24) splits in this sense.

Reinterpreting this for our original spectral sequence, we see that this means that the differential $d_{k+1} : E_2^{m-k-2,k(n-2)} \to E_2^{m-1,nk-m+1}$ is a zero-map, and that if $n = 4$, then the differential $d_{k-1} : E_2^{m-k,2k-2} \to E_2^{m-1,k}$ is an isomorphism. This proves the "B part" as well as the rest of the "A part" in this case. □

20.7 Bibliographic Notes and Conclusion

This chapter is almost entirely based on the papers [BK04] by Babson and the author, and [Ko05b] by the author, also [Ko05a] can be useful to consult. The notion of Hom$_+$ and the canonical spectral sequence connecting the Hom and Hom$_+$ complexes were introduced in [BK04]. However, a technically simpler presentation from which we have drawn here is [Ko05b], where also the combinatorial theory of mills and grinding, which was necessary to complete the computation, was developed.

We finish by mentioning that our goals for Part III of the book were limited. We wish merely to introduce the reader to the beautiful world of cell complexes associated to graph homomorphisms (and more generally associated to collections of set maps), and to develop some basic structure theory for these gadgets. Our choice of topics was in part motivated by the desire to illustrate the various combinatorial techniques that were described in Part II on a concrete set of examples. The subject of Hom complexes is very dynamic, and much more research exists and is being produced in real time. Therefore, we actively encourage the reader to consult the original sequence of papers [BK03, BK06, BK04], the author's survey [Ko05a], and perhaps most importantly the excellent papers by various authors, including but not limited to [Cs05, CL07, Cu06, CK06, CK05, Doc06, EH05, En06, Ko06a, Ko06b, Ko06c, Ko06d, Ko07, Ko05b, Pf05, Su05a, Su05b, Su06, Ziv05b].

References

[AM94]   A. Adem, J. Milgram, *Cohomology of finite groups*, Grundlehren der Mathematischen Wissenschaften [Fundamental Principles of Mathematical Sciences] **309**, Springer-Verlag, Berlin, 1994.

[Al88]   N. Alon, *Some recent combinatorial applications of Borsuk-type theorems*, Algebraic, extremal and metric combinatorics, 1986 (Montreal, PQ, 1986), pp. 1–12, London Math. Soc. Lecture Note Ser., **131**, Cambridge Univ. Press, Cambridge, 1988.

[AFL86]   N. Alon, P. Frankl, L. Lovász, *The chromatic number of Kneser hypergraphs*, Trans. Amer. Math. Soc. **298** (1986), no. 1, pp. 359–370.

[AH76]   K. Appel, W. Haken, *Every planar map is four colorable*, Bull. Amer. Math. Soc. **82**, (1976), pp. 711–712.

[AH89]   K. Appel, W. Haken, *Every planar map is four colorable*, Contemp. Math., vol. 98, Amer. Math. Soc., Providence, RI, 1989.

[AHS06]   J. Adámek, H. Herrlich, G.E. Strecker, *Abstract and concrete categories: the joy of cats*, Reprint of the 1990 original [Wiley, New York], Repr. Theory Appl. Categ. No. 17 (2006), 1–507 (electronic).

[Ba05]   E. Babson, private communication, 2005.

[BK03]   E. Babson, D.N. Kozlov, *Topological obstructions to graph colorings*, Electron. Res. Announc. Amer. Math. Soc. **9** (2003), pp. 61–68.

[BK04]   E. Babson, D.N. Kozlov, *Proof of the Lovász Conjecture*, Annals of Mathematics **165** (2007), no. 3, pp. 965–1007.

[BK05]   E. Babson, D.N. Kozlov, *Group actions on posets*, J. Algebra **285** (2005), no. 2, pp. 439–450.

[BK06]   E. Babson, D.N. Kozlov, *Complexes of graph homomorphisms*, Israel J. Math., **152** (2006), 285–312.

[Bac75]   K. Baclawski, *Whitney numbers of geometric lattices*, Advances in Math. **16** (1975), pp. 125–138.

[Bar78]   I. Bárány, *A short proof of Kneser's conjecture*, J. Combin. Theory Ser. A 25 (1978), no. 3, pp. 325–326.

[BSS81]   I. Bárány, S.B. Shlosman, A. Szűcs, *On a topological generalization of a theorem of Tverberg*, J. London Math. Soc. (2) **23** (1981), no. 1, pp. 158–164.

[BHV01]   L. Billera, S. Holmes, K. Vogtmann, *Geometry of the space of phylogenetic trees*, Advances in Applied Math. **27** (2001), pp. 733–767.

[Bj80]   A. Björner, *Shellable and Cohen-Macaulay partially ordered sets,* Trans. Amer. Math. Soc. **260**, (1980), 159–183.

[Bj96]   A. Björner, *Topological Methods,* Handbook of Combinatorics, vol. 1,2, (eds. R. Graham, M. Grötschel and L. Lovász), Elsevier, Amsterdam, 1995, pp. 1819–1872.

[BKL85]   A. Björner, B. Korte, L. Lovász, *Homotopy properties of greedoids,* Adv. in Appl. Math. **6** (1985), no. 4, pp. 447–494.

[BW83]   A. Björner, J.W. Walker, *A homotopy complementation formula for partially ordered sets,* European J. Combin. **4**, (1983), pp. 11–19.

[BoK72]   M. Bousfield, D.M. Kan, *Homotopy Limits, Completions and Localizations,* Springer Lect. Notes Math. **304**, Berlin–Heidelberg–New York 1972.

[Bre72]   G.E. Bredon, *Introduction to compact transformation groups,* Pure and Applied Mathematics **46**, Academic Press, New York–London, 1972.

[Bre93]   G.E. Bredon, *Topology and geometry,* Graduate Texts in Mathematics **139**, Springer-Verlag, New York, 1993.

[BW04]   G.R. Brightwell, P. Winkler, *Graph homomorphisms and long range action,* Graphs, morphisms and statistical physics, DIMACS Ser. Discrete Math. Theoret. Comput. Sci., **63**, Amer. Math. Soc., Providence, RI, 2004, pp. 29–47.

[BM71]   H. Bruggesser, P. Mani, *Shellable decompositions of cells and spheres,* Math. Scand. **29** (1971), pp. 197–205.

[CS04a]   G. Carlsson, V. de Silva, *A geometric framework for sparse matrix problems,* Adv. in Appl. Math. **33** (2004), no. 1, pp. 1–25.

[CS04b]   G. Carlsson, V. de Silva, *Topological estimation using witness complexes,* preprint 2004.

[CZ05]   G. Carlsson, A. Zomorodian, *Computing persistent homology,* Discrete and Computational Geometry, **33** (2) (2005), pp. 249–274.

[Cay78]   A. Cayley, *On the coloring of maps,* Proc. London Math. Soc. vol. 9, (1878), p.148.

[Co73]   M. Cohen, *A course in simple-homotopy theory,* Graduate Texts in Mathematics, Vol. 10, Springer-Verlag, New York-Berlin, 1973.

[Cs04a]   P. Csorba, *Homotopy type of the box complexes,* preprint, 11 pages, 2004. `arXiv:math.CO/0406118`

[Cs04b]   P. Csorba, private communication, 2004.

[Cs05]   P. Csorba, *Non-tidy Spaces and Graph Colorings,* Ph.D. Thesis, ETH Zürich, 2005.

[Cs07]   P. Csorba, *On simple $\mathbb{Z}_2$-homotopy type of graph complexes, and their simple $\mathbb{Z}_2$-universality,* preprint 2007, to appear in Canad. Math. Bull.

[CL07]   P. Csorba, F. Lutz, *Graph coloring manifolds,* in: *Algebraic and Geometric Combinatorics* (eds. C.A. Athanasiadis, V.V. Batyrev, D.I. Dais, M. Henk, F. Santos), Contemporary Mathematics **423**, American Mathematical Society, Providence, RI, 2007.

[CK05]   S.Lj. Čukić, D.N. Kozlov, *Higher connectivity of graph coloring complexes,* Int. Math. Res. Not. **2005**, no. 25, 1543–1562.

[CK06]   S.Lj. Čukić, D.N. Kozlov, *The homotopy type of the complexes of graph homomorphisms between cycles,* Discrete Comput. Geom. **36** (2006), no. 2, pp. 313–329.

[Cu06]   S.Lj. Čukić, *Topology of discrete structures: Graph maps and Bier spheres,* Ph.D. Thesis, ETH Zürich, 2006.

[CD06]   S.Lj. Čukić, E. Delucchi, *Simplicial shellable spheres via combinatorial blowups*, preprint, 11 pages, 2006. arXiv:math.CO/0602101

[Del06]   E. Delucchi, *Nested set complexes of Dowling lattices and complexes of Dowling trees*, preprint, 14 pages, 2006. arXiv:math.CO/0603383

[tD87]   T. tom Dieck, *Transformation groups*, de Gruyter Studies in Mathematics, 8. Walter de Gruyter & Co., Berlin, 1987. x+312 pp.

[Dir52]   G.A. Dirac, *A property of 4-chromatic graphs and some remarks on critical graphs*, J. London Math. Soc. **27** (1952), pp. 85–92.

[Doc04]   A. Dochtermann, private communication, 2004.

[Doc06]   A. Dochtermann, Hom-*complexes and homotopy theory in the category of graphs*, preprint, 30 pages, 2006. arXiv:math.CO/0605275

[Dol88]   V.L. Dol'nikov, *A combinatorial inequality* (Russian), Sibirsk. Mat. Zh. **29** (1988), no. 3, pp. 53–58, 219; translation in Siberian Math. J. **29** (1988), no. 3, pp. 375–379.

[EH05]   M.G. Eastwood, S.A. Huggett, *Euler characteristics and chromatic polynomials*, European J. Comb., to appear.

[ELZ02]   H. Edelsbrunner, D. Letcher, A. Zomorodian, *Topological persistence and simplification*, Discrete Comput. Geom. **28** (2002), pp. 511–513.

[En06]   A. Engström, *A short proof of a conjecture on the higher connectivity of graph coloring complexes*, Proc. Amer. Math. Soc. **134** (2006), no. 12, pp. 3703–3705.

[EZ50]   S. Eilenberg, J.A. Zilber, *Semi-simplicial complexes and singular homology*, Ann. of Math. (2) **51**, (1950), pp. 499–513.

[FK00]   E.-M. Feichtner, D.N. Kozlov, *On subspace arrangements of type* **D**, Formal power series and algebraic combinatorics (Minneapolis, MN, 1996). Discrete Math. **210** (2000), no. 1-3, pp. 27–54.

[FK04]   E.-M. Feichtner, D.N. Kozlov, *Incidence combinatorics of resolutions*, Selecta Math. (N.S.) **10** (2004), no. 1, pp. 37–60.

[FK05]   E.-M. Feichtner, D.N. Kozlov, *A desingularization of real differentiable actions of finite groups*, Int. Math. Res. Not. **2005**, no. 15, pp. 881–898.

[FM05]   E.-M. Feichtner, I. Müller, *On the topology of nested set complexes*, Proc. Amer. Math. Soc. **133** (2005), no. 4, pp. 999–1006.

[FS05]   E.-M. Feichtner, B. Sturmfels, *Matroid polytopes, nested sets and Bergman fans*, Port. Math. (N.S.) **62** (2005), no. 4, pp. 437–468.

[FY04]   E.-M. Feichtner, S. Yuzvinsky, *Chow rings of toric varieties defined by atomic lattices*, Invent. Math. **155** (2004), no. 3, pp. 515–536.

[Fei05]   E.-M. Feichtner, *De Concini-Procesi wonderful arrangement models: a discrete geometer's point of view*, in: Combinatorial and computational geometry, Math. Sci. Res. Inst. Publ. **52**, Cambridge Univ. Press, Cambridge, 2005, pp. 333–360.

[Fei06]   E.-M. Feichtner, *Complexes of trees and nested set complexes*, Pacific J. Math. **227** (2006), no. 2, 271–286.

[Fol66]   J. Folkman, *The homology groups of a lattice*, J. Math. Mech. **15** (1966), pp. 631–636.

[FFG86]   A.T. Fomenko, D.B. Fuks, V.L. Gutenmacher, *Homotopic topology*, Translated from the Russian by K. Mályusz. Akadémiai Kiadó (Publishing House of the Hungarian Academy of Sciences), Budapest, 1986.

[For98]   R. Forman, *Morse theory for cell complexes*, Adv. Math. **134**, (1998), no. 1, pp. 90–145.

[For03]   R. Forman, *A user's guide to discrete Morse theory*, Sém. Lothar. Combin. **48**, (2002).

[GJ76]   M.R. Garey, D.S. Johnson, *The complexity of near-optimal graph coloring*, J. Assoc. Comp. Mach. **23**, (1976), pp. 43–49.

[GJ79]   M.R. Garey, D.S. Johnson, *Computers and Intractability*, A guide to the theory of NP-completeness, A Series of Books in the Mathematical Sciences, W.H. Freeman and Co., San Francisco, 1979.

[GeM96]   S. Gelfand, Y. Manin, *Methods of homological algebra*, Translated from the 1988 Russian original, Springer-Verlag, Berlin Heidelberg, 1996.

[GR01]   C. Godsil, G. Royle, *Algebraic Graph Theory*, Graduate texts in mathematics **207**, Springer-Verlag, New York, 2001.

[GoM88]   M. Goresky, R. MacPherson, *Stratified Morse Theory*, Ergebnisse der Mathematik und ihrer Grenzgebiete, vol. **14**, Springer-Verlag, Berlin Heidelberg New York, 1988.

[Gr02]   J. Greene, *A new short proof of Kneser's conjecture*, Amer. Math. Monthly **109** (2002), no. 10, pp. 918–920.

[Gut80]   F. Guthrie, *Note on the colouring of maps*, Proc. Roy. Soc. Edinburgh, vol. 10, (1880), p. 729.

[Had43]   H. Hadwiger, *Über eine Klassifikation der Streckenkomplexe*, Vierteljschr. Naturforsch. Ges., Zürich, vol. 88, (1943), pp. 133–142.

[Han91]   P. Hanlon, *The generalized Dowling lattices*, Trans. Amer. Math. Soc. **325** (1991), pp. 1–37.

[Har69]   F. Harary, *Graph Theory*, Addison-Wesley Series in Mathematics, Reading, MA, 1969.

[Hat02]   A. Hatcher, *Algebraic topology*, Cambridge University Press, Cambridge, 2002.

[Hea90]   P.J. Heawood, *Map-Colour Theorems*, Quart. J. Math., Oxford ser., vol. 24, (1890), pp. 332–338.

[Hee69]   H. Heesch, *Untersuchungen zum Vierfarbenproblem*, Bibliog. Institut, AG, Mannheim, 1969.

[HN04]   P. Hell, J. Nešetřil, *Graphs and Homomorphisms*, Oxford Lecture Series in Mathematics and Its Applications **28**, Oxford University Press, 2004.

[HL05]   S. Hoory, N. Linial, *A counterexample to a conjecture of Lovász on the χ-coloring complex*, J. Combin. Theory Ser. B **95** (2005), no. 2, pp. 346–349.

[Il78]   N. Illies, *A counterexample to the generalized Aanderaa-Rosenberg conjecture*, Inform. Process. Lett. **7** (1978), no. 3, pp. 154–155.

[JW05]   M. Jöllenbeck, V. Welker, *Resolution of the residue class field via algebraic discrete Morse theory*, preprint 2005.
          `arXiv:math.AC/0501179`

[KSS84]   J. Kahn, M. Saks, D. Sturtevant, *A topological approach to evasiveness*, Combinatorica **4** (1984), pp. 297–306.

[KS77]   P.C. Kainen, T.L. Saaty, *The Four-Color Problem*, McGraw-Hill, New York, 1977.

[KM79]   M.I. Kargapolov, Ju.L. Merzljakov, *Fundamentals of the Theory of Groups,* Graduate Texts in Mathematics **62**, Springer-Verlag, 1979 (English translation of *Osnovy teorii grupp*, Nauka, Moscow, 1977).

[Kem79] A.B. Kempe, *On the geographical problem of four colors*, Amer. J. Math. **2**, (1879), pp. 193–204.

[KLS93] S. Khanna, N. Linial, S. Safra, *On the hardness of approximating the chromatic number*, Proc. Israel Symp. Theoretical Computer Science 1993, pp. 250–260.

[Kn55] M. Kneser, Aufgabe 360, *Jber. Deutsch. Math.-Verein.* **58**, (1955/56), 2 Abt., 27.

[Ko97] D.N. Kozlov, *General lexicographic shellability and orbit arrangements*, Annals Comb. **1** (1997), no. 1, 67–90.

[Ko98] D.N. Kozlov, *Order complexes of noncomplemented lattices are nonevasive*, Proc. Amer. Math. Soc. **126**, (1998), no. 12, 3461–3465.

[Ko99] D.N. Kozlov, *Complexes of directed trees*, J. Comb. Theory Ser. A **88** (1999), pp. 112–122.

[Ko00] D.N. Kozlov, *Collapsibility of $\Delta(\Pi_n)/S_n$ and some related CW complexes*, Proc. Amer. Math. Soc. **128** (2000), no. 8, pp. 2253–2259.

[Ko01] D.N. Kozlov, *Spectral sequences on combinatorial simplicial complexes*, J. Algebraic Combin. **14** (2001), no. 1, pp. 27–48.

[Ko02] D.N. Kozlov, *Trends in Topological Combinatorics*, Habilitationsschrift, Bern University, 2002.
http://www.math.kth.se/~kozlov/ps/main.ps

[Ko05a] D.N. Kozlov, *Chromatic numbers, morphism complexes, and Stiefel-Whitney characteristic classes*, in: *Geometric Combinatorics* (eds. E. Miller, V. Reiner, B. Sturmfels), IAS/Park City Mathematics Series **14**, American Mathematical Society, Providence, RI; Institute for Advanced Study (IAS), Princeton, NJ; in press.
arXiv:math.AT/0505563

[Ko05b] D.N. Kozlov, *Cohomology of colorings of cycles*, American J. Math., in press.
arXiv:math.AT/0507117

[Ko05c] D.N. Kozlov, *Discrete Morse theory for free chain complexes*, C. R. Math. Acad. Sci. Paris **340**, (2005), no. 12, pp. 867–872.

[Ko06a] D.N. Kozlov, *A simple proof for folds on both sides in complexes of graph homomorphisms*, Proc. Amer. Math. Soc. **134** (2006), no. 5, 1265–1270.

[Ko06b] D.N. Kozlov, *Collapsing along monotone poset maps*, International J. Math. and Math. Sciences **2006** (2006).

[Ko06c] D.N. Kozlov, *Simple homotopy types of Hom -complexes, neighborhood complexes, Lovász complexes, and atom crosscut complexes*, Topology and its Appl. **153** (2006), no. 14, 2445–2454.

[Ko06d] D.N. Kozlov, *Cobounding odd cycle colorings*, Electron. Res. Announc. Amer. Math. Soc. **12** (2006), pp. 53–55.

[Ko07] D.N. Kozlov, *Homology tests for graph colorings*, in: *Algebraic and Geometric Combinatorics* (eds. C.A. Athanasiadis, V.V. Batyrev, D.I. Dais, M. Henk, F. Santos), Contemporary Mathematics **423**, American Mathematical Society, Providence, RI, 2007.
arXiv:math.AT/0607750

[Kr92] I. Kříž, *Equivariant cohomology and lower bounds for chromatic numbers*, Trans. Amer. Math. Soc. **333** (1992), no. 2, pp. 567–577.

[Kr00] I. Kříž, A correction to *Equivariant cohomology and lower bounds for chromatic numbers*, Trans. Amer. Math. Soc. **352** (2000), pp. 1051–1052.

382   References

[dL03]   M. de Longueville, *25 Jahre Beweis der Kneservermutung der Beginn der topologischen Kombinatorik* (German), *[25th anniversary of the proof of the Kneser conjecture. The start of topological combinatorics]*, Mitt. Deutsch. Math.-Ver. 2003, no. 4, pp. 8–11.

[Lov78]   L. Lovász, *Kneser's conjecture, chromatic number, and homotopy*, J. Combin. Theory Ser. A **25** (1978), no. 3, pp. 319–324.

[Lov]   L. Lovász, private communication.

[Lu01]   F. Lutz, *Some results related to the Evasiveness Conjecture*, J. Combin. Theory Ser. B **81**, pp. 110–124.

[McL95]   S. Mac Lane, *Homology*, Reprint of the 1975 edition, Classics in Mathematics, Springer-Verlag, Berlin, 1995.

[McL98]   S. Mac Lane, *Categories for the Working Mathematician*, Second edition, Graduate Texts in Mathematics, **5**, Springer-Verlag, New York, 1998.

[Ma03]   J. Matoušek, *Using the Borsuk-Ulam theorem. Lectures on topological methods in combinatorics and geometry*, with A. Björner and G.M. Ziegler, Universitext, Springer-Verlag, Berlin, 2003.

[Ma04]   J. Matoušek, *A combinatorial proof of Kneser's conjecture*, Combinatorica **24** (2004), no. 1, pp. 163–170.

[MZ04]   J. Matoušek, G.M. Ziegler, *Topological lower bounds for the chromatic number: A hierarchy*, Jahresbericht der DMV **106**, 71–90, 2004.

[May92]   J.P. May, *Simplicial objects in algebraic topology*, Reprint of the 1967 original, Chicago Lectures in Mathematics, University of Chicago Press, Chicago, IL, 1992.

[May99]   J.P. May, *A concise course in algebraic topology*, Chicago Lectures in Mathematics, University of Chicago Press, Chicago and London, 1999.

[McC01]   J. McCleary, *A user's guide to spectral sequences*, Second edition, Cambridge Studies in advanced mathematics **58**, Cambridge University Press, Cambridge, 2001.

[MSTY]   O. Melnikov, V. Sarvanov, R. Tyshkevich, V. Yemelichev, *Lectures on Graph Theory*, BI-Wissenschaftsverlag, Mannheim, 1994. [Transl. by N. Korneenko from Russian original, Moscow, "Science," 1990].

[Mes03]   R. Meshulam, *Domination numbers and homology*, J. Combin. Theory Ser. A **102** (2003), no. 2, pp. 321–330.

[Mil57]   J.W. Milnor, *The geometric realization of semi-simplicial complex*, Ann. of Math. **65** (1957), 357–362.

[MSta74] J.W. Milnor, J.D. Stasheff, *Characteristic classes*, Annals of Mathematics Studies **76**, Princeton University Press, Princeton, 1974.

[MStu05] E. Miller, B. Sturmfels, *Combinatorial commutative algebra*, Graduate Texts in Mathematics **227**, Springer-Verlag, New York, 2005.

[Mit65]   B. Mitchell, *Theory of categories*, Pure and Applied Mathematics, Vol. XVII, Academic Press, New York-London, 1965.

[Mun84]   J.R. Munkres, *Elements of algebraic topology*, Addison-Wesley Publishing Company, Menlo Park, CA, 1984.

[Ol75]   R. Oliver, *Fixed-point sets of group actions on finite acyclic complexes*, Comment. Math. Helv. **50** (1975), pp. 155–177.

[Ore42]   O. Ore, *Theory of equivalence relations*, Duke Math. J. **9**, (1942), pp. 573–627.

[Ore67]   O. Ore, *The Four Color Problem*, Academic Press, New York, 1967.

[OS80]   P. Orlik, L. Solomon, *Combinatorics and Topology of complements of hyperplanes*, Invent. Math. **56** (1980), pp. 167–189.

[OT92]   P. Orlik, H. Terao, *Arrangements of hyperplanes*, Grundlehren der Mathematischen Wissenschaften **300**, Springer-Verlag, Berlin, 1992.

[OT01]   P. Orlik, H. Terao, *Arrangements and Hypergeometric Integrals*, MSJ Memoirs **9**, Mathematical Society of Japan, Tokyo, 2001.

[Pf05]   J. Pfeifle, *Dissections, Hom-complexes and the Cayley trick*, preprint, 21 pages, 2005.
          arXiv:math.CO/0512529

[Qu73]   D. Quillen, *Higher algebraic K-theory* I, Lecture Notes in Mathematics **341**, (1973), pp. 85–148, Springer-Verlag.

[Qu78]   D. Quillen, *Homotopy properties of the poset of nontrivial p-subgroups of a group*, Adv. Math. **28** (1978), no. 2, 101–128.

[RSST]   N. Robertson, D.P. Sanders, P.D. Seymour, R. Thomas, *The four-color theorem*, J. Combin. Theory, Ser. B **70**, (1997), pp. 2–44.

[RV75]   R.R. Rivest, J. Vuillemin, *A generalization and proof of the Aanderaa-Rosenberg conjecture*, 7th Annual ACM Symposium on Theory of Computing (Albuquerque, NM, 1975), pp. 6–11, Assoc. Comput. Mach., New York, 1975.

[Sar90]   K.S. Sarkaria, *A generalized Kneser conjecture*, J. Combin. Theory Ser. B 49 (1990), no. 2, pp. 236–240.

[Sr78]   A. Schrijver, *Vertex-critical subgraphs of Kneser graphs*, Nieuw Arch. Wiskd., III. Ser., (1978), pp. 454–461.

[Su05a]   C. Schultz, *A short proof of $w_1^n(Hom(C_{2r+1}, K_{n+2})) = 0$ for all n and a graph colouring theorem by Babson and Kozlov*, preprint, 8 pages, 2005.
          arXiv:math.AT/0507346

[Su05b]   C. Schultz, *Small models of graph colouring manifolds and the Stiefel manifolds* $Hom(C_5, K_n)$, preprint, 19 pages, 2005.
          arXiv:math.CO/0510535

[Su06]   C. Schultz, *Graph colourings, spaces of edges and spaces of circuits*, preprint, 10 pages, 2006.
          arXiv:math.CO/0606763

[Se68]   G. Segal, *Classifying spaces and spectral sequences*, Inst. Hautes Études Sci. Publ. Math. No. **34** (1968), 105–112.

[Sha02]   J. Shareshian, *On the shellability of the order complex of the subgroup lattice of a finite group*, Trans. Amer. Math. Soc. **353** (2001), no. 7, pp. 2689–2703.

[Sk06]   E. Skjöldberg, *Combinatorial discrete Morse theory from an algebraic viewpoint*, Trans. Amer. Math. Soc. **358** (2006), no. 1, pp. 115–129.

[Sta97]   R.P. Stanley, *Enumerative combinatorics*, Vol. 1, 2nd edition, Cambridge Studies in Advanced Mathematics **49**, Cambridge University Press, Cambridge, 1997.

[Ste51]   N.E. Steenrod, *The topology of fibre bundles*, Princeton University Press, Princeton, 1951; reprinted in *Princeton landmarks in mathematics and physics*, 1999.

[Th98]   R. Thomas, *An update on the Four-Color Theorem*, Notices Amer. Math. Soc., vol. 45, no. 7, August 1998, pp. 848–859.

[Va93]   V.A. Vassiliev, *Complexes of connected graphs*, The Gel'fand Mathematical Seminars, 1990–1992, pp. 223–235, Birkhäuser Boston, Boston, MA, 1993.

[Vi88]   A. Vince, *Star chromatic number*, J. Graph Theory **12**, (1988), pp. 551–559.

[Vo73]   R.M. Vogt, *Homotopy limits and colimits*, Math. Z. **134** (1973), pp. 11–52.

[Wag37]   K. Wagner, *Über eine Eigenschaft der ebenen Komplexe*, Math. Ann. **114**, (1937), pp. 570–590.

[Wa83]   J.W. Walker, *A homology version of the Borsuk-Ulam theorem*, Amer. Math. Monthly **90**, (1983), no. 7, 466–468.

[We94]   C. Weibel, *An introduction to homological algebra*, Cambridge Studies in Advanced Mathematics **38**, Cambridge University Press, Cambridge, 1994.

[WZZ99]   V. Welker, G.M. Ziegler, R. Živaljević, *Homotopy colimits – comparison lemmas for combinatorial applications*, J. Reine Angew. Math. **509** (1999), pp. 117–149.

[Wel84]   E. Welzl, *Symmetric graphs and interpretations*, J. Combin. Theory Ser. B **37** (1984), pp. 235–244.

[Wh78]   G.W. Whitehead, *Elements of homotopy theory*, Graduate Texts in Mathematics **61**, Springer-Verlag, New York, 1978.

[Yuz01]   S. Yuzvinsky, *Orlik-Solomon algebras in algebra and topology* (Russian), Uspekhi Mat. Nauk **56** (2001), no. 2(338), 87–166; translation in Russian Math. Surveys **56** (2001), no. 2, 293–364.

[Zhu01]   X. Zhu, *Circular chromatic number: a survey*, Combinatorics, graph theory, algorithms and applications, Discrete Math. **229**. (2001), no. 1-3, pp. 371–410.

[Zie02]   G.M. Ziegler, *Generalized Kneser coloring theorems with combinatorial proofs*, Invent. Math. **147**, (2002), pp. 671–691.

[ZZ93]   G.M. Ziegler, R.T. Živaljević, *Homotopy types of subspace arrangements via diagrams of spaces*, Math. Ann. **295** (1993), pp. 527–548.

[Ziv05a]   R.T. Živaljević, *WI-posets, graph complexes and $\mathbb{Z}_2$-equivalences*, J. Combin. Theory Ser. A **111** (2005), no. 2, pp. 204–223.

[Ziv05b]   R.T. Živaljević, *Combinatorial groupoids, cubical complexes, and the Lovász conjecture*, preprint, 28 pages, 2005. arXiv:math.CO/0510204

[Zo05]   A. Zomorodian, *Topology for Computing*, Cambridge Monographs on Applied and Computational Mathematics, Cambridge University Press, Cambridge, 2005.

Index